Ecohydrology

Ecohydrology is a fast-growing branch of science at the interface of ecology and geophysics, studying the interaction between soil, water, vegetation, microbiome, atmosphere, climate, and human society. This textbook gathers together the fundamentals of hydrology, ecology, environmental engineering, agronomy, and atmospheric science to provide a rigorous yet accessible description of the tools necessary for the mathematical modeling of water, energy, carbon, and nutrient transport within the soil–plant–atmosphere continuum. By focusing on the dynamics at multiple time scales, from the diurnal scale in the soil–plant–atmospheric system to the long-term stochastic dynamics of water availability, which is responsible for ecological patterns and environmental fluctuations, the book explains the impact of hydroclimatic variability on vegetation and soil microbial systems through biogeochemical cycles and ecosystems, under different socioeconomical pressures. It is aimed at advanced students, researchers, and professionals in hydrology, ecology, earth science, environmental engineering, environmental science, agronomy, and atmospheric science.

Amilcare Porporato is the Thomas J. Wu 1994 Professor of Civil and Environmental Engineering in the High Meadow Environmental Institute at Princeton University. His main research focuses on nonlinear and stochastic dynamical systems, hydrometeorology, soil–atmosphere interaction, soil moisture and plant dynamics, soil biogeochemistry, and ecohydrology. He is the author of more than 200 peer-reviewed papers, co-author of *Ecohydrology of Water Controlled Ecosystems* (Cambridge University Press, 2004), and co-editor of *Dryland Ecohydrology* (Springer, 2005). Porporato's awards include: the Arturo Parisatti International Prize of the Istituto Veneto di Scienze, Lettere e Arti; the Earl Brown II Outstanding Civil Engineering Faculty Award; the Hydrology Award from the American Geophysical Union; and the Dalton Medal from the European Geoscience Union.

Jun Yin is a Professor of Hydrometeorology at Nanjing University of Information Science and Technology, China. His research focuses on the exchange of water and energy between soil, plants, and the atmosphere. He received the Outstanding Paper Award in 2019 from the China Water Forum and the Jiangsu Specially-Appointed Faculty Award in 2020 from the Jiangsu Education Department. He has co-organized two mini-symposia sessions for the Annual Meeting of the International Society for Porous Media.

Ecohydrology
Dynamics of Life and Water in the Critical Zone

Amilcare Porporato
Princeton University, New Jersey

Jun Yin
Nanjing University of Information Science and Technology, China

CAMBRIDGE
UNIVERSITY PRESS

University Printing House, Cambridge CB2 8BS, United Kingdom

One Liberty Plaza, 20th Floor, New York, NY 10006, USA

477 Williamstown Road, Port Melbourne, VIC 3207, Australia

314–321, 3rd Floor, Plot 3, Splendor Forum, Jasola District Centre, New Delhi – 110025, India

103 Penang Road, #05–06/07, Visioncrest Commercial, Singapore 238467

Cambridge University Press is part of the University of Cambridge.

It furthers the University's mission by disseminating knowledge in the pursuit of
education, learning, and research at the highest international levels of excellence.

www.cambridge.org
Information on this title: www.cambridge.org/highereducation/isbn/9781108840545
DOI: 10.1017/9781108886321

First published 2022

Printed in the United Kingdom by TJ Books Limited, Padstow Cornwall

A catalogue record for this publication is available from the British Library.

Library of Congress Cataloging-in-Publication Data
Names: Porporato, Amilcare, author. | Yin, Jun, 1984– author.
Title: Ecohydrology : dynamics of life and water in the critical zone /
Amilcare Porporato, Princeton University, USA, Jun Yin, Nanjing
University of Information Science and Technology, China.
Description: New York : Cambridge University Press, 2021. | Includes index.
Identifiers: LCCN 2021032803 (print) | LCCN 2021032804 (ebook) | ISBN
9781108840545 (hardback) | ISBN 9781108886321 (ebook)
Subjects: LCSH: Ecohydrology.
Classification: LCC QH541.15.E19 P67 2021 (print) | LCC QH541.15.E19
(ebook) | DDC 577.6–dc23
LC record available at https://lccn.loc.gov/2021032803
LC ebook record available at https://lccn.loc.gov/2021032804

ISBN 978-1-108-84054-5 Hardback

Additional resources for this publication at www.cambridge.org/ecohydrology

Contents

Preface

With four parameters I can fit an elephant, and with five I can
make him wiggle his trunk.

John von Neumann[†]

In the last two decades, *ecohydrology* has emerged as a fast-growing branch of science at the interface of ecology and geophysics, with the goal of understanding and addressing the impending problems associated with the interactions among soil, water, vegetation, microbiome, atmosphere, climate, and human societies. Given the severity of these problems in the context of fast population growth and climate change, courses related to ecohydrology have proliferated worldwide for both undergraduate and graduate students majoring in environmental engineering as well as applied natural sciences.

Ecohydrology is an inherently interdisciplinary subject, which involves scientific fields as diverse as hydrology, fluid mechanics, ecology, biogeochemistry, plant physiology, statistical mechanics, agronomy, atmospheric science, environmental engineering, economics, and social sciences. While each field has excellent references, a textbook for ecohydrology at the upper-undergraduate or graduate level is missing. The present textbook originates from a course on ecohydrology at Duke University and Princeton University, which the authors have taught over the past 15 years, developing extensive course notes and supporting materials. The aim is to provide guidance to students in understanding the fundamental theories in ecohydrology and the connections between the processes involved, as well as a broad spectrum of applications. We also hope that the book may provide a useful reference for scientists and engineers working in this field.

We have aimed to present the tools required to formulate models for the main interactions between biotic and abiotic components in the soil–plant–atmosphere system. In order to help readers familiarize themselves with these tools we have included pedagogical features such as opening vignettes, to motivate readers to explore the chapters; chapter introductions, to briefly introduce the topics covered and to highlight the story line linking the various chapters; key terms (highlighted in italic), to which students should pay particular attention; key points, to sum up the major points of a chapter and allow readers to reflect on what has been learned; and end-of-chapter notes, which include problems and further reading, to allow readers to test their understanding of the topics covered in each chapter and to offer an entry point toward emerging topics for further research.

This textbook is intended for advanced undergraduate and graduate students, as well as researchers from a broad range of disciplines. The essential prerequisite is calculus, differential equations, and basic statistics. That said, we have paid attention to providing enough background material for readers from different backgrounds (on our courses we have had students from engineering, biology and ecology, earth sciences, as well as mathematics and physics). Between the risks of being redundant and taking concepts for granted, we have opted for the first. Accordingly, we have provided self-contained, easy-to-access

[†]Dyson, F., 2004. A meeting with Enrico Fermi. Nature, 427, pp. 297.

reviews of some useful physics and mathematics tools, painted with the same palette, style, and notation, with the hope of helping readers not to waste time dealing with frustrating notational and stylistic differences when jumping from one reference to another. We have strived to make our notation consistent as much as possible; we trust that our readers will be able to sort out the possible ambiguities that may have remained.

We have purposely stressed the temporal dimension of ecohydrological processes, illustrating the way in which rainfall intermittency propagates dynamically through the different components, from the hydrologic response at the land surface, to plant physiology and biogeochemistry, and even to social and economic systems. This is in line with the spirit of the emerging new focus on the *critical zone*, which emphasizes the connections among the critical zone layers, from "bedrock to treetop" (and beyond), as well as the coupled dynamics of natural and human systems. In no way does this mean that the spatial dimension is not interesting and important: indeed, it is a fascinating and difficult area of research and a real frontier for theoretical ecohydrology. However, even starting to deal with it explicitly would call immediately for the use of partial differential equations, the analysis of which (especially with stochastic forcing) is much more difficult. The issue is bypassed here by adopting the usual spatially lumped approach, which resorts to the *vertically averaged soil* and *big-leaf* descriptions, as well as by avoiding areas with significant topography.

We have also tried not to take for granted the theoretical foundations, which are crucial for a precise physical description of the relevant processes. On the other hand, we have privileged the use of minimalist models. While necessarily being only caricatures of reality, these minimalist models are in fact crucial to avoid a treatment that remains only descriptive, and therefore at times vague, in teaching a subject that is broad and interdisciplinary but that, at the same time, requires a quantitative analysis. Pedagogically, laying out the hypotheses, modeling assumptions, and simplifications in a mathematically clear way is also a valuable instrument to sharpen scientific questions and develop a more in-depth understanding. We have provided notes and further reading at the end of each chapter to cover more complex models, related topics, and recent developments. With the help of the related references, these notes may serve as complementary material to broaden a reader's knowledge of ecohydrology.

In summary, we hope that the result is a somewhat consistent *stick-figure description of ecohydrology*, which is easy to teach with a certain precision while provoking students to react to the necessary simplifying assumptions. Discussing the range of validity of mathematical formulations, interpreting the space and timescales at which processes take place, and challenging model assumptions to add detail while keeping a balance in the model description are very valuable pursuits that lead to a more in-depth understanding.

Book Organization

After a brief introduction (Chapter 1), the book is organized into two main parts: The first part (Chapters 2–5) focuses on short timescales to provide a *deterministic description* of the typical average daily behavior of the soil–plant–atmosphere system; it is followed by a second part (Chapters 6–10) centered on the stochastic description of soil moisture dynamics and its impacts on ecosystems, biogeochemistry, and agroecosystems.

The introduction (Chapter 1) presents the main problems of ecohydrology, especially the double two-way interactions, one between abiotic and biotic components and the other between the natural and human components. Chapter 2 provides background concepts related to thermodynamics, dimensional analysis, turbulence transport, and dynamical systems. Chapters 3–5 each describe the main topics needed to understand the three fundamental parts of the soil–plant–atmosphere continuum (SPAC). Once turbulence fluctuations are averaged out, with the help of similarity theory, reviewed in Chapter 2, and the spatial complications are lumped into a vertically averaged soil (Chapter 3) and big leaf approach (Chapter 4), the soil–plant–atmosphere dynamics can be analyzed deterministically at timescales ranging from several minutes to daily (Chapter 5). The resulting simplified descriptions have the great advantage of clearly displaying how variables "talk to each other", helping to disentangle the various pathways of soil–plant–atmosphere interaction.

The second part of the book deals with the *stochastic description* of ecohydrological processes at timescales of a growing season to several years. For such timescales, a deterministic approach is hopeless: rainfall pulses, which are crucial to ecohydrological dynamics, are virtually unpredictable and require a probabilistic description. A simple stochastic approach based on the Poisson process provides a clear and parsimonious description, offering elegant analytical results which yield theoretical insight on ecohydrological responses to future scenarios (e.g., shifts in rainfall regime). The theoretical background in Chapter 6 covers some introductory material on probability and stochastic processes, focusing on stochastic differential equations driven by marked Poisson processes. A brief introduction to these topics may be also useful for students in subjects beyond ecohydrology, including statistical physics, population dynamics, and economics (although, unlike the stock market, our jumps in ecohydrology are usually upward!). Chapter 7 brings together the methods of Chapter 6, along with the results obtained in Chapters 3–5, to arrive at a consistent model of stochastic soil moisture dynamics at different levels of complexity. Chapter 8 presents a series of applications and examples of the impact of stochastic soil moisture dynamics on plant water stress and ecosystem response. Chapter 9 plays the twofold role of a primer on soil biogeochemistry and of describing the impact of soil moisture fluctuations on soil carbon and nitrogen dynamics. Finally, Chapter 10 addresses issues related to human interactions with ecohydrological processes and the related issue of the sustainable management of soil and water resources; also in this chapter, as in the previous chapters, the emphasis is on minimalist modeling.

We normally teach this course as a one-semester graduate course open to upper undergraduates. This, however, requires focusing on the description of the soil–plant–atmosphere continuum in the first part of the course, and then concentrating on the stochastic soil moisture and plant stress dynamics to leave time to cover Chapters 9 and 10 partially. The task is facilitated if students have already some background on hydrology and environmental physics; the background material in Chapters 2 and 3 can be assigned for independent review, and similarly for the chapters on probability and stochastic processes. Alternatively one could devise a course on the soil–plant–atmosphere continuum focusing on the first part of the book, plus perhaps Chapter 9, or develop a stochastic ecohydrology course based on the second part of the book, using as a starting point the description of the soil water balance at the daily timescale (i.e., with daily averaged transpiration and carbon assimilation functions) and then introducing jump processes (Chapter 6) and delving into Chapters 7–10. It would also be logical to have two semesters, a sequence of two courses, for Chapters 1–5 and for Chapters 6–10. The open source code of the Photo3 model (Hartzell et al., 2018b, see also https://samhartz.github.io/Photo3/) can be used for simulations of soil–plant–atmosphere dynamics, as a companion while teaching the first part of the book. This may be useful to help readers to understand the interacting processes and coupling among variables, as well as a tool to develop projects and future research. The numbered notes at the end of each chapter, which include problems and further reading, may also serve as a starting point to go beyond the material covered in class and to suggest possible subjects for a student's final course presentation or to develop future research.

Acknowledgments

It is impossible to acknowledge properly all the many colleagues and friends who have helped us during the preparation of the book. We are especially indebted to Ignacio Rodríguez-Iturbe, an academic pioneer, who has been a mentor, friend, and source of inspiration for both of us. We thank John Albertson, Paolo D'Odorico, Gaby Katul, and Luca Ridolfi for their generous friendship and collegiality, as well as Roberto Revelli, Paolo Perona, Francesco Laio, Carlo Camporeale, and Lamberto Rondoni (Polytechnic of Turin), Gianluca Botter (University of Padova), Valerio Noto (University of Palermo), Francesco Viola and Nicola Montalto (University of Cagliari), Antonio Antonino and the colleagues at UFPE in Recife and UFRPE in Serra Talhada, Pierre Landolt at Fazenda Tamandua, Adjima Thiombiano (University of Ouagadougou), Juan Pedro Mellado (Universitat Politècnica de Catalunya), Joseph Santanello (NASA), Zening Wu and Caihong Hu (Zhengzhou University), Dan Li (Boston University), and Kailiang Yu (ETH Zurich).

We have benefited from the wonderful academic environment provided by our institutions, first at Duke University and now at Princeton and Nanjing; we want to thank in particular our chairs, Catherine Peters (CEE, Princeton), Mike Celia, and Gabe Vecchi (HMEI, Princeton), the President of the Nanjing University of Information Science and Technology, Beiqun Li, the Directors, Xianxin Zhou, Wei Shi, and the Deans, Xieyao Ma, Xin Yuan, Lei Tang, Zhiguo Yu, and Haibo Xu. We are grateful for stimulating research collaborations, especially with Dan Richter and the colleagues of the Calhoun CZO, as well as Jonathan Levine and Steve Pacala (CMI, Princeton), Elie Bou-Zeid, Lars Hedin, and Simon Levin. The early notes of this book started when the first author was on sabbatical at EPFL; he is grateful to Mark Parlange, Andrea Rinaldo, and Wilfried Brutsaert for the opportunities provided by the sabbatical and for stimulating conversations. We were privileged to share many research adventures with talented friends in our research groups: Edoardo Daly, Giulia Vico, Stefano Manzoni, J. R. Rigby, Xue Feng, Annalisa Molini, Federico Maggi, Tony Parolari, Yair Mau, Rodolfo Souza, Simonetta Rubol, Mark Bartlett, Sara Bonetti, Salvatore Calabrese, Norm Pelak, Samantha Hartzell, Milad Hooshyar, Shashank Anand, Sara Cerasoli, Saverio Perri, and Matteo Bertagni.

We thank our publisher, Cambridge University Press, for the opportunity to write this book and for the expert advice and guidance from Matt Lloyd, Ilaria Tassistro, Victoria Parrin, and the anonymous reviewers, who provided very useful suggestions. Last, but certainly not least, our deepest gratitude goes to our families and to Tracy and Bei for their love and support. To them we dedicate this book.

Amilcare Porporato, Princeton University
Jun Yin, Nanjing University of Information Science and Technology

1 Introduction

We forget that the water cycle and the life cycle are one

Jacques Yves Cousteau

This introductory chapter defines ecohydrology as an interdisciplinary subject, focusing on the interactions between hydrological and ecological processes. It emphasizes the feedback between biotic and abiotic elements, as well as those between the human and the environmental components. The special properties of its key ingredient – water – combined with the Earth's energy balance, define the uniqueness of the hydrological cycle and the fluxes between each "reservoir", the most important of which – the root zone – allows us to introduce the primary water fluxes within the soil–plant–atmosphere system and the main variables that will become familiar in the subsequent chapters. A brief outline of the book also presents the important connections between ecohydrology and the carbon and nutrient cycles and the implications of these connections for climate and society. The notes at the end of the chapter offer an opportunity to reflect on the manifold historical, artistic, and spiritual dimensions of ecohydrological phenomena.

1.1 Ecohydrology

Ecohydrology is the study of the *two-way interaction* between the hydrological cycle and ecosystems. More broadly, it is the science of the linkages between life and water on Earth. On the one hand, the space and time variability of the hydrological cycle controls the water availability for ecosystems; on the other hand, ecosystems, especially through transpiration by vegetation, control the main pathway by which water returns to the atmosphere from land. The terrestrial water cycle also drives some of the dynamics of soil organic matter (SOM), microbial biomass, and the related nutrient cycling. These in turn not only affect the vegetation dynamics but also impact the hydraulic and thermodynamic properties of soil, thereby directly acting on the partitioning of water and energy fluxes at the land–atmosphere interface.

The interaction between water balance and plants is responsible for some of the fundamental differences among biomes (e.g., forests, grasslands, savannas) and for the developments of their space–time patterns. Thus one of the first objectives of ecohydrology is to understand the intertwined dynamics of climate, soil, and vegetation. This interaction is especially interesting in *water-limited ecosystems*, where water is a limiting factor not only because of its scarcity but also because of its intermittent and unpredictable appearance.

Many important practical issues depend on a quantitative understanding of ecohydrological processes; these issues include environmental preservation and the proper management of soil and water resources. A

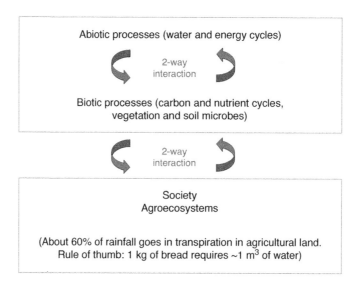

Figure 1.1 Ecohydrology is characterized by the feedbacks between hydrological and ecological processes, involving numerous biotic and abiotic components, as well as by growing interactions with social components, especially through agroecosystems.

solid scientific foundation of sustainable development and eco-agriculture must take into account the two-way interaction between biosphere and hydrosphere, as well as a quantitative description of the human interference with these processes (see Fig. 1.1).

1.2 Water and Life

Most biochemical reactions take place in water. Because of its dipolar molecular structure (see also Sec. 2.1.9), water plays the role of a universal solvent, thereby serving as the preferred vehicle for the transport of nutrients and waste in organisms and ecosystems. The presence of relatively strong hydrogen bonds means that water has a large *heat capacity* (Table 1.1 gives a comparison of the heat capacity of water with those of other substances in the soil–plant–atmosphere system) and large latent heats of melting or freezing and of evaporation or condensation.

Living beings are constituted in great part of water. For example, approximately 60% of the mass of the human body is made of water;[1] non-woody parts of plants are approximately 80% − 90% water, and similar proportions are found in the microbial world, where fungi reach about 80% and bacteria 85% − 90% water. Thus lack of water during drought conditions halts most processes in plants and soil microbes, thereby altering biogeochemical cycles and ecosystem functions.

It is believed that the origin of cellular life is related to the appearance of membranes (Morowitz, 1993), endowing cells with the ability to maintain strong gradients in nutrient concentration and water potential against the spontaneous tendency of the environment towards equilibrium. Such a strong inhomogeneity is typical of dissipative structures in nonequilibrium systems (Kondepudi and Prigogine, 2005) and requires

---

[1] Different categories of people have different percentages of their bodies made up of water (see e.g., http://water.usgs.gov/edu/propertyyou.html). Babies have the most, being born at about 78%. By one year of age, that amount drops to about 65%. In adult men, about 60% of their bodies is water, and fat tissue does not have as much water as lean tissue.

Table 1.1 Specific heat capacity of water in comparison with some other important substances in ecohydrology.

Substance	$J\,g^{-1}\,K^{-1}$
liquid water	4.186
dry air	1.0035
sandy soil	0.8
wood	0.42

a continuous influx of high-quality energy. In both animals and plants, this has required the evolution of adaptation strategies to cope with highly fluctuating environmental conditions and the development of efficient control mechanisms to maintain acceptable internal water contents while still permitting the exchange of water, nutrients, and waste with the surroundings. Plants in particular have developed fine-tuned systems to control hydrologic fluxes between the soil and atmosphere. For example, we will see in Chapter 3 how plant stomata regulate in part transpiration losses to stave off water stress, thus controlling one of the main ecohydrological processes.

1.3　Water on Earth

The energy input from the Sun provides the driving force to the climate dynamics and the hydrologic cycle, while the hydrologic cycle, in turn, exerts a strong feedback on the climate system (Webster, 1994). In contrast with the other planets of the inner solar system, the Earth has abundant water in multiple forms and receives the right amount of energy to have temperatures favorable to life. The large heat capacity and large latent heat of melting/freezing and of evaporation/condensation of water and the presence of water vapor in the atmosphere (the most important greenhouse gas) contribute to the stability of the climate (note the cloudiness of Earth in Fig. 1.2).

The frequent transitions between the water phases pervade the entire climate system. As shown on the *phase diagram of water* in Fig. 1.3, the temperature and pressure ranges of the Earth's *troposphere*[2] guarantee the presence of enough water in each phase. Figure 1.3 also shows how the Earth's hydroclimate, while developing a stable atmosphere, evolved towards the *triple point*. This is in contrast with the climate of the other neighboring planets, which instead followed runaway trajectories, either towards the only-vapor phase (Venus) or halted on the sublimation/freezing transition line (Mars).

1.4　The Hydrologic Cycle

The hydrologic cycle and the related atmospheric dynamics are fueled by solar radiation. The sequence of evaporation of liquid water from land and ocean lifts water vapor to the atmosphere and brings liquid

---

[2] The troposphere is the lower and most dynamic part of the atmosphere, where most of the air mass is present. It has a variable extent, going from an average of 6 km over the poles up to 16 km at the equator.

Table 1.2 Water content in each phase and incoming solar radiation for the Earth and its neighboring planets (Webster, 1994).

Planet	Ice (kg)	Liquid (kg)	Vapor (kg)	Distance[†] (AU)[‡]	Solar const.[*] (W m^{-2})
Venus	0	0	4.2×10^{16}	0.7184–0.7281	2647–2576
Earth	4.3×10^{19}	14×10^{21}	1.6×10^{16}	0.9833–1.017	1413–1321
Mars	1.0×10^{17}	$0^{§}$	2.0×10^{13}	1.382–1.666	715–492

[†] Perihelion and aphelion.

[‡] AU = astronomical unit = Sun–Earth distance = 150×10^6 km.

[*] Computed using the inverse squares law, by which radiation intensity is inversely proportional to the squared distance from the Sun.

[§] However, Ojha et al. (2015) reported evidence of liquid water flows on Mars.

Figure 1.2 The blue planet: NASA composition from images from remote sensing satellite instrument MODIS (Moderate Resolution Imaging Spectrometer), flying over 700 km above the Earth onboard the TERRA satellite. http:/visibleearth.nasa.gov.

water back to the surface through condensation and precipitation. Figure 1.4 shows the average *fluxes* normalized as percentages with respect to the average rainfall on land. It is apparent that the hydrologic cycle over the oceans is more than four times the size of the cycle over land. This greater water availability over the oceans, coupled to the mixing dynamics of the atmospheric circulation, gives rise to an average net circulation of the water cycle, with more atmospheric water coming towards the land. Such a flux, in turn, is balanced by the surface and subsurface flows going from the land to the oceans.

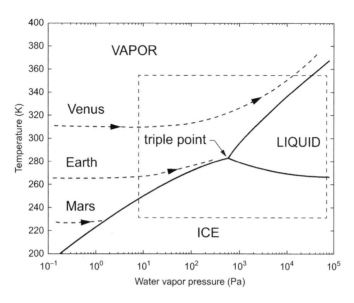

Figure 1.3 Phase diagram of water and qualitative comparison of the climate evolution of Venus, Earth, and Mars. The rectangular area refers to the typical tropospheric condition on Earth. Modified from Webster (1994).

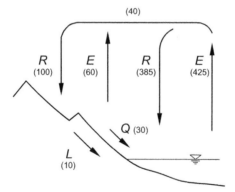

Figure 1.4 Schematic representation of the terrestrial and oceanic water cycles, with fluxes represented as percentages of rainfall on land. R, rainfall; E, evapotranspiration; L, leakage/percolation; Q, runoff.

The global averages for *stocks* and fluxes involved in the hydrologic cycle are reported in Table 1.3. Although such estimates are affected by considerable uncertainty and temporal variability, it is interesting to note the different orders of magnitude involved in the average stocks, fluxes, and residence times in the different reservoirs or *compartments* (Yin and Porporato, 2019). For example, the soil water depth of 15 cm, reported in Table 1.3, is indicative of the relatively deep soil layer considered in the estimates: as we will see in Chapter 3, the soil water depth (e.g., volume per unit surface) is given by the product of porosity, soil depth, and average soil moisture; thus, a water depth of 15 cm, divided by a typical porosity of 0.4 and an average soil moisture of 0.35, corresponds to a soil depth of about 94 cm, which also explains the relatively long mean residence time of 58 days. Comparing the fluxes referred to a unit surface, the total flux from the ocean due to evaporation, divided by the ocean surface, gives 1.20 meters (of liquid water) per year, or 3.3 mm/day, while rainfall (Fig. 1.4) is 1.1 m/yr or 2.9 mm/day. On the other hand, the total flux (rainfall) divided by 80% of the land surface (assuming 20% of desert surface with basically no evapotranspiration) gives 0.72 m/yr or 2.0 mm/day, while the evapotranspiration from the land surface,

Table 1.3 Estimates of average hydrologic stocks, fluxes, and residence times (see Webster, 1994; Brutsaert, 2005; and references therein).

Reservoir	Stock (10^{15} kg)	Depth (m)	Total flux (10^{15} kg/yr)	Residence time (days)
Oceans	1,400,000	3750[†]	434	1,170,000
Marine atmosphere	11	0.031[†]	434	9
Terrestrial atmosphere	4	0.027[‡]	107	14
Land	55,000	370[‡]	107	188,000
Ice and snow	30,000	—	—	3,240,000
Surface water (total)	360	—	—	—
Surface water (freshwater)	270	—	—	—
Soil water	15	0.15[+]	95	58
Groundwater (total)	24,000	160	—	—
Groundwater (freshwater)	10,000	—	—	—
Biota	1	0.02[+]	65	4

[†] Depth computed with reference to the ocean surface, 3.61×10^8 km^2.
[‡] Depth computed with reference to the land surface, 1.49×10^8 km^2.
[+] Excluding deserts, which are about 20% of the total land surface.

computed as 60% of rainfall (Fig. 1.4) is 0.43 m/yr or 1.2 mm/day. Such relatively low values over land may be explained by the extensive presence of deserts and arid regions.

Each compartment of the hydrologic cycle (i.e., soil, plant, atmosphere, glaciers, etc.) can be treated as a *control volume* with control surface Ω (Fig. 1.5), having a corresponding continuity equation[3]

$$\frac{dW(t)}{dt} = I(t) - O(t), \tag{1.1}$$

where $W(t)$ is the volume of liquid water within the compartment at time t, $I(t)$ is the flux of water into the control volume, and $O(t)$ is the flux of water lost from the system. This equation provides a spatially lumped description of the water balance, without giving any detail of the mechanisms responsible for the input and output water fluxes, whose specific forms (equations of motion) have to be supplemented separately on the basis of first principles (fluid mechanics and thermodynamics) or, more often, on semi-empirical formulations. In ecohydrology, it is usual to write the balance equation for a unit ground surface: dividing the previous equation by the land surface area A, the state variable $w = W/A$ becomes a depth (with dimension L) and the corresponding fluxes become depths of liquid water per unit time (with dimension L/T).

The terms in Eq. (1.1) fluctuate randomly in time as a result of complex hydroclimatic forcing. Assuming the existence of a long-term *steady state* (see Sec. 6.4.2 for a discussion of such concepts within the context of stochastic processes), one can write a long-term balance equation for the storage W, in which the inputs I and outputs O balance,

---

[3] The mass balance equation can be obtained by multiplying each term by the density of water, assumed constant (as water is incompressible fluid).

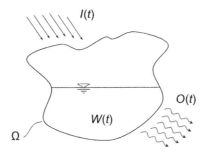

Figure 1.5 Illustration of the general continuity equation for a control volume. Ω, control surface; W, storage capacity of the control volume; I, input rate; O, output rate.

$$\overline{\frac{dW(t)}{dt}} = 0 = \bar{I} - \bar{O}, \tag{1.2}$$

where the overbars indicate long-term averages.[4] One very interesting property is that, at steady state, the *mean residence time*, τ, within a compartment is independent of the details of the transport process underlying Eq. (1.1) and can be obtained as the ratio of the mean storage and the mean input (or output),

$$\tau = \frac{\bar{W}}{\bar{I}} = \frac{\bar{W}}{\bar{O}}. \tag{1.4}$$

This equation can be used to estimate the mean residence times reported in the last column of Table 1.3.

1.5   The Soil Water Balance

The soil–plant system is the main stage of ecohydrological processes, which are controlled by the soil moisture dynamics. Using the words of Noy-Meir, "the soil is the store and regulator in the water flow system of the ecosystems, both as a temporary store for the precipitation input, allowing its use by organisms, and as a regulator controlling the partition of this input between the major outflows: runoff, evapotranspiration redistribution, and the flow between the different organisms" (Noy-Meir, 1973). The soil moisture synthesizes the interaction of climate, soil, and vegetation and their impact on plant dynamics and nutrient cycling (Fig. 1.6).

Assuming negligible lateral redistribution due to topographic effects, the rate of change of the volume of water over the rooting zone, expressed per unit ground area (in effect, a depth of water with dimension L), is given by the balance of input and output fluxes per unit ground area (with dimensions L/T):

$$w_0 \frac{ds}{dt} = R + J - (C_i + E_v + E_t + L + Q), \tag{1.5}$$

where $w_0 = nZ_r$ is the soil storage capacity per unit area, with n the depth-averaged soil porosity (i.e., the fraction of volume occupied by soil pores), Z_r the rooting depth, and s the vertically averaged *relative soil*

---

[4] For a generic quantity x,

$$\bar{x} = \lim_{T \to \infty} \frac{1}{T} \int_0^T x(t) dt. \tag{1.3}$$

Figure 1.6 Schematic representation of the various mechanisms of the soil water balance with emphasis on the role of plants. After Laio et al. (2001c).

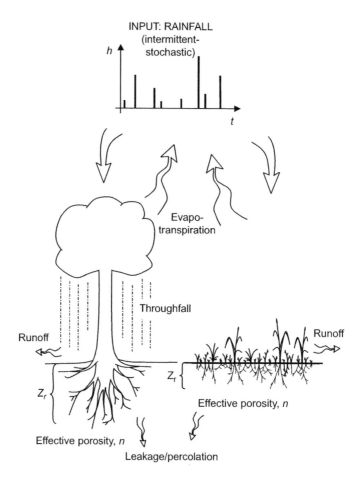

moisture, ranging between 0 and 1 for dry and saturated soils, respectively.[5] On the right-hand side of the equation, the fluxes are:

$$R = \text{rainfall rate,}$$
$$J = \text{irrigation rate,}$$
$$C_i = \text{canopy interception,}$$
$$E_v = \text{soil evaporation rate,}$$
$$E_t = \text{plant transpiration rate,}$$
$$L = \text{percolation,}$$
$$Q = \text{runoff rate.}$$

It is often useful to combine some of the previous processes: for example, $I = R + J - C_i - Q$ is the rate of infiltration into the soil, while $E = E_v + E_t$ is called the evapotranspiration (which sometimes also includes the canopy interception), while $LQ = L + Q$ refers to the combination of deep infiltration and

---

[5] These quantities will be defined and discussed in detail in Sec. 3.2.

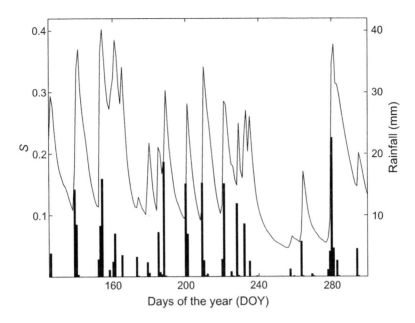

Figure 1.7 Vertically averaged soil moisture and daily rainfall during a typical growing season at the Duke Forest site (Orange County, NC, USA), measured in the rooting zone of a loblolly pine (*Pinus taeda*) plantation.

runoff. Note that the quantity L can also represent capillary rise, in which case it is negative (i.e., an input to the soil). The soil water balance equation (1.5) plays a crucial role in ecohydrology. We will come back to it many times throughout this book.

Rainfall is a highly intermittent function of time, while evapotranspiration and percolation are nonlinear functions of soil moisture. The highly unpredictable hydroclimatic variability (rainfall, in particular) makes it necessary to describe the long-term variability of soil moisture in probabilistic terms. The random, unpredictable nature of soil moisture fluctuations, with sudden jumps, is apparent in Fig. 1.7. As a result, Eq. (1.5) should be considered as a stochastic differential equation (see Chapter 6.4).

1.6   Temporal Scales of Soil Moisture and Plant Dynamics

The temporal dynamics of soil moisture involves a wide range of *timescales*. Our approach will first average out the short timescales of turbulence (e.g. seconds to minutes) to get an hourly timescale description of the soil–plant dynamics. At this level, the effects of meteorological processes (e.g., in radiation, air temperature, and humidity) and plant physiological processes (e.g., xylem cavitation and stomatal control) on the soil–plant system can be described by following their diurnal evolution (Chapters 4 and 5).

A further and important upscaling occurs when averages over a course of a day are taken, so that evapotranspiration fluxes are given in terms of their daily rates. With a focus on processes averaged at the daily timescale, the dynamics of the soil–plant system can be analyzed over a growing season and beyond. At these timescales, rainfall events may be considered instantaneous events at a point in time and their unpredictable nature injects a probabilistic connotation into the dynamics of the soil–water balance. The impacts on soil moisture distribution and plant water stress may be initially analyzed for cases with negligible seasonal and interannual rainfall components. This (statistically) steady-state assumption is a good approximation for example in the savannas of Nylsvley in South Africa or after the spring transient in temperate ecosystems, such as the Duke Forest (Fig. 1.7) and the dry-down in the dry season of a Mediterranean, monsoon, or tropical dry ecosystem. More generally, seasonal cycles in temperature and/or precipitation, especially when combined with a deep active soil layer, mean that the soil moisture dynamics are dominated by stochastic transients, as in large parts of Mediterranean, monsoon, and tropical dry ecosystem dynamics. The interannual hydroclimatic variability adds a further degree of randomness at longer timescales.

Plants heavily condition the soil water balance dynamics, which in turn impacts the plant growth. Through *stomatal closure* and other *water-use strategies* (Chapter 4), plants control transpiration directly, while other plant properties, such as changes in albedo and roughness, infiltration characteristics, and root depth and distribution, affect the terrestrial water cycle indirectly. Grossly simplifying matters, we will often group plants into functional types that have approximately similar physiological characteristics and respond similarly to environmental forcing. Thus, for modeling purposes, we will thus often speak of grasses, shrubs and trees and further distinguish their characters (e.g. evergreen, broadleaf versus needle leaf, rooting depth, photosynthetic pathway, height, leaf and root area index, etc.). In other cases we will refer to *biomes*, a biodiverse collection of plant function types forming an ecosystem type adapted to a particular environment, such as the various types of forest (e.g., temperate, tropical, drought deciduous), savannas, grasslands, etc.

Figure 1.8 shows the general patterns of vegetation productivity and biodiversity for different biomes as a function of rainfall (Shmida and Burgess, 1988). Such patterns follow closely the trends of the mean and variance of soil moisture as a function of rainfall, suggesting a strong soil moisture control and nonlinear effects of rainfall on ecosystems. Productivity increases with rainfall but tends to plateau at high rainfall rates, while the peak of biodiversity is related to the presence of ecological *niches* giving rise to the coexistence of plant functional types with different water-use strategies, adapted to either wet, dry, or intermediate conditions (Chapters 7 and 8).

1.7   Soil Moisture, Biogeochemistry, and Society

The dynamics of soil moisture is also a key factor determining the dynamics of soil *nutrient cycling*, soil carbon storage, and soil microbial processes (Chapter 9). The production of plant residues is in direct dependence with the growth of vegetation through its photosynthetic capacity, which in turn depends on water availability. Soil moisture is also a controlling factor of mineralization and uptake. Figure 1.9 presents two interesting examples of this dependence and shows how the equilibrium nitrogen-cycling rate for a site is directly related to its water availability (Aber et al., 1991). We will be especially interested in modeling the hydrologic controls on the soil nutrient cycles and the propagation of hydrologic variability into the soil carbon and nitrogen dynamics.

Through droughts, floods, and the related destruction of crops, ecohydrological fluctuations have affected social dynamics from times immemorable (see Fig. 1.10 for an interesting example of drought-induced social dynamics). As shown in the data synthesis in data synthesis in Fig. 1.11, crop yield, as a function of seasonal evapotranspiration (E_{seas}), depends strongly on water availability. The yield–E_{seas} relationship is highly species-specific: the yield of wheat (*Triticum aestivum*), a relatively drought-tolerant species, is approximately constant as long as E_{seas} remains above 300 mm, while the yield of corn

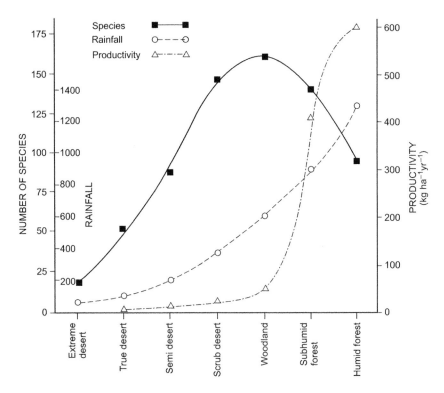

Figure 1.8 General link between precipitation, biomass, and biodiversity in water-controlled ecosystems. After Shmida and Burgess (1988).

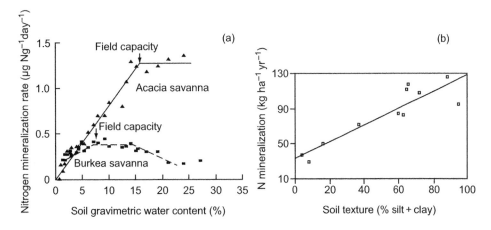

Figure 1.9 (a) Dependence of nitrogen mineralization rate on soil water content in the broad-leafed and fine-leafed savanna at Nylsvley (Scholes and Walker, 2004). (b) Measured nitrogen mineralization rates for several stands in Wisconsin in relation to soil texture, which is closely related to the soil water availability (Aber et al., 1991).

(*Zea mays*), a more drought-sensitive species, is markedly reduced by a small decrease in E_{seas}, even at relatively high values of total seasonal transpiration. As a result, wheat may adapt better to an irrigation deficit, while the drought sensitivity of corn makes stress-avoidance irrigation a better strategy to sustain

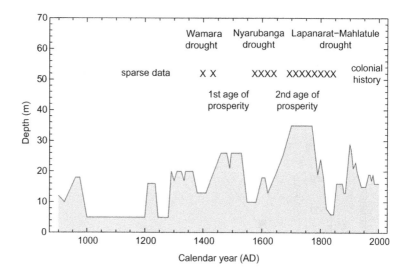

Figure 1.10 Drought-related political upheaval (X) recorded in oral tradition in East Africa with, for comparison, reconstructed levels of Crescent Island Crater lake and other related indicators of extended droughts. Redrawn after Verschuren et al. (2000).

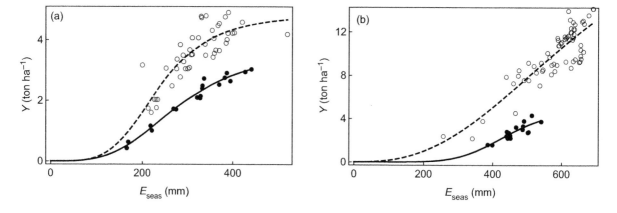

Figure 1.11 Synthesis of data for crop yield Y (Vico and Porporato, 2011a) as a function of total seasonal transpiration, for: (a) *Triticum aestivum*, and (b) *Zea mays*. The open and solid circles refer respectively to two different data sources. Points in each dataset refer to different irrigation treatments, consisting of irrigation withdrawal over different growth period and/or maintaining low, intermediate, or high soil water content throughout the season.

crop productivity. The data also show how yield depends on crop cultivar, soil features and management, nutrient availability, climatic conditions, and type of irrigation.

Recently, however, the interference of human dynamics with ecohydrological processes has also grown, because of extensive agricultural practices and the related land-use changes. As human interference with the environment escalates to worrisome levels, it is imperative to manage ecohydrological processes in a sustainable manner. Chapter 10 will analyze some of these aspects, relating to the anthropogenic acceleration of biogeochemical cycles, in view of the need for sustainable agriculture, irrigation, and management of soil water and nutrient resources.

1.8 Key Points

- Ecohydrology is characterized by the two-way interaction between hydrology (and the abiotic processes related to water and energy cycle), vegetation, and soil microbial life; these in turn impact and are influenced by human society.
- The abundance of water and the special conditions of the Earth's energy balance give rise to the kaleidoscopic variety of ecohydrological processes, centered around the triple point of the water phase diagram.
- The mass-conservation principle applied to the root zone, or active soil layer, leads to the soil–water–balance equation.
- This central equation of ecohydrology describes the dynamics of soil moisture as an alternation of intermittent jumps due to rainfall and relatively regular dry-downs, which depend on the state of the soil–plant–atmosphere continuum.
- The unpredictability and variability of environmental forcing, especially rainfall, calls for a probabilistic analysis of ecohydrological dynamics.

1.9 Notes, including Problems and Further Reading

1.1 (Elements of Ecohydrology) Draw a conceptual picture of the essential elements of ecohydrology, their interactions, and the traditional disciplines involved in their study.

1.2 (Primitive Planet Temperature) Compute the primitive surface temperature T_0 of the planets, shown on the left-hand side of Fig. 1.3, assuming that they have no atmosphere and albedo equal to 0.17.

The radiative equilibrium between incoming shortwave radiation, R_I, from the Sun and the outgoing longwave radiation, R_O, from a planet at temperature T_0 can be written as (see Note 2.14)

$$R_I = \pi r^2 S(1 - \alpha_s) = R_O = 4\pi r^2 \sigma T_0^4, \tag{1.6}$$

where S is the solar constant (Table 1.2), $\sigma = 5.67 \times 10^{-8}$ W m^{-2} K^{-4} is the Stefan–Boltzmann constant, and α_s is the albedo of the original planet without atmosphere. The solar constant is the radiation flux density measured on a surface perpendicular to the rays of the Sun, while the albedo is the ratio of the reflected irradiance and the received irradiance (see Sec. 5.1.1).

Solving for T_0 and using the values in Table 1.2 one obtains

$$T_{0,\text{Venus}} = 313 \text{ K},$$

$$T_{0,\text{Earth}} = 266 \text{ K},$$

$$T_{0,\text{Mars}} = 217 \text{ K}.$$

1.3 (Snowball Earth) Snow and ice are very efficient at reflecting solar radiation and have high surface albedo (see Table 5.1); thus, with lower temperatures and more ice surface, the Earth's albedo is larger than that of Venus or Mars. Assuming that the Earth's albedo (α_s) is a function of surface temperature, e.g.

$\alpha_s = 0.4 - 0.3 \tanh[(T_0 - 250)/10]$, use Eq. (1.6) to explore the effects on the Earth's surface temperature and the possible presence of multiple solutions. Note the low-temperature solution corresponding to a state called *Snowball Earth* (Dymnikov and Filatov, 2012).

1.4 (Hydrologic Cycle and Water Phase Diagram) Sketch the various steps of the terrestrial hydrologic cycle (see Fig. 1.4) on the phase diagram of water (Fig. 1.3). See Konings et al. (2012).

1.5 (Soil Moisture Residence Time) Using data from Table 1.3 estimate the typical residence time of water in soils. Comment on the main factors controlling its value.

1.6 (Water Use for Agriculture) Globally, approximately 16 billion acres (16 million km^2) are cultivated. The wheat for one kilogram of bread requires 1 m^3 of water for its cultivation, and the number is 15 times higher for meat (e.g., beef). Consider an ideal diet of 2,250 kcal and compute the water footprint of the current population of 7.8 billions. Compare the results with typical values of the global hydrologic cycle on land (Sec. 1.4).

1.7 (Environmental Humanities and Ecohydrology) Ecohydrological phenomena have inspired myths, religious beliefs (see Fig. 1.12), and literary works, including the Georgics of Virgil and the Grapes of Wrath of Steinbeck. In China, Shennong (神农) was a culture hero for having taught the ancient Chinese to practice agriculture and use herbal drugs. These masterpieces and mythologies may be considered *ante litteram* expressions of the modern environmental humanities in relation to ecohydrology.

1.8 (Critical Zone) The outer extent of vegetation down to the lower limits of groundwater has become known as the *critical zone* (CZ), because of its essential role in natural and managed ecosystems. Quoting Brantley et al. (2007), "the CZ is a complex mixture of air, water, biota, organic matter, and Earth

Figure 1.12 Yum-Kaax, the Mayan God of wild plants and, according to some sources, of corn, rain, and suicide, epitomizes the impact of ecohydrological processes, linked to droughts and famine, on ancient civilizations.

materials, throughout which chemical reactions proceed both abiotically and through catalysis by organisms, providing nutrients and energy for the sustenance of terrestrial ecosystems. As rocks containing high-temperature mineral assemblages re-equilibrate with fluids at the surface, environmental gradients develop. These gradients characterize the CZ both by nurturing life and by simultaneously responding to life".

2   Physics Background for Ecohydrology

> To avoid getting mired in mathematical questions beyond human capabilities, perhaps you should stay closer to physics.
>
> *Ruelle (2004)*[†]

This methodological chapter reviews some important physics concepts to quantitatively describe ecohydrological processes. The first part of the chapter presents a synthesis of thermodynamic concepts, which are necessary to define the water status (i.e., the water potential) within the soil–plant–atmosphere continuum (SPAC), which will be analyzed in detail in the first part of the book (Chapters 3–5), as well as the incessant phase transitions that characterize many aspects of the hydrologic cycle. In the second part of the chapter we summarize dimensional analysis, including the Π theorem and similarity theory. These allow us to formulate effective transport laws for turbulent fluxes between the land surface and the atmosphere (Chapter 5) along with dimensionless formulations of ecohydrological laws, such as the Budyko curve (Chapter 7) for average hydrologic partitioning. Finally, we review some basic concepts related to dynamical systems in one and two dimensions.

2.1   Review of Thermodynamics

In this section, we review some concepts needed to define water potential and describe water status within the soil–plant–atmosphere continuum, which will be important in Chapters 3–5. More in-depth descriptions of thermodynamics can be found in the following excellent books: Kestin (1979), Zemansky and Dittman (1997), Kondepudi and Prigogine (2005), Bejan (2006), and Callen (2006).

Thermodynamics is founded on two main laws, which are complementary to the mass and momentum conservation equations. They describe how energy is distributed among its different forms, transferred, and degraded during thermodynamic transformations. These laws may be applied to systems that are *isolated*, i.e., in which there is no exchange of energy or matter, or *closed*, i.e., which do not allow mass exchange but do allow energy exchange, or *open*, i.e., where both mass and energy exchange with the surroundings are possible.

2.1.1   The First Law of Thermodynamics

The first law of thermodynamics is a statement of the conservation of energy (energy is neither produced nor destroyed) and historically has been crucial in revealing the existence of a state function representing

---

[†]Ruelle, D. (2004). "Conversations on nonequilibrium physics with an extraterrestrial." *Physics Today* 57.5, pp. 48–53.

the internal energy of a system. We begin by considering an isolated system, which does not exchange mass or energy with the environment. It contains N moles[1] of a pure substance of molar mass m, and so has mass

$$M = mN \tag{2.1}$$

and volume V. For such a system, and referring to an infinitesimal transformation, the first law reads

$$dE_{\text{tot}} = d(K + P + U) = 0, \tag{2.2}$$

where E_{tot} is the total energy of the system, K is the kinetic energy, P is the potential energy, and U is the internal energy.

For closed systems, which are not energetically isolated, energy can be exchanged by heat (dQ) and/or work transfer in or out of the system (dW), i.e.,

$$dE_{\text{tot}} = d(K + P + U) = dQ + dW. \tag{2.3}$$

While still a statement of the conservation of energy, Eq. (2.3) also shows the equivalence of the work and heat modes of energy transfer. Note that in Eq. (2.3) the work done *on* the system is assumed to be positive, $dW > 0$, because it contributes energy to the system; the same convention is used for the heat flow, which is positive, $dQ > 0$, when it is an input to the system. In *simple systems* (no gravitational or dynamic effects), $K = P = 0$ so that the previous expression becomes

$$dU = dQ + dW. \tag{2.4}$$

2.1.2 The Second Law of Thermodynamics

The second law of thermodynamics tells us that entropy, unlike energy, is not a conserved quantity but is created by irreversibility in thermodynamic transformations. In particular, for an isolated system the second law says that the entropy S never decreases,

$$dS = dS_{\text{gen}} \geq 0, \tag{2.5}$$

where S_{gen} is the entropy production within the system.

The entropy change dS is zero only for ideal (reversible) transformations, while it is always positive for real (irreversible) transformations owing to entropy production. Entropy production in simple systems may be due to work dissipation (e.g., irreversible expansion and compression or friction) and/or irreversible heat flow, i.e., across finite temperature gradients. In more complex systems, other sources of entropy production may include different processes, such as mixing, nonequilibrium chemical reactions, and receiving/emitting radiation.

In agreement with the second law, when internal constraints are removed inside an isolated system, the entropy spontaneously increases owing to the consequent irreversible transformations that take place internally, until *thermodynamic equilibrium* is reached. The state of thermodynamic equilibrium is thus a state of maximum entropy for given U, V, and N (these quantities cannot change since the system is

---

[1] One mole contains an Avogadro's number of molecules of that substance, that is $N_A \approx 6.022 \times 10^{23}$ particles. The system is composed of $\tilde{N} = NN_A$ particles, each of weight $\mathcal{M}$, so that $m = \mathcal{M}N_A$ and $M = mN = \mathcal{M}NN_A$.

isolated). At equilibrium, internal transformations are no longer possible, unless mass or energy can be exchanged with the surrounding (Callen, 2006).

For a closed system, changes in entropy can be brought about by heat flux through the system boundary (dS_{heatflux}), so that the second law becomes

$$dS = dS_{\text{heatflux}} + dS_{\text{gen}}, \tag{2.6}$$

where dS_{heatflux} can be either positive or negative depending on the direction of the heat flux, while, as before, $dS_{\text{gen}} \geq 0$. It is important to note that there is no entropy flux associated with work, a fact that reveals a profound qualitative difference between heat and work fluxes, despite their quantitative equivalence stated by the first law. This is made explicit by the Carnot theorem (see the next section), which tells us that an input of work can be entirely converted into a heat flux while the opposite is not true.

2.1.3 Reversible Transformations

While real processes always produce entropy (at a rate which generally depends on the details of the transformation and on how far the system has been brought out of equilibrium), for theoretical purposes it is very useful to refer to ideal processes, called *reversible transformations*, which connect equilibrium states by quasi-static transformations that take place without entropy production. For such transformations $dS_{\text{gen}} = 0$ and the entropy flux is $dS_{\text{heatflux}} = dQ^{\text{rev}}/T$, where T is the absolute temperature of the system. For closed systems, the second law (2.6) can be written as

$$dS = \frac{dQ^{\text{rev}}}{T}. \tag{2.7}$$

In mechanical systems, the reversible work is in the form of ideal compression or expansion (Zemansky and Dittman, 1997; Kondepudi and Prigogine, 2005):

$$dW^{\text{rev}} = -pdV, \tag{2.8}$$

where p is the system pressure.[2] With Eqs. (2.7) and (2.8), the first law, for reversible transformations, becomes

$$dU = dQ^{\text{rev}} - pdV. \tag{2.9}$$

Combining the first and the second law, Eqs. (2.7) and (2.9), by eliminating dQ^{rev}, one obtains the important *Gibbs equation* for closed systems:

$$dU = TdS - pdV. \tag{2.10}$$

The *Carnot theorem* also follows by combining the two laws,[3] in the special case of cyclic, reversible processes (i.e., $\oint dU = 0$ and $\oint dS = 0$) in which an input of heat at a high T_H, $\Delta Q_H^{\text{rev}} > 0$, is used to produce work, $\Delta W^{\text{rev}} > 0$. Because reversible transformations do not produce entropy and work does not

---

[2] This equation is in agreement with the (perhaps more familiar) mechanical definition of work, which intuitively follows by thinking of a cylinder of gas in which the compression work is accompanied by a reduction in volume given by the displacement dx of the piston times the cross-sectional area A, $-dV = Adx$. In such a case $dW^{\text{rev}} = -pdV = pAdx = Fdx$, i.e., a force times a displacement.

[3] Carnot actually obtained his famous theorem before the discovery of entropy and the formulation of the second law by Clausius.

carry an entropy flux (see the discussion after Eq. (2.6)), it follows that the transformation must also be accompanied by a heat output, $\Delta Q_C^{\text{rev}} > 0$, at a cooler temperature T_C,

$$\oint dU = \Delta Q_H^{\text{rev}} - \Delta W^{\text{rev}} - \Delta Q_C^{\text{rev}} = 0, \tag{2.11}$$

$$\oint dS = \frac{\Delta Q_H^{\text{rev}}}{T_H} - \frac{\Delta Q_C^{\text{rev}}}{T_C} = 0, \tag{2.12}$$

for otherwise a complete conversion of ΔQ_H^{rev} into work (i.e., $\Delta Q_C^{\text{rev}} = 0$) would violate the second law. The *Carnot efficiency* of the transformation can then be computed, by eliminating ΔQ_C^{rev} from the previous equations, as

$$\eta \equiv \frac{\Delta W^{\text{rev}}}{\Delta Q_H^{\text{rev}}} = \left(1 - \frac{T_C}{T_H}\right). \tag{2.13}$$

Such a theoretical efficiency is attainable only for reversible processes: any irreversibility would in fact add a positive entropy production in Eq. (2.12), resulting (for given input of heat) in a larger output of heat and a reduction in the amount of extractable work.

2.1.4 Open Systems: Enthalpy and Gibbs Electrochemical Potential

When mass is reversibly added or subtracted from a system, the energy changes for two reasons: (1) the mere addition or subtraction of internal energy in the mass that is being added or subtracted; (2) the reversible work done on or by the system in exchanging matter into or out of the system at pressure p (Bejan, 2006). As a result, for a change of dN moles, the change in internal energy is $(u + pv)dN$, where $u = U/N$ is the specific molar internal energy and $v = V/N$ is the specific molar volume $dV = vdN$, and the first law becomes

$$dU = dQ + dW + (u + pv)dN. \tag{2.14}$$

Recognizing that the term in parentheses is the specific molar *enthalpy*, $h = u + pv$, Eq. (2.14) can be written as

$$dU = dQ^{\text{rev}} + dW^{\text{rev}} + hdN. \tag{2.15}$$

The specific enthalpy is also known as the *sensible heat*.

Similarly, the entropy balance equation (second law) is

$$dS = \frac{dQ^{\text{rev}}}{T} + sdN. \tag{2.16}$$

where sdN is the flux of entropy due to mass flow, and s is the specific molar entropy $s = S/N$ and sdN is the flux of entropy due to mass flow. Note again that the reversible work done by the flow of matter does not contribute to the entropy balance.

Combining these equations and using the definition of reversible work in Eq. (2.8), one obtains the *Gibbs equation for open systems*,

$$dU = TdS - pdV + \mu dN, \tag{2.17}$$

where

$$\mu = u + pv - Ts = h - Ts \tag{2.18}$$

is the Gibbs (electro)*chemical potential*, also known as the molar Gibbs free energy $\mu = G/N$ for a pure substance. After suitable extensions, as we will see in Sec. 2.1.14, this quantity will allow us to describe the water status and movement within the soil–plant–atmosphere system.

2.1.5   Extensivity and the Gibbs–Duhem Relationship

For a substance in thermodynamic equilibrium, at a given temperature, pressure, and chemical potential, the internal energy, entropy, volume, and mass are all proportional to the amount of substance present in the system, expressed by the number of moles N. For this reason, such quantities (U, S, V, M) are called *extensive quantities*, as opposed to *intensive quantities*, like T, p, and μ, whose values do not depend on the amount of matter considered. Using specific molar quantities, $U = uN$, $S = sN$, $V = \upsilon N$, and $M = mN$, we can write Eq. (2.17) as

$$u dN = T s dN - p \upsilon dN + \mu dN, \tag{2.19}$$

where u, s, and υ are the specific (molar) internal energy, entropy and volume. For given temperature, pressure, and chemical potential, Eq. (2.19) can be integrated directly with respect to N (this means that we are introducing or removing material at the same T, p, μ as that of the system), to give

$$U = TS - pV + \mu N, \tag{2.20}$$

or, in molar form,[4]

$$u = Ts - p\upsilon + \mu. \tag{2.23}$$

Considering now a generic infinitesimal reversible transformation in which p, T, and μ can change, the differential of Eq. (2.20) is

$$dU = S dT + T dS - dp V - dV p + d\mu N + dN \mu. \tag{2.24}$$

For this transformation to be consistent with the first and second laws, Eq. (2.17) must also hold. This implies that the changes in p, T, and μ cannot be completely independent but must vary according to the *Gibbs–Duhem equation*,

$$S dT - V dp + N d\mu = 0, \tag{2.25}$$

which follows from equating (2.24) and (2.17). This important equation is often used to compute how chemical potential behaves as a function of temperature and pressure when expressions for the specific molar entropy and volume are known:

$$d\mu = -s dT + \upsilon dp. \tag{2.26}$$

---

[4] Gibbs also showed that the internal energy can be expressed solely as a function of the state variables entropy, volume, and number of moles, that is $U = U(S, V, N)$, the so-called fundamental equation. This in turn provides thermodynamic definitions of temperature, pressure, and chemical potential, appearing as intensive quantities in Eq. (2.20). By comparing the differential of the fundamental equation,

$$dU(S, V, N) = \frac{\partial U}{\partial S} dS + \frac{\partial U}{\partial V} dV + \frac{\partial U}{\partial N} dN \tag{2.21}$$

with the Gibbs equation (2.17), it follows that

$$T = \frac{\partial U}{\partial S}, \quad p = -\frac{\partial U}{\partial V}, \quad \mu = \frac{\partial U}{\partial N}. \tag{2.22}$$

2.1.6   Ideal Gas

We will use the ideal gas model, which approximates well the behavior of relatively rarefied gases, when the interactions between gas molecules are not important (other than collisions). The mechanical equation of state is

$$pV = N\mathcal{R}T = MRT, \tag{2.27}$$

where $\mathcal{R}$ is the universal gas constant $8.314\ \mathrm{J\ K^{-1}\ mol^{-1}}$ and $R = \mathcal{R}/m$ is the specific gas constant with $M = mN$. The former is related to the molar specific heat capacities at constant pressure and volume,

$$\mathcal{R} = c_p - c_v, \tag{2.28}$$

reflecting the fact that the amount of heating at constant pressure is higher than at constant volume because of the extra work necessary for an isobaric reversible expansion. The state equation can also be written as

$$pv = \mathcal{R}T, \tag{2.29}$$

or, introducing the density $\rho = M/V = m/v$,

$$p = \rho RT. \tag{2.30}$$

For an ideal gas, the internal energy and, consequently, the constant-volume specific heat capacity are only functions of temperature (Kestin, 1979; Kondepudi and Prigogine, 2005). Thus, the thermodynamic equation of state for an ideal gas is

$$u = c_v T, \tag{2.31}$$

which, combined with the equation of state (2.29), gives

$$h = u + pv = c_v T + \mathcal{R}T = c_p T. \tag{2.32}$$

This shows that enthalpy and the constant-pressure specific heat capacity are functions of temperature only. Although $\mathcal{R}$ is constant, the specific heats are in principle functions of temperature, but they are often assumed constant when the considered range of temperatures is not large.

The Gibbs equation (2.10) in its molar form, with the subsequent use of Eq. (2.29), gives

$$ds = \frac{1}{T}du + \frac{p}{T}dv = c_v \frac{dT}{T} + \mathcal{R}\frac{dv}{v} = c_p \frac{dT}{T} - \mathcal{R}\frac{dp}{p}, \tag{2.33}$$

where the last equality follows after taking the logarithms of Eq. (2.29) and differentiating with respect to T so that $dv/v + dp/p = dT/T$, and then making use of Eq. (2.28). The above differential forms can be integrated to give

$$s = s_0 + c_v \ln T + \mathcal{R}\ln v = s_0 + c_p \ln T - \mathcal{R}\ln p, \tag{2.34}$$

where s_0 is the specific entropy at an arbitrary reference state of the ideal gas.

Using Eq. (2.18) for the chemical potential along with Eqs. (2.32) and (2.34) one obtains

$$\frac{\mu}{T} = c_p - s_0 - c_p \ln T + \mathcal{R}\ln p; \tag{2.35}$$

to emphasize the pressure dependence of the chemical potential, the latter is often written as

$$\mu = \mu_0(p_0, T) + \mathcal{R}T \ln \frac{p}{p_0}, \tag{2.36}$$

with reference to a given pressure p_0, typically chosen as atmospheric pressure, $p_0 = 101$ kPa.

2.1.7   Incompressible Fluids

We will also typically consider liquids as incompressible, i.e., $\rho = M/V = m/\upsilon = $ constant, which means that the reversible work $pd\upsilon$ is always zero and that the specific heat capacity is the same for constant pressure and constant volume heating,

$$c_p = c_v = c. \tag{2.37}$$

The heat capacity is often assumed constant for simplicity, so that

$$u = cT, \tag{2.38}$$

and

$$h = cT + p\upsilon. \tag{2.39}$$

For the entropy we have

$$s = c \int \frac{dT}{T} = s_0 + c \ln T, \tag{2.40}$$

while the chemical potential, $\mu = h - Ts$, becomes

$$\mu(p, T) = cT + p\upsilon - T(s_0 + c \ln T) = \mu(p_0, T) + \upsilon(p - p_0), \tag{2.41}$$

where $\mu(p_0, T) = cT + \upsilon p_0 - Ts_0 + Tc \ln T$.

2.1.8   The Clausius–Clapeyron Equation

As we saw in Secs. 1.3 and 1.4, the hydrological cycle is permeated with phase transitions. We focus here on the *vaporization–condensation* line (see Fig. 2.1), namely on the phase change between liquid and gas. Similar considerations to those we discuss below also apply for other phase changes (sublimation–deposition, and melting–freezing). The coexistence line, $p_{\text{sat}}(T)$, in the $\{p, T\}$ plane can be found by first deriving an equation for its slopes. This equation, known as the Clausius–Clapeyron equation, can be obtained by imposing the continuity of the chemical potential as follows. Consider two points A and B infinitesimally close on the coexistence curve and suppose that the system consists of one mole ($N = 1$) of vapor and liquid water coexisting in equilibrium at constant pressure and temperature (Fig. 2.1). The coexistence of the two phases implies equal chemical potentials at A and B:

$$\mu_{l_A} = \mu_{g_A} \quad \text{and} \quad \mu_{l_B} = \mu_{g_B}, \tag{2.42}$$

where the subscripts l and g refer to the liquid and gas phases of the water. It also means that

$$\mu_{l_A} - \mu_{l_B} = \mu_{g_A} - \mu_{g_B} \tag{2.43}$$

or, since the points are infinitesimally close,

$$d\mu_l = d\mu_g. \tag{2.44}$$

Using the Gibbs–Duhem equation (2.26), the infinitesimal change in chemical potential along the saturation line can be written as

$$d\mu_l = -s_l dT + \upsilon_l dp_{\text{sat}} = d\mu_g = -s_g dT + \upsilon_g dp_{\text{sat}}, \tag{2.45}$$

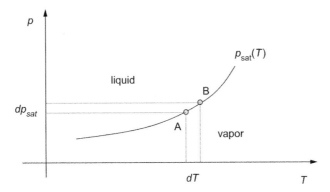

Figure 2.1 Derivation of Clausius–Clapeyron equation and saturated vapor pressure as a function of temperature.

which can be rearranged to obtain the *Clausius–Clapeyron equation*,

$$\frac{dp_{\text{sat}}}{dT} = \frac{s_g - s_l}{\upsilon_g - \upsilon_l} = \frac{\lambda_w}{T(\upsilon_g - \upsilon_l)}, \tag{2.46}$$

where we have introduced the (positive) heat input λ_w corresponding to a change in entropy of one mole of water from liquid to vapor at temperature T, called *latent heat of vaporization*:[5]

$$\lambda_w = T(s_g - s_l). \tag{2.47}$$

This can be simplified by first considering that the specific volume of gas is much larger than that of liquid, $\upsilon_g \gg \upsilon_l$, and then by assuming that the vapor behaves as an ideal gas (see Eq. (2.29)):

$$\frac{dp_{\text{sat}}}{dT} = \frac{\lambda_w}{T\upsilon_g} = \frac{\lambda_w p_{\text{sat}}}{\mathcal{R}T^2}. \tag{2.48}$$

Assuming further that λ_w is constant (i.e., independent of T), the previous equation can be easily integrated to give

$$p_{\text{sat}}(T_2) = p_{\text{sat}}(T_1) \exp\left[\frac{\lambda_w}{\mathcal{R}}\left(\frac{1}{T_1} - \frac{1}{T_2}\right)\right], \tag{2.49}$$

which is sometimes called the Antoine equation. More precisely, λ_w is in fact a weakly decreasing function of T. A useful approximation (Jones, 1992) for the water vapor saturated pressure is, using e for the partial pressure of water vapor,

$$e_{\text{sat}} = a \exp\left(\frac{bT}{c + T}\right), \tag{2.50}$$

where e_{sat} is in Pa, T is temperature in degrees Celsius, $a = 613.75$, $b = 17.502$, and $c = 240.97$. Note the strong nonlinearity of the saturation vapor pressure curve, plotted in Fig. 2.2.

---

[5] λ_w is also called the enthalpy of vaporization, because for a process at constant temperature and pressure it corresponds to a change in enthalpy of the (closed) system (the change in internal energy plus the work done on the surroundings), i.e., $\Delta Q = \lambda_w = \Delta h = \Delta u + p\Delta\upsilon$.

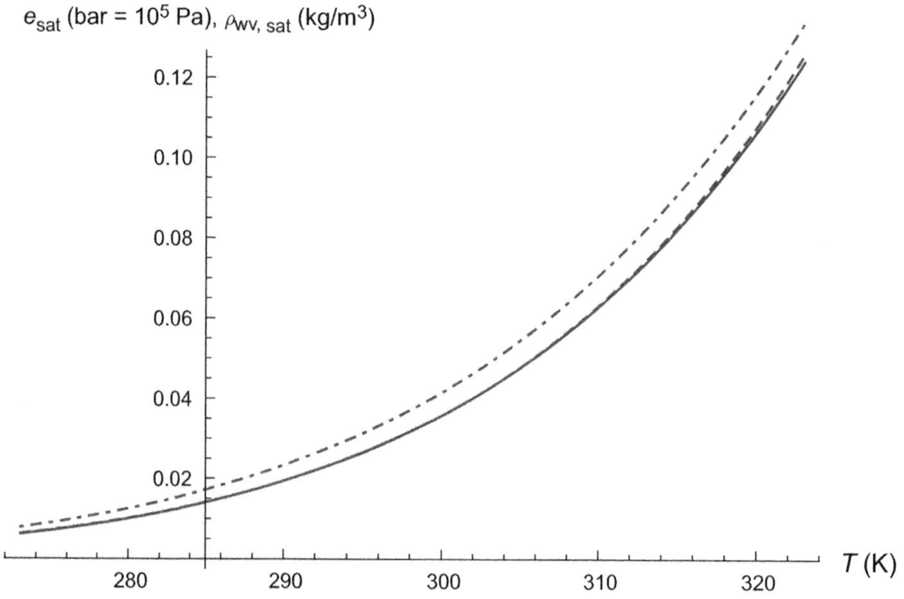

e_{sat} (bar = 10^5 Pa), $\rho_{wv, sat}$ (kg/m^3)

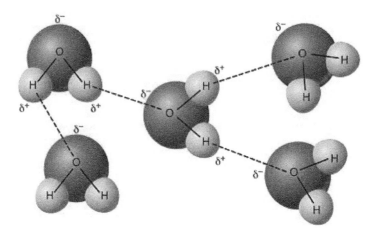

Figure 2.2 Empirical behavior of saturated water vapor pressure described by Eq. (2.49) (solid line), with for comparison the analytical approximation (2.50) (dashed line). Also shown is the absolute humidity at saturation, plotted as a dashed-dotted line according to Eq. (2.67).

Figure 2.3 Molecular structure of water and hydrogen bonds; δ^- and δ^+ are partial negative and positive charges.

2.1.9 Properties of Water and Phase Changes

The unique physico-chemical properties of water derive from its molecular structure (Fig. 2.3). In particular, the H-O-H bond of water is bent at an angle of about 105° and, because of the resulting electrical asymmetry, the hydrogen electrons are attracted by the oxygen, leaving the hydrogen atoms with an external partial positive charge, while the oxygen acquires a local partial negative charge in the zone of the unshared orbitals. As a result, water behaves as an electric dipole (Fig. 2.3), thereby making the water

Table 2.1 Latent heats of phase changes of H_2O.

Phase change	$J\,kg^{-1}$	
sublimation/deposition	2.84×10^6	at $0\,°C$
evaporation/condensation	2.50×10^6	at $0\,°C$
	2.45×10^6	at $20\,°C$
	2.25×10^6	at $100\,°C$
melting/freezing	3.34×10^5	at $0\,°C$

molecules mutually attractive with the development of *hydrogen bonds*. Such bonds, despite being about 20 times weaker than these covalent O-H bond, can give rise to significant structuring even in liquid water.

Owing to these hydrogen bonds, ice has a typical tetrahedric structure (Ice I is the normal crystalline form of ice). Only 15% of the hydrogen bonds are broken in melting, but they are however almost all destroyed in vaporization. This explains the high value of the latent heat of vaporization, which is quite close to that of sublimation, especially when compared with that of melting (see Table 2.1), which is an order of magnitude lower. Compared with other common liquids, water has very high melting and boiling points, heats of fusion and vaporization, specific heat, dielectric constant, viscosity, and surface tension. For the same reasons, liquid water is easily adsorbed on solid surfaces and combined with ions and colloids (acting as a solvent).

Water is liquid at normal temperatures, while other similar substances (e.g., H_2S) are then in the gas phase and therefore compressible; recall that the *molar volume of liquid water* is

$$\upsilon_w = \frac{V_w}{N_w} = \frac{V_w}{M_w}\frac{M_w}{N_w} = \frac{m_w}{\rho_w} = \frac{18 \times 10^{-3}\,\text{kg/mol}}{1000\,\text{kg/m}^3} = 18 \times 10^{-6}\,\frac{\text{m}^3}{\text{mol}}, \tag{2.51}$$

where m_w is known as the molar mass of water.

2.1.10 Atmospheric Air

For our purposes, atmospheric air can be considered to behave as a mixture of non-reactive ideal gases.

Mixtures of Ideal Gases

The composition of a mixture is given by the *mole fraction* x_i of each of its n constituents,

$$x_i = \frac{N_i}{N}, \tag{2.52}$$

$i = 1, \ldots, n$, where $\sum_{i=1}^{n} x_i = 1$ and $\sum_{i=1}^{n} N_i = N$, with N_i the mole number of each constituent. Under normal conditions, the mixture behaves as an ideal gas at temperature T and pressure p,

$$pV = N\mathcal{R}T = MR_{\text{mix}}T, \tag{2.53}$$

where R_{mix} is the specific gas constant for the mixture and $M = \sum_{i=1}^{n} M_i$ is the total mass of the mixture, having apparent molar weight

$$m_{\text{mix}} = \frac{\sum_{i=1}^{n} m_i N_i}{N} = \sum_{i=1}^{n} x_i m_i. \tag{2.54}$$

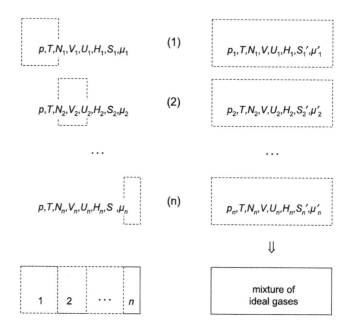

Figure 2.4 Perfect mixture of ideal gases, as a superposition of each ideal gas constituent occupying, without interactions with each other, the same volume at the same temperature (bottom right). A system consisting of a simple juxtaposition of the constituents at the same pressure without mixing (bottom left) has the same volume, internal energy and enthalpy but lower entropy and higher chemical potential.

As noted by Gibbs, a mixture of ideal gases behaves as the sum of the individual gases, as if each occupied the total mixture volume V at the same temperature T (this is known as Gibbs' theorem, Zemansky and Dittman, 1997; Callen, 2006). Thus, each gas must have a *partial pressure p_i* satisfying

$$p_i V = N_i \mathcal{R} T \tag{2.55}$$

with

$$p = \sum_{i=1}^{n} p_i, \tag{2.56}$$

which is known as Dalton's law. It is also useful to define the volume fraction, V_i, as the volume to which each gas would be reduced if it were subject to the mixture's total pressure and temperature of the mixture (see the left-hand side of Fig. 2.4), i.e.,

$$pV_i = N_i \mathcal{R} T \tag{2.57}$$

with $\sum_{i=1}^{n} V_i = V$. From the above properties, dividing (2.55) by (2.53) and (2.57) by (2.53), it follows that

$$x_i = \frac{N_i}{N} = \frac{p_i}{p} = \frac{V_i}{V}. \tag{2.58}$$

Because of Gibbs' theorem, the other thermodynamic quantities of a mixture of ideal gases can also be obtained as simple weighted sums of the properties of each gas at that volume and temperature (see the right-hand column of Fig. 2.4). In the case of internal energy and enthalpy, which for ideal gases are functions only of temperature (e.g., Eqs. (2.31) and (2.32)), this is also equal to the sums of the partial internal energy and enthalpy of each ideal gas i at the mixture values of T and p (see the left-hand column of Fig. 2.4), i.e., $U = \sum_{i=1}^{n} U_i = Nu = \sum_{i=1}^{n} N_i u_i$ and, dividing by N, $u = \sum_{i=1}^{n} x_i u_i$ and similarly $h = \sum_{i=1}^{n} x_i h_i$. Things are different, however, for the entropy and chemical potential, which also depend on pressure or volume (e.g., Eqs. (2.34) and (2.36)). As a result, if $S_i' = S_i + N_i \mathcal{R} \ln V/V_i$, where S_i is the partial entropy of the pure ideal gas i at T and p (the left-hand side of Fig. 2.4), the entropy of the mixture has an additional contribution in the sum of the partial entropies,

$$S = \sum_{i=1}^{n} S_i' = \sum_{i=1}^{n} N_i S_i + \sum_{i=1}^{n} \left(N_i \mathcal{R} \ln \frac{V}{V_i} \right), \tag{2.59}$$

where the last term is always positive and is the entropy production due to mixing of the ideal gases at constant T and p.[6] Dividing the previous expression by N and using Eq. (2.58) gives the molar entropy,

$$s = \sum_{i=1}^{n} x_i s_i'(p_i, T) = \sum_{i=1}^{n} x_i s_i(p, T) + \mathcal{R} \sum_{i=1}^{n} x_i \ln(x_i^{-1}), \tag{2.60}$$

where $\sum_{i=1}^{n} x_i \ln(x_i^{-1}) > 0$.

Finally, the chemical potential of the mixture can be obtained by first writing the chemical potential for each gas, using Eqs. (2.36) and (2.58), as

$$\mu_i'(p_i, T) = \mu_i(p, T) + \mathcal{R}T \ln \frac{p_i}{p} = \mu_i(p, T) - \mathcal{R}T \ln(x_i^{-1}). \tag{2.61}$$

It then follows that the mixture chemical potential is the sum of the partial chemical potential minus the contribution due to mixing

$$\mu = \sum_{i=1}^{n} x_i \mu_i'(p_i, T) = \sum_{i=1}^{n} x_i \mu_i(p, T) - \mathcal{R}T \sum_{i=1}^{n} x_i \ln(x_i^{-1}). \tag{2.62}$$

It is thus clear that, compared with the sum of the separate contributions at the same pressure and temperature, mixing increases entropy (see Eq. (2.60)) while decreasing the chemical potential (see Eq. (2.62)).

Dry Air

Atmospheric air (see Table 2.2) can be considered as an ideal mixture of gases of relatively constant composition, called dry air, plus water vapor, the concentration of which is highly variable and is tightly connected to the hydrologic cycle. We write the state equation for the dry air, at pressure p and temperature T, as

$$p = \rho_d R_d T, \tag{2.63}$$

---

[6] Note that the entropy production due to mixing is equal to the entropy produced by to the irreversible expansion of each gas from the situation on the left to the situation on the right in Fig. 2.4, and it is thus equal to the work that would be necessary to re-compress each gas isothermally to the "original" configuration, on the left.

Table 2.2 Atmospheric composition and relevant properties of its constituents.

Component	% (Vol. of dry air)	Molar mass (g mol^{-1})
Nitrogen (N_2)	78.1	28.02
Oxygen (O_2)	21.0	32
Argon (Ar)	0.9	39.94
Carbon dioxide (CO_2)	0.039	44.01
Methane (CH_4)	18×10^{-9}	16.04
Dry air	100	28.9†
Water vapor (H_2O)	0–0.04	18

† Apparent molar mass, given by the weighted average of each component above.
 Note that it is larger than that of water.

where ρ_d is the density,

$$R_d = \frac{8314 \text{ J kmol}^{-1}\text{K}^{-1}}{28.98 \text{ kg kmol}^{-1}} = 287.7 \text{ J kg}^{-1}\text{K}^{-1} \tag{2.64}$$

is the gas constant for dry air, and $m_d = 28.9$ is the apparent molar mass of dry air (see Table 2.2).

For future applications related to the dynamics of the atmospheric boundary layer (ABL), the layer of atmosphere next to the Earth's surface (the ABL, see Sec. 5.3), it is also useful to introduce the temperature that a parcel of dry air would have if brought adiabatically from its initial state to the standard atmospheric pressure, p_0, of 1000 mbar. This is called the *potential temperature*, and it is defined as

$$\vartheta = T \left(\frac{p_0}{p}\right)^{R_d/c_p}, \tag{2.65}$$

where $c_p = 1012 \text{ J kg}^{-1}\text{K}^{-1}$ is the mass-specific heat capacity at constant pressure. We will see in Sec. 5.3 that the ABL is characterized by a constant potential-temperature vertical profile, resulting from adiabatic mixing due to large-scale turbulence.

Humid Air

Humid air may be modeled as a mixture of dry air plus water vapor, with density

$$\rho = \rho_d + \rho_v, \tag{2.66}$$

where ρ_d and ρ_v are the densities of dry air and water vapor, respectively. The latter is also referred to as the *absolute humidity*. In conditions of saturated water vapor, the mass of water vapor per unit volume of humid air (the absolute humidity at saturation) is a strongly nonlinear function of temperature. It can be computed using the ideal gas law as (recall that we use the symbol e for water vapor pressure)

$$\rho_{v,\text{sat}} = \frac{e_{\text{sat}}(T)}{R_w T}, \tag{2.67}$$

where the gas constant $R_w = \mathcal{R}/m_w = 461.9 \text{ J kg}^{-1} \text{ K}^{-1}$. The above relationship was plotted as a function of temperature in Fig. 2.2 using Eq. (2.49).

The amount of water vapor present in the air may be quantified also in other ways. The *specific humidity* is defined as the ratio of the water vapor mass in a given volume of air and the total mass of the humid air in that volume,

$$q = \frac{M_v}{M} = \frac{\rho_v}{\rho}.$$ (2.68)

The *relative humidity* (in percent) is defined as

$$\text{RH} = 100 \frac{e}{e_{\text{sat}}},$$ (2.69)

while the *vapor pressure deficit* (VPD), which will appear in the evaporation flux equations (see Eq. (5.26)), is defined as

$$\text{VPD} = e_{\text{sat}}(T) - e.$$ (2.70)

For humid air, Dalton's law entails that the atmospheric pressure is the sum of the partial pressure of dry air and the partial pressure of water vapor,

$$p = p_d + e.$$ (2.71)

Thus the ideal gas law can be used for each component as well as for the mixture,

$$\begin{aligned} p_d &= \rho_d R_d T, \\ e &= \rho_v R_w T, \\ p &= \rho R T, \end{aligned}$$ (2.72)

where the gas constants are $R_w = \mathcal{R}/m_w = 461.9$ J kg^{-1} K^{-1} and $R_d = \mathcal{R}/m_d = 287.7$ J kg^{-1}K^{-1}, with molar masses given in Table 2.2. Combining Eq. (2.71) with Eq. (2.72), it is possible to write Eq. (2.66) as

$$\rho = \frac{p_d}{R_d T} + \frac{e}{R_w T} = \frac{p - e}{R_d T} + \frac{\epsilon \, e}{R_d T} = \frac{p}{R_d T} \left[1 - (1 - \epsilon) \frac{e}{p} \right],$$ (2.73)

where ϵ is the ratio of the molar masses, $\epsilon = m_w/m_d = R_d/R_w \approx 0.622$. Since e/p is always greater than zero, it follows that humid air is lighter than dry air at the same pressure. It is also useful to introduce the *virtual temperature*,

$$T_v = \frac{T}{1 - (1 - \epsilon)e/p} \approx T(1 + 0.61q),$$ (2.74)

which is the temperature to which dry air should be heated to have the same density as the humid air at the same pressure. The virtual temperature is always higher than the actual temperature.

The previous expressions provide a useful relationship between specific humidity and vapor pressure:

$$q = \frac{\rho_v}{\rho} = \frac{e/(R_w T)}{p \left[1 - (1 - \epsilon) \frac{e}{p} \right]/(R_d T)} = \frac{\epsilon e}{p - (1 - \epsilon)e} \approx \epsilon \frac{e}{p} \approx 0.622 \frac{e}{p},$$ (2.75)

for $(1 - \epsilon)e \ll p$. With this approximation, the virtual temperature is $T_v \approx T(1 + 0.61q)$, while the humid-air gas constant $R \approx R_d(1 + 0.61q)$, so that the equation of state for humid air may also be written as

$$p = \rho R_d T(1 + 0.61q) = \rho R_d T_v.$$ (2.76)

Finally, it will be useful to introduce the *specific humidity deficit*

$$D = q_{sat}(T) - q,$$ (2.77)

which, using (2.75), is related to the VPD in (2.70) as $D = (0.622/p)\text{VPD}$.

2.1.11 Gravitational Field and Total Potential

So far we have considered only simple systems in which thermodynamic equilibrium is characterized by homogeneous pressure, temperature, and electrochemical potential everywhere within the system. This implies negligible gravitational effects. There are some cases, however, in which such effects are important and give rise to a nonhomogeneous pressure distribution within the system even in equilibrium conditions (e.g., a hydrostatic distribution). This is important when analyzing the tendency of water and other substances to move between points at different elevations, as for example in describing the vertical water flow in soils and in tall trees, as well as when considering buoyancy effects and the related stability conditions in atmospheric dynamics. The inclusion of gravitational energy may be accounted for by adding a *gravitational potential* term to the chemical potential.

In a gravitational field with acceleration g, a mass M at an elevation z has a gravitational potential energy gzM. This corresponds to an energy gz per unit mass or to a molar energy of mgz, where m is its molar mass. Accordingly, by adding a mass $dM = mdN$ to an open system at an elevation z, the total energy of the system is increased by $mgzdN$. As a result, with the inclusion of the gravitational potential, the Gibbs equation for open systems (see Eq. (2.17)) becomes

$$dE_{tot} = d(P + U) = TdS - pdV + (\mu + mgz)dN,$$ (2.78)

or, introducing the *total potential*[7] per mole,

$$\mu' = \mu + mgz = u + pv - Ts + mgz,$$ (2.80)
$$dE_{tot} = TdS - pdV + \mu'dN.$$ (2.81)

As a consequence, using Eq. (2.36), the total water potential for an ideal gas is

$$\mu'(p, T, z) = \mu_0(p_0, T) + \mathcal{R}T \ln \frac{p}{p_0} + mgz,$$ (2.82)

while using Eq. (2.41) for incompressible fluids it is

$$\mu'(p, T, z) = \mu_0(p_0, T) + v(p - p_0) + mgz.$$ (2.83)

---

[7] It is also useful sometimes to consider the total potential divided by the molar volume, which has units of pressure (e.g., energy per mole divided by volume per mole), as it provides a measure of the amount of available energy (free energy plus gravitational) per unit volume of fluid,

$$\frac{\mu'}{v} = \frac{u - Ts}{v} + \gamma z + p,$$ (2.79)

where $\gamma = g\rho$ is the specific weight and $\gamma z + p$ is the hydrostatic pressure. The hydrostatic pressure is constant in isothermal, hydrostatic conditions (Munson et al., 2005).

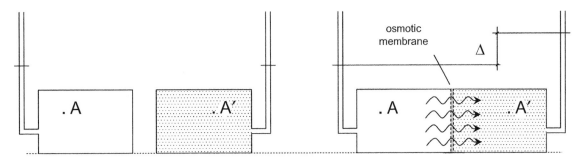

Figure 2.5 Osmometer and osmotic potential.

2.1.12 Solutions and Osmotic Potential

When a soluble compound (e.g., salt), called a solute, is added to a liquid solvent (e.g., water), the solute dissolves and forms a solution. Similarly to the case of the gas mixture considered in Sec. 2.1.10, the mixing process brings about a reduction in the chemical potential of the solution, while the entropy is increased. This reduction in chemical potential, compared to the pure solvent, is made evident by considering two points at the same elevation, temperature, and pressure (left-hand panel of Fig. 2.5). The point A is in the pure solvent, while A′ is in the solution. When the tanks are put in contact via an osmotic membrane, i.e., a semipermeable wall that blocks the passage of solute (right-hand panel), the thermodynamic disequilibrium initially present between the two compartments drives a flow of solvent towards the solution until equilibrium is reached. This causes the liquid in the right piezometer to rise to a height Δ, indicating that an excess pressure has been developed to equilibrate the change in chemical potential ($p_{A'} > p_A$). Such a pressure difference can be easily calculated using Stevin's law (Munson et al., 2005)[8]

$$p_{A'} - p_A = g\rho_w \Delta = -\Omega, \tag{2.84}$$

and its negative is called the *osmotic pressure* Ω. Since on the right-hand side of the figure the total chemical potential of water at the two points A and A′ in the right-hand panel of the figure must be the same at equilibrium, a new component of the chemical potential, μ_Ω, called the *osmotic potential*, must exist such that

$$\mu'_A = \mu_0(T) + m_w g \left(z_A + \frac{p_A}{g\rho_w} \right) = \mu'_{A'} = \mu_0(T) + m_w g \left(z_{A'} + \frac{p_{A'}}{g\rho_w} \right) - \mu_\Omega. \tag{2.85}$$

By comparing the two previous expressions,

$$\mu_\Omega = m_w g \Delta = -m_w \frac{\Omega}{\rho_w} = -\upsilon_w \Omega, \tag{2.86}$$

where $\upsilon_w = m_w/\rho_w$ is the molar volume of liquid water (see Eq. (2.51)).

To link the osmotic pressure to the amount of solute present in the solution, consider a dilute solution, in which there is a preponderant species (the solvent) and k other species present in lower amounts (solutes).

---

[8] Note that the piezometer radius is assumed to be large enough that the capillary rise is negligible compared with the effects of the osmotic pressure. Moreover the solution is assumed to be dilute so that its specific weight is the same as that of the solvent.

In this case:

$$x_{\text{solvent}} = \frac{N_{\text{solvent}}}{N_{\text{solvent}} + \sum_{i=1}^{k} N_i} = 1 - \sum_{i=1}^{k} x_i. \qquad (2.87)$$

In general, the chemical potential for a substance i in solution is related to the chemical potential of the pure substance as (Nobel, 1999; Kondepudi and Prigogine, 2005)

$$\mu_{i,\text{sol}}(p, T) = \mu_i(T, p) + \mathcal{R}T \ln a_i \qquad (2.88)$$

where a_i is the activity of the substance in solution. For ideal solutions the activity is simply equal to the mole fraction, $a_i = x_i$. Since $N_{\text{solvent}} \gg \sum_{i=1}^{k} N_i$, the chemical potential of the solution is practically equal to that of the solvent in solution, that is

$$\mu_{\text{solvent in sol.}} = \mu_{\text{pure solvent}}(p, T) + \mathcal{R}T \ln \left(1 - \sum_{i=1}^{k} x_i \right) \qquad (2.89)$$

and, using a Taylor series expansion, so that $\ln(1 - y) \approx -y$,

$$\mu_{\text{solvent in sol.}} \approx \mu_{\text{pure solvent}}(p, T) - \mathcal{R}T \sum_{i=1}^{k} x_i. \qquad (2.90)$$

Thus, referring now specifically to water as the solvent, and using Eq. (2.86),

$$\mu_{\text{water in sol.}} - \mu_{\text{pure water}}(p, T) = -\mu_{\Omega} = -\mathcal{R}T \sum_{i=1}^{k} x_i = \upsilon_w \Omega, \qquad (2.91)$$

from which

$$\Omega = -\mathcal{R}T \sum_{i=1}^{k} \frac{x_i}{\upsilon_w} = -\mathcal{R}T \sum_{i=1}^{k} c_i, \qquad (2.92)$$

where c_i is the concentration of solute i in moles per volume of water in solution (which is very close to the volume of the solution). Equation (2.92) is the *van't Hoff equation*. It is interesting to note the analogy between the previous two equations and the equation of state for mixtures of ideal gases, Eq. (2.62), which suggests that in ideal solutions the solutes behave like ideal gases diffused within an ideal gas mixture.

Finally, it will be useful to compute the reduction in the vapor partial pressure of a species i from its saturation value in a gas mixture in contact with a liquid solution consisting of the same species. Assuming equilibrium, the chemical potential for species i is the same in the ideal solution (i, sol) as in the mixture of ideal gases (i, mix), that is,

$$\mu_{i,\text{sol}} = \mu_{i,\text{liq}}(p, T) + \mathcal{R}T \ln x_i = \mu_{i,\text{mix}} = \mu_{i,\text{gas}}(p, T) + \mathcal{R}T \ln \frac{p_i}{p}. \qquad (2.93)$$

Considering the case of $x_i \to 1$, that is, when in practice the species i is preponderant, the pure vapor is in equilibrium with its liquid, $\mu_{i,\text{liq}}(p, T) = \mu_{i,\text{gas}}(p, T)$. Assuming further that the pressure p is equal to the saturated pressure, $p_{i,\text{sat}}$, the equality above immediately yields *Raoult's law*,

$$p_i = p_{i,\text{sat}} x_i \qquad (2.94)$$

which allows us to compute the reduction in partial pressure of a vapor due to the presence of a solute in the liquid phase (see Note 2.9).

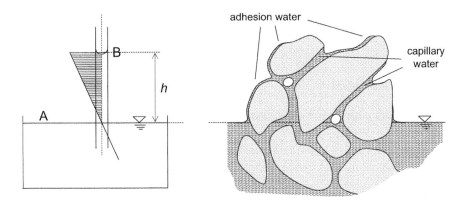

Figure 2.6 Capillary rise in a small tube and related interface phenomena in unsaturated soils, giving rise to suction (negative pressure with respect to the atmosphere) in the liquid and resulting in negative matric potentials. The shading in the left-hand panel indicates that the pressure is negative compared with atmospheric.

2.1.13 Capillarity and Interface Phenomena

The difference in free surface energy between the solid–gas interface and the liquid–gas interface of different substances produces a tensile force at the liquid-gas interface, called the *surface tension*, which is responsible for the well-known phenomenon of *capillary rise* (e.g., Fig. 2.6) and similar interface phenomena in unsaturated porous media, such as soil and a plant xylem (see Secs. 3.2.2 and 4.4). We begin by recalling that capillary rise is accompanied by a reduced pressure in the water phase. For a circular pipe, the extent of the capillary rise is related to the surface tension of the liquid σ and the contact angle θ_c as well as the radius of the capillary tube r by the expression

$$h = \frac{2\sigma \cos \theta_c}{\gamma_w r}, \tag{2.95}$$

which can easily be obtained by writing the balance of vertical forces in the pipe (Munson et al., 2005). From the hydrostatic pressure distribution inside the pipe (Fig. 2.6, left) one can immediately compute the reduction of pressure from the atmospheric pressure p_0 for a point just below the meniscus. Indicating this pressure difference as Ψ, one has that

$$\Psi = p_{\text{cap}} - p_0 = -\gamma_w h, \tag{2.96}$$

which corresponds to a state of suction (i.e., a negative pressure with respect to atmospheric).

Thermodynamic equilibrium implies a constant total chemical potential within the entire liquid mass. Thus, considering two points A and B, one at the free surface and the other just below the meniscus, and using Eq. (2.83),

$$\mu'_A = \mu_0(p_0, T) = \mu'_B = \mu_0(p_0, T) + \upsilon_w(p_{\text{cap}} - p_0) + m_w g h, \tag{2.97}$$

where the reduction in chemical potential due to capillary phenomena, called the *matric potential*, is

$$\mu_\Psi = \upsilon_w \Psi = -g m_w h. \tag{2.98}$$

This component is very important in unsaturated soils (Fig. 2.6) and accounts for the progressive reduction in the Gibbs free energy of water as water–air–solid interfaces appear at increasingly smaller pores while the soil moisture is depleted. As a result, the lower soil water chemical potential makes it more difficult for plants to extract water and corresponds to increased risk of cavitation inside the plant xylem (Sec. 4.4).

2.1.14 Total Water Potential

To describe the tendency of water to move within the soil–plant–atmosphere system, it is convenient to define the *total water potential*, ψ. It is the sum of the Gibbs free energy of water and the gravitational energy, expressed per unit volume of liquid water and relative to standard conditions. Thus for a unit volume of liquid water, the molar chemical potential (Gibbs free energy per unit mole) needs to be multiplied by the number of moles of water and then divided by the total volume of water, that is, divided by the liquid-water molar volume υ_w, i.e.,

$$\psi = \frac{\mu' - \mu_0}{\upsilon_w}. \tag{2.99}$$

With this normalization, the total water potential has units of pressure, e.g., Pa or MPa. Pure liquid water and saturated vapor, at the same standard pressure and temperature and zero elevation, both have total water potential equal to zero.

For liquid water, its total water potential can be obtained from Eq. (2.83) with the inclusion of possible osmotic and matric potential effects:

$$\psi_w = \frac{[\mu_0(p_0, T) + \upsilon_w(p_0 + \Psi + \Omega) + m_w gz] - [\mu_0(p_0, T) + \upsilon_w p_0]}{\upsilon_w}. \tag{2.100}$$

Denoting the gravitational potential per unit volume as $G = m_w gz/\upsilon_w = \gamma_w z$, the previous expression becomes

$$\psi_w = G + \Psi + \Omega, \tag{2.101}$$

which makes the three contributions to the total water potential evident: the gravitational, the matric, and the osmotic components. Note that the sum of the gravitational and matric (pressure) components is equal to the hydrostatic gauge pressure, $G + \Psi = \gamma_w z + (p - p_0)$.

For water vapor, the total water potential can be obtained by considering the chemical potential of a component of a mixture of ideal gases: Eq. (2.61) along with Eq. (2.35) can be written as

$$\mu_i'(p_i, T) = \mu_i(T, p_0) + \mathcal{R}T \ln \frac{e}{p_0} + m_w gz, \tag{2.102}$$

so that

$$\psi = \frac{\mu_i(p_0, T) + \mathcal{R}T \ln \frac{e}{p_0} + m_w gz - \left[\mu_i(p_0, T) + \mathcal{R}T \ln \frac{e_{\text{sat}}(T)}{p_0}\right]}{\upsilon_w}, \tag{2.103}$$

or[9]

$$\psi = \frac{\mathcal{R}T}{\upsilon_w} \ln \frac{e}{e_{\text{sat}}(T)} + \gamma_w z. \tag{2.104}$$

[9] Assuming that e_{sat} is practically independent of pressure, as was done when deriving the Clausius–Clapeyron equation.

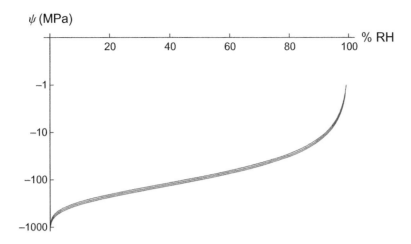

Figure 2.7 Atmospheric water potential ψ versus relative humidity for different values of air temperature (from top to bottom $T = 0, 20, 40\,°C$).

Finally, introducing the gravitational potential G and the relative humidity of water vapor, i.e.,

$$\psi = \frac{\mathcal{R}T}{\upsilon_w} \ln \frac{\text{RH}(\%)}{100} + G, \tag{2.105}$$

For normal air conditions the value of ψ for water vapor is usually very low: Fig. 2.7 shows the behavior of Eq. (2.105) as a function of relative humidity for different temperatures ($T = 0, 20, 40\,°C$). The water-vapor potential drops very quickly to low values as soon as the relative humidity is reduced from saturation conditions. Notice that the direct effect of temperature is minimal when compared with that of the relative humidity.

2.2   Dimensional Analysis and Similarity

Dimensional analysis and similarity are very useful for obtaining general relationships in complex systems when solutions from first principles are not available, as in the case of the hydrologic partitioning (e.g., the Budyko diagram, Sec. 7.6), turbulent fluxes (Sec. 2.3), and the similarity solutions of Richards' equation (Sec. 3.4.1).

2.2.1   Units of Measurements and Dimensions

A quantitative description of a physical problem requires first identifying the nature or type of the physical quantities involved (e.g., length, force, time, etc.) and comparing them with a "standard" (e.g., the gram for a mass, the meter for a length, etc.) through a measurement. This standard is called a unit of measurement, or simply a unit. There are two types of units: *fundamental units*, which require only a direct comparison with the unit of measurement, and *derived units*, for which only an indirect comparison with the unit of measurement based on the definition of the physical quantity is possible. In this second case, for example, the unit of force follows from an indirect comparison with units of length, mass, and time according to the definition $F = ma$; this means that the unit of force is a derived unit. The same is true for density, which is obtained as mass divided by volume, and for most other quantities that are encountered.

The number k of fundamental units in a physical problem depends on the type of problem under examination: it is equal to 2 in kinematic problems, 3 in dynamics, and 4 in thermodynamics. Such a number plays a key role in the Π theorem (see Sec. 2.2.2 below). There is some arbitrariness in choosing which types of units are fundamental, in the sense that any set is admissible as long as the chosen quantities are dimensionally independent (we will come back later to the meaning of this statement). Once a set of dimensionally independent units has been selected, their dimensions define the class of system of units; within the same class, the different systems differ only in magnitude, not in their physical nature. For example, the LMT (length, mass, time) class is typically used in mechanics; within it, one may use the meter, kilogram, second (SI) system or the cm, gram, second (cgs) system.

The function that determines the factor by which the numerical value of a physical quantity changes upon passage from the original system of units to another, within a given class, is called the *dimension function* or simply the dimension of that quantity. Taking velocity as an example, if the unit of length is decreased/increased by a factor L and the unit of time is decreased/increased by a factor T, the velocity is changed by a factor LT^{-1} compared with the original unit. We use Maxwell's convention, according to which the dimension of the variable φ is indicated as $[\varphi]$. Thus for velocity, the dimension is indicated as $[v] = LT^{-1}$.

For fundamental units, such as a height h, the dimension function is simple:

$$\text{height, } h \qquad [h] = L,$$

while for derived quantities the dimensions are more complex, e.g,

$$
\begin{aligned}
\text{weight, } w & \qquad [w] = F = MLT^{-2}, \\
\text{density, } \rho & \qquad [\rho] = ML^{-3}, \\
\text{dynamic viscosity, } \mu & \qquad [\mu] = ML^{-1}T^{-1}, \\
\text{rainfall frequency, } \lambda & \qquad [\lambda] = T^{-1};
\end{aligned}
\tag{2.106}
$$

in the first of these relations the quantity F is the unit of force.

It can be shown (Barenblatt, 1996) that dimension functions are always power-law monomials. For example, in a problem in mechanics a generic variable q has dimension function $[q] = L^{\alpha}M^{\beta}T^{\gamma}$, but it is never a transcendental function such as $\ln(M)$. It follows that the arguments of transcendental functions must be dimensionless (quantities whose dimension function is unity), so that their numerical values are identical in all systems of units.

It is important to have a practical criterion to determine whether some quantities, say $a_1, a_2, \ldots, a_k$, are dimensionally independent and so can be used to form a new class of systems of units, for a problem that requires k fundamental units. One can demonstrate (Barenblatt, 1996) that a set of variables is dimensionally independent if and only if none of their dimension functions can be represented in terms of a product of powers of the dimensions of the remaining quantities. In turn, it is not difficult to show that this is true only if the determinant formed with the exponents of the dimension functions is different from zero. Note 2.11 explains this important concept in practice for a problem related to fluid mechanics.

2.2.2 Buckingham or Π Theorem

This fundamental theorem formalizes the logical fact that the equations describing a physical problem must remain valid independently of the chosen system of units. Its practical power lies in providing a mathematical structure among the fundamental dimensional groups underlying the dynamics of a physical problem, thereby offering a logical way to reduce the number of independent variables needed in a description of the problem. We will state here only the main steps of the theorem, referring to Barenblatt (1996) for details.

The starting point assumes that we can functionally relate a quantity of interest, a, to other n *governing quantities*, $a_1, \ldots, a_n$,

$$a = f(a_1, \ldots, a_n); \tag{2.107}$$

suppose also that the type of problem admits k dimensionally independent quantities, listed among the governing quantities,[10] so that the previous equation can be formally rearranged as

$$a = f(a_1, \ldots, a_k; a_{k+1}, \ldots, a_n). \tag{2.108}$$

The Π *theorem* states (Barenblatt, 1996) that it is always possible to rewrite the previous expression in dimensionless form as a function of $n - k$ dimensionless groups,

$$\Pi = \varphi(\Pi_1, \ldots, \Pi_{n-k}), \tag{2.109}$$

where

$$\Pi = \frac{a}{a_1^{\alpha_1}, a_2^{\alpha_2}, \ldots, a_k^{\alpha_k}}$$

$$\Pi_1 = \frac{a_{k+1}}{a_1^{\alpha_1^1}, a_2^{\alpha_2^1}, \ldots, a_k^{\alpha_k^2}} \tag{2.110}$$

$$\ldots$$

$$\Pi_{n-k} = \frac{a_n}{a_1^{\alpha_k^{n-k}}, a_2^{\alpha_k^{n-k}}, \ldots, a_k^{\alpha_k^{n-k}}},$$

with exponents $\alpha_1, \ldots, \alpha_k^{n-n}$ chosen in such a way to make the groups in Π dimensionless.

By comparing Eqs. (2.107) and (2.109), it is clear that the Π theorem allows us to re-express a general mathematical relationship between some dimensional quantity and several (n) governing quantities as a new relationship between some dimensionless quantity and fewer ($n - k$) dimensionless combinations of the governing quantities. The number of dimensionless quantities, $n - k$, and thus the degree of dimensionality reduction, is equal to the number of the governing quantities (n) minus the number of governing quantities with independent dimensions (k).

The main steps for a successful application of the Π theorem may be summarized as follows:

- list all important variables, but not more than necessary, and write a formal dimensional relation of the type (2.107);
- select the right independent variables among the governing quantities ($a_1, \ldots, a_n$);

---

[10] While there may be different possible choices for the k dimensionally independent quantities, each choice will lead to equivalent, if formally different, results.

- construct the dimensionless quantities (Π groups) by computing the appropriate exponents $(\alpha_1, \ldots, \alpha_k^{n-n})$.

When one or more of the Π groups attain(s) very large or very small values, the function φ in (2.109) may reach an asymptotic form, in which case the problem is said to be self-similar in this regime. In the simplest form of *self-similarity*, called complete or of the first kind (see Barenblatt, 1996), the function φ in (2.109) reaches a constant plateau for either very small or very large values of the governing groups. As a result, the physical problem does not change (i.e., it remains self-similar) even if the values of these groups change, and the self-similar group can then be eliminated from Eq. (2.109), allowing for further simplification. For example, assuming complete self-similarity with respect to the group Π_1, one can write Eq. (2.109) as

$$\Pi = \varphi \left(\begin{matrix} 0 \\ \infty \end{matrix} , \Pi_2, \ldots, \Pi_{n-k} \right) = \varphi'(\Pi_2, \ldots, \Pi_{n-k}). \tag{2.111}$$

Thus complete self-similarity allows us to further reduce the dimensionality of the problem by as many dimensions as there are self-similar groups. This type of similarity is frequently encountered in near-wall turbulence (Sec. 2.3), where the global Reynolds number and the dimensionless distance from the wall appear as self-similar groups.

2.3 Turbulent Fluxes

The similarity theory of near-wall turbulence may be used to quantify the fluxes of heat, water vapor, and other quantities such as CO_2 between the land surface and the atmosphere (Tennekes and Lumley, 1972; Kaimal and Finnigan, 1994; Barenblatt, 1996; Brutsaert, 2005). The resulting expressions will be used in subsequent chapters, especially in Chapters 4 and 5. It will be assumed that the land surface and the overlying turbulent atmosphere are horizontally homogeneous with a preferential horizontal mean wind speed in the x direction, so that only vertical profiles along the vertical direction z are of interest. It will be further assumed that sources and sinks of momentum, sensible heat, water, and CO_2 are at the surface, and that the turbulent fluctuations in the ABL are quasi-stationary (i.e., they fluctuate at a faster timescale) compared to the diurnal trends. While buoyancy is essential for the diurnal growth of the ABL, it has only a secondary effect on the turbulence structure of the ABL and the related fluxes. Thus, for simplicity, no stability corrections will be considered and, to a first approximation, heat and moisture will be treated as passive scalars (i.e., scalar quantities that do not have a dynamic effect on the flow and are therefore passively advected).[11] The land surface is always a sink of momentum and, during the day, normally a source of moisture (evapotranspiration) and sensible heat, as well as a sink of CO_2 because of plants' photosynthetic activity.

Given the role of temporal fluctuations in turbulent transport, it is important to distinguish them explicitly from the mean values, by performing a so-called Reynolds' decomposition (Tennekes and Lumley, 1972; Brutsaert, 2005). For the longitudinal velocity in the horizontal direction such a decomposition is

$$u = \bar{u} + u', \tag{2.112}$$

---

[11] The reader may consult Brutsaert (2005) and Katul et al. (2011) for a discussion of the corrections to the turbulence profiles due to non-neutral atmospheric conditions.

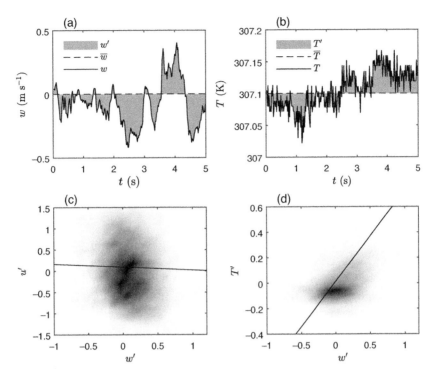

Figure 2.8 Wind and temperature measurements sampled at 56 Hz rate in July 1995 at 5 m above a grass clearing at a Duke Forest site (Katul et al., 1997). Reynolds' decomposition of (a) vertical velocity w and (b) temperature T over a 5-second period; density maps of (c) (w', u') and (d) (w', T') during 20-minute observations, where regions with a darker color have a higher density of samples. The prime on a quantity indicates the fluctuation from the average value (compare Eq. (2.112)). Data courtesy of G. Katul.

where the overbar stands for long-time averaging:

$$\bar{u} = \frac{1}{T} \int_{t-T/2}^{t+T/2} u(s)ds. \tag{2.113}$$

Here T refers to a time much larger than the timescales at which turbulent fluctuations are correlated (Tennekes and Lumley, 1972). Examples of Reynolds' decomposition are given in Fig. 2.8 for the near-surface vertical wind velocity and temperature over a grassland.

In the ABL the sharp gradients in the average profiles of velocity, temperature, humidity, CO_2, etc., coupled to the turbulent fluctuations of vertical velocity, give rise to net fluxes of these quantities. To describe these profiles, we refer to a generic passive scalar B with dimensions $[B] = $ B and to its corresponding (intensive) quantity per unit mass of air, $b = B/M$. Because an air mass M has dimension M, the dimensions of b are $[b] = $ BM^{-1}; the latter can be decomposed as

$$b = \bar{b} + b'. \tag{2.114}$$

In future applications, b will stand for the mass specific enthalpy (sensible heat), or $c_p T$ (J/kg), or the specific humidity, q (kg of water per kg of air), or the CO_2 concentration, c (kg of CO_2 per kg of air). All the mean quantities evolve slowly compared with the turbulent fluctuations and may be considered constant over a time period of few minutes, while the fluctuations (primed quantities) evolve very rapidly and form a quasi-stationary process around their mean values. We will also use w' to indicate the vertical velocity fluctuations (positive upward), with zero average vertical velocity, $\bar{w} = 0$. Finally, we will consider the density fluctuations as negligible, $\rho' = 0$.

Neglecting molecular diffusion compared with turbulent dispersion, and assuming horizontally homogeneous conditions, the time-averaged continuity equation for the generic conserved quantity b can be written as

$$\rho \frac{\partial \bar{b}}{\partial t} = -\frac{\partial}{\partial z} F_b, \tag{2.115}$$

where the average vertical flux is

$$F_b = \rho \, \overline{w'b'}. \tag{2.116}$$

This follows from considering the instantaneous flux ρbw, and noting that $\overline{w} = 0$ and $\rho' = 0$ (e.g., $\rho = \bar{\rho}$) so that its time average is $\rho \overline{w'b'}$. The latter expresses the important fact that turbulent fluxes are linked to correlations between velocity fluctuations and the fluctuations in the concentration of the transported variable (see Figs. 2.8c and d). These fluxes are regarded as positive if directed from the surface to the atmosphere.

The turbulent fluctuations in a near-neutral atmosphere are driven by shear stress, which in turn is related to the vertical turbulent transfer of longitudinal momentum (Munson et al., 2005). Since the concentration of longitudinal momentum (e.g., momentum per unit mass of air) is u, the vertical fluxes of horizontal momentum (which is in the opposite direction to the horizontal shear stress τ) are

$$F_u = -\tau = \rho \, \overline{w'u'}, \tag{2.117}$$

where $\overline{w'u'}$ has units of force per unit surface (stress). The slight negative correlation between the fluctuations of horizontal and vertical velocity in Fig. 2.8c indicates that the momentum flux is towards the surface.

For the *sensible heat flux* the intensive quantity is the specific enthalpy per unit mass of air, $c_p T$, so that

$$F_{c_p T} = H = \rho c_p \, \overline{w'T'}. \tag{2.118}$$

This flux, usually indicated as H, has units of energy per unit time (power) per unit ground surface (e.g., W m^{-2}). It is typically positive during the day because the land surface tends to be warmer than the atmosphere. Figure 2.8d shows an example of positive correlation between the fluctuations of temperature and vertical velocity, suggesting the presence of a surface-to-atmosphere heat flux.

The mass flux of water vapor is due to correlations between the fluctuations of the vertical velocity and those of specific humidity,

$$F_q = \rho_w E = \rho \, \overline{w'q'}. \tag{2.119}$$

While in the previous equation F_q is the mass of water per unit time per unit surface, E is the flux of water vapor in terms of the volume of liquid water per unit surface area per time, e.g. mm/day. Such a flux is positive during the day, when it corresponds to evapotranspiration, but may become negative at night (i.e., giving rise to dew), when the atmosphere is saturated and the surface is colder (see Note 5.15). The energy flux corresponding to this water flux is the *latent heat flux*, LH $= \lambda_w \rho_w E$. Note that c_p and λ_w are mass-specific parameters in this section.

Finally, the CO_2 flux is

$$F_{CO_2} = \rho \, \overline{w'c'}, \tag{2.120}$$

where c' is the fluctuation in the CO_2 concentration and the flux is in mass of CO_2 per unit time per unit surface. Such a flux is generally positive at night and during periods when the soil and plant respiration is dominant, while it is negative when plants are active photosynthetically (during the day[12] and in well-watered conditions). In Chapter 5 we will analyze in detail the temporal behavior of the fluxes defined above.

2.3.1 Turbulence Similarity Theory

Because turbulent flows cannot be solved analytically, we can only hope to get approximate expressions linking turbulent fluxes and their governing quantities. To this purpose we can apply dimensional analysis and the Π theorem (Sec. 2.2) and resort to the observed self-similar behavior of turbulent fluctuations in the logarithmic layer of the atmosphere (Sec. 2.3.2). We first consider the shear stress in the longitudinal direction. Experimental evidence shows that close to the Earth's surface (in the so-called *surface layer*), the shear stress (the negative of the momentum flux) is approximately constant and equal to the value at the surface, the wall shear stress, indicated as τ_0. In this layer, τ_0 can be related to the main properties of the ABL by the general relation

$$\tau_0 = f_{\tau_0}(h, \rho, \Delta \bar{u}, \mu, \epsilon^*), \tag{2.121}$$

where h is the typical thickness of the boundary layer, $\Delta \bar{u}$ is the velocity difference which drives the momentum flux over the surface layer (it is equal to the mean velocity $\bar{u}_\infty$ in the upper part of the ABL (Fig. 2.9), because the velocity is zero at the surface), μ is the dynamic viscosity of air, and ϵ^* is a length representing the effective height of the surface roughness, called the protrusion height, for momentum transfer. Because of the constancy of the shear stress in the vertical direction and the horizontal homogeneity, there is no spatial dependence in Eq. (2.121). Using $h, \rho, \Delta \bar{u}$ as dimensionally independent quantities, the Π theorem gives (Sec. 2.2)

$$\frac{\tau_0}{\rho \Delta \bar{u}^2} = \varphi_{\tau_0}\left(\frac{\mu}{\rho \Delta \bar{u} h}, \frac{\epsilon^*}{h}\right), \tag{2.122}$$

where the first dimensionless group in parentheses is the inverse of a global *Reynolds number*, $\mathrm{Re} = \rho \bar{u}_\infty h/\mu$, and the other is the relative roughness.

Further simplifications are possible because of the self-similarity of turbulent fluctuations in the region where the shear stress is almost constant (see Sec. 2.2.2). In fact, observations show that in this region the statistics of turbulent fluctuations are self-similar with respect to the global Reynolds number (Barenblatt, 1996; Munson et al., 2005) so that the dimensionless group obtained from the Π theorem tends to a constant as $\mathrm{Re} = \rho \bar{u}_\infty h/\mu \to \infty$. As a result, the dimensionless function in Eq. (2.122) depends only on the relative roughness ϵ^*/h,

$$\frac{\tau_0}{\rho \Delta \bar{u}^2} = \varphi_{\tau_0}\left(0, \frac{\epsilon^*}{h}\right) \sim \varphi_{\tau_0}'\left(\frac{\epsilon^*}{h}\right) \sim \text{constant}; \tag{2.123}$$

---

[12] Apart from CAM (crassulacean acid metabolism) plants, which instead uptake CO_2 at night (see Chapter 4).

this constant is called the drag coefficient,

$$\mathrm{Cd} = \frac{\tau_0}{\rho \, \Delta \bar{u}^2}. \tag{2.124}$$

The previous scaling relationship can be used to define a typical velocity scale, u_*, called the friction velocity, as

$$u_* = \left(\frac{\tau_0}{\rho} \right)^{1/2} = \Delta \bar{u} \, \mathrm{Cd}^{1/2}. \tag{2.125}$$

For the generic scalar b, a general expression similar to Eq. (2.121) must include not only a measure of the gradient of such quantity, e.g., $\Delta \bar{b} = (\bar{b}_\infty - b_0)$ (see Fig. 2.9), but also the shear or velocity gradient, which drives the flux of b. Accordingly,

$$F_b = f_b(h, \rho, \Delta \bar{b}, \Delta \bar{u}, \mu, \epsilon_b^*), \tag{2.126}$$

where ϵ_b^* is the effective roughness height for b (see Fig. 2.9). Using $h, \rho, \Delta \bar{u}, \Delta \bar{b}$ as dimensionally independent quantities, so that $[F_b] = \mathrm{BL}^{-2}\mathrm{T}^{-1}$, the Π theorem (Sec. 2.2) gives

$$\frac{F_b}{\rho \, \Delta \bar{u} \Delta \bar{b}} = \varphi_{F_b} \left(\frac{\mu}{\rho \, \Delta \bar{u} h}, \frac{\epsilon_b^*}{h} \right). \tag{2.127}$$

Complete self-similarity with respect to the global Reynolds number, as in Eq. (2.123), also exists for the flux of any passive scalar. This provides a way to further simplify the previous expression, allowing us to write

$$- \frac{F_b}{\rho \, \Delta \bar{u} \Delta \bar{b}} \sim \mathrm{constant} = \mathrm{Cb}, \tag{2.128}$$

where the minus sign keeps the proportionality coefficient Cb positive for the case of upward flux. This in turn allows the introduction of a typical scale, b_*, for the passive scalar, through

$$- \frac{F_b}{\rho} = u_* b_* \tag{2.129}$$

which, using Eq. (2.125), can be written as

$$b_* = \mathrm{Cb} \, \mathrm{Cd}^{-1/2} \, \Delta \bar{b}. \tag{2.130}$$

In the specific cases of the sensible heat and evaporation fluxes, Eqs. (2.118) and (2.119), similar considerations that led to Eqs. (2.126)–(2.130) allow us to introduce temperature and humidity scales, T_*, and q_*, and relate them to the fluxes through

$$u_* T_* = \overline{w'T'} = \frac{H}{\rho c_p} \tag{2.131}$$

and

$$u_* q_* = \overline{w'q'} = \frac{\rho_w E}{\rho}. \tag{2.132}$$

The above relationships between turbulent fluxes and typical turbulent scales for the different transported quantities are not yet usable in practice because they depend on undefined characteristic turbulent scales; by linking them to the mean profiles of velocity and the transported quantity, in the coming sections we will obtain the required gradient-flux expressions.

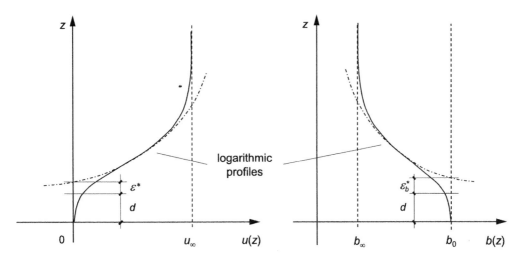

Figure 2.9 Logarithmic profiles of longitudinal mean velocity, $\bar{u}(z)$ and of the mean concentration, $\bar{b}(z)$, of a passive scalar.

2.3.2 Logarithmic Profiles

We show now that the mean concentrations of horizontal momentum (and thus the longitudinal velocity) and the generic variable b attain a logarithmic profile in the self-similar region in the ABL (see Fig. 2.9). This result is very useful for obtaining practical expressions for the turbulent fluxes.

We begin with the longitudinal mean velocity profile, first noting that the vertical profiles of mean turbulent quantities can be obtained in terms of their vertical gradients, expressed as a function of a vertical coordinate y with an arbitrary origin within the rough surface. We can then proceed to construct a general relationship of the type (2.107), by listing all the relevant governing variables. Thus, besides y, one should include also the height of the ABL, h, the representative roughness height, ϵ^*, the fluid density and dynamic viscosity, ρ and μ, a measure of the global velocity difference across the ABL, such as $\Delta \bar{u} = \bar{u}_\infty$ (see Fig. 2.9), and the wall shear stress τ_0. However, looking at Eq. (2.125), it is clear that the last two variables ($\Delta \bar{u}$ and τ_0) are linked through the turbulent velocity scale u_*, which can thus be used as a single governing quantity instead of two. On the basis of these considerations, as a starting point of our dimensional analysis we can write an equation linking the gradient of the velocity profile to a function of the variables discussed above:

$$\frac{d\bar{u}(y)}{dy} = f_u(y, h, \rho, u_*, \mu, \epsilon^*). \tag{2.133}$$

Choosing ρ, y, and u_* as the governing, dimensionally independent, quantities, we can apply the Π theorem (Sec. 2.2) to get

$$\frac{y}{u_*} \frac{d\bar{u}(y)}{dy} = \varphi_u \left(\frac{h}{y}, \frac{\mu}{\rho u_* y}, \frac{\epsilon^*}{y} \right). \tag{2.134}$$

Laboratory and field experiments support a self-similarity assumption (Barenblatt, 1996) for high-Reynolds-number boundary layers (see Sec. 2.2.2). Specifically, $y/u_* \times d\bar{u}(y)/dy$ reaches an asymptotic shape in a region which is at the same time close to the wall, compared with the boundary layer height h, so that $h/y \to \infty$, and far enough from it that the local Reynolds number $\rho u_* y / \mu \to \infty$ and the relative roughness $\epsilon^*/y \to 0$. In this region, turbulence self-similarity (see Sec. 2.2.2) allows us to write

$$\frac{y}{u_*} \frac{d\bar{u}(y)}{dy} = \varphi_u(\infty, 0, 0) \sim \text{constant} = \frac{1}{k}, \tag{2.135}$$

where k is the von Karman constant, approximately equal to 0.4. The previous expression is easily integrated, providing a *logarithmic profile*,

$$\bar{u}(y) = \frac{u_*}{k} \ln y + \mathcal{C}, \tag{2.136}$$

where the integration constant is $\mathcal{C} = (-u_*/k) \ln \epsilon^*$, obtained by imposing that the zero of the logarithmic velocity is at $y = \epsilon^*$, where ϵ^* is the momentum *roughness height*. Measured data show that the effective roughness height for the velocity profile is less than the average tree canopy height (see Table 2.3 and Fig. 2.9). It is more convenient to use a vertical axis $z = y + d$, where d is the *displacement height*, starting at the surface level, so that

$$\bar{u}(z) = \frac{u_*}{k} \ln \left(\frac{z - d}{\epsilon^*} \right). \tag{2.137}$$

This profile is found to fit experimental data quite well in the self-similar region above the canopy, typically going from the top of the canopy to a height of about one tenth of the entire ABL height.

Regarding the passive scalar $\bar{b}$, one can proceed similarly to Eq. (2.133) with the proviso of including both the turbulent scales u_* and b_* among the governing quantities, i.e.,

$$\frac{d\bar{b}(y)}{dy} = f_b(y, h, \rho, u_*, b_*, \mu, \epsilon_b^*). \tag{2.138}$$

Using y, ρ, u_*, b_* as dimensionally independent variables, the Π theorem and the self-similarity property (Sec. 2.2.2) yield

$$\frac{y}{b_*} \frac{d\bar{b}(y)}{dy} \sim \text{constant} = \frac{1}{k} \tag{2.139}$$

and, in turn,

$$\bar{b}(z) - b_0 = \frac{b_*}{k} \ln \left(\frac{z - d}{\epsilon_b^*} \right), \tag{2.140}$$

where b_0 is the value of $\bar{b}$ at the surface (see Fig. 2.9).

Typical values for the height parameters of the logarithmic profiles are given in Table 2.3 for vegetated surfaces. The parameters ϵ_T^* and ϵ_q^* are the roughness heights for sensible heat and water vapor. These enter in the expressions for the vertical profiles of mean temperature and specific humidity, respectively.

2.3.3 Gradient-Flux Expressions

Using the logarithmic profiles (2.137) and (2.140), we can now specify typical values for the turbulent scales appearing in the flux expressions (2.125) and (2.130). For this purpose, we consider a level, z_a, within the logarithmic layer, and solve Eq. (2.137) for u_* to obtain

$$u_* = \frac{k\bar{u}(z_a)}{\ln[(z_a - d)/\epsilon^*]} \tag{2.141}$$

Table 2.3 Typical values of the height parameters for logarithmic profiles (Eqs. (2.150) and (2.151)). Data from Kaimal and Finnigan (1994).

Symbol	Parameter	Value
d	displacement height	~ 0.75 of canopy height, h_c
ϵ^*	momentum roughness height	$\sim d/10$
ϵ_T^*	sens. heat roughness height	$\sim 20\%$ of ϵ^*
ϵ_q^*	w. vapor roughness height	$\sim 20\%$ of ϵ^*

and similarly for b_*, using Eq. (2.140):

$$b_* = \frac{k(\bar{b}(z_a) - b_0)}{\ln[(z_a - d)/\epsilon_b^*]}. \tag{2.142}$$

The fluxes can now be obtained, using Eqs. (2.125) and (2.129), as

$$\frac{\tau_0}{\rho} = u_*^2 = \frac{k^2[\bar{u}(z_a)]^2}{\{\ln[(z_a - d)/\epsilon^*]\}^2} \tag{2.143}$$

and

$$-\frac{F_b}{\rho} = u_* b_* = \frac{k^2 \bar{u}(z_a)[\bar{b}(z_a) - b_0]}{\ln[(z_a - d)/\epsilon^*]\ln[(z_a - d)/\epsilon_b^*]}. \tag{2.144}$$

In terms of the dimensionless coefficients introduced in Eqs. (2.124) and (2.128) at level z_a,

$$\mathrm{Cd}_a = \frac{k^2}{\{\ln[(z_a - d)/\epsilon^*]\}^2} \tag{2.145}$$

and

$$\mathrm{Cb}_a = \frac{k^2}{\ln[(z_a - d)/\epsilon^*]\ln[(z_a - d)/\epsilon_b^*]}, \tag{2.146}$$

the previous fluxes can also be written as

$$\tau_0 = \mathrm{Cd}_a\,\rho\bar{u}_a^2 \tag{2.147}$$

and

$$-F_b = \mathrm{Cb}_a\bar{u}_a\,\rho(\bar{b}_a - b_0), \tag{2.148}$$

where $\bar{b}_a = \bar{b}(z_a)$.

Taking into account that for sensible heat $F_b = H$ and $\bar{b} = c_p\bar{T}$, and for evaporation $F_b = \rho_w E$ and $\bar{b} = \bar{q}$, the following expressions are obtained:

$$\begin{aligned}
-H &= \mathrm{Ch}_a\,\rho c_p \bar{u}_a(\bar{T}_a - T_0), \\
-\rho_w E &= \mathrm{Ce}_a\,\rho \bar{u}_a(\bar{q}_a - q_0).
\end{aligned} \tag{2.149}$$

These important results show that the turbulent fluxes are characterized by the differences between values of the variables at the surface and at a level z_a within the logarithmic layer, with $\bar{u}(z_a) = \bar{u}_a$, $\bar{T}(z_a) = \bar{T}_a$, and $\bar{q}(z_a) = \bar{q}_a$. For sensible heat and evaporation the bulk-transfer coefficients are called the Stanton and Dalton coefficients:

$$Ch_a = \frac{k^2}{\ln[(z_a - d)/\epsilon_T^*]\ln[(z_a - d)/\epsilon^*]}, \tag{2.150}$$

$$Ce_a = \frac{k^2}{\ln[(z_a - d)/\epsilon_q^*]\ln[(z_a - d)/\epsilon^*]}. \tag{2.151}$$

Since the roughness heights for sensible heat and water vapor are similar, $\epsilon_T^* \approx \epsilon_q^*$ and so $Ch_a \approx Ce_a$. Therefore, it is convenient to introduce a single *atmospheric conductance*, having the dimension of velocity, for both the sensible heat and water-vapor fluxes,

$$g_a = \frac{\bar{u}_a\, k^2}{\ln[(z_a - d)/\epsilon^*]\ln[(z_a - d)/\epsilon_q^*]}, \tag{2.152}$$

so that the previous fluxes become

$$-H = \rho c_p g_a(\bar{T}_a - T_0), \tag{2.153}$$

$$-\rho_w E = \rho g_a(\bar{q}_a - q_0). \tag{2.154}$$

Using the approximation (2.75), the water-vapor flux is also written as

$$- \rho_w E = \frac{\epsilon}{p}\rho g_a(\bar{e}_a - e_0), \tag{2.155}$$

where p is the near-surface atmospheric pressure, $\bar{e}_a$ and e_0 are the water vapor pressures at level z_a and near the evaporating surface, and ϵ (not the surface roughness ϵ^*) is approximately 0.622.

The previous expressions allow us to compute the main fluxes between the land and the atmosphere, assuming that the values of mean velocity, temperature, and humidity in the logarithmic layer and at the surface are known. Unfortunately, it is seldom the case that all these quantities are measured, thus requiring further considerations. For example it is especially difficult to know the surface temperature. Moreover, the surface conditions, as in the case of surface humidity at the leaf level, may depend on the moisture conditions underlying the surface and thus involve additional resistance to transport beneath the surface. We will discuss the application of these flux expressions and the additional modeling assumptions relating to the soil–plant–atmosphere continuum and the surface energy balance in Chapters 4 and 5.

2.4   Dynamical Systems

The mathematical description of ecohydrological processes very often leads to systems of differential equations. These equations form a dynamical system, which can be written synthetically as

$$\frac{dx}{dt} = f(x, t) \tag{2.156}$$

for $x(0) = x_0$, where t is time, $x(t) = (x_1(t), x_2(t), \ldots, x_n(t)) \in \mathbb{R}^n$ is the vector of the state variables, and $f(t) = (f_1(t), f_2(t), \ldots, f_n(t)) \in \mathbb{R}^n$ are the rate functions. When the latter contain random terms, the equations are called stochastic differential equations (see Sec. 6.4); otherwise the system is said to be *deterministic*.

The evolution of the solution $x(t)$ as a function of time gives rise to a multivariate time series, which can be studied with several specific techniques; however, $x(t)$ can also be considered as a point in the n-dimensional space of coordinates $(x_1(t), x_2(t), \ldots, x_n(t))$, called *phase space*, where it describes a trajectory driven by the field $f(x(t), t)$, so that the dynamical system can be studied geometrically. We refer to Strogatz (2001), Logan (2013), and Argyris et al. (2015) for more information. In what follows we discuss some simple cases in one and two dimensions, setting $x_1 = x$ and $x_2 = y$ to simplify the notation.

2.4.1 One-Dimensional Case

The simple case of a time-independent one-dimensional dynamical system is of the form

$$\frac{dx}{dt} = f(x), \tag{2.157}$$

for $x(0) = x_0$. The zeros (for $x = x_e$) of the rate function,

$$f(x_e) = 0, \tag{2.158}$$

are called the *equilibrium* or fixed points of the system, because if the system starts there it stays there forever since $dx/dt = 0$.

The slope of the rate function at the fixed points determines their linear stability: if the slope is positive then the equilibrium point is unstable, while it is stable if the slope is negative. This can be understood by considering small perturbations around x_e and the shape of the related potential function $V(x)$; the latter is related to the rate function by

$$f(x) = -\frac{\partial V(x)}{\partial x} \quad \rightarrow \quad V(x) = -\int f(x)dx + \text{const.} \tag{2.159}$$

Logistic Equation with Harvesting

We illustrate the previous concepts with the example of a normalized logistic equation with a harvesting term (Logan, 2013),[13]

$$\frac{dx}{dt} = x(1 - x) - h, \tag{2.160}$$

where h is the harvest rate.

The equilibrium point is obtained by solving $x(1 - x) - h = 0$, which gives

$$x_e = \frac{1 \pm \sqrt{1 - 4h}}{2}. \tag{2.161}$$

The rate function, split into $x(1 - x)$ and h, has the form depicted in Fig. 2.10a, which shows there is no steady state for $h > 1/4$ and there are two equilibrium points for $h < 1/4$. For the smaller x_e (○), the rate function is negative on the left side of the equilibrium point ($dx/dt < 0$) and positive on the right side ($dx/dt > 0$), moving the system away from equilibrium (unstable point). For the larger x_e (●), the

---

[13] The logistic equation $dx/dt = rx(1 - x)$ should not be confused with the logistic map, a difference equation in discrete time, $x_{n+1} = rx_n(1 - x_n)$, with much richer dynamics (Strogatz, 2001; Argyris et al., 2015); the difference between these two models, i.e., the logistic equation and the logistic map, is a good reminder of the profound distinction between continuous and discrete equations.

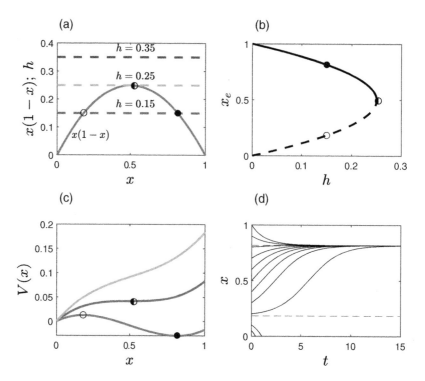

Figure 2.10 Logistic equation with harvest, as a case of bistable dynamics. (a) Rate function split into $x(1-x)$ and h; (b) equilibrium points as a function of the harvest rate h (stable solution, solid line; unstable solution, dashed line); (c) potential function for different harvest rates ($h = 0.15$ (bottom), 0.25 (middle), 0.35 (top)); and (d) time series for $h = 0.2$ starting from different initial conditions.

rate function is positive on the left side ($dx/dt > 0$) and negative on the right side ($dx/dt < 0$), moving the system toward equilibrium (stable point). The corresponding plot representing the equilibrium points and their stability as a function of the equation parameters, is called a bifurcation diagram and is shown in Fig. 2.10b. Figure 2.10c shows the potential function for different values of the harvest rate, confirming the bistable behavior and the fact that for $h > 1/4$ the solution does not have any equilibrium point and decays to minus infinity.

The solution as a function of time completes the picture of the system. In this case, it can be obtained explicitly as

$$x(t) = \frac{1}{2}\left\{1 - \sqrt{-1 + 4h}\, \tan\left[\frac{1}{2}\sqrt{-1 + 4h}\, t - \arctan\left(\frac{-1 + 2x_0}{\sqrt{-1 + 4h}}\right)\right]\right\}. \tag{2.162}$$

From the solutions plotted in Fig. 2.10d, one can see the evolution for a range of initial conditions. Below the unstable solution, the trajectories decay to zero (before plunging to minus infinity) in finite time, while above it the time evolution of the system tends toward the stable state, which for this reason is called an attractor. Without harvesting ($h = 0$), Eq. (2.162) becomes

$$x(t) = \frac{x_0}{(1 - x_0)e^{-t} + x_0}, \tag{2.163}$$

which describes a logistic growth (a slow start followed by a fast growth period and a plateau). A one-dimensional model with a similar harvest term, but slightly more complicated dynamics, will be considered in Sec. 10.4.2 in relation to crop growth, fertilization, and harvest.

2.4.2 Two-Dimensional Case

A much greater variety of cases becomes possible in two dimensions, including point attractors of different types as well as limit cycles, related to self-sustained oscillations and Hopf bifurcations (e.g., Strogatz, 2001); the existence of a potential function becomes an exception. The general form of a two-dimensional dynamical system is

$$\frac{dx}{dt} = f(x, y), \tag{2.164}$$

$$\frac{dy}{dt} = g(x, y). \tag{2.165}$$

By eliminating t with the chain rule one obtains

$$\frac{dy}{dx} = \frac{g(x, y)}{f(x, y)}, \tag{2.166}$$

the solutions of which, $y(x)$, trace out a trajectory in phase plane with time as an implicit parameter. A visual inspection of this *streamplot* often reveals the global behavior of the solutions. The streamplot is easily constructed by interpreting the system as a fluid flow with the velocity $v(x, y) = (f(x, y), g(x, y))$.

As a next step in analyzing these systems, it is useful to determine the *nullclines*, which are the curves given by the conditions $dx/dt = f(x, y) = 0$ and $dy/dt = g(x, y) = 0$. The intersections of the nullclines give the critical (or equilibrium) points, which are the solutions (x_e, y_e) of the system

$$\begin{cases} f(x_e, y_e) = 0, \\ g(x_e, y_e) = 0. \end{cases} \tag{2.167}$$

The stability of the critical points is found by considering the evolution of $x(t) = x_e + u(t)$ and $y(t) = y_e + v(t)$, where $u(t)$ and $v(t)$ are small perturbations from the equilibrium state, and then assessing whether these perturbations grow or decay. Since the perturbations are small, in normal conditions one can limit the analysis to a linearized system for the perturbations near the critical point, which is

$$\begin{aligned} \frac{du}{dt} &= Au + Bv, \\ \frac{dv}{dt} &= Cu + Dv. \end{aligned} \tag{2.168}$$

The matrix

$$\mathcal{J}(x_e, y_e) = \begin{pmatrix} A & B \\ C & D \end{pmatrix} = \begin{pmatrix} \frac{\partial f}{\partial x}\big|_{x_e, y_e} & \frac{\partial f}{\partial y}\big|_{x_e, y_e} \\ \frac{\partial g}{\partial x}\big|_{x_e, y_e} & \frac{\partial g}{\partial y}\big|_{x_e, y_e} \end{pmatrix} \tag{2.169}$$

is called the *Jacobian* matrix at equilibrium.

Table 2.4 Linear stability analysis: nomenclature, geometry, and stability of equilibrium points in terms of eigenvalues.

Eigenvalues	Stability	Name
$(+,+)$	unstable	unstable node
$(-,-)$	stable	stable node
$(+,-)$	unstable	unstable saddle point
$a \pm bi; a > 0$	unstable	unstable spiral
$a \pm bi; a < 0$	stable	stable spiral
$a \pm bi; a = 0$	neutral	circle

Because of the linearity of the system (2.168), the evolution around the critical point is of the form $x = v e^{\lambda t}$, which substituted into Eq. (2.168) transforms the *linear stability analysis* into an algebraic eigenvalue problem,

$$\det \begin{pmatrix} A - \lambda & B \\ C & D - \lambda \end{pmatrix} = \lambda^2 - (\mathrm{tr}\mathcal{J})\lambda + \det\mathcal{J} = 0, \tag{2.170}$$

where $\mathrm{tr}\mathcal{J} = A + D = \lambda_1 + \lambda_2$ is the trace of $\mathcal{J}$ and $\det\mathcal{J} = AD - CB = \lambda_1\lambda_2$ is its determinant. The eigenvalue with larger absolute value is referred to as the dominant eigenvalue, λ_d. In terms of asymptotic stability, when $\mathrm{Re}(\lambda_d) > 0$ the critical point is unstable, while $\mathrm{Re}(\lambda_d) < 0$ the point is stable. The names and the local geometry of the critical points are summarized in Table 2.4.

Substrate–Microbe Dynamical System

As an example we consider a minimalist model of substrate–microbe interaction in soils, which will be discussed in depth in Sec. 9.2,

$$\begin{aligned} \frac{dx}{dt} &= a - kxy + by, \\ \frac{dy}{dt} &= (1 - r)kxy - by, \end{aligned} \tag{2.171}$$

where a, k, b, and r are constant coefficients. The first equation is the rate equation for the substrate x, which is increased by vegetation litter and root decay given by the term a; the term kxy gives the decomposition of substrate by microbial biomass (k is the decomposition constant) and by expresses microbial death (b is a mortality coefficient); the dead biomass is assumed to be recycled into the substrate. The equation for the microbial biomass y has a growth term given by the decomposition minus the fraction of the decomposition that is respired (r is the respiration fraction), and a decay rate due to microbial death. We first determine the equilibrium point, which is

$$x_e = \frac{b}{(1 - r)k}, \quad y_e = \frac{(1 - r)a}{rb}. \tag{2.172}$$

Next, the Jacobian is obtained by linearizing around this point, leading to

$$\mathcal{J}(x_e, y_e) = \begin{pmatrix} -ky_e & -kx_e + b \\ (1-r)ky_e & (1-r)kx_e - b \end{pmatrix} = \begin{pmatrix} \dfrac{ak(1-r)}{br} & -\dfrac{br}{1-r} \\ \dfrac{ak(1-r)^2}{br} & 0 \end{pmatrix}. \tag{2.173}$$

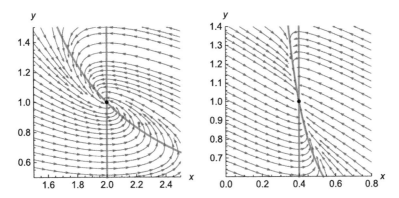

Figure 2.11 Phase diagrams for the substrate–microbe dynamics of Eq. (2.171) for $k = 1$ (left) and $k = 5$ (right).

To keep the expressions simple, we focus on the role of the decomposition constant $k > 0$ and assume $a = b = 2r = 1$. In this case, the Jacobian simplifies to

$$\mathcal{J} = \begin{pmatrix} -k & -1 \\ k/2 & 0 \end{pmatrix},$$ (2.174)

from which, solving the characteristic equation (2.170), one obtains the eigenvalues

$$\lambda_{1,2} = \frac{1}{2}\left(-k \pm \sqrt{k(k-2)}\right).$$ (2.175)

Since the real parts of both eigenvalues are negative, the critical point is always stable; the imaginary parts are nonzero for $0 < k < 2$, so the critical point is a spiral, while it is a stable node for $k > 2$ (see Table 2.4). Phase diagrams for the nonlinear model (2.171) are shown in Fig. 2.11 for two different values of k, $k = 1$ for the spiral case and $k = 5$ for the node case.

More complicated dynamics become possible due to nonlinearities, including long-term solutions with self-sustained oscillations; see Note 2.15. These periodic oscillations in phase space form a closed loop and represent a stable attractor called *limit cycle*. Non-periodic oscillatory solutions, referred to as deterministic chaos, are only possible in dimensions higher than 2 (this includes two-dimensional systems with periodic forcing), when the system has sensitivity to initial conditions (Strogatz, 2001; Argyris et al., 2015). In chaotic conditions, the system trajectory never visits exactly the same point and describes a fractal object called strange attractor.

2.5   Key Points

- The Gibbs chemical potential, Eq. (2.18), is obtained from the combination of the first and second laws of thermodynamics for reversible processes in open systems.
- The Gibbs–Duhem equation (2.26), coupled to the condition of equal chemical potential at phase transitions, leads to the Clausius–Clapeyron equation for the saturated vapor pressure as a function of temperature, $p_{sat}(T)$.

- The vapor pressure deficit (VPD) and the specific humidity deficit are connected by the relationship between specific humidity and vapor pressure, Eq. (2.75).
- The total water potential, defined in Eq. (2.101), is the sum of matric, osmotic, and gravitational components.
- The main two pillars of dimensional analysis are the Π theorem (2.109) and self-similarity (2.111); while the former is a rigorous mathematical deduction, the latter is an assumption that requires empirical or numerical verification.
- Applying the Π theorem and complete self-similarity twice, turbulent fluxes can be written as the product of bulk transfer coefficients and the gradient in the mean quantity being transported (Sec. 2.3.3).
- In one-dimensional dynamical systems, the stability of the fixed points is given by the slope of the rate function at those points; in the two-dimensional case, the main fixed points are centers, nodes, foci, and saddles (depending on the eigenvalues of the Jacobian); however, more complicated cases, including limit cycles related to self-sustained oscillations, are also possible in the presence of nonlinear terms.

2.6 Notes, including Problems and Further Reading

2.1 Determine which forms of specific energy have units of pressure, length, or velocity squared.

2.2 Derive the expression for the latent heat of vaporization λ_w using Eq. (2.47) with the entropy expressions for perfect gas and incompressible fluids. Use the expression obtained in Eq. (2.48) to derive an equation for $p_{sat}(T)$.

2.3 Compute the density of the dry air at one standard atmospheric pressure ($p_0 = 101.325$ kPa) for $T = 10$ and $30\,°C$, as well as the corresponding volumes for one mole of dry air.

2.4 Compare the final pressure, temperature, and work done by a mole of dry air for three different types of expansion: isothermal, adiabatic, and free expansion, starting from the same initial condition.

2.5 Derive the expression for the potential temperature, Eq. (2.65), as the final temperature for an adiabatic reversible compression of an ideal gas from a lower pressure p to the standard atmospheric pressure p_0, and link it to the molar entropy of dry air (considered as a perfect gas).

2.6 Compute the difference in density between dry air and saturated humid air at 20 °C and standard atmospheric pressure. Compute the temperature at which the dry air has the same density as saturated air at temperature T and pressure p.

2.7 Consider isothermal hydrostatic equilibrium for dry air (treated as a perfect gas) and liquid water (treated as an incompressible fluid). Obtain the vertical profiles of the main thermodynamic quantities in both cases.

2.8 Seawater is a solution of water with dissociated salts (Nobel, 1999): each cubic meter of solution contains approximately 546 mol of Cl^-, 470 mol of Na^+, 53 mol of Mg^{2+}, 28 mol of SO_4^{2-}, and 10 mol of K^+ and Ca^{2+}, with lesser concentrations of other ions. Compute the osmotic potential of seawater at 20 °C using the van't Hoff equation (2.92).

Solution:

$$\Omega = -8.314 \, \frac{J}{K \, mol} \times 293 \, K \times 1107 \, \frac{mol}{m^3} = -2.7 \, MPa. \tag{2.176}$$

2.9 Use Raoult's law to compute the relative humidity of the atmosphere in equilibrium with seawater (see the previous problem).

2.10 Compute the total water potential of air (treated as a perfect gas) at relative humidities 35% and 75% and temperatures 20 and 30 °C, at ground level, at 100 m, and at 1000 m (take the gravitational acceleration as constant).

2.11 (Class of System of Units) For a generic problem in fluid mechanics ($k = 3$), show that a density ρ, a velocity v, and a length l can be chosen to form an alternative class of system of units, instead of the usual LMT class, which adopts a length l, a mass m, and a time t, with $[l] = L$, $[m] = M$ and $[t] = T$.

Solution: Compute the determinant of the exponents of the dimension functions,

$$\begin{aligned}
&\text{density, } \rho & &[\rho] = ML^{-3}, \\
&\text{velocity, } v & &[v] = LT^{-1}, \\
&\text{length, } l & &[l] = L,
\end{aligned}$$

that is

$$\begin{vmatrix} 1 & -3 & 0 \\ 0 & 1 & -1 \\ 0 & 1 & 0 \end{vmatrix} = 1 \neq 0, \tag{2.177}$$

which proves the dimensional independence of the chosen set of variables and their suitability to form a new class of system of units.

In this new class, which may be indicated as RVL, the units of density, velocity, and length become fundamental, i.e.,

$$\begin{aligned}
&\text{density, } \rho & &[\rho]_{RVL} = R, \\
&\text{velocity, } v & &[v]_{RVL} = V, \\
&\text{length, } l & &[l]_{RVL} = L,
\end{aligned}$$

and all derived quantities can be expressed using the new class. Thus, for example, the unit of viscosity μ (see Eq. (2.106)), while still remaining a derived one in the new RVL system, changes as follows:

$$\text{viscosity, } \mu \qquad [\mu]_{RVL} = RVL,$$

and similarly for the previously fundamental quantities in the LMT system, which become derived quantities:

$$\begin{aligned}
&\text{length, } l & &[l]_{RVL} = L, \\
&\text{mass, } m & &[m]_{RVL} = RL^3, \\
&\text{time, } t & &[t]_{RVL} = LV^{-1}.
\end{aligned}$$

It is also interesting to consider explicitly the effect of this change of class of system of units, from LMT to RVL, on the Reynolds number,[14] $\text{Re} = \rho u l / \mu$, whence

$$[\text{Re}]_{\text{MLT}} = \frac{\text{ML}^{-3}\text{LT}^{-1}\text{L}}{\text{ML}^{-1}\text{T}^{-1}} = [\text{Re}]_{\text{RVL}} = \frac{\text{RVL}}{\text{RVL}} = 1. \tag{2.178}$$

Thus, clearly, while a quantity that is fundamental in one class may not be so in another one, dimensionless groups remain unchanged upon a change of class of units.

2.12 Plot the atmospheric conductance as a function of wind velocity for three different types of vegetation: grassland ($h_c = 0.4$ m), shrubland ($h_c = 1.5$ m), forest ($h_c = 20$ m).

2.13 (Normal Form of Bistable Systems) In the following chapters, we will see several cases of bistable dynamics in a soil–plant system (e.g., Note 8.9 on vegetation, Sec. 10.1.3 on land degradation, and Note 10.16 on salinity). Each model has it own specific nonlinearities, but their common behavior is captured by the same *normal form* (Strogatz, 2001; Argyris et al., 2015). For a state variable h, the normal form of bistable dynamics is

$$\frac{dh}{dt} = f(h) = h - \frac{h^3}{3} + r - w, \tag{2.179}$$

where r is a constant term which makes h grow and w is a negative term which makes it decrease.

Determine the steady-state solutions of Eq. (2.179) and their linear stability. Show that, when plotting them as a function of the parameter $\epsilon = r - w$, the system undergoes a pitchfork bifurcation. Plot the potential, $\mathcal{V}(h)$, defined by $f(h) = -\partial \mathcal{V}(h)/\partial h$ as a function of ϵ.

2.14 (Temperature Dynamics of the Snowball Earth) Reconsider the simplified problem of the temperature of a planet (Notes 1.2 and 1.3). The planet's energy balance is

$$\frac{dU}{dt} = R_I - R_O, \tag{2.180}$$

where U is the planet's internal energy, R_I is the incoming shortwave radiation, and R_O the outgoing longwave radiation. For $U = CT$, where C is a representative constant heat capacity of the planet, and using the radiation functions (1.6), one obtains a one-dimensional dynamical system,

$$\frac{dT}{dt} = f(T), \tag{2.181}$$

for the temporal evolution of the Earth's temperature. Show that including the snow albedo effect (Note 1.3) produces a bistable system. Plot the solution of the system for different values of the heat capacity and initial temperature conditions, as well as the function $f(T) = -\partial \mathcal{V}(T)/\partial T$ and the corresponding potential function $\mathcal{V}(T)$ as a function of temperature.

2.15 (Excitable Dynamics of Socio-Environmental Systems) Let us consider again the normal form of the bistable system (2.179) for an environmental variable h and assume that the loss term w is now a variable

---

[14] Recall that the Reynolds number is the dimensionless group (formed from a typical velocity and lateral dimension of the flow along with the fluid density and viscosity), which describes the relative importance of inertial and viscous terms in fluid flows (Munson et al., 2005).

representing the pressure from a socio-economical component. Assuming that the latter "benefits" from h, according to the equation

$$\frac{dw}{dt} = h - a - bw, \tag{2.182}$$

the coupled system of Eqs. (2.179) and (2.182) is then the well-known Fitzhugh–Nagumo model, an extension of the van der Pol equations (see Murray, 2002 and Lindner et al., 2004).

Solve numerically the above system and plot the solutions both as functions of time and as phase-space orbits in the plane (h, w) for different values of the parameters. Show that the system undergoes a Hopf bifurcation, giving rise to oscillations that could be interpreted as cycles of exploitation–degradation–collapse–recovery (see Sec. 10.1.2).

3   The Soil

Pinguis item quae sit tellus, hoc denique pacto
discimus: haud umquam manibus iactata fatiscit,
Sed picis in morem ad digitos lentescit habendo.

The fatter earth by handling we may find,
With ease distinguished from the meagre kind:
Poor soil will crumble into dust; the rich
Will to the fingers cleave like clammy pitch.

Virgil, Georgics, L248–250 (tr. Dryden)

> This chapter deals with the first component of the soil–plant–atmosphere continuum (SPAC). We begin by classifying the different types of soils and conceptualizing the soil as a porous medium, to define the main soil hydrologic variables, including the porosity and the relative soil moisture. We proceed by focusing on the soil's hydraulic properties: the soil water-retention curve, which links the soil moisture content to the soil water potential, and the hydraulic conductivity, which relates water flow to the water-potential gradients. Using simple models of pore hydraulic behavior, we emphasize the statistical nature of these curves and discuss how organic matter and biotic components may alter their form. Finally, after reviewing the continuum approach to unsaturated water flow, given by Richard's equation, we present its sorption solutions for infiltration modeling and its integration over the root zone to arrive at a vertically averaged water-balance equation that will be used in subsequent chapters.

3.1   Soil Types and Classification

The soil is the store of water and nutrients for most plants and living organisms on land (Noy-Meir, 1973), and acts as an important hydrologic "valve" for the partitioning of the land-surface water and energy fluxes. While soil is often idealized as a mixture of mineral, liquid, and gaseous components, it is in reality a very complex biomaterial (Brady and Weil, 1996; Richter and Markewitz, 2001; Paul, 2006), which provides different environments for plants and microbial life. Owing to this complexity, there are several ways in which soils have been classified and analyzed.

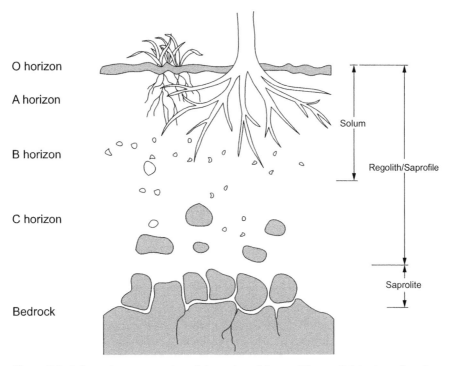

Figure 3.1 Schematic representation of the main soil layers. The regolith is also referred to as saprofile, while the weathered part of the bedrock is called saprolite.

3.1.1 Soil Layers and Taxonomy

Soils develop slowly from the underlying bedrock through physico-chemical processes, collectively called weathering. Such processes are strongly impacted by hydroclimatic conditions and usually mediated by plants and soil organisms. As a result, soils tend to develop horizontal layers or *horizons* (see Fig. 3.1):

- O horizon: the most surficial organic layers, resulting from the recent accumulation of plant litter. It is often also referred to as the litter layer.
- A horizon: the layers near the surface, dominated by a mixture of mineral particles darkened by consistent accumulation of organic matter; the site of most roots.
- B horizon: the layer with less organic matter, with typical accumulation of clays and varying amounts of silicates, iron and aluminum oxides, calcium carbonate, and other materials percolating from the surface or produced by the weathering processes.
- C horizon: the least weathered part of the soil profile (also called the regolith, see Fig. 3.1), with rock intrusions and less abundant plant roots and microorganisms.

In what follows we will focus on the A horizon, where most roots are located, often assuming that its properties are relatively homogeneous to allow for a vertically averaged description.

Soils are also very different, depending on their composition and age. The commonly used United States Department of Agriculture (USDA) soil taxonomy is based on 12 *soil orders* (Brady and Weil, 1996), summarized here by increasing degree of weathering and soil development (the numbers in parentheses give the approximate coverage worldwide, the rest being covered by rocks and sands):

- Entisols (16%): recent soils at the earliest stages of formation, with little profile development.
- Histosols (1%): dark soils typical of humid areas (peats and bogs) with high levels of accumulation of organic matter (more than 20% of the total).
- Gelisols (9%): young soils characterized by layering and churning caused by the weathering action of freezing and thawing in the permafrost layer.
- Inceptisols (10%): young soils with barely developed soil horizons.
- Andisols (< 1%): moderately weathered, fertile[1] soils formed by volcanic deposits and rich in amorphous glass and iron and/or aluminum minerals that impart a high water and organic-matter retention capacity.
- Aridisols (12%): light-colored soils with low organic-matter content typical of arid regions and characterized by the possible accumulation of salt deposits due to incomplete leaching, surface armoring by rocks and pebbles, and surface salinization.
- Vertisols (2%): dark soils with high levels of very active (shrinking and swelling) clays, typical of warm mesic regions and supporting grassland vegetation.
- Alfisols (10%): soils rich in silicate clay typical of relatively humid areas, with a possible tendency to accumulate sodium (natric soils). They typically host deciduous forests or savannas.
- Mollisols (7%): well-developed, calcium- and humus-rich soils formed under prairie grasses.
- Ultisols (9%): acidic soils formed from the clays typical of forested areas with low-to-moderate fertility.
- Spodosols (3%): acidic sandy soils typically supporting forests (especially conifers).
- Oxisols (8%): highly weathered soils found in warm and humid areas (e.g., tropical forests) that are easily washed by leaching and thus with mild acidity and low fertility and with abundant, quasi-inert clays and fine-textured materials.

3.1.2 Soil Texture

The solid phase of soils is mostly mineral, made up of a mixture of sand, silt, clay, and to a lesser extent organic matter. For classification purposes, it is conventionally assumed that all soil particle sizes can be sorted into three textural fractions, namely *sand, silt, and clay*. According to the USDA classification, sand particles have sizes between 2 and 0.05 mm (gravel has particle sizes >2 mm), silt particles are 0.05-0.005 mm, and clays have sizes less than 0.005 mm.[2]

Depending on the fractions of sand, silt, and clay, the soil can be divided into textural classes. The most important and general class is loam, a mixture of sand, silt, and clay in more or less equal portions.

---

[1] Note however that the vegetation in andisols tends to be phosphorous limited because of the very high phosphorous retention capacity of its constituents.

[2] For their size and composition, clays tend to behave as colloids with active chemical properties and often display plastic behavior. This is due to their irregular crystalline structure, consisting of more or less regular arrangements of silica and alumina sheets (see, Brady and Weil, 1996 or Hillel, 1998 for an introduction to clay structure, classification, and properties). Such sheet structures or clay platelets may be disrupted by water (some clays shrink when dry and swell when water is added), chemicals (clays tend to adsorb cations like Ca^+ and Na^+), or organic matter.

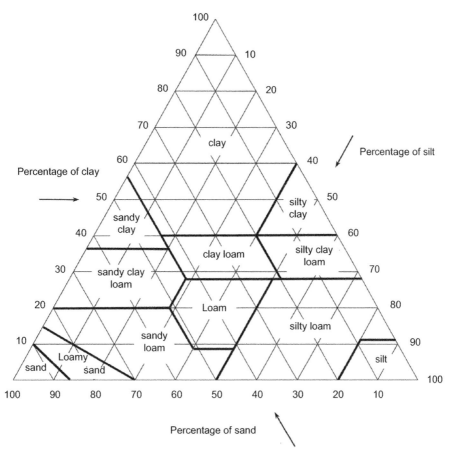

Figure 3.2 The USDA soil textural triangle. The arrows indicate the directions in which the axes should be read. Thus, for example, the percentage of sand is given by the numbers down the right-hand sloping side of the triangle. Image from Soil Survey Division Staff (1993), Soil Survey Manual. Credit: US Department of Agriculture.

Real soils are typically loams of some sort. More specifically the USDA soil-texture classification distinguishes 12 textural classes (Fig. 3.2). Soil texture is often a good descriptor of a soil's physical properties.

Despite amounting only to 5% in volume and 3% in weight in fertile soils, soil organic matter (SOM) is a very important factor determining the hydraulic and thermal properties of soils (Sec. 3.2.4) and the dynamics of nutrients (see Chapter 9).

3.2   Hydraulic Properties of Soil

From a hydraulic point of view, a soil may be idealized as an *unsaturated porous medium*. This means that, despite being made of an aggregate of irregular particles, for modeling purposes it may be treated as a continuum. This description is achieved by averaging its properties over volumes that are small enough to be mathematically considered as a point, but large enough to achieve the statistical convergence of such averages (Hillel, 1998). Under this assumption, all soil properties (soil water potential, soil porosity, soil moisture hydraulic conductivity, etc.) can be defined at a point in space. The spatial scale at which this averaging is done is known as the *Darcy scale* and the averaging volume is called the "representative elementary volume" (e.g., Bear, 1972). Depending on the soil texture, the Darcy scale can range from a few centimeters to decimeters, and defines the smallest scale at which the results of the continuum scheme can be applied. In natural, vegetated soils, the Darcy scale may be quite large and, because of rock intrusions, roots, and macropores, it may even exceed the depth of the A horizon or the rooting depth. In such conditions, the use of soil layers or vertically averaged descriptions may offer better approximations than a vertically continuous description.

3.2.1 Soil Moisture

Assuming the soil to be composed of solid (mostly mineral), liquid (water), and air phases, the total representative volume of soil around a point (x, y, z) is given by the sum of the volumes of air, water, and mineral components, i.e., $V_s = V_a + V_w + V_m$, where V_s is defined by the Darcy scale. The *porosity* is defined as

$$n = \frac{V_a + V_w}{V_s} \tag{3.1}$$

and typically ranges between 0.35 and 0.45, with clays having higher values than silts and sands. Typical values are reported in Table 3.1.

The volumetric water content, θ, is the ratio of water volume to soil volume, so that the local *relative soil moisture* is

$$s = \frac{V_w}{V_a + V_w} = \frac{\theta}{n}, \tag{3.2}$$

which is the fraction of pore volume containing water. The relative soil moisture is a dimensionless, normalized variable, which varies between zero and one (saturation).

The previous formulae define fields in space and time, $n(x, y, z, t)$, $\theta(x, y, z, t)$, and $s(x, y, z, t)$. While both $\theta(x, y, z, t)$ and $s(x, y, z, t)$ vary on relatively short timescales with fast jumps and slower decays (see Fig. 1.7), the porosity varies only at timescales related to vegetation growth and decay (years and longer) and soil formation; for this reason it will be considered constant in time, $n(x, y, z)$.

3.2.2 Soil Water-Retention Curve

The total water potential (Sec. 2.1.14) in the soil serves as a fundamental quantity to describe the water flow in porous media. As with the porosity and soil moisture, the water potential is interpreted as a spatial average over a representative volume, of a size determined by the Darcy scale. In unsaturated conditions and with low solute concentration, the most important component of the soil water potential, ψ_s, is the *soil matric potential*, Ψ_s, due to the surface tension of water in the soil pores, which act as capillary tubes (see Sec. 2.1.14 and Eq. (2.101)). Because of the curvature of the water menisci in the pores, the liquid water in unsaturated soils (i.e., when $s < 1$) is under a negative water potential, or suction. The water potential, becomes more negative as the soil moisture level drops, and liquid water, confined to the smallest pores, becomes progressively less available to roots and microorganisms (i.e. lower free energy).

Typical values of the soil water potential range from around −0.01 MPa, for well-watered soils, to about −10 MPa or lower for very dry conditions (at the so-called *hygroscopic point*, $\Psi_{s,h}$, when the soil water is in equilibrium with the atmospheric water vapor). The soil moisture content at this point is also called the residual soil moisture content.

The mathematical relationship linking the soil matric potential to the relative soil moisture is the soil *water-retention curve*, $\Psi_s = \Psi_s(s)$. This curve is a static one-to-one relation representing the macroscopic (i.e., averaged over the Darcy scale) link between the amount of water in the soil pores and its energetic (e.g., water potential) content. It entails a quasi-equilibrium assumption, which neglects the dynamic process of wetting and drying (which is responsible, among other things, for the hysteretic behavior of retention curves; e.g., De Gennes, 1985; Brutsaert, 2005).

Several mathematical forms have been proposed in the literature for the retention curve (Brutsaert, 2005). A simple and convenient representation was proposed by Brooks and Corey (1966) in the form of a power law,[3]

$$\Psi_s = \overline{\Psi}_s s^{-b}, \tag{3.3}$$

where $\overline{\Psi}_s$ and b are experimentally determined parameters (see Table 3.1). In particular, $\overline{\Psi}_s$ represents the *air-entry suction* point (i.e., the level of suction needed for air to overcome the capillary forces and enter the soil pores under saturated conditions[4]), whereas b is related to the soil pore-size distribution (see Sec. 3.2.3).

Soil texture plays an important role in determining the range of values of soil water potential for typical soil moisture values, as can be seen from Fig. 3.3. Thus, different types of soil may yield very different levels of relative soil moisture corresponding to the same value of soil matric potential. In particular, through the retention curves we can derive the soil moisture values that correspond to some water potential levels that are important in plant physiology, such as the *wilting point*, s_w, and the *point of incipient stomatal closure*, s^*. So, for example, values of s_w corresponding to a soil water potential of −3 MPa are found around 0.1 for loamy sand, while for loam they are near 0.25. Similarly, s^* related to a potential of −0.03 MPa is found around 0.3 for loamy sand and moves above 0.55 in the case of loam (see Table 3.1). A reference value of −10 MPa is used for the hygroscopic point, Ψ_{s,s_h}, (see Fig. 3.3), and in general it can be related to the relative humidity assuming equilibrium with the surrounding air (see Eq. (2.104)). These concepts will be discussed in more detail and linked to plant physiology in Chapter 4.

3.2.3 Interpretation of the Soil Water-Retention Curve

As porous media made of irregular grains and aggregates, soils are often suitably described by statistical models. Thus, one can formulate the averaging of soil properties over the Darcy scale as an ensemble average over a statistical distribution, such as the *pore-size distribution*. Here we assume that our reader is

---

[3] For soils having non-unimodal pore-size distributions, because of their more irregular grain aggregation, the retention curve may be modeled by superposing unimodal retention curves (Durner, 1994; Coppola, 2000).

[4] It is worth noting that in order to obtain values of soil moisture in reasonable agreement with measured field data, one should use the value of $\overline{\Psi}_s$ given by the geometric mean of the measured values (Clapp and Hornberger, 1978) instead of the arithmetic mean, which would yield unrealistically high values of soil moisture.

Table 3.1 Parameters describing the soil characteristics used in the model in Sec. 3.2 for four different soil textures. After Laio et al. (2001c).

	$\overline{\Psi}_s$ (kPa)	b	c	K_s (cm/d)	n	β	s_h	s_w	s^*	s_{fc}
sand	−0.34	4.05	11.1	> 200	0.35	12.1	0.08	0.11	0.33	0.35
loamy sand	−0.17	4.38	11.7	≈ 100	0.42	12.7	0.08	0.11	0.31	0.52
sandy loam	−0.70	4.90	12.8	≈ 80	0.43	13.8	0.14	0.18	0.46	0.56
loam	−1.43	5.39	13.8	≈ 20	0.45	14.8	0.19	0.24	0.57	0.65
clay	−1.82	11.4	25.8	< 10	0.5	26.8	0.47	0.52	0.78	≈ 1

s_h, s_w, and s^* correspond to Ψ_s values of -10, -3, -0.03 MPa, respectively; s_{fc} is the s level at which leakage becomes negligible compared with evapotranspiration.

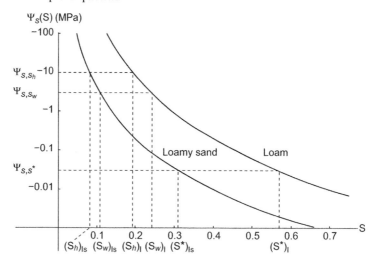

Figure 3.3 Soil moisture-retention curves for two different soil types, a loamy sand and a loam. The subscripts l and ls on the horizontal refer to loam and loamy sand, respectively. Such curves may be used to derive values of the soil moisture levels s_h, s_w, and s^* from the corresponding values of the soil matric potential, $\Psi_{s,s_h} = -10$ MPa, $\Psi_{s,s_w} = -3$ MPa, and $\Psi_{s,s^*} = -0.03$ MPa, using the parameters reported in Table 3.1. After Laio et al. (2001c).

familiar with the basic concept of the probability density function (PDF) and its integral, the cumulative distribution function (CDF); we refer to Chapter 6 for a review of probabilistic tools for ecohydrology.

In this subsection, we use the pore-size distribution and an idealized representation of the matric potential due to capillary action in order to deconstruct the basic physical and statistical ingredients that give rise to the retention curves. While a full statistical description of the soil's three-dimensional aggregation structure is very difficult, the physical and statistical meaning of the shapes and the parameters of the retention curves discussed in Sec. 3.2.2 can be illustrated using a simple conceptual model known as the bundle-of-tubes model (see Fig. 3.4). It idealizes the soil pores as an ensemble of random tubes of radius r. The pore-size distribution, $p(r)$, is defined as the relative abundance of the pores in terms of the pore volume rather than the number of pores. The cumulative distribution function $P(r)$ can then be expressed as the ratio of the combined volume of pores whose effective radius is less than r to the total volume

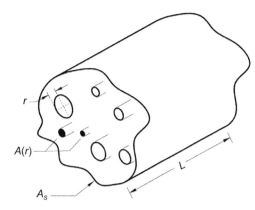

Figure 3.4 Graphical representation of pore size, pore volume, and cross-sectional pore area for the bundle-of-tubes model. The solid regions indicate pores filled with water.

of all pores. Since the soil water tends to occupy the smaller of the available pores, where it is held by capillary forces, the relative soil moisture is equal to the CDF of the pore radii (Brutsaert, 2005; Pelak and Porporato, 2019),

$$s(r) = P(r). \tag{3.4}$$

Both s and P vary between 0 and 1. One can invert the previous equation to get the maximum radius filled with water for relative soil moisture s:

$$r = P^{-1}(s), \tag{3.5}$$

where P^{-1} is the inverse function of the pore-size CDF.

The pore radius r is related to the soil's water matric potential Ψ_s as follows. Recalling that the capillary rise h in a circular pipe is inversely proportional to the pipe radius (see Eq. (2.95) and Sec. 2.1.13), and that the matric potential must compensate for the negative pressure due to the capillary rise, i.e., $\Psi_s + g\rho_w h = 0$, results in

$$\Psi_s(r) = -\frac{C_s}{r}, \tag{3.6}$$

where the proportionality constant C_s depends on the contact angle at the water–grain–air interface and the surface tension.[5] Using Eqs. (3.5) and (3.6), one can finally relate the retention curve to the statistical properties of the pore sizes (see Fig. 3.5) and the soil's physical properties:

$$\Psi_s(s) = -\frac{C_s}{P^{-1}(s)}. \tag{3.7}$$

Figure 3.5 illustrates graphically the steps needed to derive $\Psi_s(s)$ from $P(r)$. For power-law distributed pore radii,[6] the PDF is

---

[5] The contact angle for quartz–water–air is almost zero, but such an angle can considerably increase when hydrophobic substances are present as may occur following a fire (Rockhold et al., 2002).

[6] Several other mathematical forms of retention curves have been proposed in the literature, each of which can be linked to a different assumption on the statistical distribution of soil pores (Brutsaert, 2005). A power-law distribution is scale-free, which reflects the typical statistically self-similar geometry of soil structure (Feder, 2013).

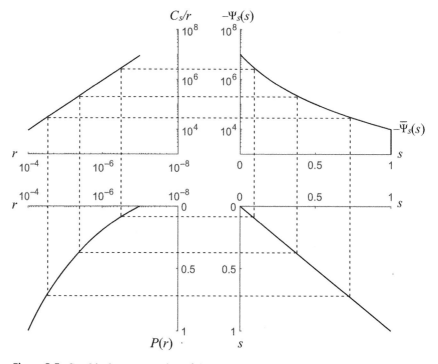

Figure 3.5 Graphical representation of the steps linking the cumulative pore-size distribution $P(r)$ (lower left) to the soil water-retention curve $\Psi_s(s)$ (upper right) for a power-law distribution of pore sizes, Eq. (3.8). The upper left panel (giving the reciprocal function) and the lower right panel link $P(r)$ to $\Psi_s(s)$. The parameters used in this figure are $\alpha = 1 - 1/b$, $b = 4.9$, $r_m = 0.0001$ mm, $r_M = 0.1$ mm, $C_s = 1$.

$$p(r) = \frac{1-\alpha}{r_M^{1-\alpha} - r_m^{1-\alpha}} r^{-\alpha}, \tag{3.8}$$

where α is the exponent, and the r_m and r_M are the minimum and maximum pore radii, respectively ($r_m < r < r_M$). The corresponding CDF can be expressed as

$$P(r) = \frac{1}{r_M^{1-\alpha} - r_m^{1-\alpha}} \left(r^{1-\alpha} - r_m^{1-\alpha} \right). \tag{3.9}$$

The inverse function of this CDF is

$$P^{-1}(s) = \left[(r_M^{1-\alpha} - r_m^{1-\alpha})s + r_m^{1-\alpha} \right]^{1/(1-\alpha)}. \tag{3.10}$$

Then the soil matric potential can be derived from Eq. (3.7) as

$$\Psi_s(s) = -C_s \left[(r_M^{1-\alpha} - r_m^{1-\alpha})s + r_m^{1-\alpha} \right]^{-1/(1-\alpha)}. \tag{3.11}$$

When r_m approaches zero (e.g. fine-grained soils) and $\alpha = 1 - 1/b$, the retention curve has the form of Eq. (3.3) and the air-entry suction point (see Sec. 3.2.2) can be expressed as (Pelak and Porporato, 2019)

$$\overline{\Psi}_s = -\frac{C_s}{r_M}. \tag{3.12}$$

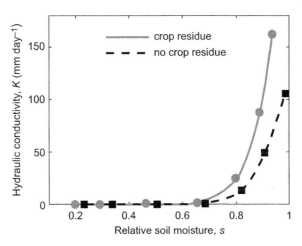

Figure 3.6 Measured behavior of unsaturated hydraulic conductivity in silty loam soils with crop residue incorporation (solid line) and without crop residue incorporation (dashed line). Adapted from Hassan et al. (2013).

3.2.4 Soil Hydraulic Conductivity

The soil hydraulic conductivity, $K(s)$, is the average water flux per unit cross-sectional area of soil for a unitary gradient of water potential. It describes how conductive a soil is to water flow, as a function of soil saturation levels. It depends on the type of soil, in particular the soil texture, and is a strongly nonlinear function of soil moisture. Figure 3.6 shows measured data for hydraulic conductivity and the role of organic matter and amendments in agricultural soils. It is evident that the hydraulic conductivity quickly decreases from its value at saturation K_s to very low values at intermediate to low saturation levels. Also clear is the effect of organic matter addition and soil amendments, typical in agricultural practices.

Following Brooks and Corey (1966), a simple and popular model for the decay of the hydraulic conductivity with decreasing soil moisture assumes a power-law behavior (e.g., Clapp and Hornberger, 1978; Hillel, 1998; Dingman, 2015),

$$K(s) = K_s s^c, \tag{3.13}$$

where K_s is the *saturated hydraulic conductivity* (see Table 3.1). A re-analysis of several datasets by Brutsaert (2005) showed that a good fit with measurements is obtained with $c = 2 + 2.5b$, where b is the exponent of the retention curve (3.3).

It is important to note the particularly strong nonlinear behavior of the unsaturated hydraulic conductivity and the fact that it represents an average behavior of the soil under the continuum hypothesis, resulting from the slow water flow through increasingly smaller pores as the soil dries. In addition, since the flow refers to an average over a complex porous matrix of the size of the Darcy scale, it is not unexpected that measurements have shown that the exponent c of Eq. (3.13) is related to the exponent b of the retention curve, which, as we saw in Sec. 3.2.3, also depends on the soil pore statistics. A conceptual interpretation of the hydraulic conductivity curve, illustrating its physical and statistical underpinning, is discussed using a simple conceptual model in Sec. 3.2.5.

For reasons of mathematical tractability, exponential decay from a value equal to the saturated hydraulic conductivity K_s at $s = 1$ to a value of zero at the *field capacity*, s_{fc}, may also be employed as a model (Laio et al., 2001c). The field capacity is a conventional soil-moisture level at which the soil hydraulic conductivity can be assumed to be practically zero. Here it is operationally defined as the value of soil moisture at which the hydraulic conductivity according to Eq. (3.13) becomes negligible (10%) compared to the

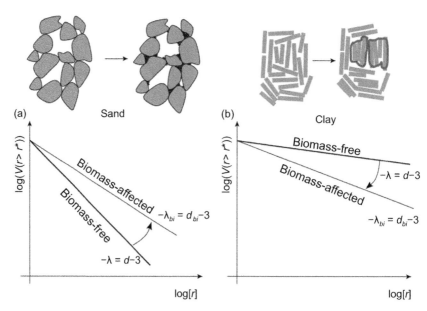

Figure 3.7 Role of organic matter in altering aggregate and pore-size distributions in soils. (a) Sand and (b) clay. After Maggi and Porporato (2007).

maximum daily evapotranspiration losses, E_{max} (for this purpose fixed at 0.5 cm/day). This exponential model can be expressed as

$$K(s) = \begin{cases} \dfrac{K_s}{e^{\beta(1-s_{fc})} - 1} \left[e^{\beta(s-s_{fc})} - 1 \right] & s_{fc} < s \le 1, \\ 0 & s \le s_{fc}. \end{cases} \tag{3.14}$$

A possible criterion to minimize the discrepancies between the two expressions (3.13) and (3.14) is to impose the condition that they subtend the same area between s_{fc} and $s = 1$. This provides $\beta = 2b + 4$, so that the value of β depends on the type of soil, varying from $\simeq 12$ for sand to $\simeq 26$ for clay. Using the typical values reported in Table 3.1, the two expressions have very similar behavior.

Soil texture tends to account for most of the discernible patterns in K_s, with clay having a smaller K_s than sand (Table 3.1). Soil organic matter, roots, and soil fauna may also strongly impact a soil's hydraulic properties (Wilcox et al., 2003; Thompson et al., 2010; Hassan et al., 2013), although their effects are not easy to quantify. In particular, soil organic matter (SOM) tends to homogenize the pore-size distributions (Maggi and Porporato, 2007) and thus make the sand and clay behaviors more similar to that of an intermediate loamy soil (Fig. 3.7). Figure 3.8 shows the role of different vegetation types in modifying a soil's structural and hydraulic properties after 50 years of imposed chaparral and pine vegetation (Johnson-Maynard et al., 2002). In this Californian site with a Mediterranean climate, populations of burrowing fauna and earthworms in the A horizon create distinctive patterns of *macroporosity*, which result in high hydraulic conductivity and preferential flows (see also the discussion of fingering in Note 3.11 for another example of preferential flow).

Changes in soil chemistry and in particular soil salinization and sodicity (see Chapter 10) can dramatically alter the structure of soil aggregates and thus cause strong reductions in soil hydraulic conductivity (McNeal and Coleman, 1966; Mau and Porporato, 2015). When experiencing prolonged drought,

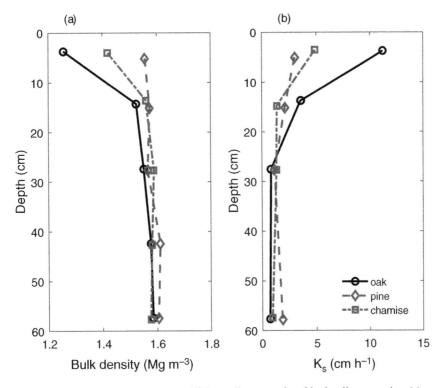

Figure 3.8 Role of vegetation in modifying soil structural and hydraulic properties. (a) Bulk density and (b) saturated hydraulic conductivity. Adapted from Johnson-Maynard et al. (2002).

especially after a relatively wet period, clayey soils tend to shrink and crack, forming macropores and increasing the soil hydraulic conductivity (Tang et al., 2008). This phenomenon of soil cracking and shrinking may change the soil hydraulic conductivity with potential implication for water resources and ecohydrology (Ritchie and Adams, 1974; Arnold et al., 2005).

3.2.5 Interpretation of the Unsaturated Hydraulic Conductivity

The statistical and physical significance of the unsaturated hydraulic conductivity (e.g., Eq. (3.13)) may be understood by following again the simple analogy of the bundle-of-tube model (Brutsaert, 2005), already used in Sec. 3.2.3 for the interpretation of the retention curve.

On the basis of the considerations made in Sec. 3.2.3, we may assume that the cumulative pore size distribution is related to the soil moisture by Eq. (3.4) and that the pores are approximately circular tubes with random radius r (see Fig. 3.4), so that there is a one-to-one relation between the pore size (r), pore volume (V), and cross-sectional pore area (A) distributions,

$$\frac{A(r)}{A_s} = \frac{A(r)L}{A_s L} = \frac{V(r)}{V_s} = ns(r) = nP(r). \tag{3.15}$$

Differentiating with respect to r,

$$\frac{1}{A_s}\frac{dA(r)}{dr} = n\frac{dP(r)}{dr} = np(r), \tag{3.16}$$

where $p(r)$ is the probability density function for the pore radii.

We can then consider that for a given soil moisture s all the pores of radius $r' \leq r$ will be filled with water because of capillary effects. Moreover, within these pores, the water flux takes place as a laminar (creeping) flow (see Sec. 4.4). As a result (see Eq. (4.14)), the flow rate per unit cross-sectional area can be written as

$$\Phi_{\text{lam}}(r') = \frac{\gamma_w r'^2}{8\mu}\frac{\Delta h}{l}. \tag{3.17}$$

Since the hydraulic conductivity k is the flux per unit cross-sectional area and per unit gradient, dividing Eq. (3.17) by $\Delta h/l$ yields

$$k(r') = \frac{\gamma_w r'^2}{8\mu}. \tag{3.18}$$

The total flux per unit cross section of soil is then given by integrating across all pores that are filled with water, using Eq. (3.16),

$$\begin{aligned}
K(r) &= \frac{1}{A_s}\int_0^{A(r)} k(r')dA \\
&= \frac{1}{A_s}\int_0^r \frac{\gamma_w r'^2}{8\mu}A_s np(r')dr' \\
&= \frac{n\gamma_w}{8\mu}\int_0^r r'^2 p(r')dr'.
\end{aligned} \tag{3.19}$$

Finally, a change of variable using Eq. (3.6), so that $r'(s)^2 = C_s^2\Psi_s(s')^{-2}$, and the fact that, according to Eq. (3.4), $p(r)dr = dP(r) = ds$, gives the final expression,

$$K(s) = \frac{nC_s^2\gamma_w}{8\mu}\int_0^s \frac{1}{\Psi_s(s')^2}ds', \tag{3.20}$$

which provides the desired link between the retention curve and the soil hydraulic conductivity.

With a power-law form of the retention curve (3.3), Eq. (3.20) can be integrated to get a corresponding power-law form of the conductivity function (see Eq. (3.13)), with $c = 1 + 2b$. In order to make things more realistic, Burdine et al. (1953) added a multiplicative tortuosity function, which gives an exponent $c = 2b + 3$. A re-analysis of several datasets by Brutsaert (2005), however, showed that a better fit with measurements is obtained with $c = 2.5b + 2$.

3.3   Unsaturated-Soil Water Flow

Having described the main soil hydraulic properties, in this section we derive the equations for unsaturated water flow for a continuum mechanics description based on porous medium averaging over the Darcy scale (Sec. 3.2). We first introduce the continuity equation based on the principle of mass conservation (Sec. 3.3.1). We then describe Richards' equation modeling the soil water flow (Sec. 3.3.2) and its vertically averaged formulation (Sec. 3.3.3).

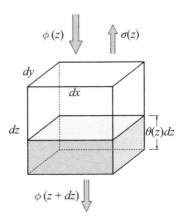

Figure 3.9 Infinitesimal soil element used as a control volume in the derivation of the continuity equation (3.23). The volume fraction of water in the soil element is denoted by θ.

3.3.1 Soil Water Continuity Equation

We assume that the soil hydraulic properties are horizontally homogenous and focus on the water flow in the vertical direction. The *continuity equation* for the conservation of water in the vertical direction can be obtained by considering the balance of inputs and outputs in an infinitesimal element of soil during an infinitesimal time interval dt (Fig. 3.9). Thus, the change in the water volume is equal to the difference in the vertical fluxes between z and $z + dz$ minus the local sink,

$$d\theta\, dxdydz = \phi(z)dxdydt - \phi(z + dz)dxdydt - \sigma(z)dxdydzdt, \tag{3.21}$$

where z is the vertical direction (pointing downward), $\phi(z)$ is the volumetric water flux at the given level z, and $\sigma(z)$ is the sink of water due to root uptake per unit volume of soil per unit time.

Expanding in a Taylor series and neglecting infinitesimal terms of higher order,

$$d\theta\, dxdydz = \phi(z)dxdydt - \left[\phi(z) + \frac{\partial\phi}{dz}dz\right]dxdydt - \sigma(z)dxdydzdt. \tag{3.22}$$

Simplifying and rearranging the terms in the previous equation, we obtain the continuity equation

$$\frac{\partial\theta}{\partial t} = -\frac{\partial\phi}{\partial z} - \sigma. \tag{3.23}$$

3.3.2 Richards' Equation

Richards' equation results from combining the continuity equation with the equation of water flow, the generalized Darcy's law. As before, we will restrict our analysis to one-dimensional flow in the vertical direction.

In unsaturated soils, the water flux depends on both the gradient of the total water potential and the soil hydraulic conductivity. Buckingham was the first to generalize Darcy's law (Brutsaert, 2005), originally formulated for saturated soils, to unsaturated porous media. The equation may be written in terms of the gradient of the total water potential ψ_s, where

$$\psi_s = G_s + \Psi_s + \Omega_s. \tag{3.24}$$

Here G_s is the gravitational potential, Ω_s is the soil water osmotic potential, and Ψ_s is the soil matric potential (see Eq. (2.101) in Sec. 2.1.14 for a review of the definitions). Assuming zero osmotic potential, we can express the extended Darcy's law as

Figure 3.10 Soil diffusivity calculated from Eq. (3.28) with parameters of soil characteristics from Table 3.1.

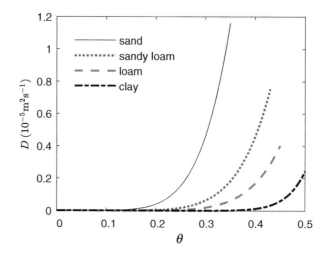

$$\phi = -K(\theta)\frac{\partial \psi_s(\theta)/\gamma_w}{\partial z} = -K(\theta)\frac{\partial}{\partial z}\left[-z + \frac{\Psi_s(\theta)}{\gamma_w}\right] = K(\theta)\left(1 - \frac{\partial \Psi(\theta)}{\partial z}\right), \tag{3.25}$$

where z is taken positive downward, and $\Psi = \Psi_s/\gamma_w$ is the suction head ($[\Psi] = L$). The latter is also denoted as h or H in the literature (e.g., Brutsaert, 2005). It is important to stress the empirical nature of this equation of motion and the fact that it is a one-to-one link between the Gibbs free energy gradients (through the water potential) and the water flux. This implicitly means that the dynamic (i.e., inertial) components are neglected in the resulting formulation.

Combining the continuity equation (3.23) with the equation of motion (3.25), we obtain *Richards' equation*,

$$\frac{\partial \theta}{\partial t} = -\frac{\partial K(\theta)}{\partial z} + \frac{\partial}{\partial z}\left[\mathcal{D}(\theta)\frac{\partial \theta}{\partial z}\right] - \sigma, \tag{3.26}$$

where

$$\mathcal{D}(\theta) = K(\theta)\frac{\partial \Psi(\theta)}{\partial \theta} \tag{3.27}$$

is the *soil diffusivity*, a strongly nonlinear function of the soil moisture θ. For a power-law soil (see Eqs. (3.3) and (3.13)), the soil diffusivity in Eq. (3.27) is expressed as

$$\mathcal{D}(\theta) = -bK_s\overline{\Psi}_s\gamma_w^{-1}n^{b-c}\theta^{c-b-1}, \tag{3.28}$$

which is illustrated in Fig. 3.10 for different soil types. This power-law diffusivity covers a variety of soil types.

Richards' equation (3.26) is one of the most important equations of hydrology, resulting from the combination of the continuity equation (water-mass conservation) and an equation of motion (the extended Darcy's law), which in turn links empirically the two curves of unsaturated soil hydrology, the retention curve and the soil hydraulic conductivity curve. From the mathematical point of view, it is a nonlinear partial differential equation of the parabolic type. The first term on the right-hand side of the equation represents advection and the second is a diffusion term. Analytical solutions are available only for specific conditions (e.g., the sorption solution will be described in Sec. 3.4.1 and 3.4.2). The fact that the flux is directly related to the gradient of the moisture fraction in the extended Darcy's law (3.25) gives rise to the diffusive term and implies a one-to-one and instantaneous response between the gradients and the moisture flux. Sharper gradients propagate faster, and reach infinite speed in the limit of step changes (like the one that will be used in the sorption solution). This unphysical feature is in general not a real problem[7] and is related to the fact that the extended Darcy's law neglects dynamic components, such as those related to the propagation of capillary fronts in irregular pores during fast wetting events after intense rainfall. More generally, these limitations serve as good reminders of the fact that Richards' equation, like any other mathematical model, has its own assumptions and limits of validity. For ecohydrology, we especially emphasize those related to the Darcy-scale averaging and the continuum approach to porous media, as well as the assumption of local homogeneity and the lack of macroporosities and preferential flows at those scales (see Sec. 3.2.4).

3.3.3 Vertically Averaged Soil Moisture Equation

For general conditions, a vertically explicit description of soil moisture dynamics via Richards' equation almost always requires numerical solutions. These are computationally demanding and not readily amenable to theoretical considerations. Thus, for long-term analyses with stochastic rainfall inputs it is often convenient to consider vertically averaged soil moisture dynamics over the rooting zone. This approximation is similar in spirit to the big-leaf approximation that will be adopted in Sec. 5.4 and is quite appropriate for soils that have well-defined horizons or soils that have been made vertically homogenous by agricultural practices; the presence of macroporosity and roots, with their enhanced soil water redistribution action, may also contribute to the validity of this approximation (see Guswa et al., 2002 and Sec. 4.3.1 on hydraulic redistribution).

Consider the average behavior of soil water within a layer of depth Z_r comprising most of the roots (e.g., 75%, see Table 4.1) and roughly corresponding to the depth of the A horizon (see Fig. 3.1). Averaging Richards' equation (3.26) with respect to z between 0 and Z_r and writing the result in terms of the relative soil moisture yields (see Eq. (3.2))

$$nZ_r\frac{d\bar{s}}{dt} = K(s)\left(1 - \frac{\partial\Psi}{\partial z}\right)\bigg|_{z=0} - K(s)\left(1 - \frac{\partial\Psi}{\partial z}\right)\bigg|_{z=Z_r} - \int_0^{Z_r}\sigma(z)dz, \tag{3.29}$$

where, for simplicity, we have assumed constant porosity with depth and define the vertically averaged soil moisture

$$\bar{s} = \frac{1}{Z_r}\int_0^{Z_r} s(z)dz. \tag{3.30}$$

To emphasize the physical meaning of each term, the previous equation (3.29) can be rewritten as

$$nZ_r\frac{d\bar{s}}{dt} = I(t) - L(t) - \Phi(t) \tag{3.31}$$

and compared to Eq. (1.5).

The first term is the water flux at the soil surface,

$$I(t) = K(s)\left(1 - \frac{\partial\Psi}{\partial z}\right)\bigg|_{z=0}. \tag{3.32}$$

---

[7] This issue is present also in the well-known diffusion or heat equation and has resulted in numerous extensions, including the Maxwell–Cattaneo heat equation (Christov, 2009).

When positive, this water flux represents surface *infiltration* due to rainfall $R(t)$ and/or irrigation $J(t)$; when negative, it is the soil water evaporation $E_v(t)$. In fact, during rain or irrigation events, the surface becomes almost saturated so that Ψ is close to zero or even positive (e.g., ponding), resulting in a water potential gradient $\partial\Psi/\partial z$ less than one and thus a positive surface water flux. When dry atmospheric air causes large near-surface negative Ψ values, the water potential gradient $\partial\Psi/\partial z$ is greater than one (that is, it exceeds the gravitational potential), causing water to evaporate into the atmosphere. Thus the infiltration $I(t)$ equals the rainfall (plus irrigation, if present) minus the sum of the interception and runoff. The irrigation term $J(t)$ will be discussed further in Sec. 10.2; we assume here that it is applied at the soil surface, although in microirrigation it could come from below-ground irrigation pipes.

The second term, depending on whether it is negative or positive, represents either the percolation output or the capillary input from lower layers,

$$L(t) = K(s(z)) \left(1 - \frac{\partial\Psi}{\partial z} \right) \Bigg|_{z=Z_r}. \tag{3.33}$$

In places such as wetlands where the water table is shallow, the water potential gradient $\partial\Psi/\partial z$ can be large, inducing capillary rise and increasing soil water content (Eagleson, 1978a; Ridolfi et al., 2006). This process, also known as *exfiltration*, could play an essential role in the long-term water balance and influence the plant dynamics. When the water table is deep, the water potential gradient tends to be close to zero and the flux is downward (leakage or percolation), dominated by gravity. In such cases, the leakage is given by

$$L(t) = K(s(t)), \tag{3.34}$$

a regime called *free gravity drainage*. This version will be used for deriving the analytical solutions for stochastic soil moisture dynamics in Sec. 7.2.2.

The last term in (3.31) is the plant uptake via transpiration,

$$\Phi(t) = \int_0^{Z_r} \sigma \, dz, \tag{3.35}$$

where Z_r was defined in the text above Eq. (3.29). In the absence of plant capacitance (i.e., ability to absorb water), it is equal at any instant to the flux of water through the plant and into the atmosphere, i.e., $\Phi(t) = E_t(t)$. Such a transpiration flux is modulated by multiple factors within the SPAC (see Sec. 5.4). The distribution of root density also influences the water uptake rate $\sigma(z, t)$ at different levels z, and plants may extract more water in the wet soil layers to compensate for the reduced water uptake in the dry soil layers (Guswa et al., 2002); see also the discussion of hydraulic redistribution by roots in Sec. 4.3.1.

3.4   Infiltration

The term I on the right-hand side of Eq. (3.31) is extremely important as its dynamics primarily controls the *rainfall–runoff partitioning*. While the temporal dynamics of infiltration can be obtained by solving Richards' equation numerically, analytical approximations can be very useful to get insight into the underlying processes and provide practical expressions for applications.

Figure 3.11 illustrates the typical temporal dynamics of infiltration for a relatively long and intense rainfall event, taking place after a dry spell. The process can be divided into three main phases. At the

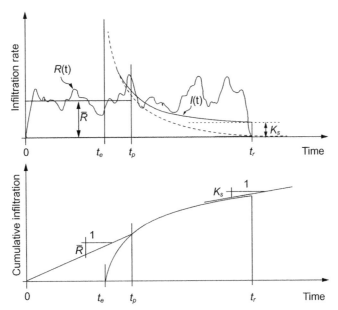

Figure 3.11 Upper graph: rainfall and infiltration rate as a function of time. Lower graph: cumulative infiltration before and after the ponding time t_p; the parameter t_e is used to match the transition between phases I and II (see Sec. 3.4.3).

beginning (phase I), because of the sharp gradient in the water potential between the wet surface and the underlying dry layers, the infiltration rate is very high and the soil can accommodate all the incoming rainfall.[8] As the wetting front propagates downward, however, the suction may be reduced to levels that do not allow all the rainfall to infiltrate. At this point, ponding of water occurs on the soil surface, quickly followed by surface runoff driven by surface heterogeneities and microtography. This second phase (phase II) is characterized by a constant zero value of the water potential at the surface and a progressive deepening and diffusion of the wetting front. If rainfall continues then, as the soil becomes uniformly saturated, the vertical percolation flux tends to a rate equal to the saturated hydraulic conductivity of the soil, K_s (phase III).

The infiltration dynamics in phase II can be idealized as a sorption problem and treated analytically in an elegant way (Sec. 3.4.1). The actual matching of the solution with phase I, however, can in general be done only approximately and requires a readjustment of the times at which the sorption solution starts, called the time compression approximation (Sec. 3.4.3).

3.4.1 Sorption Solution of Richards' Equation

In the first two phases of infiltration (Fig. 3.11), the advection term in Richards' equation is small and the process is dominated by the diffusion term arising from the matric potential gradients at the surface (see Eqs. (3.26) and (3.27)). Once ponding occurs, the surface boundary condition switches from a flux boundary condition dictated by the throughfall rate (a Neumann boundary condition) to a constant-head boundary condition (a Dirichlet boundary condition). Such conditions can be approximated quite well analytically by the so-called *sorption problem* presented in this section.

---

[8] It should be clear that here we are neglecting canopy interception and simply assuming that all rainfall actually reaches the soil surface.

For negligible advection, the starting equation of the sorption problem is (Brutsaert, 2005),

$$\frac{\partial \theta}{\partial t} = \frac{\partial}{\partial z} \left[\mathcal{D}(\theta) \frac{\partial \theta}{\partial z} \right]. \tag{3.36}$$

We can idealize the boundary and initial conditions by assuming that the soil moisture is initially uniform at a level θ_i while the surface value is suddenly raised to a new level θ_f and kept constant thereafter, i.e.,

$$\begin{aligned} \theta(z) &= \theta_i, \quad t = 0, \\ \theta(z = 0) &= \theta_f, \quad t > 0. \end{aligned} \tag{3.37}$$

It is clear that for the phase II infiltration, taking place after ponding, θ_f is equal to the value at saturation. Moreover, since these boundary conditions initially produce unlimited water flux into the surface layer, the corresponding solutions refer only to the infiltration capacity; to match the actual infiltration rates, the time origin of phase II will have to be readjusted (see Sec. 3.4.3).

To obtain the evolution of the profile of the moisture fraction in space and time for $z \geq 0$, it is useful to start with dimensional considerations, first relating θ to other governing quantities (Sec. 2.2.2). In our case here, the task is facilitated by having the model equation available, i.e., Eq. (3.36), and also by the simplicity of the initial and boundary conditions. As a result, for our problem here we can write

$$\theta = f_\theta(z, t, \mathcal{D}(\theta), \theta_i, \theta_f), \tag{3.38}$$

where, dimensionally, $[\theta] = [\theta_i] = [\theta_f] = 1$, $[z] = L$, $[t] = T$, and $[\mathcal{D}] = L^2 T^{-1}$. As the problem belongs to the LT class, it follows that $k = 2$; choosing t and $\mathcal{D}(\theta)$ as dimensionally independent quantities, with the Π theorem (Sec. 2.2.2) one can transform Eq. (3.38) into the dimensionless equation

$$\theta = \varphi_\theta \left[\frac{z}{\sqrt{\mathcal{D}(\theta)}\sqrt{t}}, \theta_i, \theta_f \right]. \tag{3.39}$$

This equation provides an implicit function of θ, which shows us that the latter can be written as $\theta = \theta(\xi)$, meaning that it depends only on a single variable ξ defined as

$$\xi = \frac{z}{\sqrt{t}}. \tag{3.40}$$

This relation is known as the Boltzmann transformation, as it was first introduced by Boltzmann in 1894 to analyze nonlinear diffusion (Philip, 1957a; Brutsaert, 2005).

An important consequence of the previous result is that the problem admits a similarity solution. As a result, our nonlinear partial differential equation (3.36) can be written for the initial and boundary conditions (3.37) as an ordinary differential equation (ODE). In fact, writing

$$\frac{\partial \theta}{\partial t} = \frac{d\theta}{d\xi} \frac{\partial \xi}{\partial t} = -\frac{1}{2} z t^{-3/2} \frac{d\theta}{d\xi} \tag{3.41}$$

and

$$\frac{\partial}{\partial z} \left[\mathcal{D}(\theta) \frac{\partial \theta}{\partial z} \right] = \frac{d}{d\xi} \left[\mathcal{D}(\theta) \frac{d\theta}{d\xi} \frac{\partial \xi}{\partial z} \right] \frac{\partial \xi}{\partial z} = \frac{d}{d\xi} \left[\mathcal{D}(\theta) \frac{d\theta}{d\xi} \right] t^{-1}, \tag{3.42}$$

and substituting the right-hand sides into Eq. (3.36), one obtains the second-order ODE

$$\frac{d}{d\xi} \left[\mathcal{D}(\theta) \frac{d\theta}{d\xi} \right] + \frac{\xi}{2} \frac{d\theta}{d\xi} = 0, \tag{3.43}$$

with boundary conditions $\theta = \theta_i$ for $\xi \to \infty$ and $\theta = \theta_f$ for $\xi = 0$.

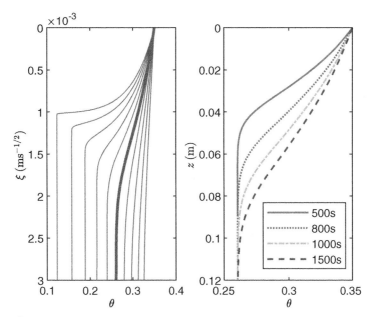

Figure 3.12 (a) Self-similar solutions of $\theta(\xi)$ for a sandy loam soil. (b) Volumetric water content as a function of soil depth at different times for the thick line in (a). The soil parameters are listed in Table 3.1 and the diffusivity is a power-law function as in Eq. (3.28).

The ODE in Eq. (3.43) can be solved analytically only for special forms of the soil diffusivity. For most cases a numerical solution is necessary. Figure 3.12(a) shows examples of numerical solutions for a power-law diffusivity (see Eq. (3.28)) for different boundary conditions. Figure 3.12(b) further shows the evolution of the wetting front at different times for a specific set of boundary conditions. For very dry initial soil, the wetting front tends to be rectangular and a further simplification for this infiltration process was given by Green and Ampt (1911) (see Notes 3.6 and 3.7).

3.4.2 Sorptivity and Infiltration

A second and very important consequence of Eq. (3.39) is that, for the sorption problem, the total or cumulative infiltration scales in time as $t^{1/2}$, independently of the specific form of the diffusivity function.

To demonstrate this, we begin by noting that, in general, the cumulative infiltration capacity $\mathcal{I}(t)$ up to time t is represented by the area between the curve $\theta(z, t)$ and the axes. This area can be expressed as an integral with respect to either z or θ,

$$\mathcal{I}(t) = \int_0^\infty [\theta(z, t) - \theta_i]dz = \int_{\theta_i}^{\theta_f} z(\theta, t)d\theta. \tag{3.44}$$

For the specific case of the sorption problem, Eq. (3.40) can be inverted:

$$z = \xi(\theta)t^{1/2} \tag{3.45}$$

and substituted in Eq. (3.44) to give

$$\mathcal{I}(t) = t^{1/2} \int_{\theta_i}^{\theta_f} \xi(\theta)d\theta. \tag{3.46}$$

Figure 3.13 Schematic diagram illustrating calculation of the cumulative infiltration $\mathcal{I}$ as the area between the $\theta(z, t)$ curve and the axes. This area can be found by integrating elemental area $zd\theta$ over the θ coordinate or $(\theta - \theta_i)\,dz$ over the z coordinate.

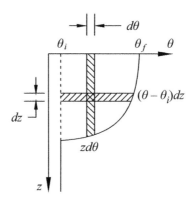

The integral in Eq. (3.46) is an important quantity in soil physics, called the *sorptivity*,

$$A_0 = \int_{\theta_i}^{\theta_f} \xi(\theta)d\theta, \tag{3.47}$$

and it is equal to the cumulative infiltration in unit time. It depends on the soil type via the diffusivity function $\mathcal{D}(\theta)$, which contains both the retention and the hydraulic conductivity curves, as well as the boundary conditions θ_i and θ_f.

With this definition of sorptivity, the cumulative infiltration becomes

$$\mathcal{I}(t) = A_0 t^{1/2}, \tag{3.48}$$

and its time derivative becomes

$$I(t) = \frac{d\mathcal{I}(t)}{dt} = \frac{1}{2} A_0 t^{-1/2}. \tag{3.49}$$

This important result shows that the theoretical infiltration rate is initially infinite and quickly decays in time with a characteristic $-1/2$ power-law dependence (see Fig. 3.11). Interestingly, this type of decay is general and does not depend on the form of the diffusivity function.

3.4.3   Matching Phases I, II, and III of Infiltration

Recall that in phase I of infiltration, the dry soil absorbs all the incoming rainfall, $I = R$ and $\mathcal{I} = \bar{R}t$. It is clear from the solutions of (3.36) that phase II of infiltration is dominated by the diffusion term, while phase III is dominated by the advection term in Richards' equation (3.26). A good approximation that describes both these phases of infiltration can be expressed as (Brutsaert, 2005)

$$I(t) = K_s + A_0(t - t_e)^{-1/2}, \quad t > t_p \tag{3.50}$$

where t_p is the time to ponding, which marks the transition of phase I to phase II and t_e is a parameter, the compression time. used for matching this transition. Subtracting the time t from t_e means that the starting times of phases I and II are compressed; for this reason the approach is called the *time-compression approximation* (TCA).[9]

---

[9]   The TCA was first introduced by Sherman (1943) and Holtan (1945); however, several variants of this approximation have been presented in the literature (see Brutsaert, 2005).

To find the compression time t_e, one can follow the schematic diagram shown in Fig. 3.11 and assume approximately rectangular rainfall, that is, $R \approx \bar{R}$, where $\bar{R}$ is the mean rainfall rate before ponding. At the ponding time t_p, the infiltration capacity equals the rainfall intensity (Fig. 3.11a),

$$\bar{R} = I(t_p). \tag{3.51}$$

Moreover, the cumulative infiltration equals the total rainfall at the ponding time t_p (Fig. 3.11b),

$$\bar{R}t_p = \int_{t_e}^{t_p} I(t)dt. \tag{3.52}$$

Equations (3.51) and (3.52) can be used to derive t_e and t_p for matching phases II and III with phase I of the infiltration. In case of the infiltration modeled by Eq. (3.50),

$$t_p = \frac{(\bar{R} - K_s)(2A_0^2 - K_s^2 + K_s\bar{R})}{A_0^2\bar{R}} \tag{3.53}$$

and

$$t_e = \frac{(\bar{R} - K_s)(2A_0^2 - K_s^2 + 2K_s\bar{R} - \bar{R}^2)}{A_0^2\bar{R}}. \tag{3.54}$$

With these equations, the TCA gives quick estimations of the vertical infiltration with reasonable accuracy.

3.4.4 From the Point Scale to the Plot Scale

The previous developments offer a way to calculate infiltration in idealized conditions of homogeneous soils, where Richards' equation is a good approximation. In reality, however, soil heterogeneities, roots, micro-topography and the related drainage network make the modeling of infiltration more difficult. For vegetated soils without considerable topography and soil crusting, infiltration remains a highly nonlinear process, where the ponding threshold, after which infiltration is greatly reduced, plays a crucial role.

The difficulties of realistic infiltration modeling are somewhat alleviated when considering vegetated soils at the *plot scale* (say, a land area of a thousand square meters). In such a case, it is often sufficient to know the total amount of rainfall in order to compute the so-called *event-scale infiltration* rather than the detailed temporal dynamics of infiltration itself. The typical timescale of the duration of a rainfall event is on the order of several minutes to a few hours for convective storm and frontal or tropical cyclone events. We can compare the total runoff Q with the total rainfall $\mathcal{R}$ per event, although at these scales the relationship between rainfall and runoff is not one-to-one, because the same total rainfall depth can be produced by different hyetographs (graphs of rainfall intensity against time). To this purpose, it is useful to define a runoff curve,

$$\frac{Q}{\mathcal{R}} = \varphi(\mathcal{R}, nZ_r(1 - \bar{s}_-), K_s, a, b, \ldots), \tag{3.55}$$

where $\bar{s}_-$ is the average soil moisture before rainfall. The runoff ratio, $Q/\mathcal{R}$, is a function of the total rainfall, of the available storage in the soil before the rainfall, represented by $nZ_r(1 - \bar{s}_-)$, of the rate of percolation after ponding, which linked to K_s, of a quantity a related to the initial abstraction (e.g., regarding canopy interception; see Sec. 7.3.2), and of the fraction b of impervious area (due to crusts, etc.), and so on.

Roughly speaking, the many semi-empirical formulations of Eq. (3.55) can be classified according to their shape, as in Fig. 3.14. Typically, for vegetated soils without considerable topography and soil

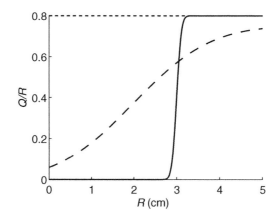

Figure 3.14 Schematic representation of different types of runoff ratios for rainfall-event infiltration characterization. The solid line represents a threshold-type or saturation excess runoff, corresponding to an antecedent soil moisture condition that is far from saturation. The dashed line is an intermediate case with higher infiltration at small rainfall depths due to the presence of Hortonian-type runoff (when the throughfall rate exceeds the infiltration capacity and the excess part is converted into surface runoff) or previously saturated parts of the soil. The dotted line is the extreme case of constant runoff ratio, adopted in the so-called rational method.

crusting, the infiltration displays a threshold-like behavior, giving rise to the so-called *infiltration-excess runoff*, where the rainfall is practically all infiltrated if saturation and ponding do not occur. This corresponds to the continuous line, where the location of the sharp increase marks a rainfall depth roughly equal to the volume of soil available for water storage before the rainfall event. This effect of the antecedent soil moisture conditions is accounted for by the presence of the average soil moisture fraction just before the rainfall event, $\bar{s}_-$. To first order, this case is dominated by a strong initial sorption phase and may be approximated as a step function. These considerations will be used to justify the simple runoff model used in Chapter 7 for the stochastic soil moisture fraction.

In plots with topography, crust, and impervious surfaces, the runoff increases more progressively with rainfall depth, because some rainfall is already lost as runoff before soil saturation (the dashed line in Fig. 3.14); the extreme case of proportional response with a constant runoff coefficient is shown with a dotted line in Fig. 3.14 and corresponds to the assumption made in the rational method. At larger spatial scales (e.g., a watershed) the event runoff is typically the result of the averaging of a mixture of the different behaviors represented above (the dashed line in Fig. 3.14), which gives rise to smoother and less nonlinear runoff curves (see Note 3.12).

3.5   Key Points

- Soils have a natural tendency to form layers, called horizons; the A horizon contains most of the roots and is ecohydrologically the most active.
- The soil texture triangle classifies soils on the basis of the percentages of their main soil textures: sand, silt, and clay. At its center, loam is a mixture of roughly equal parts of these components.

- The relative soil moisture s is the volume fraction of soil pores filled with water, calculated over a representative elementary volume of soil (at the Darcy scale).
- The soil water-retention curve links the volumetric soil moisture to the soil matric potential; it describes how water is energetically bound to the soil matrix.
- The soil hydraulic conductivity is a very nonlinear function of the soil moisture. Like many other soil hydraulic properties, it reflects the average behavior at the Darcy scale and depends on soil texture as well as organic matter and soil heterogeneities.
- In the self-similar sorption solution of Richards' equation, the infiltration decays as $t^{-1/2}$ for any type of soil diffusivity.
- The saturation excess runoff typical of relatively flat, vegetated plots stems from the threshold-like behavior of infiltration before and after ponding.

3.6   Notes, including Problems and Further Reading

3.1 (Van Genuchten Model)   Following the bundle-of-tubes model used in Sec. 3.2.3 to link the pore-size distribution $p(r)$ to the retention curve $\Psi_s(s)$, find the form of $p(r)$ for the van Genuchten (1980) model with zero residual moisture, $s = 1/(1 + a|\Psi_s|^n)^m$, where a, n, and m are parameters.

3.2   Besides the idealized conceptual model introduced in Secs. 3.2.3 and 3.2.5, other conceptual models have been used for interpreting the links between soil moisture, retention curve, and hydraulic conductivity (see, e.g., Mualem, 1976; Reeves and Celia, 1996).

3.3   Owing to the constitutive relationships between θ and Ψ, Richards' equation (3.26) can be written in another form:

$$\frac{\partial \theta}{\partial t} = -\frac{\partial K}{\partial z} + \frac{\partial}{\partial z}\left[K(\Psi)\frac{\partial \Psi}{\partial z}\right] - \sigma,$$

which has mixed dependent variables θ and Ψ. Such a form may be convenient for some numerical purposes (see, e.g., Celia et al., 1990).

3.4 (Two-Stage Evaporation)   The soil drydown process can be classified into two stages (Brutsaert, 2005; Fisher, 1923). In the first stage, the soil surface is sufficiently wet that evaporation is at its maximum potential rate; in the second stage, evaporation is limited by the water available from the soil. The second stage, when neglecting gravity effects, can be simplified as a reverse sorption problem. Use the sorption solution from Sec. 3.4.1 to analyze the soil moisture profiles during the soil drydown process.

3.5   The water uptake σ in Fig. 3.9 is the source of water flux into the plants and is directly controlled by multiple factors such as root density and the soil–root resistance (see Sec. 4.3). Philip (1997) formulated an analytical treatment of Richards' equation with the uptake term σ (also see Sec. 4.3 and Note 4.3).

3.6 (Green–Ampt Approximation)   In the infiltration process, the wetting fronts tend to be sharper when the initial soil is dry. It can be shown that the wetting front is exactly rectangular when the diffusivity is a delta function (e.g., $\mathcal{D}(\theta) = \mathcal{D}_0 \delta(\theta - \theta_f)$) (Philip, 1957b; Philip, 1973; Milly, 1985). On the basis of this idealized soil moisture profile, Green and Ampt (1911) introduced approximate solutions for the vertical infiltration process, which are appealing for their parsimonious representation of infiltration.

3.7 Several approximate analyses of Richards' equation are available in the literature. A quasi-linear analysis treats the K–Ψ relationship as an exponential function to describe steady flows (Wooding, 1968; Philip, 1968). A linear absorption assumes that the soil diffusivity is a constant or a linear function of soil moisture (Liu et al., 1998; Parlange et al., 2000). Neglecting the gradient of Ψ gives the kinematic wave approximation describing flows that are dominated by gravity (Lighthill and Whitham, 1955). Neglecting advection corresponds to the sorption problem (Brutsaert, 2005) (see Sec. 3.4.1).

3.8 Solve Eq. (3.43) analytically for a so-called linear soil, corresponding to constant D (see, e.g., Liu et al., 1998).

3.9 Work out the details of the time compression approximation for the Green–Ampt model. (See Salvucci and Entekhabi, 1994 and Note 3.6.)

3.10 The vertical infiltration discussed in Sec. 3.4 can be adjusted to describe the infiltration over a hillslope with homogenous isotropic soils. Philip (1991) showed how to rotate the coordinates to obtain the infiltration rates normal and parallel to the slope.

3.11 (Infiltration Instability and Fingering) In several field and laboratory conditions, preferential flow (fingering) has been observed, especially in sandy soils during infiltration into dry soils (Selker et al., 1992; De Rooij, 2000). These dynamic instabilities are related to those occurring in viscous flows (the Saffman–Taylor instability); they are not only very interesting from a theoretical point of view (Bensimon et al., 1986; Homsy, 1987; Cueto-Felgueroso and Juanes, 2008; DiCarlo, 2013), but may also play a role in the transport of water and nutrients to deeper layers, especially in arid and semi-arid regions. Their role in ecohydrology and their relations to other preferential flows due to macropores (see Sec. 3.2.4), created by soil heterogeneities, decaying roots, and fauna, are interesting open problems.

3.12 (Rainfall-Runoff Response) A runoff event at the watershed scale is the result of the combination of runoff generation events at various points (i.e., at the plot scale) of the watershed and thus it reflects an averaging of different responses similar to those shown in Fig. 3.14. In some cases the rainfall-runoff response remains very nonlinear and is reminiscent of the saturation excess response at the plot scale (see Fig. 3.15a); this has been called the violent response by Hawkins (1993) and is associated with a threshold-like behavior called *filling-and-spilling* (McDonnell, 2013). In other cases, some runoff has been produced already by small rain events, owing to the presence of wetland and/or impervious areas; this behavior, characterized by a more linear runoff coefficient, has been called the complacent response (Hawkins, 1993). The typical response described by the SCS-CN method (called the standard response by Hawkins, 1993) is somewhat in between these two extremes.

Several semi-distributed rainfall-runoff models predict runoff from a distribution of watershed properties. Each model, such as VIC (Wood et al., 1992; Liang et al., 1994), PDM (Moore, 1985), and TOPMODEL (Beven, 2006), differs in the way that these distributions are parameterized and linked to physical attributes at a point. When their event-based versions are compared, the runoff curves reproduce

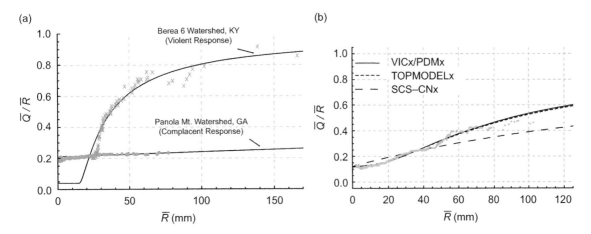

Figure 3.15 Comparison of rank-order data for the models. Panel (a) shows the extended SCS-CN method for the complacent response of the Panola Mountain Watershed (gray dots) and the violent response of the Berea 6 watershed (gray crosses). After Bartlett et al. (2017). Panel (b) shows the observations (gray dots) and different modeling results (lines) with similar average antecedent potential retention and pre-threshold runoff index. After Bartlett et al. (2016b).

the different runoff-response types described above (see Fig. 3.15b), depending on the topographic and soil properties as well as on the antecedent watershed saturation levels.

3.13 (Groundwater Coupling) At the scale of river basins involving several nested watersheds, streamflow routing and groundwater coupling become important. While a certain degree of empirical modeling inevitably remains, owing to a lack of resolved physics and of sufficient space–time numerical resolution and to uncertainty in the boundary conditions, considerable advances in the numerical simulation of hydrologic response have been made (e.g., the model intercomparisons in Refsgaard and Knudsen, 1996; Maxwell et al., 2014).

4 Plants

The sun flared down on the growing corn day after day until a line of brown spread along the edge of each green bayonet.

J. Steinbeck, The Grapes of Wrath

Plants are crucial players in ecohydrological processes. On the one hand, they are the site of transfer of most of the water from the soil to the atmosphere. On the other hand, water is necessary for plants to perform photosynthesis and to maintain turgor and growth. In this chapter, we discuss the main plant characteristics related to modeling the fluxes of water, carbon, and energy, as part of the soil–plant–atmosphere continuum, with the goal of characterizing ecohydrological dynamics and in particular plant water stress. Using an electrical analogy for water transport, we analyze the resistance to flow from the soil to the leaves and into the atmosphere; we also review the processes and modeling of photosynthesis and the parameterization of stomatal conductance. Since the calculation of transpiration requires a knowledge of leaf temperature and thus leaf energy balance, a full description of transpiration is postponed to Chapter 5.

4.1 Plant Water

Plant physiologists usually distinguish two kinds of water in vascular plants:[1] *apoplastic water*, located outside the plasma membranes and relatively free to move from roots to leaves through the xylem conduits, and *symplastic water*, which is contained in the protoplast of the living cells (see Fig. 4.1). We discuss here only the main issues related to modeling water transport and refer to specialized books for more details on plant physiology and ecology (e.g., Salisbury and Ross, 1969; Jones, 1992; Lambers et al., 1998; Nobel, 1999; Larcher, 2003).

We will indicate by ψ_x and ψ_p the water potentials of the xylem and protoplast respectively. The apoplastic water potential in the xylem can be approximated as the sum of the matric and gravitational potentials (see Sec. 2.1.14 for a definition of water potential):

$$\psi_x(z, t) = \Psi_x(z, t) + \Omega_x(z, t) + G_x(z), \tag{4.1}$$

where we have explicitly indicated the dependence on time t (during soil moisture evolution) and elevation z. The most important component of the xylem water potential is the matric potential, Ψ_x. Xylem sapwood

---

[1] Roughly speaking, these are plants with a hydraulic apparatus, i.e., lignified tissues for conducting water, minerals, and photosynthetic products through the plant; non-vascular plants include moss and algae.

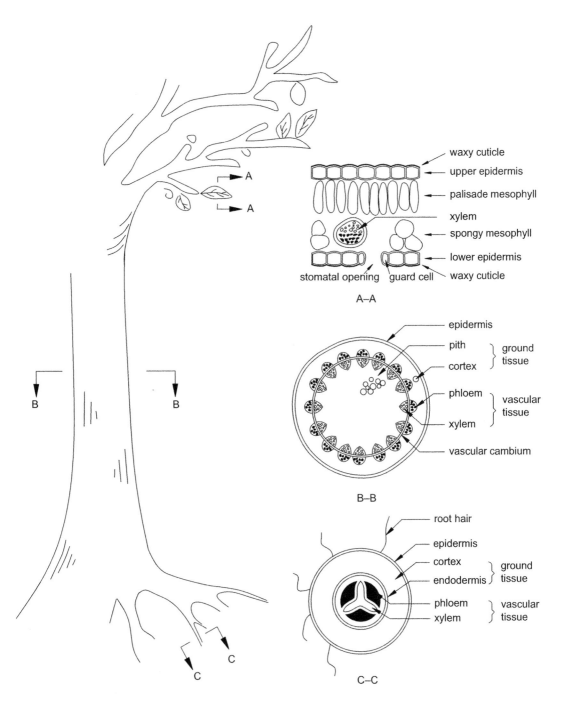

Figure 4.1 Elements of plant anatomy in relation to the water movement from the roots to the leaves. Modified after Salisbury and Ross (1969), Rand (1983), and Jensen et al. (2016).

is, in fact, quite diluted along its whole path, so that the osmotic potential is modest (~ -0.1 MPa) and practically constant everywhere (e.g., Nobel, 1999), whereas the gravitational potential is relevant only for tall plants.

The main components of the symplastic water potential are the matric, osmotic, and gravitational potentials,

$$\psi_p(z, t) = \Psi_p(z, t) + \Omega_p(z, t) + G_p(z). \tag{4.2}$$

For tissues at a given height, the apoplastic and symplastic water may be assumed to be in local equilibrium, so that

$$\Psi_p(z, t) + \Omega_p(z, t) = \Psi_x(z, t) + \Omega_x(z, t). \tag{4.3}$$

Figure 4.2 presents a schematic representation of the plant hydraulic system. Under normal transpiration conditions in well-watered soils, the typical values of plant water potential are only slightly negative, ranging from -0.1 MPa to -1 MPa. Living cells contain high concentrations of solute within them and consequently the osmotic potential, Ω_p, may be quite low, with typical values from -0.5 MPa to -3 MPa (Salisbury and Ross, 1969).[2] Low values of osmotic potential are very important for plant cells. According to Eq. (4.3), when the xylem water potential, ψ_x, is very low, only a low value of Ω_p can ensure that the matric potential, Ψ_p, is maintained positive, thus providing the essential turgor for plant growth ("turgor" refers to the state of turgidity and resulting rigidity of cells or tissues, typically due to the absorption of fluid and the positive internal pressure that pushes the membrane against the cell wall). A positive matric potential is often referred to as a pressure potential and is then denoted by Π (see Sec. 2.1.14). We will return to the relationship between the symplastic water potential components in the next section.

Typically, a soil–water potential value of -0.3 MPa corresponds to relative soil moisture values of 0.2 and 0.4 for a loamy sand and a loam respectively (e.g., Fig. 4.2). Such values correspond to conditions of moderate water stress, where the stomata are partially closed during the central part of the day. For plants in semi-arid ecosystems, typical values of the soil water potential for the start of stomata closure are of the order of -0.03 MPa (Scholes and Walker, 2004). The atmospheric water potential is usually much lower than that of the soil water, except during rainfall events. For average daytime conditions, it has a typical value of -100 MPa, corresponding to a relative humidity of about 40% (see Sec. 2.1.14).

Figure 4.2 also shows the symplastic water connected to the apoplastic water through an osmotic membrane. Its total potential, ψ_p, follows the same behavior of that of the apoplastic water, but the existence of a more concentrated solution (i.e., more negative osmotic potentials, Ω_p) ensures positive matric potentials, Ψ_p, and thus positive turgor pressure. The potential gradient drives the water flow from roots to xylem and to the leaves, as introduced in the following sections (also see Fig. 4.1).

4.2   The Soil–Plant–Atmosphere Continuum

The water flow within the *soil–plant–atmosphere continuum* (SPAC) takes place along the gradient of decreasing water potential from the soil to the atmosphere. The SPAC is often simplified as a discrete series of compartments connected by resistors. A schematic representation of this electrical analogy is

---

[2] As a reference value, the osmotic potential of seawater is of the order of -2.7 MPa. See Note 2.8.

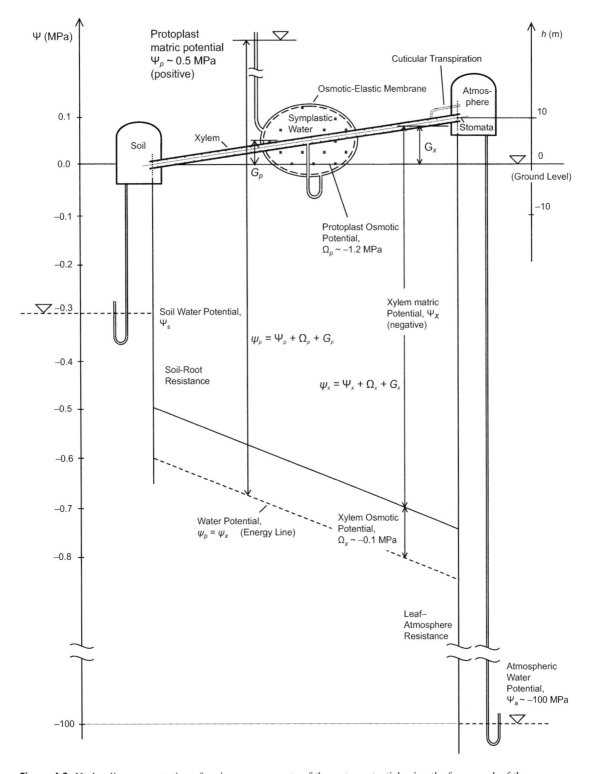

Figure 4.2 Hydraulic representation of various components of the water potential using the framework of the soil–plant–atmosphere continuum. The soil and leaf compartments are represented as tanks connected by a pipe representing the xylem. The symplastic water (oval tank) is idealized as being connected to the apoplastic water of the xylem at the center via a conduit and an osmotic membrane, while the matric potential in each tank is visualized by means of piezometers. The example refers to a 10-meter tall plant with specific values of the water potential components taken from Nobel (1999). After Porporato et al. (2001).

presented in Fig. 4.3. The first part of this pathway takes place in the liquid phase, from the soil at water potential ψ_s to the leaf tissues at ψ_l, at a low Reynolds number (i.e., laminar flow) through the soil pores and the plant's xylem conduits (Sec. 4.4). Once it reaches the leaves, the water then evaporates, passing from the leaf tissues into the stomatal cavities. The water vapor then diffuses out of the stomata into the atmosphere, where it is then transported by turbulence above the canopy in the so-called logarithmic layer (see Sec. 2.3) at lower specific humidity q_a.

In order to simplify the mathematical description, the water flow within the SPAC may be assumed to take place as a succession of steady states between compartments where water is in local thermodynamic equilibrium, a reasonable assumption considering that the system adjusts very quickly to time-varying conditions. With these simplifications, the water flow from soil to atmosphere is analogous to a current through a series of nodes connected by resistors (Fig. 4.3). Each node is characterized by its water potential, and the transport between them is controlled by their gradients, in proportion to their resistances (or their inverses, i.e., *conductances*). Thus in each node the concepts and definitions of equilibrium thermodynamics (reviewed in Chapter 2) can be used to unambiguously define the water status in each of the SPAC compartments. The water transport is then described by means of gradient transport formulae linking the conductances to the gradients in the driving force (see Note 4.5). While in each compartment water is assumed to be in local thermodynamic equilibriums, the transport through the resistors is a nonequilibrium process, as it gives rise to irreversible free-energy dissipation and thus a drop in water potential.

We consider the soil–root conductance, g_{sr} (Sec. 4.3), the plant conductance, g_p (Sec. 4.4), the stomatal conductance, g_s (Sec. 4.8), and the atmospheric conductance g_a (Sec. 2.3). Typically, such conductances are functions of the water potential inside the system, making the SPAC a highly nonlinear system. For the sake of simplicity, the soil moisture may be vertically averaged over the root zone, of depth Z_r (Sec. 3.3.3). Similarly, the canopy may be lumped into a single compartment, using the standard big-leaf approach (Raupach and Finnigan, 1988), which assumes that all the leaf stomata and exchange surfaces are concentrated at a single representative height (Fig. 4.3).

The soil water potential (ψ_s) and the atmospheric humidity (q_a) provide the boundary conditions to the SPAC. Going from the soil to the leaves, the liquid water flux is proportional to the gradient of the water potential, according to the so-called *van den Honert* equation,

$$\Phi = g_{srp}(\psi_s - \psi_l), \tag{4.4}$$

where Φ is the volumetric water flux per unit ground surface (LT^{-1}), g_{srp} is the soil–root–plant conductance, expressed per unit ground area. The soil–root–plant conductance (the inverse of the soil–root–plant resistance) is obtained as the series addition of the soil–root conductance and the plant–xylem conductance, so that

$$g_{srp} = \frac{g_{sr}g_p}{g_{sr} + g_p}, \tag{4.5}$$

all expressed in terms of unit ground area. Since the driving quantity is the water potential, with the dimensions of pressure, $[\psi] = FL^{-2}$, and thus having units of pascals (Pa), all these conductances have dimensions $[g_{sr}] = [g_p] = [g_{srp}] = L^3T^{-1}F^{-1}$ and may thus be expressed in $m\,s^{-1}\,Pa^{-1}$.

If one neglects the changes in plant water storage (see Sec. 4.4.2), the flux inside the plant must equal the transpiration rate E through the stomata,

$$\Phi = E \tag{4.6}$$

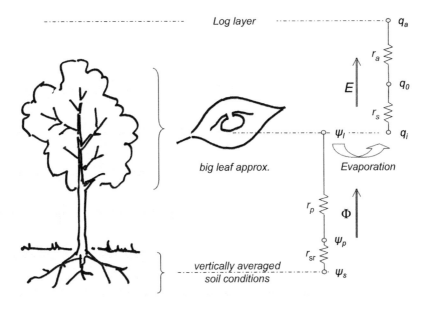

Figure 4.3 Electrical analogy of the resistances to water movement in the SPAC in the big-leaf approximation.

(note that here we neglect bare soil evaporation). The latter is due to diffusion and may be assumed to be proportional to the gradient in the specific humidity between the inside of the stomatal cavities and just outside the leaf:

$$\rho_w E = g_s \rho [q_i(T_l, \Psi_l) - q_0], \qquad (4.7)$$

where ρ_w is the liquid water density, ρ is the air density, q_i and q_0 are the specific humidities inside and outside the leaf, T_l is the leaf temperature, and g_s is the stomatal conductance (per unit ground area). The water vapor inside the stomatal cavities can be assumed to be in equilibrium with the liquid water in the surrounding leaf tissues (the mesophyll), which is at the same water potential ψ_i(water vapor) $= \psi_l$(liquid) and the same temperature, $T_i = T_l$. Consequently, using Eq. (2.104), i.e.,

$$\psi_i = \frac{\mathcal{R} T_l}{\upsilon_w} \ln \frac{e_i}{e_{\text{sat}}(T_l)} + g\rho_w z_i = \psi_l, \qquad (4.8)$$

and Eq. (2.75), we obtain

$$q_i \approx \frac{\epsilon}{p_0} e_i = \frac{\epsilon}{p_0} e_{\text{sat}}(T_l) \exp\left[\frac{\upsilon_w(\psi_l - g\rho_w z_i)}{\mathcal{R} T_l} \right]. \qquad (4.9)$$

Finally, for the last step, from outside the leaves to the logarithmic layer, the transpiration flux due to turbulent transport can be written as (Sec. 2.3.3, Eq. (2.154))

$$\rho_w E = g_a \rho (q_0 - q_a), \qquad (4.10)$$

where g_a is the atmospheric conductance and q_a is the specific humidity at a height z_a within the logarithmic velocity profile. Since the transpiration rate E is the volumetric flux per unit (ground) area, $[E] = \text{L T}^{-1}$, the stomatal and atmospheric conductances too have dimensions $[g_s] = [g_a] = \text{L T}^{-1}$ and are typically expressed in m s^{-1} (note there are differences in units among g_s, g_a, g_{sr}, and g_p).

In the following we will discuss each compartment and the conductances which connect them. This will allow us to express the water flux through the plant as a function of the boundary conditions, the soil water potential and the specific humidity in the atmosphere. However, since Eq. (4.9) also contains the leaf temperature, a complete determination of the transpiration flux will only be possible when the leaf energy balance is included (Sec. 5.1.1).

4.3   Roots

Water moves from soils to plants through root hairs and mycorrhizae that are very intimately intertwined with the grains and organic matter of the soil matrix; it passes through the cell wall of the root hair and gets into the root, where it crosses the epidermis and cortex to reach the xylem vessels located in the axial part of the root (Fig. 4.1).

Root characteristics are difficult to model and often their parameterization requires a great deal of empirical assumption. Moreover, roots differ considerably depending on the type of plant; Table 4.1 reports the main root characteristics for different biomes. For example, the *rooting depth*, Z_r, reported in Table 4.1 as the level where 75% of the roots are found, is one of the important parameters defining the volume of water accessible to plants, thus defining the timescale of soil moisture dynamics (see Chapter 3). A second important parameter is the *root area index* (RAI), defined as the ratio of the surface root area per unit ground area, which may change by an order of magnitude, from about 5 in arid and cold biomes up to 80 in temperate grasslands (see Table 4.1).

It is practically impossible to account for the details of the complex soil–plant interface connecting the porous-medium structure of the soil with the root network. As a first approximation, and in line with the simplifications made in modeling the soil and other parts of the plant, the soil–root conductance may be assumed to be proportional to the soil hydraulic conductivity divided by the typical distance, Λ_{sr}, traveled by the water from the soil to the root surface (Katul et al., 2003; Manzoni et al., 2013)

$$g_{sr} = \frac{K(s)}{\gamma_w \Lambda_{sr}}, \tag{4.11}$$

where γ_w is the specific weight of water. The distance Λ_{sr} depends on many factors such as the root geometry and density (Tardieu and Davies, 1993; Katul et al., 2003). A useful expression for the latter may be obtained by assuming parallel cylindrical roots of average diameter d_r, uniformly distributed over the root zone depth Z_r, with a total RAI given by (Manzoni et al., 2013)

$$\Lambda_{sr} = \sqrt{\frac{d_r Z_r}{RAI}}. \tag{4.12}$$

The root area index and Z_r vary across biomes but also depend on the type of soil and the depths of the soil horizons (Table 4.1), while the root diameter d_r is more constrained, ranging from around 0.2 mm in grasses to 0.6 mm in trees (Jackson et al., 1997).

Using a schematic conceptualization of the bulk behavior of roots and again in line with a parsimonious approach to modeling, the previous expression (4.11) treats the root zone as an averaged representative layer of given soil and root characteristics. It is based on averaging out much of the root complexity (including the complex soil–root interface; see Note 4.4 on mycorrhizae), similarly to the big-leaf approximation for canopies (see Sec. 5.4).

There are very interesting approaches in the literature that add increasingly more complexity and detail to root modeling. The simplest allow only for some sort of vertical root distribution, while the more sophisticated contain dynamic branching, growth, and decay in a three-dimensional soil–root domain with

Table 4.1 Fine-root (FR) characteristics averaged by biome (Jackson et al., 1997) and typical values of the soil–root conductance ($g_{sr,sat}$) computed for a loam soil (see Table 3.1) using Eqs. (4.11) and (4.12)).

Biome	FR biomass (kg m^{-2})	RAI	$Z_{r,75\%}$ (cm)	$g_{sr,sat}$ (m s^{-1} MPa^{-1})	LAI
Boreal forest	0.60	4.6	24	0.042	4.8
Desert	0.27	5.5	45	0.058	2
Sclerophyll, shrubs and trees	0.52	11.6	27	0.063	4
Temperate coniferous forest	0.82	11.0	69	0.039	5
Temperate deciduous forest	0.78	9.8	41	0.047	5.2
Temperate grassland	1.51	79.1	24	0.303	3
Tropical deciduous forest	0.57	6.3	76	0.028	4.8
Tropical evergreen forest	0.57	7.4	49	0.038	8.5
Tropical grassland/savanna	0.99	42.5	49	0.156	5
Tundra	0.96	5.2	15	0.098	1

complex root architectures (see Note 4.3 on root modeling). While these efforts are interesting, they are also very difficult to parameterize.[3]

4.3.1 Hydraulic Redistribution

Hydraulic redistribution, or *hydraulic lift*, is the movement of water from moist to dry soil through plant roots (Caldwell et al., 1998). Figure 4.4 offers a schematic representation of the soil–plant water fluxes due to hydraulic redistribution. Unlike transpiration, which results in net water movement from the soil to the roots and then to the leaves, with hydraulic redistribution, which typically happen at night, the potential in the root may exceed the potential in the drier part of the soil, with the result that moisture may also flow from the roots to the dry soil. Both laboratory and field experiments have shown that hydraulic redistribution moves water through the soil profile at a much faster rate than could have been possible by gravity and diffusion in the soil matrix alone (Amenu and Kumar, 2008).

Hydraulic redistribution occurs within a range of different ecosystems and plant species (Neumann and Cardon, 2012), potentially influencing both water use and carbon assimilation (Domec et al., 2010). It may also enhance seedling survival and maintain overstory transpiration during summer drought (Brooks et al., 2006) by bringing water from deeper layers to surface layers, where nutrients and microbial life are typically more abundant. Hydraulic redistribution has been shown to be beneficial in rainfed agriculture in arid and semi-arid regions. Bogie et al. (2018) reported improved soil moisture and nutrient conditions for millet when intercropped with native woody shrub in the Sahel, providing a novel approach for farmers in semi-arid regions to fight against drought.

Models of hydraulic redistribution are necessarily quite complex and require resolving the gradients in water potential between soil and root at different depths (Amenu and Kumar, 2008; Siqueira et al., 2008). Several aspects of hydraulic redistribution remain elusive; the effect of specific plant controls, including

---

[3] As often in environmental modeling, extreme care should be exercised when adding model complexity; it is very important not to overemphasize some processes compared with others, which could lead to even more unrealistic outcomes.

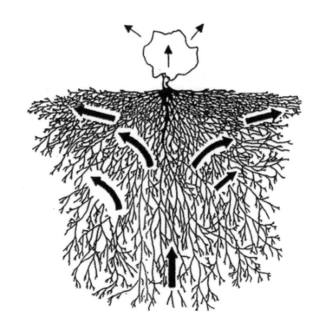

Figure 4.4 Schematic representation of water flow due to hydraulic lift. At night, with a low transpiration rate, water moves from moist soil through the root system to drier soil layers. After Caldwell et al. (1998).

possible osmotic adjustments to actively induce water flows, has not been investigated even in the most recent models. Allowing for a greater water redistribution and a more active role of the plant in extracting water from different soil depths, the presence of hydraulic redistribution also challenges the traditional approach of plant uptake based on Richards' equation (see Sec. 3.3.2 as well as Note 4.3).

4.4   Xylem

The xylem presents itself as a series of interconnected and mostly parallel conduits interrupted by pits and walls (Fig. 4.1), going from the roots, through the trunk, all the way to the leaves, and forming what is called the apoplastic pathway (see Sec. 4.1). Such conduits tend to be rather long in deciduous trees (vessels), while they are shorter in conifers (tracheids).

In the xylem, water flux takes place as a laminar, creeping flow, driven by negative pressure gradients (Nobel, 1999). Along the upward path, energy losses progressively reduce the level of the plant water potential, so that the apoplastic water potential attains large negative values, especially in the highest parts of the plant (see Fig. 4.2). Accordingly, the well-known Hagen–Poiseuille law for *laminar flow* in pipes can be used to analyze the water flow in the xylem. This law (Munson et al., 2005) can be written as a generalized transport formula of the type (4.48): recalling from fluid mechanics that the volumetric flow rate Q for the laminar flow of a Newtonian fluid of viscosity μ in a pipe of radius r and length L is given by (see also the bundle-of-tube model in Sec. 3.2.5)

$$Q = \frac{\pi r^4 \gamma_w}{8\mu} \frac{\Delta h}{L},$$

(4.13)

where Δh is the hydrostatic head. The mean velocity or flux per unit cross-sectional area of the pipe is thus

$$\Phi_{\text{lam}} = \frac{Q}{\pi r^2} = \frac{r^2}{8\mu} \frac{\Delta \psi}{L}, \tag{4.14}$$

where $h = \psi/\gamma_w$, as seen in Eqs. (2.96) and (3.25). As a result, we can define the hydraulic conductivity for the laminar flow as

$$k_{\text{lam}} = \frac{\gamma_w r^2}{8\mu}, \tag{4.15}$$

where, consistently with its definition in soil water flow (see Sec. 3.3), the hydraulic conductivity is defined here as the water flux per unit cross-sectional area and per unit gradient of hydrostatic head. Other references may define it as the volumetric flow rate per unit pressure gradient, resulting in different forms and units (e.g. Cruiziat et al., 2002; Manzoni et al., 2013).

The Hagen–Poiseuille law helps us to understand the essential factors that control the flow in the xylem under well-watered conditions. Under such conditions, water can ascend the tree like a continuous column, a concept that is the base of the so-called *cohesion theory* for plant-water ascent (e.g., Salisbury and Ross, 1969). For example, considering that the radius of a xylem vessel is typically $r = 2 \times 10^{-5}$ m and the flow may have velocity $\Phi = 10^{-3}$ m s^{-1} and viscosity $\mu = 10^{-3}$ Pa s (Rand, 1983), the Reynolds number ($R_e = \rho \Phi r/\mu$, see Sec. 2.2 on dimensional analysis) is 0.02 and the hydraulic conductivity is 0.5×10^{-3} m s^{-1}. This suggests that the flow takes place at a very low Reynolds numbers (i.e., creeping flow). Equation (4.15) is useful in showing how xylem hydraulic properties (radius and viscosity) affect the xylem resistance to water flow. However, given the irregular geometry and interconnectedness of the vessels, an estimate of the actual overall values of k_p entails an averaging of the complex flow network and is usually done empirically.

4.4.1 Xylem Embolism

As the plant water potential drops during a soil drying phase, *xylem embolisms* become more frequent, typically starting from vessels of larger section. With *embolisms* filling the xylem conduits with gas, the plant conductivity to water flow is also strongly reduced, in a manner which is reminiscent of the loss of hydraulic conductivity in unsaturated soils (see Sec. 3.2.5). As a result, the plant conductivity k_p decreases almost to zero with the reduction in plant water potential. The functions for different parts of the plants are shown in Fig. 4.5. As can be seen, the root xylem in woody plants is generally more vulnerable to cavitation than are shoots, although the upper branches are those at a lower water potential. This may be a mechanism to confine cavitation to a more easily replaceable part of the plant. This behavior is often modeled using a *vulnerability curve* (e.g. Manzoni et al., 2013),

$$k_p = k_{p,\max} \frac{1}{1 + (\psi_p/\psi_{50})^a}, \tag{4.16}$$

where $k_{p,\max}$ is the maximum xylem hydraulic conductivity, a is a shape parameter, and ψ_{50} is the leaf water potential at 50% loss of conductivity (see Fig. 4.6 for typical values). The value of $k_{p,\max}$ can be linked to sapwood-specific conductivity or leaf-specific conductivity when these data are available ($k_{p,\max} = k_{p,\text{leaf}}g\text{LAI}$, or $k_{p,\max} = k_{p,\text{sap}}g\text{SAI}$, where LAI and SAI are the leaf area index and the sapwood area index).

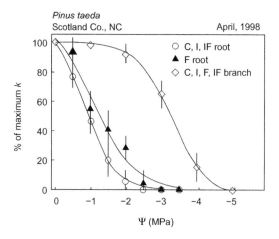

Figure 4.5 Curves of vulnerability to embolism of loblolly pines (*Pinus taeda*), showing the reduction in xylem conductance as a function of xylem water potential (Ewers et al., 2000). C, I, F, IF refer to control, irrigated, fertilized and irrigated/fertilized experiments, respectively.

The plant conductance, g_p, is related to the hydraulic conductivity, k_p, by

$$g_p = \frac{k_p}{\gamma_w h_c}, \tag{4.17}$$

where h_c is the canopy height and γ_w is the specific weight of water. Combining Eqs. (4.16) and (4.17) yields an expression for $g_p(\psi_p)$. Other forms of vulnerability curves have been used to relate g_p to the xylem water potential; for example (Sperry et al., 2002; Katul et al., 2003)

$$g_p = g_{p,\max} \exp\left[-\left(\frac{-\psi_p}{d}\right)^c\right], \tag{4.18}$$

where typical values of d and c are listed in Table 4.3.

The conductance from soil to root and plant can then be expressed as a series combination of the soil–root resistance (Sec. 4.3) and the plant resistance (Sec. 4.4), so that in terms of the conductances we have

$$g_{\rm srp} = \frac{g_{\rm sr} g_p}{g_{\rm sr} + g_p}. \tag{4.19}$$

Note that if g_p is the leaf-specific conductance, it needs to be multiplied by LAI to convert it to conductance per unit ground area (Daly et al., 2004a). The water flux from the roots to the leaves can be modeled as (van den Honert, 1948)

$$\Phi = g_{\rm srp}(\psi_s - \psi_l). \tag{4.20}$$

This water flux will transport water to the leaves for photosynthesis and then escape from their stomates, as discussed in the following sections.

4.4.2 Plant Capacitance

Many plants rely on water storage capacity to increase the accessibility of water for transpiration, reduce competition for water with neighboring plants, and buffer water supply during dry periods. This results

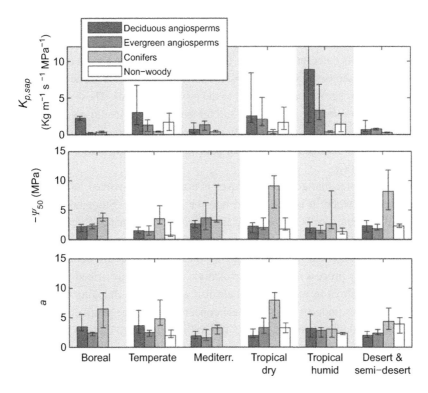

Figure 4.6 Median, 25%, and 75% quartiles (upper and lower error bars in each case) for $k_{p,\mathrm{sap}}$, ψ_{50}, and a for various plant functional types and biomes (Manzoni et al., 2013).

in a decrease in plant water stress and increase in productivity (Meinzer, 2009). The volume of water stored in plants can be significant relative to that stored in the soil (e.g., on the order of 10% for a softwood tree; Waring and Running, 1978). The stored water is utilized by the plant on both daily and seasonal timescales. For example, the volume of water stored in a Douglas Fir (*Pseudotsuga menziesii*) may fluctuate 3%–5% over the course of a day, supplying up to 25% of the transpired water and being recharged at night, when transpiration is near zero (Cermak et al., 2007). The stored water volume may drop by as much as 40% during a long drought. Plant capacitance is typical in succulents (often in association with the CAM photosynthetic pathway, see Note 4.11) as well as in seasonally dry climates (e.g., Baobab trees can store up to 400 liters by plant capacitance; this water is used for seasonal regulation).

The use of the stored sapwood water is principally governed by the water storage capacity (the total volume of water which may be stored in the plant tissue), the hydraulic capacitance (the change in stored water per unit change in water potential), and the storage conductance (the conductance for the water flow between the water storage and transport pathways). These traits describe a plant's water storage strategy – how much water it will withdraw from storage tissue when under stress and how quickly this transfer takes place. Various models have been used to describe the dynamics of plant water capacitance, including the porous-medium model, the unit pipe model, and resistance–capacitance models with varying degrees of detail. We refer to Hartzell et al. (2017) and references therein for more details.

4.5   Leaves

The xylem conduits enter into the leaves (see Fig. 4.1) to transport water into the mesophyll, where photo-synthesis takes place (see Sec. 4.6). Substantial hydraulic resistances occur both in the leaf xylem as well as in the flow paths across the mesophyll to the evaporation sites. These components respond differently to the ambient conditions (e.g., irradiance and temperature), indicating an involvement of aquaporins.[4] The maximum leaf conductance varies widely across species, and typically scales with plant conductance; the leaf is a substantial resistance in the plant pathway, 30% and upward of the whole-plant resistance (Sack and Holbrook, 2006).

Most of the water that gets to the leaves is not directly involved in photosynthesis, but moves to the stomatal cavities where it is evaporated into the atmosphere. Stomatal closure is controlled by the tur-gor level of the guard cells surrounding the *stomata* (or stomates),[5] see Fig. 4.1: high guard-cell turgor produces stomatal opening, while low turgor induces stomatal closure (e.g., Salisbury and Ross, 1969; Larcher, 2003). When the water stress is not too strong, the active control of guard-cell turgor is mainly an osmoregulatory process similar to the osmotic adjustment described above. The decrease in guard-cell turgor is brought about by an increase of abscisic acid, which, changing the permeability characteristics of the cell membranes, induces a lower solute concentration in the guard cells (Orcutt, 2000; Larcher, 2003). Stomatal regulation first evolved around 400 million years ago (Chater et al., 2017), and then became an essential component to control transpiration and photosynthesis. The sensitivity of the stomatal response to water loss varies widely from species to species, ranging from isohydric (approximately constant leaf water potential) or anisohydric (more variable leaf water potential); see Sec. 8.6.

The light presence controls the diurnal cycle of stomatal opening and closure in the morning and at dusk.[6] Stomatal closure can also be induced as a preventive measure to reduce internal water losses and risk of cavitation, before the plant water potential is seriously lowered. In fact, roots have been shown to be able to induce a feed-forward control on stomata by producing abscisic acid in conditions of incipient water stress (e.g., Schulze, 1993). Similarly, a reduction in relative atmospheric humidity can induce stomatal closure to prevent water stress.

The process of *stomatal closure* covers the entire scale of *water stress* (Fig. 4.14): On the one hand, incipient stomatal closure is among the first effects of water deficit on plant physiology, while, on the other hand, complete stomatal closure only takes place at the very end of the sequence of the effects on physiology, when the plant starts wilting. These properties make stomatal closure the ideal candidate to delimit both the starting and the maximum point of water stress. Moreover, the use of incipient and complete stomatal closure as indicator of water stress is also very convenient because these points are frequently measured in field experiments. These properties will be used in defining plant water stress thresholds in the stochastic models that will be introduced later in this book.

---

[4] Aquaporins or "water channels" are membrane proteins, which form pores to help transport water between cells.

[5] The word comes from the Greek $\sigma\tau o\mu\alpha$, which means mouth. They are small openings typically located on the lower side of the leaves (see Fig. 4.1).

[6] Only desert plants with Crassulacean acid metabolism (CAM) open their stomata at night, when transpiration demand is lower (Larcher, 1995; Nobel, 1999). During the day, when the stomata are closed, these acids are hydrolyzed and the resulting CO_2 is used by photosynthesis. CAM photosynthesis is the most evolved photosynthetic pathway, and plants in this group include cacti, agave, opuntia, pineapple, many orchids, and other aerial and succulent plants (see Note 4.11).

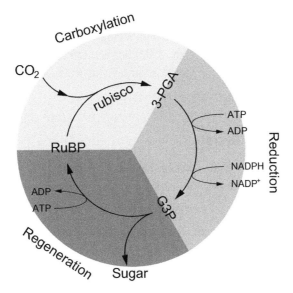

Figure 4.7 Schematic representation of the three phases of dark reactions; rubisco is an enzyme. See Sec. 4.6.2 for an explanation of terms.

At the stomatal level, through photosynthesis and transpiration control, the carbon and water cycles are intimately linked and, in turn, interact with the surface energy balance. For this reason, modeling stomatal conductance is a crucial step in ecohydrology; however, before doing this (see Sec. 4.8) it is important first to review briefly the process of photosynthesis.

4.6   Photosynthesis

Photosynthesis is one of the most fundamental processes for life, as its products are then used by all heterotrophic life which feeds directly or indirectly on plants. It employs water to transform solar energy into chemical energy. The process occurs in the chloroplasts of leaves through two subsequent processes. The first process involves light directly and is thus called a "light reaction", whilst the second is called a "dark reaction" as light is not directly involved. We will review schematically the basic functioning of photosynthesis, while defining both the notation and the terminology with reference to the most ancient and most common type, the so-called C_3 type of photosynthesis (see the modern types of photosynthesis pathways C_4 and CAM in Notes 4.10 and 4.11).

4.6.1   Light Reaction

Light energy is first absorbed by pigment molecules (mostly chlorophyll) and transferred to reaction centers, where a series of biochemical reactions is initiated to create reducing power in the form of NADPH (reduced nicotinamide adenine dinucleotine phosphate) and chemical energy in the form of ATP (adenosine triphosphate). Water is necessary in this phase as it is oxidized into two protons, two electrons and one half of an oxygen molecule (photolysis). For this reason, leaf cells need to be well hydrated and the leaf water potential ψ_l must be high, otherwise photosynthesis is reduced. This is important in water-controlled ecosystems (e.g., Larcher, 2003; Bonan, 2002) but is often neglected in models. The water abundance in the leaf cells is also the cause for the great amount of water lost by evaporation when the stomata are opened to allow a CO_2 influx.

4.6.2 Dark Reaction (Calvin Cycle)

NADPH and ATP from the light reaction are used to convert the CO_2 (taken up from the atmosphere through the stomata) into carbohydrates through the Calvin cycle. This consists of three phases (see Fig. 4.7):

1. *Carboxylation*: The five-carbon sugar RuBP (ribulose-1,5-biphosphate) combines with CO_2 and water to form three-carbon compounds (i.e. 3-PGA in C_3 plants). This reaction is catalyzed by the rubisco enzyme (ribulose biphosphate carboxylase–oxygenase). The rate at which CO_2 is fixed by carboxylation is indicated as A (Fig. 4.8).
2. *Reduction*: Next the three-carbon compounds (3-PGA) are modified using ATP and NADPH and then they are in part transformed into carbohydrates (e.g. sugar).
3. *Regeneration*: The remaining part of the modified three-carbon compounds (G3P) is combined with additional ATP to regenerate RuBP.

During the Calvin cycle, rubisco also catalyzes the oxygenation of RuBP, consuming oxygen and producing CO_2, in a process called either *photorespiration* or *RuBP oxygenation*, which thus reduces by 30% to 50% the net CO_2 uptake. The rate of photorespiration is indicated by P (Fig. 4.8).

The opposite of photosynthesis is respiration, by which organic compounds are oxidized to produce energy to maintain plant functions and grow new tissues. Respiration takes place day and night in almost all the plant parts; in the leaf cells at daytime it occurs simultaneously during photosynthesis through the stomata. This type of respiration is called either *dark respiration* or *daytime respiration* (e.g., Farquhar et al., 1980; Campbell and Norman, 1998) and is distinct from photorespiration. The rate of dark respiration is indicated by R_d (see Fig. 4.8).

4.7   Modeling Net Assimilation

During the day, plants uptake CO_2 by turbulent transport through the surface layer of the atmosphere and then by diffusion through the stomata (see Fig. 4.8). Accordingly, assuming quasi-steady-state conditions, we can express the net CO_2 flux per unit ground area (i.e., the net assimilation) as

$$A_n = g_{sa,CO_2}(c_a - c_i), \tag{4.21}$$

where c_a and c_i are the CO_2 concentrations in the atmosphere and in the stomatal cavities, respectively, and g_{sa,CO_2} is given by a series combination of the stomatal and atmospheric resistances to CO_2 (see Fig. 4.8), which in term of conductances becomes

$$\frac{1}{g_{sa,CO_2}} = \frac{1}{g_{s,CO_2}} + \frac{1}{g_{a,CO_2}}. \tag{4.22}$$

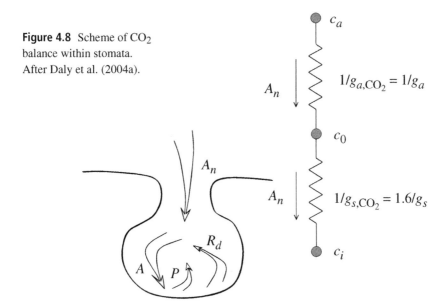

Figure 4.8 Scheme of CO_2 balance within stomata. After Daly et al. (2004a).

These conductances are expressed in units of moles of CO_2 m^{-2} s^{-1}. Owing to the fact that CO_2 is a heavier molecule than water, its diffusion coefficient is lower and this results in a reduction in the stomatal conductance for CO_2 as compared with the stomatal conductance for water, g_s (Jones, 1992):

$$g_{s,CO_2} = \frac{1}{1.6}g_s. \tag{4.23}$$

In contrast, the transport of carbon dioxide and the transport of water vapor are driven by the same turbulent advection process and are thus characterized by the same value of the conductance,

$$g_{a,CO_2} = g_a, \tag{4.24}$$

as discussed in Sec. 2.3.3.

Because the equilibration of CO_2 concentration inside the stomatal cavities is much faster than the timescale of stomatal adjustment, the CO_2 balance inside the stomata may also be described as a quasi-steady-state balance (Fig. 4.8),

$$A_n = A - P - R_d, \tag{4.25}$$

where A is the rate of CO_2 fixation by carboxylation, P is the photorespiration, and R_d is the dark respiration. Since R_d is typically a small fraction of the assimilation, it will be neglected for the sake of simplicity. The quantities A and P primarily depend on ϕ, c_i, T_l, and ψ_l (the absorbed photon irradiance, the CO_2 concentration inside the leaf pore, the leaf temperature, and the leaf water potential).

The dependence of assimilation on the leaf potential can be assumed to act multiplicatively, with a function similar to that used for Jarvis' stomatal conductance (e.g. Daly et al., 2004a),

$$A_n \simeq A - P = f_{\psi_l} A_{\phi,c_i,T_l}(\phi, c_i, T_l). \tag{4.26}$$

The function f_{ψ_l} is normalized between one in well-watered conditions and zero in fully stressed (e.g., wilting) conditions and will also be used for modeling leaf–potential effects on stomatal conductance (see Eq. (4.41)). It is important to note that, even if the effects are similar, here the water–potential limitation refers to the metabolic impairment of photosynthesis by water stress, which acts directly on the photosynthetic process,[7] as opposed to a reduction due to stomatal closure by water stress.

The dependence of $A_n = A - P$ on ϕ, c_i, and T_l may be modeled following the classic approach of Farquhar et al. (1980) as well as the approach given in the review by Von Caemmerer (2013), originally proposed for C_3 photosynthesis in well-watered conditions (the model can be extended easily to include the C_4 pathway; see Note 4.10). This semi-empirical model assumes that, when the water potential is not limiting, the rate of CO_2 fixation is the minimum of the two potential capacities to fix carbon, i.e.,

$$A_n(\phi, c_i, T_l) = \min(A_c, A_q),\tag{4.27}$$

where A_c is the assimilation rate limited by rubisco activity (i.e., CO_2 concentration), which depends on c_i, T_l, and the oxygen concentration, o_i, while A_q is the assimilation rate when the photosynthetic electron transport limits RuBP regeneration (light limitation), which thus depends on ϕ, c_i, and T_l.

The Rubisco-limited rate of photosynthesis is given by

$$A_c = V_{c,\max} \frac{c_i - \Gamma^*}{c_i + K_c(1 + o_i/K_o)},\tag{4.28}$$

where $V_{c,\max}$ is the maximum carboxylation rate, K_c and K_o are the Michaelis–Menten coefficients for CO_2 and O_2 respectively (see Note 4.9), and Γ^* is the CO_2 compensation point, at which the rate of photosynthesis equals the rate of photorespiration. The Michaelis–Menten coefficients, the maximum carboxylation rate, and the T_l-dependence of the CO_2 carboxylation point are modeled using the following empirical functions (Leuning, 1995):

$$K_c = K_{c_0} \exp\left[\frac{H_{K_c}}{\mathcal{R}T_0}\left(1 - \frac{T_0}{T_l}\right)\right],\tag{4.29}$$

$$K_o = K_{o_0} \exp\left[\frac{H_{K_o}}{\mathcal{R}T_0}\left(1 - \frac{T_0}{T_l}\right)\right],\tag{4.30}$$

$$\Gamma^* = \gamma_0[1 + \gamma_1(T_l - T_0) + \gamma_2(T_l - T_0)^2],\tag{4.31}$$

$$V_{c,\max} = V_{c,\max_0} \frac{\exp\left[\frac{\mathcal{H}_{vV}}{RT_0}\left(1 - \frac{T_0}{T_l}\right)\right]}{1 + \exp\left(\frac{S_v T_l - H_{dV}}{\mathcal{R}T_l}\right)},\tag{4.32}$$

the parameters of which are reported in Table 4.2.

The assimilation rate when the photosynthetic electron transport limits RuBP regeneration (light limitation) is

$$A_q = \frac{J}{4}\frac{c_i - \Gamma^*}{(c_i + 2\Gamma^*)},\tag{4.33}$$

where J is the electron transport rate, given by

$$J = \min(J_{\max}, \phi),\tag{4.34}$$

---

[7] The direct effects of these metabolic limitations were investigated in detail in Vico and Porporato (2008).

Table 4.2 Parameters for the model of C_3 photosynthesis. After Daly et al. (2004a).

Parameter	Value	Units	Description
H_{K_c}	59,430	J mol^{-1}	Activation energy for K_c
H_{K_o}	36,000	J mol^{-1}	Activation energy for K_o
H_{vV}	116,300	J mol^{-1}	Activation energy for $V_{c,max}$
H_{dV}	202,900	J mol^{-1}	Deactivation energy for $V_{c,max}$
H_{vJ}	79,500	J mol^{-1}	Activation energy for J_{max}
H_{dJ}	201,000	J mol^{-1}	Deactivation energy for J_{max}
J_{max_0}	75	μmol m^{-2} s^{-1}	Eq. (4.35)
K_{c_0}	302	μmol mol^{-1}	Michaelis constant for CO_2 at T_0
K_{o_0}	256	mmol mol^{-1}	Michaelis constant for O_2 at T_0
o_i	0.209	mol mol^{-1}	Oxygen concentration
$\mathcal{R}$	8314	J kmol^{-1} K^{-1}	Universal gas constant
S_v	650	J mol^{-1}	Entropy term
T_0	293.2	K	Reference temperature
V_{c,max_0}	50	μmol m^{-2} s^{-1}	Eq. (4.32)
γ_0	34.6	μmol mol^{-1}	CO_2 compensation point at T_0
γ_1	0.0451	K^{-1}	Eq. (4.31)
γ_2	0.000347	K^{-2}	Eq. (4.31)

where the maximum electron transport, J_{max}, can be expressed as (Leuning, 1995)

$$
J_{max} = J_{max_0} \frac{\exp\left[\dfrac{H_{vJ}}{\mathcal{R}T_0}\left(1 - \dfrac{T_0}{T_l}\right)\right]}{1 + \exp\left(\dfrac{S_v T_l - H_{dJ}}{\mathcal{R}T_l}\right)},
\tag{4.35}
$$

and ϕ (mol photons m^{-2}s^{-1}) is the absorbed photon irradiance and may be computed as

$$
\phi = 0.5(1 - \alpha_l)\frac{R_{sw}\bar{\lambda}}{N_A c \hbar}\kappa_2,
\tag{4.36}
$$

where α_l is the albedo of the leaf, 50% of the absorbed shortwave radiation $(1 - \alpha_l)R_{sw}$ (W/m^2) is considered photosynthetically active radiation (PAR) (Jones, 1992), $\bar{\lambda}$ is the average wavelength in meters for PAR (assumed to be 550×10^{-9} m), κ_2 is the quantum yield of photosynthesis (0.20 mol electrons mol^{-1} photons (Leuning, 1995), c is the speed of light (m s^{-1}), N_A is Avogadro's constant (mol^{-1}), and $\hbar$ is Planck's constant (J s).

A simplified approach to calculating the net assimilation may be obtained by linearizing the Rubisco-limited photosynthesis kinetics (Manzoni et al., 2011a). Assuming the ratio of internal and atmospheric CO_2 concentration $R_c = c_i/c_a$ in the denominator of A_c is constant, one can express the net assimilation as

$$
A_n = a_w f_{\psi_l} \frac{\phi}{\phi + \gamma_k} \frac{\eta_c c_i - \Gamma^*}{\eta_c R_c c_a + K_c(1 + o_i/K_o)} - R_d
\tag{4.37}
$$

where γ_k is a kinetic constant that depends on leaf temperature, η_c is the efficiency of the CO_2 pump ($\eta_c = 1$ for C_3 species), and a_w represents the carboxylation rate under well-watered conditions. Since

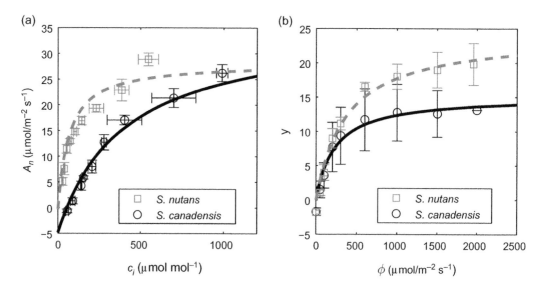

Figure 4.9 (a) CO_2 response and (b) light response of photosynthesis. According to Eq. (4.37), the y axis in panel (b) represents the assimilation rate under constant c_i (Manzoni et al., 2011a).

the resulting behaviors are similar to those in the full Farquhar model (Eqs. (4.27)–(4.36)), the simplified model (4.37) may be employed as an agile alternative. For example, it has been used to simulate the CO_2 and light responses of photosynthesis in the Lysimeter CO_2 Gradient (LYCOG) facility (Manzoni et al., 2011a). The results are shown in Fig. 4.9, which gives the measured and modeled assimilation rates for comparison.

4.8   Stomatal Conductance

While the details of plant hydraulics are quite complex and not completely understood (Rockwell and Holbrook, 2017), to a first approximation the water flux from the leaves to the atmosphere just outside the leaf surface can be regarded as a one-dimensional diffusion flux through the stomatal openings. The corresponding transport equation is given by Eq. (4.7), and is repeated here for convenience:

$$\rho_w E = g_s \rho [q_i(T_l, \Psi_l) - q_0]. \tag{4.38}$$

It is important to note that this equation assumes that all the water vapor fluxes take place through the stomata, thus neglecting the so-called cuticular transpiration. The latter is the vapor flux through the leaf cuticle, which happens even if the stomata are fully closed. If necessary for modeling purposes, a *cuticular conductance* can be assumed to act in parallel with the stomatal conductance.

The presence of Ψ_l in Eq. (4.38) reflects the hypothesis of the SPAC of the continuity of the water potential. It should be noted however that, for simplicity, the water vapor inside the stomatal cavities is often considered to be at saturation (i.e., on the Clausius–Clapeyron curve, Sec. 2.1.8); this assumption is made for example in the Penman–Monteith approximation; see Sec. 5.2.4. While it is technically in disagreement with the SPAC, in general it involves only a small overestimation of the evaporation flux, because of the overwhelming role of the stomatal conductance which markedly drops as the water potential is reduced, with a corresponding, reduction in transpiration. This may not be the case when leaf and stomatal damage caused by drought increases the cuticular conductance.

Because of the complex physiological and environmental factors that control stomatal movement and plant hydraulics, no complete model for g_s has been developed so far. In general, the functional dependence of stomatal conductance is of the type

$$g_s = g_s(\phi, T_l, \psi_l, D, c_i), \tag{4.39}$$

since it depends on the available light (ϕ is the photosynthetically active photon flux density; see Eq. (4.36)), the leaf temperature T_l the leaf water potential ψ_l, the specific humidity deficit D (see Eq. (2.77)), and the leaf internal CO_2 concentration c_i. For simplicity, some of these variables may be surrogated with more readily available quantities: for example, the atmospheric temperature and CO_2 concentration may be used instead of the corresponding leaf values, which are more difficult to measure or model.

Several models of stomatal conductance have been proposed in the literature. They can be roughly subdivided into (i) empirical and data-based models (e.g., Jarvis, 1976; Jones, 1992; Lhomme et al., 1998; Lhomme, 2001); (ii) semi-empirical models linked to assimilation (e.g., Norman, 1982; Ball et al., 1987; Leuning, 1990, 1995); (iii) mechanistic models (e.g, Delwiche and Cooke, 1977; Dewar, 1995; Gao et al., 2002); (iv) the optimization of stomatal functioning for maximum carbon uptake (e.g., Cowan and Farquhar, 1977; Hari et al., 1986; Medlyn et al., 2011, see also Note 4.8). Here, we will only focus on (i) and (ii), presenting an empirical, multiplicative formulation by Jarvis (Sec. 4.8.1) and a more physiological approach (Sec. 4.8.2) which involves the computation of the photosynthetic CO_2 assimilation flux.

4.8.1 Jarvis' Approach

Jarvis' empirical formulation (Jarvis, 1976; Jones, 1992; Lhomme et al., 1998; Lhomme, 2001) uses a multiplicative relationship of functions of the main factors affecting stomatal movement, such as the photosynthetically active radiation (PAR), the ambient temperature (T_a), the leaf water potential (ψ_l), the saturation deficit (D), and the CO_2 concentration:

$$g_s = g_{s,\max} f_{PAR}(PAR) f_{T_a}(T_a) f_{\psi_l}(\psi_l) f_D(D) f_{CO_2}(c_a), \tag{4.40}$$

where $g_{s,\max}$ is the maximum stomatal conductance per unit ground area when none of the factors is limiting.[8] A typical value of $g_{s,\max}$ is around 25 mm s^{-1} (Jones, 1992; Kelliher et al., 1995b). The functions in Eq. (4.40) are obtained from controlled environment studies and account separately for the influence of each variable (Fig. 4.10). The formulation is simple and has the advantage of allowing more direct calculations without having to couple it to a photosynthesis model.

The direct effect of increasing light is usually expressed as an exponential function $f_{PAR}(PAR) = 1 - \exp(-k_1 P_{PAR})$ (Fig. 4.10a), where k_1 is equal to 0.005 m^2 W^{-1} (Jones, 1992). In terms of temperature dependence, in general the stomata open as the ambient temperature increases up to an optimum value, T_{opt}, after which they start closing (Fig. 4.10b). A simple function commonly used is $f_{T_a} = 1 - k_2(T_a - T_{opt})^2$, with $k_2 = 0.0016$ K^{-2} and $T_{opt} = 298$ K (Lhomme et al., 1998).

As for f_{ψ_l}, we assume no control of leaf water potential up a certain point ψ_{l_1} (i.e., under well-watered conditions). As the leaf water potential drops below ψ_{l_1}, the stomatal conductance begins to decrease, reaching zero when ψ_l drops to ψ_{l_0} (Fig. 4.10d) (Daly et al., 2004a):

---

[8] Note that Jarvis' original formulation did not include limitations on ψ_l and CO_2 concentration.

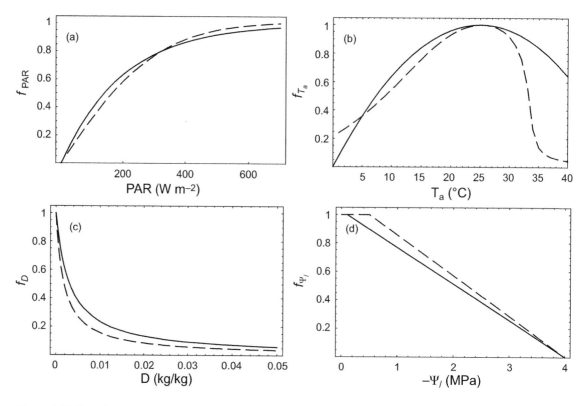

Figure 4.10 Functions controlling the stomatal response to the environment: (a) solar irradiance, (b) ambient temperature, (c) water vapor deficit, and (d) leaf water potential, following Jarvis' (continuous line) and Leuning's (dashed line) approaches. After Daly et al. (2004a).

$$
f_{\psi_l}(\psi_l) = \begin{cases} 0 & \psi_l < \psi_{l_0}, \\[2mm] \dfrac{\psi_l - \psi_{l_0}}{\psi_{l_1} - \psi_{l_0}} & \psi_{l_0} \leq \psi_l \leq \psi_{l_1}, \\[2mm] 1 & \psi_l > \psi_{l_1}, \end{cases}
\tag{4.41}
$$

where $\psi_{l_1} = -0.07$ MPa and $\psi_{l_0} = -4$ MPa are typical values in agreement with measurements of C_3 plants in semi-arid ecosystems (Scholes and Walker, 1993; Bonan, 2002). A sigmoidal function can also be used instead of the piecewise function (4.41).

A linear relation for the function $f_D(D)$ was suggested by Jarvis (1976), but other forms have also been used, such as polynomial (Shuttleworth et al., 1989), hyperbolic (Fig. 4.10c), power-law (Hari et al., 1999; Katul et al., 2009; Medlyn et al., 2011; see also Eq. (4.54)), or exponential functions (Oren et al., 1999; Ewers et al., 2000). A hyperbolic relationship, suggested by Leuning (1995), is usually recommended (see also Lohammar et al., 1980; Daly et al., 2004a):

$$
f_D(D) = \frac{1}{1 + \dfrac{D}{D_x}},
\tag{4.42}
$$

where D_x is species-dependent, with a typical value of 7.7 g/kg (Leuning, 1995).

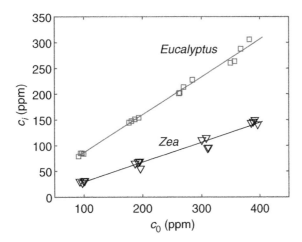

Figure 4.11 Measured relationship between the internal and ambient CO_2 concentrations for *Eucalyptus pauciflora*, a C_3 plant, at four different irradiances, and for *Zea mais* (corn), a C_4 plant, grown at four different nitrogen levels. After Wong et al. (1979).

Finally, stomata are sensitive to the atmospheric CO_2 concentration, c_a, because when c_a decreases they open to maximize photosynthesis. However, as the CO_2 concentration is almost constant during the day, $f_{CO_2}(c_a)$ is seldom included in Eq. (4.40). As we will see, this dependence is more naturally included in the physiological approach described next.

4.8.2 Physiological Model of Stomatal Conductance

The so-called physiological model of stomatal conductance was introduced by Norman (1982) and Ball et al. (1987) and improved by Leuning (1990, 1995). It provides a valuable alternative to the purely empirical stomatal conductance model of Jarvis (Sec. 4.8.1). The model is based on the observed constancy of the ratio of the internal (stomatal) and ambient CO_2 concentrations for different values of the assimilation rate, A_n, under normal environmental conditions (see Fig. 4.11).

Making the ratio of the internal (c_i) and ambient (just outside the stomata, c_0) CO_2 concentrations explicit in the equation (similar to Eq. (4.21)) for diffusion through the stomata,

$$A_n = g_{s,CO_2} c_0 \left(1 - \frac{c_i}{c_0} \right),\tag{4.43}$$

and solving for the stomatal conductance, after including all the constant terms into a single parameter κ and adding a dependence on saturation deficit D (this step is due to Leuning, 1995), we obtain

$$g_{s,CO_2} = \kappa A_n \frac{f_D(D)}{c_0},\tag{4.44}$$

where $\kappa \approx 15$ (for C3 plants). The same functional form for $f_D(D)$ in Eq. (4.42) as that used in Jarvis' equation (4.40) may be used in Eq. (4.44) also, but the value of D_x needs to be adjusted to 1.8 g/kg (Leuning, 1995) in order to obtain similar values of transpiration and assimilation in well-watered conditions with both approaches. At low CO_2 concentration levels (e.g., $c_0 = \Gamma^*$), a plant increases the stomatal openings so that the CO_2 uptake by photosynthesis is compensated by the release of CO_2 from respiration, resulting in a zero net assimilation rate. Such a stomatal behavior cannot be described by Eq. (4.44), and a modified form is given as (Leuning, 1990, 1995)

$$g_{s,CO_2} = g_0 + \kappa A_n \frac{f_D(D)}{(c_s - \Gamma^*)}, \tag{4.45}$$

where g_0 is the stomatal conductance at compensation point as $A_n \to 0$.

Unlike Jarvis' approach, which completely determines g_s independently of the photosynthesis model, in the physiological approach Eq. (4.44) requires the net carbon assimilation A_n to be known and thus needs to be linked to an assimilation model, such as the Farquhar model of Sec. 4.7 or its simplified version (4.37). The resulting stomatal conductance behaves very similarly to Jarvis' equation (4.40), as shown by Fig. 4.10. As a result, with Jarvis' approach, Eqs. (4.23) and (4.26) may be used to directly evaluate the assimilation once the stomatal conductance has been determined. With the physiological approach, the SPAC equations must first be solved for the leaf temperature and leaf water potential. Then the carbon assimilation and transpiration can be calculated. Finally, Eq. (4.44) can be used to compute the stomatal conductance.

4.8.3   A First Analysis of Transpiration and Plant Carbon Assimilation

Having introduced the stomatal conductance models, we can now couple all the water-transport equations and take stock of our progress in modeling plant transpiration. The main parameters of the SPAC and the photosynthesis model are reported in Tables 4.2 and 4.3.

According to the SPAC with no plant capacitance (see Sec. 4.4.2), the flux of liquid water from soil to leaves (mesophyll) is equal to the diffusion and turbulent flux of water vapor from stomatal pores to the logarithmic layer. Accordingly, the soil-to-leaf water flux is given by Eq. (4.20), namely

$$E = g_{srp}(\psi_l, s)[\psi_s(s) - \psi_l], \tag{4.46}$$

and the leaf-to-atmosphere flux is given by the series of (4.7) and (4.10),

$$\rho_w E = \rho g_{sa}(T_a, e_a, \psi_l) \frac{\epsilon}{p_0}[e_i(\psi_l, T_l) - e_a], \tag{4.47}$$

where the partial pressure of water vapor inside a leaf, e_i, is related to the leaf water potential ψ_l and leaf temperature T_l.

Equations (4.46) and (4.47) cannot be solved as a system for E and ψ_l given the soil and atmosphere boundary conditions, because the leaf temperature (T_l) remains undefined. For this full problem, we will have to wait until the next chapter. where we will introduce the leaf energy balance. For now we can only solve a subset of it, by first using Eq. (4.46) for the soil–root–plant part to plot the relationship between E and ψ_l under different soil moisture conditions, s. These curves are reported as dashed lines in Fig. 4.12. In the same figure, we also show the curves resulting from Eq. (4.47) for the plant–atmosphere flux, assuming various different values of leaf temperature (see the solid lines in Fig. 4.12).

The intersections of dashed and solid lines satisfy the soil–root–plant–atmosphere continuum constraints and give the transpiration as a function of leaf temperature. A dramatic decrease in transpiration with lower soil-moisture values is evident, due to the reduction in the soil–plant conductance; this decrease corresponds to a reduction in leaf water potential. In contrast, for the same value of soil moisture, higher leaf temperatures lead to higher transpiration, while the leaf water potential decreases. In Sec. 5.2.3, we will use the leaf energy balance to obtain the transpiration as a function of leaf temperature; this will give

Table 4.3 Parameter values used in the model of the hourly dynamics of the SPAC. Modified from Daly et al. (2004a).

Parameter	Value	Units	Description
κ	15	—	Eq. (4.44)
c	2	—	Eq. (4.18)
c_a	350	μmol mol^{-1}	Atmospheric carbon concentration
c_p	1012	J kg^{-1} K^{-1}	Air specific heat at constant pressure
d	2	MPa	Eq. (4.18)
$D_{x,\text{Jarvis}}$	7.7	g/kg	Eq. (4.42)
$D_{x,\text{Leuning}}$	1.8	g/kg	Eq. (4.44)
$g_{p,\max}$	16.4	μm MPa^{-1} s^{-1}	Maximum plant conductance
g_a	20	mm s^{-1}	Atmospheric conductance
g_b	20	mm s^{-1}	Leaf boundary layer conductance
$g_{s,\max}$	25	mm s^{-1}	Maximum stomatal conductance
h_0	50	m	Boundary layer height at night
k_1	0.005	m^2 W^{-1}	Eq. (4.40)
k_2	0.0016	K^{-2}	Eq. (4.40)
p_a	1.013×10^5	Pa	Air pressure
T_{opt}	298	K	Eq. (4.40)
λ_w	2.5×10^6	J kg^{-1}	Latent heat of water vaporization
ρ	1.2	kg m^{-3}	Air density
$\phi_{\max}$	500	W m^{-2}	Maximum leaf available energy
Ψ_{l1}	−0.05	MPa	Eq. (4.41)
Ψ_{l0}	−4.5	MPa	Eq. (4.41)

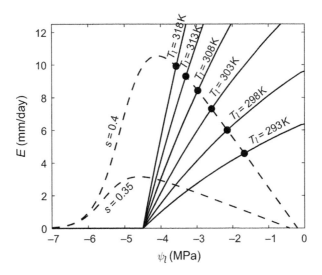

Figure 4.12 Relationship between ψ_l and E from the soil-to-leaf liquid water flux in Eq. (4.46) (dashed lines), and the leaf-to-atmosphere water vapor flux in Eq. (4.47) (solid lines). The air temperature was set as 293 K, the relative humidity was 30%, and the other SPAC parameters are listed in Table 4.3.

the extra condition to find the value of leaf temperature and obtain a complete solution for E, ψ_l, and T_l. Regarding plant carbon assimilation, the equations for carbon flux and assimilation, Eqs. (4.21) and (4.26) respectively, need to be coupled to the water-transport equations in order to determine the leaf water potential, which is required to compute A_n and g_{s,CO_2}. Similarly to the case of transpiration, however, for net assimilation the leaf temperature remains undetermined and requires coupling with the surface energy balance (see Sec. 5.2.3).

4.9   Plant Water Stress

The reduction in soil moisture content during droughts lowers the plant water potential. This in turn causes a reduction in cell turgor and relative water content in plants and drives a decrease in transpiration and photosynthesis. Low water potentials and carbon starvation by reduced photosynthesis bring about a sequence of damage of increasing seriousness as the drought intensifies. Here we discuss some of the physiological effects of water stress; this will help set the stage for mathematical modeling of plant water stress in Chapter 8.

4.9.1   Physiological Effects of a Reduction in Plant Water Potential

We can begin analyzing the effects of soil moisture changes on plant water potential with the aid of Fig. 4.2. Assuming for simplicity that during the development of a drought the level of the atmospheric water potential does not change, consider first a condition with higher soil moisture than that shown in Fig. 4.2. In such conditions, the soil water potential ψ_s is noticeably increased, whereas the resistance at the soil–root interface is reduced (Nobel, 1999). As a consequence, the plant water potential is relatively high, so that the negative pressure in the xylem is reduced and the plant tissues are in full-turgor conditions. At the leaf–atmosphere interface, the resistance is reduced by an increase in the stomatal opening. Because of both the decreased resistances to flow and the increased soil–atmosphere water potential gradient, the water flux is considerably increased. This produces higher energy losses along the xylem, thereby steepening the energy line.

When the soil moisture level is lower than that of the example in Fig. 4.2, the situation is opposite to that discussed above. The soil water potential quickly drops to very negative values and the soil–root resistance increases considerably. The stomata are now almost completely closed and the resistance to water vapor efflux is very high. The water flux is scanty and the energy line, which is now shifted down, becomes almost flat. In the xylem the resulting low pressure values make increasingly more likely the appearance of cavitation, starting from the highest parts of the plant, while in the cells the low plant water potential induces a decrease in turgor and relative water content. If the soil moisture is further depleted, the stomata close completely: at this point, the soil water potential is so low that the plant is no longer able to take up water. However, since small water losses keep going on through cuticular transpiration, plant cells progressively lose their turgor and drought damage sets in.

A reduction in the plant water potential has serious impacts on the cell matric (pressure) and osmotic potentials, which are related to the osmotic and elastic properties of the cell membranes in a combined way that is well synthesized by the Höfler diagram shown in Fig. 4.13 (see also Jones, 1992; Nobel, 1999). At full-turgor conditions, the plant water potential is quite high (slightly below zero) and the relative water content of the cells at a maximum value. As the soil moisture is depleted and the plant water potential

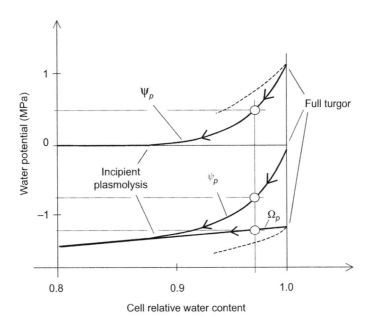

Figure 4.13 Höfler diagram linking the cell relative water content to the protoplast water potential ψ_p, the protoplast turgor potential Ψ_p, and the protoplast osmotic potential Ω_p. The dashed lines show the variations in Ψ_p and Ω_p as osmotic adjustment takes place. The open circles correspond to the situation depicted in Fig. 4.2. After Porporato et al. (2001).

lowers, the osmotic potential decreases only slightly because of the solute concentration increase caused by water losses. According to Eq. (4.2), a reduction in water potential, if not accompanied by an equivalent decrease in osmotic potential, produces a decrease of matric potential and turgor loss. The three circles on the curves in Fig. 4.13 correspond to the situation depicted in Fig. 4.2. As the drought continues, the water potential and turgor keep decreasing until the plasma membranes of the cells start pulling away from the cell walls (i.e., incipient *plasmolysis*). At this point, the plant conditions have become very critical and permanent damage is produced.

Figure 4.14 reproduces a typical sequence of effects caused by a decrease in cell water potential. As can be seen, during drought conditions this decrease triggers a series of harmful events on plant physiology, whose seriousness grows with the intensity and duration of the water deficit. As Ψ_p is reduced, the first effect is a reduction in cell growth and wall-cell synthesis; then the nitrogen uptake from the soil is diminished owing to the sharp decrease in nitrate reductase, an enzyme which catalyzes the reduction of nitrate to nitrite (this is the first internal step for nitrogen assimilation after nitrate has been uptaken by the plant).

Stomatal closure and the related reduction in CO_2 assimilation also begin very soon: at first, stomatal closure is mostly a way to optimize water loss and carbon assimilation and does not produce permanent damage to the plant. Soon after, however, respiration (O_2 uptake and CO_2 release) is affected. Changes in patterns of resource allocation are then induced, including the accumulation of the amino acid proline, whose drastic increase is a characteristic sign of disturbance in protein metabolism. Sugar accumulation, along with possible flowering reduction, inhibition of seed production, and fruit abortion are possible further effects of higher levels of water stress (Nilsen and Orcutt, 1998). At the same time, the reduction in the cooling effect of transpiration makes the insurgence of heat and radiation stress more likely. Finally,

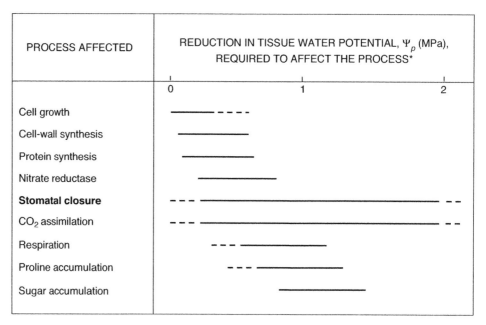

*With Ψ_p of well-watered plants under mild evaporative demand as the reference point

Figure 4.14 Typical sequence of the effects on plant physiology caused by a decrease in cell water potential. Redrawn from Hsiao (1973) and Lawlor and Tezara (2009).

at very low water potentials, complete stomatal closure takes place, and this usually also marks the point of complete turgor loss and wilting (Bradford and Hsiao, 1982).

4.9.2   Drought Tolerance and Escape Strategies: Osmotic Adjustment

Many species in semi-arid ecosystems have adapted to frequent and prolonged periods of water deficit, developing various strategies to delay the appearance of water stress (e.g., Ehleringer and Monson, 1993; Larcher, 1995; Nilsen and Orcutt, 1998). The ways by which plants compensate for water limitations range from mechanisms for drought escape, such as rapid phenological development, drought deciduousness, or extended dormancy, to drought tolerance by osmotic adjustment, desiccation tolerance, and reduction in water loss with enhancement of water accumulation.

Osmotic adjustment is a particularly widespread process adopted by plants. As shown by the dashed lines in Fig. 4.13, plants in semi-arid ecosystems may adjust their levels of cell osmotic potential by increasing the cell solute concentration. This allows them to maintain higher *turgor* (from Eq. (4.2), a decrease in Ω_p causes an increase of Ψ_p, when ψ_p is kept fixed) and to reduce the level of water stress (see the dashed lines in Fig. 4.13). This phenomenon is called osmotic adjustment, and is considered to be an important form of adaptation to water stress (Nilsen and Orcutt, 1998).

With such adjustments, the stomata remain open longer, leaving more time for carbon assimilation, and the plants continue to acquire water from the soil even at low water potentials (Larcher, 2003; Lambers et al., 1998). Turgor maintenance through osmotic adjustment in response to water stress is widespread, occurring in the leaves, roots, and reproductive organs of many species (see Fig. 4.15 and Turner and

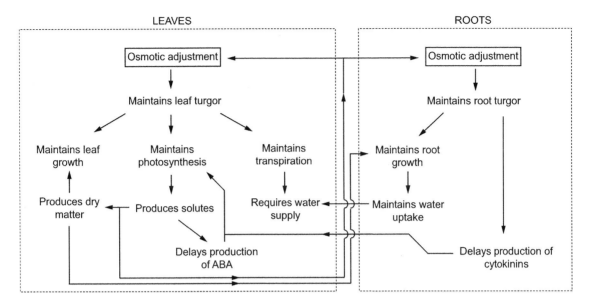

Figure 4.15 Effects produced by osmotic adjustment in roots and leaves. Cytokinins are plant phytohormones, which, along with absisic acid (ABA), are involved in the process of adjustment of the stomatal regulatory system's growth and development. Redrawn after Turner (1986).

Jones, 1980). Even a partial turgor maintenance can be advantageous and may be accomplished by a passive concentration effect as the cell water decreases. We will see in Chapter 8 how plants can achieve maximum transpiration and assimilation rates under conditions of unpredictable soil moisture availability by adjusting their osmotic status and adapting stomatal control strategies.

4.10 Key Points

- The apoplastic (xylem) and symplastic (plant cell) water may be assumed to be in local equilibrium, with the same water potential; compared with the water in the xylem, the cell water has higher pressure and lower osmotic potential, owing to a higher solute concentration.
- The water flux through the plant from the soil to the atmosphere follows a path of decreasing water potential. The losses in water potential are modeled using an electrical analogy in terms of conductances whose values depend on the water potential at the nodes.
- The dramatic decrease in the bulk soil–root conductance is modeled as proportional to the unsaturated soil hydraulic conductivity and depends also on the total root density and depth.
- The laminar flow through the xylem vessels can be disrupted by embolism caused by strong suction during drought; the disruption of the water column by air bubbles causes a dramatic loss of conductivity in the xylem.
- The Farquhar et al. (1980) photosynthesis model reproduces empirically the light and CO_2 limitations of the light and dark reactions, as well as their dependences on leaf temperature;

however, it does not contain the dependence on the leaf water potential, ψ_l, which needs to be included separately as an additional limitation.

- Guard cells control the leaf stomatal opening to maximize carbon uptake while minimizing transpiration losses. This is reflected in the dependence of the stomatal conductance on the main environmental variables including the leaf water potential.
- The transpiration flux can only be computed as a function of the unknown leaf temperature. Determining the latter requires coupling the water transport equations with the equations for energy exchanges with the atmosphere (see Chapter 5).
- The sequence of physiological damage induced by a reduction in water potential during droughts can in part be alleviated by osmotic adjustments, i.e., the active control of the solute concentration in plant cells to increase turgor.

4.11 Notes, including Problems and Further Reading

4.1 Assuming that roots of diameter d_r, evenly distributed within the root zone of depth Z_r, behave like a bundle of tubes (see Fig. 3.4), derive an expression for the distance between each root, Λ_{sr}, and compare it with Eq. (4.12).

4.2 Compute the typical soil–root resistance for temperate deciduous forest and a loamy sand soil, using data from Table 4.1.

4.3 (Root Modeling) Numerous root models have been proposed in the literature. Following an order of increasing complexity, Laio (2006) provides an elegant analysis of the role of root distribution with suitable simplifications that still allow for a stochastic soil moisture analysis (Chapter 7). Guswa et al. (2002) considered a root distribution within a vertically distributed soil model based on Richards' equation (see Eq. (3.26) and Chapter 3); their model contains an interesting tunable parameter to account for preferential water uptake by the roots. This in turn leads to an averaging behavior, which implies some form of hydraulic redistribution, with results similar to those obtained by vertically averaged schemes. More detailed models of root functioning are reviewed in Javaux et al. (2013).

4.4 (Mycorrhizae) Mycorrhizae are symbiotic associations between a fungus and a plant. The term mycorrhiza most often refers to the role of the fungus in the plant's root system; it improves the connections between plant and soil water and nutrients. The extraradical hyphae (the branching filaments that make up the fungus) effectively increase the volume of soil explored for nutrients and water, whereas the small diameter of the fungal hyphae allows them to penetrate small soil pores and access microsites that roots cannot reach (Allen et al., 2003). The hyphae may only be 2 to 10 μm in diameter, but individuals can extend across many hectares.

Augé (2004) observed a slight but significant mycorrhizal effect on the soil moisture characteristic curve, as well as considerable reduction in the overall soil–plant resistance to water flow. Similarly, mycorrhizae were found to be involved in direct and inverse hydraulic redistribution (from wet patches in soils to plants and vice versa to keep fungi alive), as well as in reducing the tortuosity of water paths from soils to plants (Allen, 2007).

4.5 (Steady-State Diffusion and Discrete Transport Equations)   In several instances in the text (e.g., Secs. 3.2.5, 4.4, 5.2.3, 5.4), we have seen that fluxes can be written as a discrete transport equation of the type

$$\Phi_0 = g\,(c_i - c_{i-1})\,, \tag{4.48}$$

where g is the conductance, with dimensions of velocity, $L\,T^{-1}$, between compartments i and $i-1$ in the SPAC, and c_i and c_{i-1} are the concentrations of the quantity being transported. Several mechanisms lead to this form of equation. For example, this equation may be obtained by considering one-dimensional diffusion,

$$\frac{\partial c}{\partial t} = \mathcal{D}\frac{\partial^2 c}{\partial x^2}, \tag{4.49}$$

where $\mathcal{D}$ is a constant diffusion coefficient with dimension $L^2 T^{-1}$. For steady-state conditions, (4.49) becomes

$$0 = \mathcal{D}\frac{\partial^2 c}{\partial x^2}. \tag{4.50}$$

Integrating (4.49) with respect to x gives

$$-\mathcal{D}\frac{\partial c}{\partial x} = \Phi_0. \tag{4.51}$$

Considering now a one-dimensional interval of length L and boundary conditions $c(0)$ and $c(L)$, and integrating (4.51) with respect to x, one obtains

$$-\mathcal{D}\,(c(L) - c(0)) = \Phi_0 L. \tag{4.52}$$

Isolating the flux term in Eq. (4.52) yields

$$\Phi_0 = \frac{\mathcal{D}}{L}\,(c(0) - c(L))\,, \tag{4.53}$$

which can be compared to Eq. (4.48) to obtain $g = \mathcal{D}/L$, with the dimensions of velocity, $L\,T^{-1}$. A similar linear equation is also obtained for laminar flow in circular pipes when solving the Navier–Stokes equations; see Eq. (4.14).

In nonlinear diffusion, as is the case described by Richards' equations (see Sec. 3.3.2), and xylem flow in conditions of cavitation, the same procedure leads to state-dependent conductance functions. Note that in near-wall turbulence, a similar equation is obtained for turbulent transport (see Sec. 2.3.3).

4.6 (Phloem and Sugar Transport in Plants)   The phloem is a pipe-like network of living cells, which runs parallel and mostly opposite to the xylem, to transport sap, the solution of water and sugars from the sites of photosynthesis to the rest of the plant. As in the xylem, the sap flow in the phloem takes place as a laminar flow; however, while the water flow in the xylem is driven by negative pressures (tension), the phloem flow is driven by positive hydrostatic pressures generated by osmotic effects (the Munch mechanism). Such effects are due to the loading and unloading of sugars in the phloem, as well as to the transfer of water from and to the xylem, from which the phloem is separated by a semipermeable membrane (allowing water to pass relatively freely while preventing the passage of larger molecules such as sugars). Phloem sap is also thought to play a role in sending informational signals throughout vascular plants. Modeling of the coupled water–carbon flow in the xylem–phloem system is an interesting open area of research (Jensen et al., 2016).

4.7 Link each term of the Farquhar et al. (1980) model of photosynthesis to the corresponding process of photosynthesis (light reactions and the phases of the Calvin cycle). Plot the various equations of the model as a function of the main limiting factors, c_i, ϕ, and T_l.

4.8 (Stomatal Optimality) As mentioned in Sec. 4.8, an interesting approach to modeling stomatal conductance is based on an optimization hypothesis whereby plants maximize carbon gain while minimizing water loss (Cowan and Farquhar, 1977). Following this theory, an expression for the stomatal conductance is obtained which scales with the square root of the vapor pressure deficit (Hari et al., 1986). This relationship has been shown to be consistent with observed experimental data (see e.g. Katul et al., 2009). When combining this stomatal optimization hypothesis with the Farquhar model of photosynthesis, the following analytical expression for stomatal conductance is obtained (Medlyn et al., 2011):

$$g_{s,CO_2} = g_1 \frac{A_n}{c_0 \sqrt{D}}. \tag{4.54}$$

Note the similarity with Eq. (4.44). The slope parameter g_1 is proportional to the square root of the CO_2 compensation point Γ^* and the marginal carbon cost of water, λ, i.e.,

$$g_1 \propto \sqrt{\Gamma^* \lambda}. \tag{4.55}$$

The slope parameter g_1 can be found by fitting Eq. (4.54) to stomatal conductance measurements (Medlyn et al., 2011).

4.9 (Michaelis–Menten Kinetics) Michaelis–Menten kinetics, named after Leonor Michaelis and Maud Menten, is often used for modeling enzyme dynamics in biochemistry; here it is used to model photosynthesis (Sec. 4.7) and the decomposition of soil organic matter (Sec. 9.2). This kinetics describes the rate of enzymatic reactions as follows:

$$E + S \underset{k_r}{\overset{k_f}{\rightleftharpoons}} ES \overset{k_c}{\longrightarrow} E + P, \tag{4.56}$$

where an enzyme, E, and a substrate, S, bind in a reversible process to form ES, which then releases a product, P, and regenerates the enzyme; k_f, k_r, and k_c are the forward, reverse, and catalytic rate constants. Since the rate of a reaction is proportional to the product of the concentrations of the reactants, the system can be described by four differential equations (Logan, 2013):

$$\frac{dC_E}{dt} = -k_f C_E C_S + k_r C_{ES} + k_c C_{ES}, \tag{4.57}$$

$$\frac{dC_S}{dt} = -k_f C_E C_S + k_r C_{ES}, \tag{4.58}$$

$$\frac{dC_{ES}}{dt} = k_f C_E C_S - k_r C_{ES} - k_c C_{ES}, \tag{4.59}$$

$$\frac{dC_P}{dt} = k_c C_{ES}, \tag{4.60}$$

where the coefficients C denote the concentrations of the corresponding chemical components. Combining Eqs. (4.57) and (4.59) yields $dC_E/dt + dC_{ES}/dt = 0$, which suggests the conservation of the total enzyme:

$$C_E + C_{ES} = C_{E,tot} = \text{const.} \qquad (4.61)$$

In most systems, the ES concentration rapidly approaches a steady state (see Sec. 6.4.2), such that

$$dC_{ES}/dt = 0. \qquad (4.62)$$

Rearranging Eqs. (4.57)–(4.60) with (4.61) and (4.62) yields the rate of product formation as

$$\frac{dC_P}{dt} = k_c C_{ES} = k_c C_{E,tot} \frac{C_S}{(k_r + k_c/k_f) + C_S} = V_{max} \frac{C_S}{K_M + C_S}. \qquad (4.63)$$

Clearly, when $C_S \to \infty$, $dC_P/dt \to V_{max}$; when $C_S = K_M$, $dC_P/dt = \frac{1}{2}V_{max}$. Therefore, $V_{max} = k_c C_{E,tot}$ is referred to as the reaction rate at saturation substrate concentration, whereas $K_M = (k_r + k_c)/k_f$, known as Michaelis–Menten coefficient, is the substrate concentration at which the reaction velocity is $\frac{1}{2}V_{max}$. Expressions with a similar format also have been used for modeling plant vulnerability curves (Eq. (4.16)) and crop yields (Eq. (10.39)).

4.10 (C$_4$ Photosynthesis)   In C$_3$ plants the primary product of photosynthesis is a three-carbon sugar, while in C$_4$ species it is a four-carbon compound. As opposed to the situation in C$_3$ plants, in C$_4$ plants carbon dioxide and oxygen do not compete for the same enzyme, so that they have higher CO_2 concentrations inside the stomata and higher CO_2 fixing rates. This is generally associated with the higher photosynthetic rates and water-use efficiencies in C$_4$ plants (Campbell and Norman, 1998).

The biochemical and physiological differences between C$_3$ and C$_4$ plants have important consequences in ecological and hydrological processes. C$_3$ plants tend to be more active during colder and humid growing seasons, whereas C$_4$ plants are more active for warmer and drier conditions. Regions with warm-season rainfall tend to have greater C$_4$ abundance than do regions with cool-season rainfall (Lambers et al., 1998; Larcher, 2003).

4.11 (CAM Photosynthesis)   A second variation on the C$_3$ pathway is the Crassulacean acid metabolism (CAM) photosynthetic pathway. Like the C$_4$ pathway, the CAM pathway relies on an additional carbon fixation step, but it is unique in its temporal regulation. During the night, when transpiration drivers are low, CAM plants open their stomata to take up and store carbon dioxide. During the day, this carbon dioxide is fixed as sugars in the Calvin cycle. In order to do this, CAM plants control the timing of chemical reactions and stomatal opening via an endogenous circadian rhythm (Lüttge and Beck, 1992; Bartlett et al., 2014; Hartzell et al., 2015).

The CAM pathway is present in many cacti and succulent species, including common crops such as agave, pineapple, and prickly pear. Because of their very high water-use efficiency, these plants make up almost 50% of the biomass in certain arid and semi-arid regions of the world (Syvertsen et al., 1976).

4.12 (Evolution of Photosynthetic Pathways)   The evolution of photosynthesis pathways (C$_3$, C$_4$, CAM) is a fascinating topic. Interestingly, the evolution of car engines has undergone a similar path (combustion, turbo-charger, and hybrid engines) (Hartzell et al., 2018a). In both cases the main core process has remained the same. In plants, the basic process is the Calvin cycle in all photosynthesis pathways, while in cars it is the internal combustion engine (typically an Otto or Diesel cycle). With the so-called carbon pump, C_4 plants add a CO_2-concentrating mechanism and turbo-charged engines add an oxygen-concentrating mechanism via a turbo-charger. Finally, CAM plants store carbon dioxide as malic acid in vacuoles at night, while hybrid vehicles use a battery to store energy under favorable conditions.

4.13   Solve graphically and numerically the coupled system formed by the soil–root–plant water transport equation $E = g_{srp}(\psi_s - \psi_l)$ and the stomatal water transport equation $\rho_w E = g_s \rho[q_i(T_l, \Psi_l) - q_0]$, as shown in Fig. 4.12, but with constant g_{srp} and g_s.

4.14   Sketch graphically the implications of a reduction in g_{srp} and g_s with leaf water potential on the graphical solution of the previous problem.

5 The Atmosphere

Chi ha provato il volo camminerà guardando il cielo, perché là è stato e là vuole tornare.

You will forever walk the earth with your eyes turned skyward, for there you have been, and there you will always long to return.

Leonardo da Vinci

The primary goal of this chapter is to describe and model the fluxes of water and energy between the land and the atmosphere, with a focus on the role of atmospheric conditions on the soil–plant system. From a modeling viewpoint, the surface energy balance provides the extra constraint to obtain the surface temperature, which is necessary to compute the sensible and latent heat fluxes and therefore the evapotranspiration. We also extend our description to the energy and water balances of the whole atmospheric boundary layer (ABL). With suitable simplifying assumptions, this provides a dynamical system for the evolution of the ABL and the water and energy fluxes between the surface and the atmosphere as a function of the radiation input as well as of the surface and free atmosphere conditions. In this manner, the full extent of land–atmosphere feedbacks is taken into account in defining the evolution of diurnal hydrometeorological conditions.

5.1 Land–Atmosphere Exchanges of Water and Energy

During the day, the input of energy from the sun drives the uptake of carbon by photosynthesis and the water-vapor losses by evapotranspiration from the vegetated land surface; the latent heat from the land to the atmosphere, related to the evapotranspiration, is coupled to the sensible heat flux, which typically is also positive because of surface warming by solar radiation (e.g., Fig. 5.1). Nighttime conditions, typically characterized by water and energy exchanges of lower intensity as well as by CO_2 emissions because of soil and plant respiration, are not considered here (see Note 5.15).

In the first part of this chapter we will focus mostly on transpiration from land uniformly covered by a vegetation canopy, which will be modeled according to the big-leaf approximation. In these conditions, the soil evaporation is less important and we will simply refer to transpiration and evapotranspiration without distinction as E. When the atmospheric conditions are known within the logarithmic layer, just above the surface, the evapotranspiration fluxes can be computed using the SPAC formulation of Chapter 4, as long as the evaporative surface temperature is known via the surface energy balance.

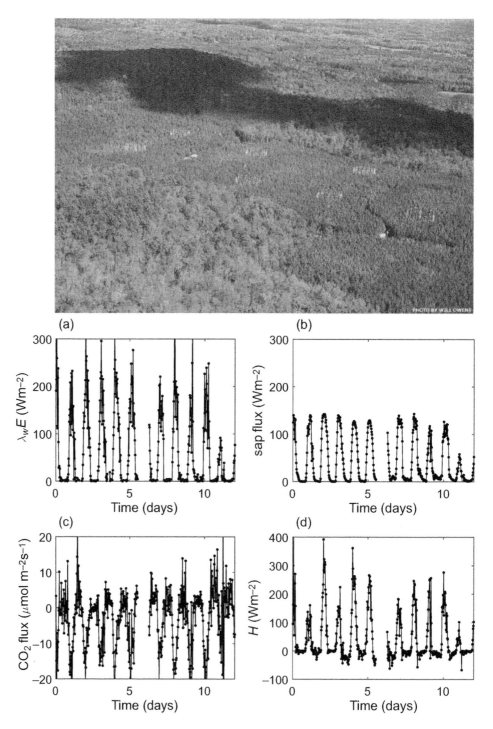

Figure 5.1 Upper panel: Loblolly pine (*Pinus taeda*) forest surrounded by hardwood forest in the fall, at the Duke Forest research site. The rings of the CO_2-enriched atmosphere (the FACE experiment) are visible (http://face.env.duke.edu/main.cfm). Driven by turbulent winds during the relatively warm day, the forested land surface is emitting energy as sensible and latent heat fluxes, as well as water vapor into the atmosphere while absorbing carbon dioxide from it. Photograph courtesy of Roni Avissar. Lower panels: The daily evolution of surface fluxes measured by the eddy-covariance method at the average canopy height in Duke Forest during a sequence of mild-temperature days with no rainfall. After Lai et al. (2000).

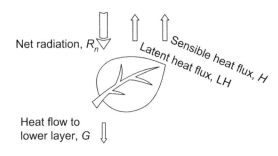

Figure 5.2 Main fluxes in the surface energy balance.

5.1.1 Surface Energy Balance

A simplified version of the surface energy balance (Fig. 5.2), assuming instantaneous-equilibrium conditions (i.e., no heat storage at the surface), is given by the balance of the net radiation input, R_n, the loss of sensible heat, H, the latent heat flux, $\mathrm{LH} = \lambda_w \rho_w E$, and conduction to the lower soil layers, G, as

$$R_n = H + \mathrm{LH} + G = H + \lambda_w \rho_w E + G, \tag{5.1}$$

where all the fluxes are in units of power per unit surface, e.g., W m^{-2}.

The *net radiation* is given by the balance of shortwave and longwave components:

$$R_n = (1 - \alpha_s) R_s + R_{ld} - R_{lu}, \tag{5.2}$$

in which α_s is the surface albedo, R_s is the shortwave radiation flux reaching the surface (roughly 50%–60% of the extraterrestrial solar radiation at the particular time of the day and year), R_{ld} is the longwave radiation flux downward from the atmosphere, and R_{lu} is the upward longwave radiation flux from the surface. The latter fluxes tend to be relatively constant during the day and night (e.g., Fig. 5.3) and the corresponding emissivity and absorptivity of the Earth's surface are assumed to be unity.

The time variability of the incoming shortwave radiation, $R_s(t)$, drives the temporal trends of transpiration and photosynthesis. For clear sky, it can be computed on the basis of astronomical and topographic considerations (Dingman, 2015); for simplicity, it may be modeled as a parabolic function of time:

$$R_s(t) = \begin{cases} -R_{s,\max} \dfrac{4}{(t_s - t_r)^2}(t - t_s)(t - t_r) & t_r \leq t \leq t_s, \\ 0 & \text{otherwise,} \end{cases} \tag{5.3}$$

where $R_{s,\max}$ is the maximum daily irradiance, and t_r and t_s are the times of sunrise and sunset.

The heat flow G in Eq. (5.1) tends to be less important for canopies, owing to their relatively low heat capacity, but becomes more relevant for soil and is a dominant term in the surface energy balance of lakes and seas. For simplicity, G can be assumed to be $\sim 0.1 R_n$ for grass, $\sim 0.3 R_n$ for bare soil, and $\sim 0.2 R_n$ for crops like maize (Brutsaert, 2005). When it is incorporated into the *net radiation* term, $Q = R_n - G$, the surface energy balance, Eq. (5.1), becomes simply

$$Q = H + \mathrm{LH} = H + \lambda_w \rho_w E. \tag{5.4}$$

The net available energy could also be approximated as a parabolic function by replacing $R_{s,\max}$ in Eq. (5.3) with $Q_{\max}$. Note that λ_w and c_p are mass specific parameters in this chapter.

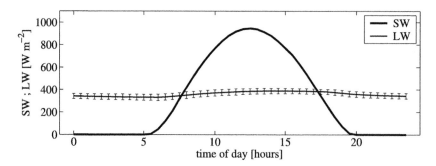

Figure 5.3 Diurnal evolution of the short- and longwave (SW, LW) radiation fluxes. The vertical bars on the longwave curve represent one standard deviation and are derived from measurements above the canopy at the Duke Forest site during July and August. The shortwave radiation is the clear-sky theoretical value for the Duke Forest latitude. After Siqueira et al. (2009).

An important indicator of how the incoming radiation is partitioned at the surface is provided by the *Bowen ratio*,

$$\mathrm{Bo} = \frac{H}{\mathrm{LH}} = \frac{H}{\lambda_w \rho_w E}. \tag{5.5}$$

For well-watered conditions, it is on the order of 0.2–0.5, but it can be greater and even much larger than one in conditions of water stress and drought, when stomatal closure and increased resistance to water flow from the soil to the atmosphere may dramatically reduce the denominator in Eq. (5.5). In general, the Bowen ratio varies from low values in deciduous forest and agricultural sites to high values in the dry regions within Mediterranean climates (see Fig. 5.5). Evaporative fraction is another useful indicator, which is defined as the ratio of the latent heat flux and the net available energy and is related to the Bowen ratio as $1/(\mathrm{Bo} + 1)$ (Brutsaert, 2005).

5.1.2 Albedo

The surface albedo (α_s), appearing in the surface energy balance (5.2), is an important property affecting evapotranspiration and plant carbon assimilation through its control of leaf temperature. It is the ratio of the global reflected solar radiative fluxes and the flux of the corresponding incident radiation, both integrated over the entire spectrum of solar radiation.

The surface albedo is quite strongly related to the solar zenith angle as well as the type of vegetation cover. As illustrated in Fig. 5.4, forests usually have lower albedo and thus absorb more solar radiation than the corresponding grassland. This provides an interesting example of how plants may affect their environmental conditions. Typical values of albedo for some other types of surfaces are reported in Table 5.1.

5.2   Evapotranspiration

Having described the surface energy balance, we can now proceed with the analysis of evapotranspiration. We begin with the case of evaporation from a wet surface, where the evaporation at zero water potential

Table 5.1 Surface albedo (data from Webster, 1994; Brutsaert, 2005).

Surface	Albedo (α_s)
Earth (average global)	0.33
Clouds (cirrus–stratocumulus)	0.1–0.9
Ice/snow (fresh)	0.6–0.9
Ice/snow (old)	0.4–0.6
Deep water	0.04–0.08 ‡
Soil – moist dark (ploughed fields)	0.05–0.15
– gray soil (bare fields)	0.15–0.25
– dry soil (desert)	0.20–0.35
Vegetation (see Fig. 5.4)	0.1–0.3
Venus	0.71
Mars	0.17

‡ Oceans have low albedo, which combined with their high water thermal capacity, determines a high absorption and storage of heat.

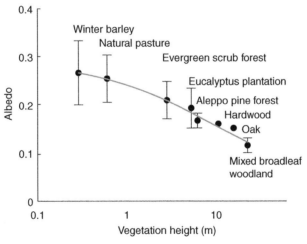

Figure 5.4 Albedo as a function of vegetation height. Adapted from Stanhill (1970).

allows a simple and instructive mathematical formulation. Then we deal with transpiration from vegetation, where evaporation takes place in conditions of negative leaf water potential and, because of the stomata, there is a different overall resistance to the sensible and latent heat fluxes.

5.2.1 Evaporation from Wet Surfaces

We consider first the special case of wet surfaces, for which the air at the surface may be assumed to be in saturated conditions and in quasi-equilibrium with the underlying liquid water surface at temperature T_0. Moreover, unlike transpiration through the leaves, the conductances to evaporation and to sensible heat flux are the same since there is no stomatal resistance. Thus, the evaporation flux is (see Eq. (2.155) in Sec. 2.3)

$$\rho_w E = \frac{\epsilon}{p_0}\rho g_a[e_{\text{sat},0}(T_0) - e_a], \qquad (5.6)$$

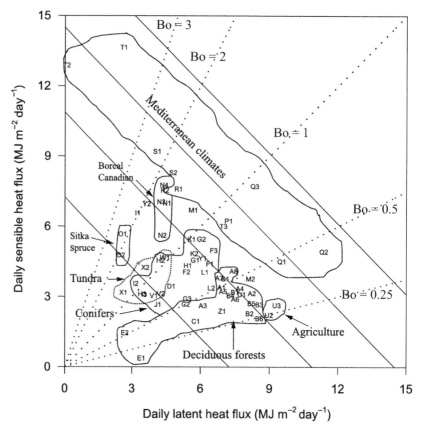

Figure 5.5 The daily cumulative sensible heat flux in MJ m^{-2} day^{-1} versus the daily cumulative latent heat flux in MJ m^{-2} day^{-1} between days 165 and 235 for the FLUXNET sites. Lines of constant Bowen ratio (dotted lines) and lines of constant total available energy (solid diagonal lines) are shown. Approximate delineations around groups of sites indicate the different vegetation types and climates. After Wilson et al. (2002).

where ϵ (≈ 0.622) is the ratio of the gas constant for dry air to that of water vapor (see Eq. (2.75)), and $e_{\text{sat},0}(T_0)$ is the saturation vapor pressure at T_0, given by the Clausius–Clapeyron equation (see Eq. (2.48)) in Sec. 2.1.8).

The energy balance (5.4) requires that

$$\lambda_w \rho_w E = Q - H, \tag{5.7}$$

which combines with sensible heat flux (see Eq. (2.153) in Sec. 2.3) to set a energy constraint on the evaporation process,

$$E = \frac{Q - \rho c_p g_a (T_0 - T_a)}{\lambda_w \rho_w}. \tag{5.8}$$

Equations (5.6) and (5.8) furnish a closed system of equations in E and T_0 once the available energy Q and the atmospheric conditions e_a and T_a are known.

Figure 5.6 shows the behavior of the evaporation, as a function of surface temperature for fixed atmospheric temperature, obtained by numerically solving the coupled system (5.6) and (5.8) for different values

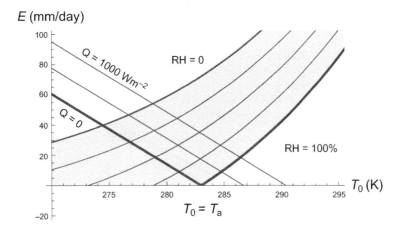

Figure 5.6 The evaporation from a wet surface, given by the intersection of energy balance and water vapor flux, plotted as a function of surface temperature for fixed $T_a = 283$ K. The curves with positive slopes are the turbulent transfer water vapor fluxes for different values of relative humidity (RH = 100%, 75%, 50%, 25%, 0% in the shaded area), while the decreasing straight lines are the evaporation flux from the energy balance equation (5.8) for different values of the net energy: $Q = 0, 500$, and 1000 W m^{-2}.

of relative humidity and available energy. According to Eq. (5.6), a higher surface temperature (T_0) tends to correspond to more evaporation owing to the larger saturation vapor pressure (e_0) (see the curves with positive slopes in Fig. 5.6). However, according to Eq. (5.8), higher T_0 also tends to correspond to larger sensible heat ($H = \rho c_p g_a (T_0 - T_a)$), so that less energy is available for evaporation (see the lines with negative slopes in Fig. 5.6). The crossing points of these two groups of curves and lines, which satisfy both the continuity and energy constraints (i.e., Eqs. (5.6) and (5.8)), provide both the evaporation rate and surface temperature for the given atmospheric conditions.

The system of Eqs. (5.6) and (5.8) can also be used to explain the behavior of the Bowen ratio. Dividing these two equations by the Bowen ratio, defined in Eq. (5.5), one obtains

$$\text{Bo} = \frac{c_p (T_0 - T_a)}{\lambda_w (q_0 - q_a)} = \frac{c_p p_0 (T_0 - T_a)}{\epsilon \lambda_w (e_0 - e_a)}. \tag{5.9}$$

Note that a multiplicative factor of g_s / g_{sa} would be needed in Eq. (5.9) to account for the stomatal resistance in non-wet or vegetated surfaces. Figure 5.7 shows contour plots of the Bowen ratio as a function of air temperature and humidity for a wet surface. The Clausius–Clapeyron equation (see Sec. 2.1.8) is used to obtain the saturation water vapor pressure under varying atmospheric temperature ($e_{\text{sat},a}(T_a)$, solid line) and for a fixed wet-surface temperature ($e_{\text{sat},0}(T_0)$, dashed line). The bold solid line sets the upper limit of the atmospheric vapor pressure ($e_a \leq e_{\text{sat},a}$), while the dashed line provides a boundary between the condensation ($e_a > e_0$) and evaporation ($e_a < e_0$) processes.

5.2.2 Combination Approach: the Penman Equation

As discussed in the previous subsection, the system of equations (5.6) and (5.8) only requires a knowledge of the incoming radiation and the atmospheric temperature and humidity, but does not need direct knowledge of the surface temperature, which is determined by the system itself. However, because of the nonlinearity of the $e_{\text{sat}}(T)$ curve, the system requires numerical solution. To avoid this, an analytical

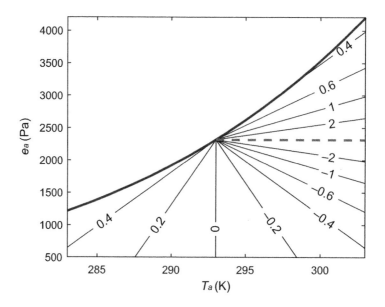

Figure 5.7 Contour plot of Bowen ratio as a function of air temperature, T_a, and water vapor pressure, e_a, for a wet (i.e., saturated) surface with a fixed temperature of 293 K. The solid lines indicate the saturation water vapor pressure at the given atmospheric temperature; on these lines are marked the values of the equilibrium Bowen ratio (see Eq. (5.21) and Note 5.1). The dashed line marks the saturation water vapor of the wet surface, which separates condensation (above the line) from vaporization (below the line).

approximation was proposed by Penman (1948), who combined Eqs. (5.6) and (5.8) into a single equation for the evaporation flux. For this purpose, Eq. (5.6) can be written as

$$\rho_w E = g_a \rho \frac{\epsilon}{p_0} [e_{\text{sat},0}(T_0) - e_{\text{sat},a}(T_a) + (e_{\text{sat},a}(T_a) - e_a)], \tag{5.10}$$

or, introducing the vapor pressure deficit (VPD) (see Eq. (2.70) in Sec. 2.1.10),

$$\rho_w E = g_a \rho \frac{\epsilon}{p_0} [e_{\text{sat},0}(T_0) - e_{\text{sat},a}(T_a) + \text{VPD}]. \tag{5.11}$$

Penman eliminated the dependence on T_0 by introducing a local linearization of the Clausius–Clapeyron curve (Sec. 2.1.8) around the known atmospheric temperature, which is indicated as the slope of the saturation vapor pressure curve,

$$\Delta(T_a) \approx \frac{e_{\text{sat},0}(T_0) - e_{\text{sat},a}(T_a)}{T_0 - T_a}. \tag{5.12}$$

Equation (5.11) can be approximated as

$$\rho_w E = g_a \rho \frac{\epsilon}{p_0} [\Delta(T_a)(T_0 - T_a) + \text{VPD}]. \tag{5.13}$$

The surface temperature can then be eliminated as follows. We find the difference between the surface and atmospheric temperatures from Eq. (5.8),

$$T_0 - T_a = \frac{Q - \lambda_w \rho_w E}{g_a \rho c_p}.$$ (5.14)

This difference can be substituted into Eq. (5.13) to eliminate the surface temperature:

$$\rho_w E = g_a \rho \frac{\epsilon}{p_0} [\Delta(T_a) \frac{Q - \lambda_w \rho_w E}{g_a \rho c_p} + \text{VPD}].$$ (5.15)

Finally, solving for E and introducing the so-called psychrometric constant

$$\gamma^* = \frac{c_p p_0}{\epsilon \lambda_w},$$ (5.16)

one obtains the Penman equation (Penman, 1948)

$$\rho_w E = \frac{\Delta}{\lambda_w (\Delta + \gamma^*)} Q + \frac{\gamma^*}{\Delta + \gamma^*} E_A,$$ (5.17)

where E_A is usually referred to as the *drying power of air*:

$$E_A = \frac{\epsilon}{p_0} \rho g_a \text{VPD}.$$ (5.18)

The first term in Eq. (5.17) is also called the *equilibrium evaporation, E_{eq}*, and is the evaporation that would occur in conditions when the air near the surface and the air in the atmosphere are both saturated and the surface temperature of the system is kept higher than the atmospheric temperature by the incoming radiation,

$$\rho_w E_{\text{eq}} = \frac{\Delta}{\lambda_w (\Delta + \gamma^*)} Q.$$ (5.19)

In summary, the total evaporation from a wet surface,

$$\rho_w E = \rho_w E_{\text{eq}} + \frac{\gamma^*}{\Delta + \gamma^*} E_A,$$ (5.20)

is the sum of the equilibrium evaporation and the evaporation due to the drying power of the air. For well-watered conditions and at the daily timescale, Priestley and Taylor (1972) found that E is approximately 1.26 times E_{eq}; lower values have been reported in forests, while larger values have been observed in dry and cold regions.

Recalling the definition of the Bowen ratio in Eqs. (5.5) and (5.9), one can further define an equilibrium Bowen ratio as

$$\text{Bo}_{\text{eq}} = \frac{H}{\lambda_w \rho_w E_{\text{eq}}} = \frac{c_p}{\lambda_w} \frac{T_0 - T_a}{q_{\text{sat},0} - q_{\text{sat},a}} = \frac{c_p p_0}{\lambda_w \epsilon} \frac{T_0 - T_a}{e_{\text{sat},0} - e_{\text{sat},a}},$$ (5.21)

where the coefficient of the last term is the pychrometric constant γ^* (see Eq. (5.16)). Note that a multiplicative factor of g_s/g_{sa} is needed in Eq. (5.21) to account for the stomatal resistance in non-wet or vegetated surfaces (see Eq. (5.43)). This Bowen ratio is the upper limit as $Q \rightarrow \infty$ (Porporato, 2009) and it is also the value of the Bowen ratio when both surface and atmospheric air become saturated (see the solid line in Fig. 5.7 and Note 5.1).

Figure 5.8 Different pathways and conductances for the water and sensible heat fluxes, from the soil through the leaf to the atmosphere.

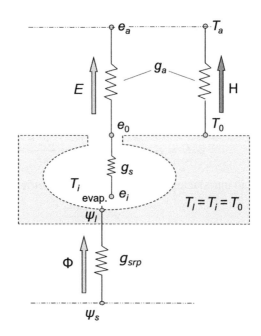

5.2.3 Transpiration from Vegetation

We are now ready to resume our program of calculating the transpiration that we started in Sec. 4.8.3, where we saw that the water transport laws through the SPAC only constrain transpiration to a function of the leaf temperature, which remains undetermined. Here we close the problem by coupling the water transport law to the energy balance.

Unlike the wet-surface case of the previous section, in the case of evapotranspiration from a non-wet surface, the water vapor at the surface, e_0 in Eq. (5.6), is not at saturation (i.e., it lies below the Clausius–Clapeyron curve, Sec. 2.1.8) and its value must be computed while taking into account the hydration state underlying such a surface. As seen in Chapter 4, the water vapor has to pass through the stomata, and thus through an additional conductance in series, before reaching the atmosphere. Therefore, the water vapor pressure at the surface depends on the value of the water potential in the leaves and its computation requires a coupling with the full SPAC dynamics. Figure 5.8 shows the different pathways for the sensible and latent heat fluxes (and thus for the transpiration) in the case of leaf transpiration.[1]

As seen in Chapter 4, the flux of liquid water from the soil to the leaves equals the diffusion and the turbulent flux of water vapor from the stomatal pores to the logarithmic layer (recall that for vegetated surfaces we assume $E \sim E_t$). Accordingly (the equations are repeated here for convenience), the soil-to-leaf water flux is given by Eq. (4.20),

$$E = g_{\text{srp}}[\psi_s(s) - \psi_l] \tag{5.22}$$

and the leaf-to-atmosphere flux is given in Eq. (2.155),

---

[1] The leaf surface – also called the cuticle – and the closed stomata themselves are not perfectly impervious to water fluxes, so that a small amount of transpiration may still take place even with the stomata "completely" closed. To account for this, as already noted in Sec. 4.8, it may be useful to add a so-called *cuticular conductance* in parallel with the stomatal conductance.

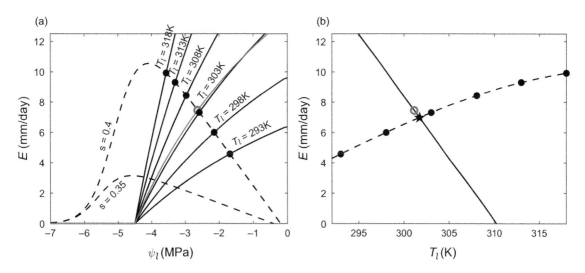

Figure 5.9 Graphical representation of the steps in the numerical calculation of the transpiration. In (a), the dashed lines are plotted from Eq. (5.22), the bold solid lines are from Eq. (5.23), and the thin solid line is from Eq. (5.27). In (b), the solid line is from Eq. (5.24), and the dashed line connects the dots obtained in (a). The star shows the final result for the transpiration rate from leaves as the solution of Eqs. (5.22), (5.23), and (5.24). The open circle is the solution using the Penman–Monteith approach with Eqs. (5.22) and (5.27). The air temperature is set as 293 K, the relative humidity is 30%, the net radiation is 400 Wm^{-2}, the soil moisture is 0.4, and the other SPAC parameters used in the calculation are listed in Table 4.3.

$$\rho_w E = \rho g_{sa} \frac{\epsilon}{p_0} [e_i(\psi_l, T_l) - e_a], \qquad (5.23)$$

where the partial pressure of water vapor inside the leaf, e_i, is related to the leaf water potential ψ_l and the leaf temperature T_l (see Eq. (5.70) and also Sec. 2.1.14). As shown in Fig. 4.12, these two equations are not closed because of the presence of the undetermined leaf temperature. Closure is now obtained by coupling them with the leaf energy balance (see Eq. (5.8));

$$E = \frac{Q - \rho c_p g_a(T_l - T_a)}{\rho_w \lambda_w}. \qquad (5.24)$$

Equations (5.22), (5.23), and (5.24) can be solved as a system for E, T_l, ψ_l given the conditions of the soil (ψ_s), atmosphere (T_a, e_a), and radiation (Q) and the conductances within the SPAC (g_{srp}, g_{sa}, g_a).

As we previously did for Fig. 4.12, we use Eq. (5.22) for the soil–root–plant system to link E and ψ_l under different soil moisture conditions, s (see Table 4.3 for the SPAC parameters); this is shown by the dashed lines in Fig. 5.9a. The relationship is then redrawn by using Eq. (5.23) for the plant–atmosphere part under different leaf temperatures (see the solid lines in Fig. 5.9a). The intersections of these two sets of lines, satisfying the soil–root–plant–atmosphere continuum constraints, give the solutions of E under various values of T_l and of the constant s. These solutions are reported in Fig. 5.9b (dashed lines), where the energy balance equation (5.24) is also used to plot the relationship between E and T_l (solid lines). The intersection obtained by satisfying both the continuity and energy constraints finally gives the solutions in terms of the evaporation, E, and leaf temperature, T_l (the star in Fig. 5.9b), for the specific conditions $s = 0.4$, $T_a = 293$ K, and RH$_a = 30\%$.

5.2.4 Penman–Monteith Equation

Monteith applied Penman's combination approach (see Sec. 5.2.2) to the case of non-wet surfaces considering, for simplicity, the water vapor inside the stomates to be in a saturated condition (Jones, 1992). Following the steps leading to Penman's equation (see Sec. 5.2.2), one can write Eq. (5.23) as

$$\rho_w E = \rho g_{sa} \frac{\epsilon}{p_0}[e_i(\psi_l, T_l) - e_{sat,a} + (e_{sat,a} - e_a)]. \tag{5.25}$$

While g_{sa} still depends on ψ_l, we set $\psi_l = 0$ for the function $e_i(\psi_l = 0, T_l)$, so that $e_i = e_{sat,i}(T_l)$. Although this assumption breaks the SPAC continuum, it introduces only a slight overestimation of the transpiration, which, combined with the underestimation due to the Penman linearization, is not a big concern (see an example in Fig. 5.9).[2] The reduction in transpiration due to the reduction in plant water potential is mainly through stomatal closure and the related drop in stomatal conductance, which is still retained in g_{sa}.

   The assumption of saturated water vapor inside the stomata allows us to follow the Penman combination approach, which couples Eq. (5.25) to the energy balance equation (5.24). Similarly to (5.15), we have

$$\rho_w E = g_{sa} \rho \frac{\epsilon}{p_0} \left[\Delta(T_a) \frac{Q - \lambda_w \rho_w E}{g_a \rho c_p} + \text{VPD} \right], \tag{5.26}$$

which provides the *Penman–Monteith equation* on solving for E:

$$\rho_w E = \frac{\Delta}{\lambda_w \left[\Delta + \gamma^* \left(1 + \frac{g_a}{g_s} \right) \right]} Q + \frac{\gamma^*}{\Delta + \gamma^* \left(1 + \frac{g_a}{g_s} \right)} E_A, \tag{5.27}$$

where E_A is given by Eq. (5.18). For $g_s \gg g_a$, the case of no stomata and a wet surface, (5.27) corresponds to the Penman equation (5.17), as expected.

   The Penman–Monteith equation (5.27) combines the water flux equation (5.23) with the energy balance equation (5.24). If the stomatal conductance, g_s, is given, such an equation can speedily be used to estimate the transpiration (Dingman, 2015). If g_s is not available, the Penman–Monteith equation still needs to be coupled to the SPAC equation (5.22). As shown in Fig. 5.9, the solution of the Penman–Monteith approach (open circle) satisfies the energy balance constraint (the open circle is on the solid line in Fig. 5.9b) and it is close to the full numerical solution presented in Sec. 5.2.3 (star). Thus, in normal conditions the Penman–Monteith equation gives satisfactory solutions in spite of the assumption of vapor saturation.[3] However, this approximation should be used with care under conditions of leaf or stomatal damage, which may result in nonnegligible conductances even at low water potentials (Sec. 4.8).

   With the determination of leaf temperature, allowed by the coupling of the leaf energy balance and the SPAC equations, the calculation of plant carbon assimilation also becomes possible (during the day CO_2 goes from the atmosphere to the leaves, so the carbon flux from the soil to the atmosphere is the

---

[2] In fact, the ratio of the partial pressure of water vapor, e_i, and its value at saturation, $e_{sat,i}$, is given by Eq. (5.70) (see also Sec. 2.1.14). This ratio is only 0.95 for $\psi_l = -6.92$ MPa at 20 °C (Jones, 1992, p. 157). Moreover, at low ψ_l, when $e_i(T_l, \psi_l) \leq e_{sat,i}(T_l)$, the transpiration is reduced and an effective reduction can be performed by an earlier stomatal closure.

[3] Note that saturation here refers to saturation of the water vapor as in the Clausius–Clapeyron equations in the stomatal cavities (Sec. 2.1.8) and not necessarily to unsaturation (i.e., the presence of air) in the mesophyll, considered as a porous medium. Interaction with cell walls, capillary interfaces, and the presence of osmotic components make it possible to have unsaturated vapor in contact with an unsaturated porous medium (see Secs. 2.1.12 and 2.1.13).

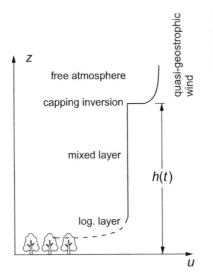

Figure 5.10 Schematic representation of the lower atmosphere with the mean velocity profile.

negative of the net carbon assimilation; see Fig. 5.1). Thus, the equations for carbon flux and assimilation, Eqs. (4.21) and (4.26) respectively, along with the water-transport equations, can be closed with the leaf energy balance to compute A_n and g_{s,CO_2} once the radiation and the soil and atmospheric conditions are known.

5.3 Dynamics of the Atmospheric Boundary Layer (ABL)

The atmospheric conditions (velocity, temperature, and humidity) in the logarithmic layer were considered as known quantities in the previous formulae for the heat and water vapor fluxes. Their diurnal evolution is controlled by the dynamics of the *atmospheric boundary layer* (ABL), also known as the planetary boundary layer (PBL). When these quantities are not available from direct observation, the ABL dynamics can be modeled in a relatively simple way by adding suitable conservation equations for energy and water vapor in the entire ABL. Not only does this allow us to reconstruct the diurnal evolution of the ABL from the sole knowledge of the surface and free atmosphere conditions (including the net energy input R_n), but it also provides us with a mathematical model to analyze the most fundamental land–atmosphere feedbacks, including the role of ecohydrological processes in the partitioning of the incoming radiation energy, the growth of the ABL, and the related meteorological dynamics (e.g., convective precipitation).

5.3.1 ABL Characteristics

The atmospheric boundary layer can be schematically divided into two parts (Fig. 5.10): a surface layer comprising the canopy and the logarithmic layer (see Sec. 2.3.2) and a *mixed layer* above it. The ABL is separated from the free atmosphere by a relatively thin capping inversion, or *inversion layer*, which can often be simplified as a sudden jump in the atmospheric state (e.g., in the velocity, potential temperature, humidity).

The ABL grows during the day, driven by inputs of energy provided by turbulent shear and the sensible heat flux (e.g., buoyancy). During a warm clear day at mid-latitude continental regions, the ABL may reach heights of the order of 1500–2000 m in the mid-afternoon. It then typically collapses in the early evening, leaving a residual layer which evolves into a shallow, stable boundary layer during the night (Garratt, 1994). When the atmosphere is very moist and unstable, instead of collapsing the ABL may evolve into towering cumulus, which could give rise to convective precipitation.

From a fluid dynamic point of view, the ABL is dominated by turbulence, which is responsible for strong mixing, resulting in a flat velocity profile in the mixed layer (see Fig. 5.10). In contrast, the *free atmosphere* is only very slightly turbulent and can be thus treated as an ideal potential flow in quasi-geostrophic conditions (apart from when fronts are developing; see Holton, 2004). The free atmosphere determines the boundary conditions of wind, humidity, and temperature for the ABL and in turn for the SPAC. Its evolution is controlled by synoptic weather patterns, changing on timescales of the order of two days to weeks. Apart from conditions of strong convective instability,[4] the hydrostatic approximation is usually valid for both the free atmosphere and the ABL (Holton, 2004).

5.3.2   Evolution Equations for the ABL

In the absence of water phase changes, the potential temperature (ϑ, see Eq. (2.65)) and specific humidity (q) are conserved quantities, which may be assumed to be uniform throughout a well-mixed slab of air of thickness h (see Fig. 5.11). The governing equations for these quantities were first developed by Tennekes (1973), Betts (1973), Carson (1973) and others; see Garratt (1994).

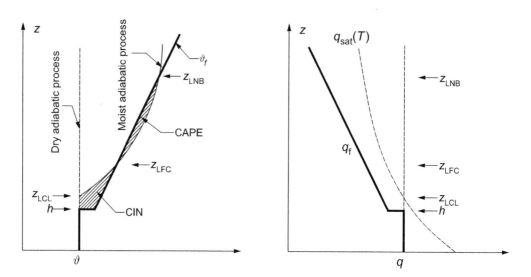

Figure 5.11 Schematic representation of profiles of the ABL and adiabatically lifted air parcels. The shaded areas represent the convective available potential energy (CAPE) and the convective inhibition energy (CIN) (see Sec. 5.3.4). Note that a more precise evaluation of CAPE and CIN requires the use of the virtual potential temperature (or virtual temperature, see Eq. (2.74)) to quantify the density of the moist air (Doswell III and Rasmussen, 1994). Adapted from Yin et al. (2015).

---

[4] We assume that the ABL is neutrally buoyant in order to compute the surface fluxes using a logarithmic profile (see Sec. 2.3). The corrections to the logarithmic profile for stable or unstable conditions are discussed, for example, in Brutsaert (2005) and Katul et al. (2011).

Replacing the conserved variable b in Eq. (2.115) with ϑ and q and integrating it with respect to z throughout the boundary layer, one obtains

$$h\frac{d\vartheta}{dt} = \overline{(\vartheta'w')}_0 - \overline{(\vartheta'w')}_h,$$
$$h\frac{dq}{dt} = \overline{(q'w')}_0 - \overline{(q'w')}_h, \tag{5.28}$$

where the overbars for ϑ and q are not indicated, following conventional practice. The left-hand side terms in Eq. (5.28) are the storage terms, the first terms on the right-hand sides are the surface sensible and latent heat fluxes,

$$\rho c_p \overline{(\vartheta'w')}_0 = H,$$
$$\rho \overline{(q'w')}_0 = \rho_w E, \tag{5.29}$$

and the last terms are the entrainment fluxes, which can also be obtained by integrating Eq. (2.115) with respect to z across the inversion layer:

$$\frac{dh}{dt}[\vartheta_f(h) - \vartheta] = -\overline{(\vartheta'w')}_h,$$
$$\frac{dh}{dt}[q_f(h) - q] = -\overline{(q'w')}_h, \tag{5.30}$$

where $\vartheta_f(h)$ and $q_f(h)$ are the potential temperature and the specific humidity in the free atmosphere at altitude h.

To close the system, one needs to parameterize the entrainment flux by known variables. In a cloud-free atmosphere, the boundary-layer turbulence is often driven by surface buoyancy flux and shear stress so that, to a good approximation, the entrainment flux is given as[5] (Tennekes, 1973; Garratt, 1994)

$$-\overline{(\vartheta'w')}_h = A\frac{T_0 u_*^3}{gh} + \beta\overline{(\vartheta'w')}_0, \tag{5.31}$$

where T_0 is the surface temperature, u_* is the friction velocity, the coefficient A is approximately 2.5, and β has a typical value of 0.2. The first term on the right-hand side is due to mechanical turbulence and the second term is due to buoyancy-generated turbulence. The former is less important in warm seasons and typically may be neglected.

With these approximations, Eqs. (5.28)–(5.31) for the ABL can be reorganized as follows:

$$\rho c_p h\frac{d\vartheta}{dt} = H + \rho c_p[\vartheta_f(h) - \vartheta]\frac{dh}{dt}, \tag{5.32}$$

$$\rho h\frac{dq}{dt} = \rho_w E + \rho[q_f(h) - q]\frac{dh}{dt}, \tag{5.33}$$

$$\beta\frac{H}{\rho c_p} = [\vartheta_f(h) - \vartheta]\frac{dh}{dt}. \tag{5.34}$$

Such equations, when coupled to the surface energy balance and evapotranspiration equations in the previous sections, form a closed system, which succinctly describes the impacts of vegetation and the

---

[5] To account for the effect of water vapor on the buoyancy of air, the potential temperature in Eq. (5.31) needs to be replaced by the virtual potential temperature $\vartheta_v = \vartheta[1 + (R_w/R_d - 1)q]$, which is the temperature of dry air with the same density and pressure as that of the given mixed air (also see the virtual temperature in Eq. (2.74)). This correction, however, is not very important and can also be partially effected by adjusting the value of β for certain atmospheric conditions (see Rigby et al., 2015).

hydrological processes on the ABL dynamics. For example, taller vegetation with its lower albedo (see Fig. 5.4) tends to have higher available energy (see Eq. (5.2)) along with lower aerodynamic resistance because of larger surface roughness, facilitating the transport of latent and sensible fluxes (see Eqs. (5.23) and (5.24)), and these effects impact the dynamics of the ABL through Eqs. (5.32)–(5.34). One can also trace the role of soil moisture through the leaf water potential, assimilation rate, and stomatal conductance (see Chapter 4), which in turn controls the evaporative flux and the water content in the ABL (see Eqs. (5.23) and (5.33)). This SPAC–ABL coupling is analyzed in Sec. 5.4.

5.3.3 Analytical Approximation to the ABL Dynamics

Approximate analytical solutions of the ABL evolution provide useful insight. Following Rigby et al. (2015), we begin with the assumption of linear free atmospheric profiles (e.g. Tennekes, 1973; Garratt, 1994):

$$\vartheta_f(z) = \gamma_\vartheta z + \vartheta_{f0}, \tag{5.35}$$

$$q_f(z) = \gamma_q z + q_{f0}, \tag{5.36}$$

where γ_ϑ and γ_q are the slopes of the potential temperature and specific humidity in the free atmosphere (γ_ϑ is also known as the lapse rate), and ϑ_{f0} and q_{f0} are the profile intercepts. Under this assumption and neglecting the morning transition ($\vartheta(h = 0) = \vartheta_{f0}$), the growth rate of the boundary layer can be derived from Eq. (5.31) as (Tennekes, 1973; Garratt, 1994)

$$\frac{dh}{dt} = \frac{(1 + 2\beta)H(t)}{\rho c_p \gamma_\vartheta h}. \tag{5.37}$$

Integrating Eq. (5.37) with respect to t, one can immediately find a solution for $h(t)$ as (Porporato, 2009)

$$h(t) = \sqrt{\frac{2(1 + 2\beta)}{\rho c_p \gamma_\vartheta} \int_0^t H(x)dx}, \tag{5.38}$$

where x is an integration variable.

The solution shows that at a given time the height of the ABL depends on the integrated history of the sensible heat flux. Explicit solutions for ϑ and q can also be obtained by combining Eqs. (5.37), (5.32), and (5.33) with the initial condition $\vartheta(t = 0) = \vartheta_{f_0}$ (Porporato, 2009; Rigby et al., 2015):

$$\tilde{\vartheta}(t) = \gamma_\vartheta \frac{1 + \beta}{1 + 2\beta} h(t) \tag{5.39}$$

$$\tilde{q}(t) = \frac{1}{\lambda_w \rho h(t)} \int_0^t Q(u)du + \frac{1}{2}\left[\gamma_q - \frac{c_p \gamma_\vartheta}{\lambda_w(1 + 2\beta)}\right] h(t), \tag{5.40}$$

where $\tilde{x} = x - x_{f_0}$ for any variable x. As can be seen, the potential temperature at any given time is proportional to the ABL height, which includes the integrated information about the sensible heat flux as shown in Eq. (5.38). The specific humidity is not only related to this integrated information about the sensible heat flux in terms of $h(t)$ but also to the latent heat flux, in the expression for the integrated available energy. These expressions show how the available energy and its partition into latent and sensible heat fluxes drive the growth and state of the ABL.

To make the system analytically tractable, g_s and g_a may be assumed to be approximately constant during the day (see examples of the diurnal variations in g_s in Fig. 5.14). The transpiration and leaf energy balance equations (5.23) and (5.24) can then be reorganized as

$$Q = H + \rho_w \lambda_w E = \rho c_p g_a(\vartheta_0 - \vartheta) + \lambda_w \rho g_{sa}[q_0 - q]$$
$$\approx \rho g_a c_p(\tilde{\vartheta}_0 - \tilde{\vartheta}) + \lambda_w \rho g_{sa}(\tilde{q}_{sat,0} - \tilde{q}) + \lambda_w \rho g_{sa} D_0, \tag{5.41}$$

where $q_0 \approx q_{sat,0}$ (see the Monteith approximation in Sec. 5.2.4), $D_0 = q_{sat}(\vartheta_{f_0}) - q_{f0}$ is the initial specific humidity deficit, and the near-surface potential temperature is equivalent to the near-surface temperature because $p = p_0$ (see Eq. (2.65)).

Equation (5.41) implicitly provides a solution for the surface temperature ϑ_0. One can solve it explicitly by adapting the Penman–Monteith approach (Sec. 5.2.2). Accordingly, by linearizing the saturation specific humidity curve around a suitable reference temperature,

$$\tilde{q}_{sat,0} \sim \frac{\epsilon \Delta}{p_0} \tilde{\vartheta}_0 \tag{5.42}$$

where Δ is the slope of the saturation vapor pressure curve (see Eq. (5.12)), one can rewrite the equilibrium Bowen ratio in Eq. (5.21) as

$$\mathrm{Bo}_{eq} = \frac{g_a c_p(\vartheta_0 - \vartheta)}{g_{sa} \lambda_w(q_{sat,0} - q_{sat})} = \frac{g_a c_p p_0}{g_{sa} \lambda_w \epsilon \Delta}. \tag{5.43}$$

Substituting Eqs. (5.42) and (5.43) into (5.41), one can find an explicit approximation for the surface temperature (Rigby et al., 2015):

$$\tilde{\vartheta}_0 \approx \frac{\mathrm{Bo}_{eq}}{1 + \mathrm{Bo}_{eq}}\left(\frac{Q}{\rho g_a c_p} + \frac{\lambda_w g_{sa}}{c_p g_a}\tilde{q} + \tilde{\vartheta} - \frac{\lambda_w g_{sa} D_0}{g_a c_p}\right). \tag{5.44}$$

Substituting this temperature approximation into Eqs. (5.37) and (5.41) along with the expression $\tilde{\vartheta}(h)$ and $\tilde{q}(h)$ in Eqs. (5.39) and (5.40) yields the following nonlinear ordinary differential equation for the ABL growth rate:

$$\frac{dh}{dt} = \underbrace{\frac{f(t)}{h}}_{\mathrm{I}} \underbrace{}_{\mathrm{II}} + \underbrace{\frac{g(t)}{h^2}}_{\mathrm{III}} + \underbrace{C}_{\mathrm{IV}}, \tag{5.45}$$

where

$$f(t) = \frac{1 + 2\beta}{\rho c_p \gamma_\vartheta} \frac{\mathrm{Bo}_{eq}}{1 + \mathrm{Bo}_{eq}}(Q(t) - \rho \lambda_w g_{sa} D_0), \tag{5.46a}$$

$$g(t) = g_{sa} \frac{1 + 2\beta}{\rho c_p \gamma_\vartheta} \frac{\mathrm{Bo}_{eq}}{1 + \mathrm{Bo}_{eq}} \int_0^t Q(u)du, \tag{5.46b}$$

$$C = \left(-\frac{1 + \beta}{1 + \mathrm{Bo}_{eq}}\right)g_a + \left(-\frac{c_p \gamma_\vartheta - \lambda_w \gamma_q(1 + 2\beta)}{2c_p \gamma_\vartheta} \frac{\mathrm{Bo}_{eq}}{1 + \mathrm{Bo}_{eq}}\right)g_{sa}. \tag{5.46c}$$

While Eq. (5.45) is difficult to solve analytically, one can gain some insight by numerically investigating each term. Figure 5.12 shows the result for the parameters listed in Table 5.2, representing typical summertime atmospheric sounding profiles under well-watered conditions in the central facility at Southern Great Plains (Atmospheric Radiation Measurement Program; http://www.arm.gov/). As can be seen, the

Table 5.2 Typical parameters for model testing derived from observations in the Central Facility, Southern Great Plains, USA. Adapted from Rigby et al. (2015).

Variable	Value	Unit
β	0.2	—
λ_w	2.45×10^6	$J\,kg^{-1}$
ρ	1.225	$kg\,m^{-3}$
c_p	1005	$J\,kg^{-1}\,K^{-1}$
g_s	0.014	$m\,s^{-1}$
g_{sa}	0.009	$m\,s^{-1}$
γ_ϑ	0.0065	$K\,m^{-1}$
ϑ_{f0}	293	K
γ_q	-3.3×10^{-6}	$kg\,kg^{-1}\,m^{-1}$
q_{f0}	0.014	$kg\,kg^{-1}$
Q_{max}‡	493	$W\,m^{-2}$

‡ $Q(t)$ is then modeled as a parabolic function, as in Eq. (5.3).

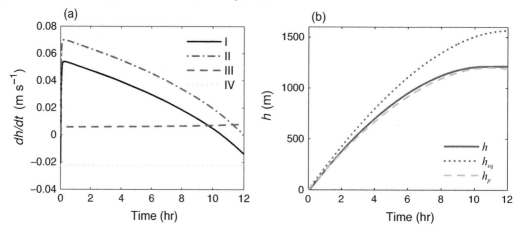

Figure 5.12 Simulation of boundary-layer dynamics with parameters from Table 5.2. (a) Evolution of each term in Eq. (5.45). (b) Evolution of ABL height predicted by the full numerical model (h), equilibrium solution of Eq. (5.48) (h_{eq}), and approximate solution of Eq. (5.50) (h_p). Adapted from Rigby et al. (2015).

diurnal variation of dh/dt, term I, is mostly captured by term II, while term III does not have a noticeable diurnal variation and term IV is constant as per Eq. (5.46c).

On the basis of the insights obtained from Fig. 5.12, we can initially assume that the terms III and IV are negligible, leading to a solution corresponding to the case of equilibrium evaporation. In such a case, Eq. (5.45) reduces to

$$\frac{dh}{dt} = \frac{f(t)}{h}. \tag{5.47}$$

Assuming that there is no morning transient ($h(0) = 0$) and no initial humidity deficit under well-watered condition ($D_0 = 0$), the solution is

$$h_{eq}(t) = \left[2 \int_0^t f(u)\mathrm{d}u\right]^{1/2}. \tag{5.48}$$

This "equilibrium" solution has exactly the same form as the solution for a mixed layer under constant equilibrium Bowen ratio (i.e. H in Eq. (5.38) is calculated as $\mathrm{Bo}_{eq}/(1 + \mathrm{Bo}_{eq})Q$; see Tennekes, 1973; Garratt, 1994; Porporato, 2009). The equilibrium Bowen ratio is the algebraic limit of the Bowen ratio as $Q \to \infty$ (Porporato, 2009). The overestimation of the Bowen ratio results in more surface sensible heat flux and a fast-growing boundary layer (the dotted lines in Fig. 5.12b).

To improve the equilibrium solution of Eq. (5.48), the sum of terms III and IV can be approximated by a constant C' after acknowledging the relatively invariant behavior of term III (the dashed line in Fig. 5.12a). With this, Eq. (5.45) becomes (Rigby et al., 2015)

$$\frac{dh}{dt} = \frac{f(t)}{h} + C'. \tag{5.49}$$

From a physical point of view, $f(t)/h$ contributes to the ABL growth through the equilibrium evaporation (E_{eq}), while the constant C' in Eq. (5.49) influences the ABL growth by evaporation due to the drying power of air (E_A, see Eq. (5.18)). From a mathematical point of view, Eq. (5.49) is a particular form of the Abel equation of the second kind, which is itself a generalization of the Riccati equation (Zaitsev and Polyanin, 2012). Unfortunately no general solution exists for this equation with arbitrary radiative forcing $f(t)$.

However, we can further assume that C' is a small term and write the solution $h(t)$ of Eq. (5.49) as a first-order perturbation, that is

$$h_p(t) = h_{eq}(t) + C't, \tag{5.50}$$

shown with a dashed line in Fig. 5.12b. The inclusion of C' adds the contribution from the vapor pressure deficit, which efficiently reduces the overestimation in the equilibrium solution (5.48).

The approximations in Eq. (5.49) extend the Penman approach to the entire ABL dynamics. The resulting ABL growth has two parts, similarly to the Penman equation (5.17): the first part is related to the equilibrium evaporation and the second part is linked to the drying power of the air. Recalling that the evaporation is roughly proportional to the equilibrium evaporation with the so-called Priestley–Taylor coefficient of 1.26 at a daily timescale (see Secs. 5.2.2 and 7.2.1), one may expect that this coefficient is closely related to h/h_p or the constant C', which essentially accounts for the entrainment of dry air from the top of the atmosphere and its effects on the boundary-layer states (De Bruin, 1983). Such an approximation allows us to better understand the roles of surface hydrology and atmospheric conditions in the coupled SPAC system, providing new perspectives on the land–atmosphere interaction.

5.3.4 Thermodynamics and Atmospheric Convection

When discussing the coupling between surface heat flux partitioning and the boundary-layer dynamics analyzed in the previous sections, we considered only clear-sky conditions. Here, we extend the analysis to the coupling between the ABL and atmospheric convection, thus providing a more comprehensive picture of land–atmosphere interactions. We follow Yin et al. (2015) as we review the thermodynamics of pseudoadiabatic processes; this is then used to derive the lifting condensation level (LCL) and the convective available potential energy (CAPE). These indices are critical for understanding the surface controls on the initiation and intensity of atmospheric moist convection (see Sec. 5.4.3).

Dry and Moist Adiabatic Processes

We begin by considering an air parcel of ideal gas adiabatically lifted through the surrounding atmosphere with pressure and temperature profiles $p_{\text{srd}}(z)$ and $T_{\text{srd}}(z)$, respectively. The lifted air parcel is set to keep the same pressure as its surroundings (i.e., $p(z) = p_{\text{srd}}(z)$) but may have a different temperature profile $T(z)$ depending on its location. Within the boundary layer, the air is well mixed so that $T_{\text{srd}}(z < h) = T(z)$; above the boundary layer the surrounding air is the free atmosphere so that $T_{\text{srd}}(z > h) = T(z)$.

The enthalpy of the air parcel is determined by Eq. (2.32). With constant mass specific heat capacity c_p, an infinitesimal change in enthalpy is

$$c_p dT = du + p dv + v dp. \tag{5.51}$$

Note that u and v are also mass specific variables. In the case of *dry adiabatic processes*, the air parcel is isolated without heat exchange. Thus, the work done in changing the volume of the air parcel against the surrounding pressure directly reduces the parcel's internal energy ($du = -p dv$), and Eq. (5.51) becomes (Emanuel, 1994; Tsonis, 2002; Yin et al., 2015)

$$c_p dT = v dp. \tag{5.52}$$

This defines the temperature variation (dT/dp) of an air parcel without water phase change, a process often referred to as a dry adiabatic process. For such a process, assuming ideal gas behavior, the potential temperature as defined in Eq. (2.65) remains constant (see Note 5.9).

In addition to the work done by changing the volume and pressure of the air, for an isolated system the enthalpy can also be increased by a phase change (Emanuel, 1994; Tsonis, 2002; Yin et al., 2015); therefore,

$$c_p dT = v dp + \lambda_w dq_L, \tag{5.53}$$

where λ_w is the latent heat of vaporization and q_L is the liquid water content. Here, the heat capacity of liquid water or ice is neglected, and such an approximation, causing only a 1% variation in the lapse rate, is referred to as pseudoadiabatic (Emanuel, 1994).

In an isolated system with water phase change, water vapor and liquid coexist and the total water content is conserved, so that

$$d(q + q_L) = d(q_{\text{sat}} + q_L) = dq_{\text{sat}} + dq_L = 0, \tag{5.54}$$

where $q_{\text{sat}} \approx \epsilon e_{\text{sat}}/p$ is the saturation water vapor content (see Eq. (2.75)). The change in e_{sat} follows the saturation curve as described by the Clausius–Clapeyron equation (see Eq. (2.48)). Combining it with Eqs. (5.53), (5.54), and the ideal gas law (2.72), one finds

$$\frac{dT}{dp} = \frac{1}{p} \frac{RT + \lambda_w q_{\text{sat}}}{c_p + \dfrac{\lambda_w^2 q_{\text{sat}}}{R_w T^2}}, \tag{5.55}$$

where R and R_w are the gas constants of air and water vapor, respectively. This defines the temperature variation (dT/dp) of an air parcel with water phase change, referred to as a *moist adiabatic process*.

The dry and moist lapse rates (dT/dp) are often expressed in terms of altitude (dT/dz). To find these expressions, one can assume the surrounding atmosphere to be in hydrostatic equilibrium,

$$\frac{dp}{dz} = -\frac{pg}{RT_{\text{srd}}(z)}. \tag{5.56}$$

Assuming $T_{\text{srd}}(z)/T(z) \approx 1$, the dry adiabatic lapse rate can be derived by substituting Eq. (5.56) into Eq. (5.52):

$$\Gamma_{\text{dry}} = \frac{dT}{dz} = -\frac{g}{c_p}, \tag{5.57}$$

and the moist adiabatic lapse rate can be obtained by substituting Eq. (5.56) into Eq. (5.55):

$$\Gamma_{\text{moist}} = \frac{dT}{dz} = \Gamma_{\text{dry}} \frac{1 + \dfrac{\lambda_w q_{\text{sat}}}{RT}}{1 + \dfrac{\lambda_w^2 q_{\text{sat}}}{c_p R_w T^2}}, \tag{5.58}$$

where the last expression shows the differences between Γ_{dry} and Γ_{moist}. The lapse rate can also be expressed in terms of the potential temperature; the corresponding dry adiabatic lapse rate is zero and the moist adiabatic lapse rate is positive (see Fig. 5.11 and Note 5.10).

Lifting Condensation Level (LCL) and Convective Available Potential Energy (CAPE)

The dry and moist adiabatic processes discussed above can be used to derive some critical indices for *atmospheric convection*. We consider an adiabatically lifted unsaturated air parcel, which first experiences a dry adiabatic process and cools down at the rate given in Eqs. (5.52) or (5.57). The cooling reduces its saturation vapor pressure, and the air parcel becomes saturated at the *lifting condensation level* (LCL). Above this level, the continuously released latent heat from condensation warms the air parcel and reduces the lapse rate to the value given by Eqs. (5.55) or (5.58).

The altitude of the LCL can be found by tracking the state of the air parcel from its initial condition (T_0 and q_0). At the LCL, the air just becomes saturated but its specific humidity still equals its initial value:

$$q_{\text{LCL}} = q_0 = q_{\text{sat}}(T_{\text{LCL}}, P_{\text{LCL}}) = \epsilon \frac{e_{\text{sat}}(T_{\text{LCL}})}{P_{\text{LCL}}}, \tag{5.59}$$

where $e_{\text{sat}}(T_{\text{LCL}})$ can be estimated by the Antoine equation (2.49) with $T_2 = T_{\text{LCL}}$ and a suitable reference temperature T_1. The temperature at the LCL can be linked to T_0 by integrating Eq. (5.52) using the ideal gas law (2.30), obtaining

$$T_{\text{LCL}} = T_0 \left(\frac{P_{\text{LCL}}}{p_0} \right)^{R/c_p}, \tag{5.60}$$

or it can be directly obtained by using the dry adiabatic lapse rate:

$$T_{\text{LCL}} = T_0 + \Gamma_{\text{dry}} z_{\text{LCL}}. \tag{5.61}$$

The location of the LCL can be found by analytically solving Eqs. (5.59)–(5.61) together with (2.49) (Yin et al., 2015):

$$z_{\text{LCL}} = -\frac{T_0}{\Gamma_{\text{dry}}} + \frac{\lambda_w R}{g R_w W_{-1} \left[\dfrac{\lambda_w R \Gamma_{\text{dry}}}{g R_w T_0} \left(\dfrac{p_0 q_0}{\epsilon e_{\text{sat},0}} \right)^{R/c_p} \exp \left(\dfrac{\lambda_w R \Gamma_{\text{dry}}}{g R_w T_0} \right) \right]}. \tag{5.62}$$

where $W_{-1}[\cdot]$ is the lower branch of the Lambert W function (Corless et al., 1996) and the value of the reference temperature T_1 is taken as the initial air parcel temperature T_0. Equation (5.62) provides an analytical solution for the LCL as function of the initial temperature T_0, pressure p_0, and humidity q_0 of

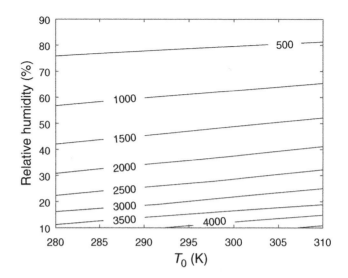

the lifted air parcel (a similar formula for the LCL was derived by Romps (2017)). Figure 5.13 shows the contour plot of the LCL as a function of near-surface temperature and relative humidity, from which it becomes clear that the LCL is more sensitive to the humidity than to the air temperature.

Above the LCL, the latent heat release due to condensation warms up the air parcel, possibly making it warmer than the surrounding environment. The location at which the air parcel temperature just becomes larger than the surrounding air temperature is the level of free convection (LFC) (see Fig. 5.11). Below the LFC, adiabatic lifting results in negative buoyancy and inhibits convection, while above the LFC the situation is reversed, resulting in positive buoyancy. Still further above, the moist adiabatic process again crosses the surrounding temperature profile at the level of neutral buoyancy (LNB), above which the buoyancy is again negative (see Fig. 5.11).

The difference in temperature between the adiabatically lifted air parcel and the surrounding atmosphere is a measure of the buoyant force on an air parcel. The total area of positive buoyancy (see the shaded area in Fig. 5.11) between the LFC and the LNB is thus a measure of the buoyant potential energy, termed the *convective available potential energy* (CAPE) (Emanuel, 1994; Yin et al., 2015), formally given by

$$\text{CAPE} = \int_{z_{\text{LFC}}}^{z_{\text{LNB}}} g \frac{T_p(z) - T_{\text{srd}}(z)}{T_{\text{srd}}(z)} dz, \tag{5.63}$$

where T_{srd} is the surrounding air temperature, and T_p is the adiabatically lifted air parcel temperature, which needs to be calculated using Eqs. (5.57) and (5.58). Similarly, the total negative buoyancy below the LFC will restrain the rising air parcel. This is called the convective inhibition (CIN) (Emanuel, 1994; Yin et al., 2015):

$$\text{CIN} = -\int_{z_0}^{z_{\text{LFC}}} g \frac{T_p(z) - T_{\text{srd}}(z)}{T_{\text{srd}}(z)} dz. \tag{5.64}$$

One needs to replace the temperature by the virtual temperature (see Eq. (2.74)) in Eqs. (5.63) and (5.64) to quantify the density of moist air more accurately (Doswell III and Rasmussen, 1994).

The LCL gives an indication of the location of the cloud base; this level is often seen in the flat bottom of summer afternoon clouds. CAPE and CIN are useful to estimate the intensity of atmospheric moist convection. As we will see in Sec. 5.4.3, the previous quantities, when coupled into the ABL model, may be used to assess the likelihood of convective rainfall.

5.4 Coupling of the Soil–Plant–Atmosphere Continuum (SPAC) with the ABL

Coupling the components of the soil–plant–atmosphere continuum (SPAC) to the ABL, we may simulate the diurnal evolution of evapotranspiration and carbon assimilation; these may then be integrated over the daily timescale to obtain useful parameterizations for the stochastic analysis to be presented in the following chapters.

5.4.1 A Brief Review of the SPAC

Before proceeding further, it is useful at this point to recapitulate the main equations of the SPAC. The water flow within the SPAC was assumed to take place as a succession of steady states between compartments connected by resistors (Fig. 4.3). We considered the soil–root conductance, g_{sr} (Sec. 4.3), the plant conductance, g_p (Sec. 4.4), the stomatal conductance, g_s (Sec. 4.8), and the atmospheric conductance g_a (Sec. 2.3). The soil water potential (ψ_s) and the atmospheric humidity (q_a) provide boundary conditions for the SPAC. Going from the soil to the leaves, the liquid water flux is proportional to the gradient of the water potential (Sec. 4.2),

$$\Phi = g_{srp}(\psi_s - \psi_l), \tag{5.65}$$

where Φ is the volumetric water flux in terms of volume per unit ground surface (LT^{-1}), g_{srp} is the soil–root–plant conductance per unit ground area (the inverse of the soil–plant–root resistance). The soil–root–plant conductance was obtained as the series combination of the reciprocal soil–root conductance and plant xylem conductance,

$$g_{srp} = \frac{g_{sr}g_p}{g_{sr} + g_p}, \tag{5.66}$$

all expressed in terms of unit ground area (see Eq. (4.19). Since the driving quantity is the water potential, with dimensions of pressure, $[\psi] = FL^{-2}$, and thus having units of pascals (Pa), all these conductances have dimensions $[g_{sr}] = [g_p] = [g_{srp}] = L^3 T^{-1} F^{-1}$ and may thus be expressed in m s^{-1} Pa^{-1}.

Neglecting any changes in plant water storage, the flux inside the plant must equal the transpiration rate through the stomata,

$$\Phi = E. \tag{5.67}$$

The latter may be assumed to be proportional to the gradient in the specific humidity between the interior of the stomatal cavities and just outside the leaf:

$$\rho_w E = g_s \rho [q_i(T_l, \Psi_l) - q_0], \tag{5.68}$$

where ρ_w is the liquid water density, ρ is the air density, q_i and q_0 are the specific humidity valus inside and outside the leaf, T_l is the leaf temperature, and g_s is the stomatal conductance (per unit ground area).

The water vapor inside the stomatal cavities is at the same water potential, ψ_i(water vapor) $= \psi_l$(liquid), and temperature $T_i = T_l$. Consequently, using Eq. (2.104),

$$\psi_i = \frac{\mathcal{R}T_l}{\upsilon_w} \ln \frac{e_i}{e_{\text{sat}}(T_l)} + g\rho_w z_i = \psi_l, \tag{5.69}$$

from which, using Eq. (2.75),

$$q_i \approx \frac{\epsilon}{p_0} e_i = \frac{\epsilon}{p_0} e_{\text{sat}}(T_l) \exp\left[\frac{\upsilon_w(\psi_l - g\rho_w z_i)}{\mathcal{R}T_l}\right]. \tag{5.70}$$

From outside the leaves to the atmosphere, the transpiration flux is

$$\rho_w E = g_a \rho(q_0 - q_a), \tag{5.71}$$

where g_a is the atmospheric conductance and q_a is the specific humidity at a height z_a within the logarithmic velocity profile. Since E is the volumetric flux per unit (ground) area, $[E] = \text{L T}^{-1}$, the stomatal and atmospheric conductances also have dimensions $[g_s] = [g_a] = \text{L T}^{-1}$ and are typically expressed in m s^{-1} (note that there are differences in units among g_s, g_a, g_{sr}, and g_p).

5.4.2   Diurnal Evolution of the SPAC–ABL Model

The system of equations (5.65)–(5.71) for the SPAC, coupled to the surface energy balance and partitioning equations (5.22)–(5.24), allows us to solve for the water flux and leaf temperature once the soil water potential ψ_s, air temperature T_a, and humidity q_a are available (see Sec. 5.2.3). If these quantities are unknown, the SPAC system can be coupled to the ABL model introduced in Sec. 5.3, Eqs. (5.32)–(5.34)), to simulate the full dynamics of land–atmosphere interaction.

Here we present the results of a simulation of the entire system forced by net radiation and based on the soil moisture and free atmosphere boundary conditions. Following Daly et al. (2004a) we use the parameters given in Table 4.3, which are typical for woody plants, and assume the leaf available energy as a parabolic function of time, as in Eq. (5.3). The stomatal conductance, g_s, is calculated by both the Jarvis approach (Sec. 4.8.1) and the physiological model (Sec. 4.8.2). This calculated stomatal conductance (through either the Jarvis or the physiological approach) can be used to evaluate the net carbon assimilation, A_n, using the model described in Sec. 4.7, and the boundary-layer dynamics, using the model introduced in Sec. 5.3.

Figure 5.14 shows results for the diurnal evolution of g_s, E, and A_n, using both Jarvis' and physiological approaches. For each case the evolution of the ABL height is similar to that shown in Fig. 5.12, reaching higher values with decreasing soil moisture (i.e., higher sensible heat flux). The stomatal conductance has a maximum in the morning, then reaches a plateau in the afternoon, and finally goes to zero at night when the stomata are closed. The stomata quickly open in the morning when radiation starts and then partially close due to the effect of D and ψ_l; the plateau in the afternoon is related to the fact that the reduction in ϕ and the increments in D and T_a are compensated by the positive effect of higher leaf water potentials. With the Jarvis formulation, the daily course of E is moderately symmetrical, while with the physiological formulation it more closely follows that of g_s, so that in the afternoon it tends to be lower.

Figure 5.15 shows the evolution of the *water use efficiency* (WUE), also using both the Jarvis and the physiological approaches. This efficiency is defined as the ratio of A_n/A_{max} over E/E_{max}, in which A_{max}

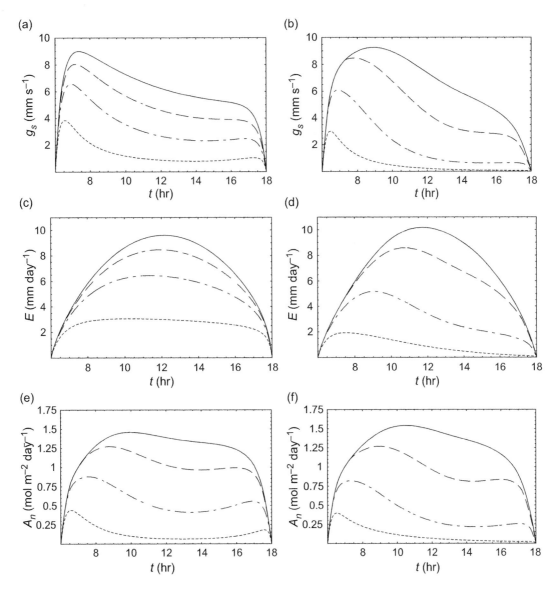

Figure 5.14 Results at an hourly timescale following Jarvis' (left column) and Leuning's (right column) formulations, for different initial soil moisture levels: $s = 1$ (solid line), 0.45 (dashed line), 0.4 (dashed and dotted line) and 0.35 (dotted line). Parameters as in Fig. 5.20. After Daly et al. (2004a).

and E_{max} are the net assimilation and transpiration rates averaged over a day in well-watered conditions. In both cases, the WUE reaches its peak value in the early morning when transpiration is hindered by the moister air and assimilation is growing owing to the increasing solar irradiance. As the air humidity declines during the day so does the WUE, until the late afternoon when the situation is reversed by the slower decay of assimilation compared with transpiration. The fact that the maximum is reached in the morning is probably due to the dominance of the increase in solar radiation compared with the reduction in the water potential that follows stomatal opening.

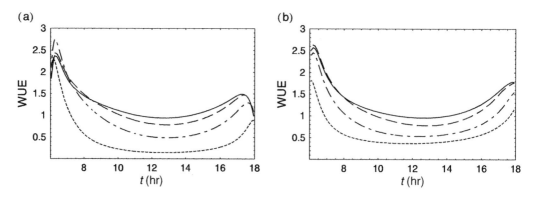

Figure 5.15 Daily behavior of the WUE following (a) Jarvis' and (b) Leuning's approach, for different initial soil moisture levels, as in Fig. 5.14. After Daly et al. (2004a).

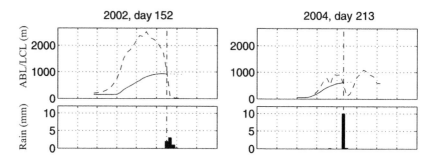

Figure 5.16 Modeled ABL height (solid line), LCL height (dashed line), and measured rainfall at Duke Forest near Durham (NC). After Juang et al. (2007).

5.4.3 Soil Moisture Control on Atmospheric Convection

The soil moisture control on evapotranspiration affects the energy partitioning and, through the boundary-layer dynamics, may influence the dynamics of atmospheric convection. We will track these impacts through an analysis of the dynamics of the LCL (see Sec. 5.3.4) and in particular the timing of when the ABL height crosses the LCL, assuming that at that time the air at the top of the ABL becomes saturated, with the formation of the first convective clouds. If then the energetic conditions of the ABL allow for the development of deep convection (i.e., if the CAPE is large enough), one can expect that the LCL crossing may be closely followed by a convective rainfall event (see Fig. 5.16; Juang et al., 2007).

The statistical analyses of Juang et al. (2007) showed that convective rainfall occurred in only 45% of the cases of LCL crossing, suggesting that LCL crossing is a necessary but not sufficient condition for the initiation of deep convection and thunderstorms. The occurrence of convective precipitation is in fact also controlled by the intensity of atmospheric convection, as measured by CAPE (see Fig. 5.11 and Sec. 5.3.4). Here we follow Yin et al. (2015) to explore the surface controls on both the initiation and intensity of atmospheric convection. Using the SPAC parameters in Sec. 5.4.2 and the free atmospheric conditions obtained from measurements in the Southern Great Plains, Yin et al. (2015) obtained the diurnal evolution of ABL, LCL, LNB, and LFC under various soil moisture conditions. Figure 5.17 shows that, during the day, both ABL and LCL increase and end up crossing in the afternoon when convective clouds are formed. To further induce convective rainfall, the ABL height needs to be close to the LFC to initiate

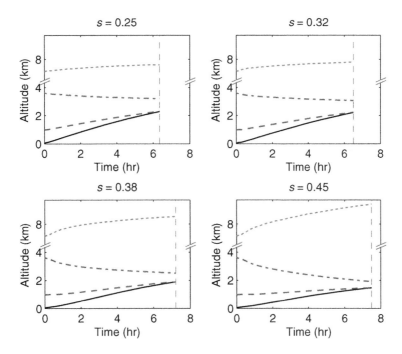

Figure 5.17 Evolution of the ABL (solid lines), LCL (dashed lines), LFC (dashed and dotted lines), and LNB (dotted lines) under different soil moisture conditions as functions of time after sunrise ($t = 0$). The vertical thin dashed lines mark the LCL crossing time. Note that the y axis has been cut between 4 and 6 km to facilitate comparison of the evolution at the different altitudes. After Yin et al. (2015).

free convection as well as far from the LNB to have a large value of CAPE. As can be seen in Fig. 5.17, the LFC decreases while the LNB increases, so that the gap between the two gradually widens. Under drier soil conditions, more sensible heat flux is available for the growth of the ABL, which in turn induces an earlier LCL crossing. However, the distance between the LFC and LNB increases more slowly than under wetter soil conditions.

Figure 5.18 shows further the diurnal evolution of CAPE and CIN as calculated from Eqs. (5.63) and (5.64) in Sec. 5.3.4. Under drier soil conditions, more sensible heat flux is used to extend the ABL, which may then cross the LCL earlier. However, the limited water supply from the dry surface also retains a higher LCL, causing a slower increase in CAPE. For this typical example, the wet soil tends to induce afternoon thunderstorms, although the timing of convective cloud formation, following the LCL crossing time, is later.

The LCL crossing and CAPE can be analyzed under various atmospheric conditions by changing the slopes of the free atmospheric temperature and humidity profiles (i.e., γ_{ϑ} and γ_q). The results in Fig. 5.19 show the existence of four regimes for the conditions of LCL crossing and CAPE at the time of the crossing (or at the end of the day if there is no crossing). When the atmospheric conditions are within regime I, the CAPE value is larger than 400 J kg^{-1} at the end of the day, but the ABL does not cross the LCL. Within regime II, the CAPE value is too low at the end of the day, while the ABL still does not cross the LCL. In regime III, the ABL crosses the LCL, but the value CAPE is too low (< 400 J kg^{-1}) at the crossing time. Only when the atmospheric conditions are within regime IV is deep convection likely to be triggered, since the CAPE value is large enough at the time when the ABL crosses the LCL. In general, the regimes I, II,

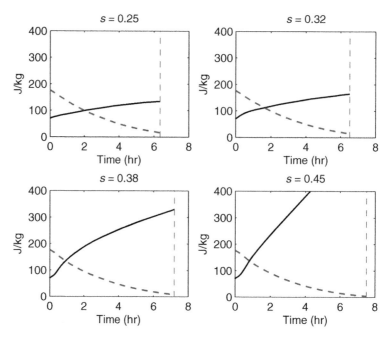

Figure 5.18 CAPE (solid line) and CIN (dashed line) evolution under different soil moisture conditions as functions of time after sunrise ($t = 0$). The vertical thin dashed lines mark the LCL crossing time. After Yin et al. (2015).

III, and IV map onto four quadrants, corresponding to atmospheres in "near-neutral-and-dry", "stable-and-dry", "stable-and-moist", and "near-neutral-and-moist" conditions. Such regimes change under different soil moisture conditions, demonstrating how land–surface conditions influence atmospheric convection through the whole SPAC–ABL dynamics.

5.4.4   Multi-Day Dry-Down

Evapotranspiration and boundary-layer dynamics can be modeled over a dry-down period without rainfall by coupling the SPAC–ABL model to a soil water balance model. In the absence of rainfall, the water losses are due to percolation (L) and evapotranspiration (E) and the soil moisture equation (1.5) simply becomes

$$nZ_r\frac{ds}{dt} = -E - L, \tag{5.72}$$

where s is the relative soil moisture averaged over the root depth Z_r and n is porosity (see also Sec. 3.3.3).

Following Daly et al. (2004a), Fig. 5.20 shows the temporal dynamics of some of the main variables characterizing the SPAC during a dry-down of several days after the drainage become negligible and with E calculated from the full SPAC–ABL model. It is apparent how the soil moisture decays more slowly during the night for lack of transpiration. The leaf water potential is equal to the soil water potential during the night, while during the day it is much lower because of transpiration (Fig. 5.20b). Through the energy balance, transpiration also influences the boundary layer, affecting leaf temperature (not shown) and VPD (Fig. 5.20c). Both the daily minimum leaf water potential and the maximum assimilation decrease with

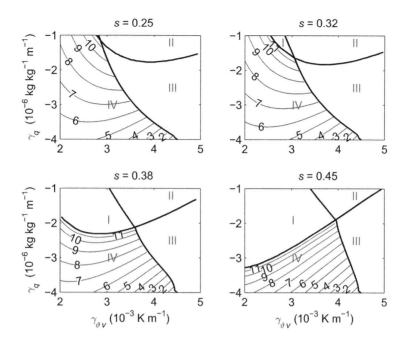

Figure 5.19 LCL crossing time under different soil moisture and atmospheric conditions as a function of the time after sunrise ($t = 0$). Labeled contour lines represent the hour after sunrise at which LCL crossing occurs. LCL crossing occurs only in regime IV. After Yin et al. (2015).

time, while the maximum VPD increases (Fig. 5.20b, c, and f). Figure 5.20 also highlights some differences due to the two stomatal functions. The higher leaf water potential from the physiological approach leads to lower transpiration and assimilation rates in the late stage of the dry-down (Fig. 5.20d). These diurnal trends in the surface fluxes are evident in the field observations shown in Fig. 5.1.

5.4.5 Integration to the Daily Level

Figure 5.20 shows that the total daily transpiration and assimilation are modulated mostly by the changes in soil moisture. One can thus obtain more straightforward relations between the total daily transpiration, stomatal conductance, and assimilation with soil moisture dynamics, as was done in Daly et al. (2004b).

Figure 5.21 shows for comparison the model results and the measured leaf conductance of the C_3 plant *Nerium oleander* as a function of the so-called extractable soil moisture (Gollan et al., 1985). As the measurements were conducted under constant irradiance, the model results during a phase of dry-down are sampled at a given hour of the day to ensure constant irradiance. The time of sampling was chosen so that the modeled stomatal conductance matched the measurements at high soil water content. The general behavior is very well reproduced for both approaches with a plateau for well-watered conditions and a regular (almost linear) decay at low soil moisture values. Such results are quite useful and from them one can obtain the soil moisture dependence of transpiration and assimilation at a daily timescale, which in turn can be used in stochastic models of soil moisture dynamics, as will be seen in the following chapters.

Examples of the daily processes obtained from integration of the results over an hourly timescale are shown in Figs. 5.22 and 5.23. The mean daily values of soil moisture during the drying phase are observed

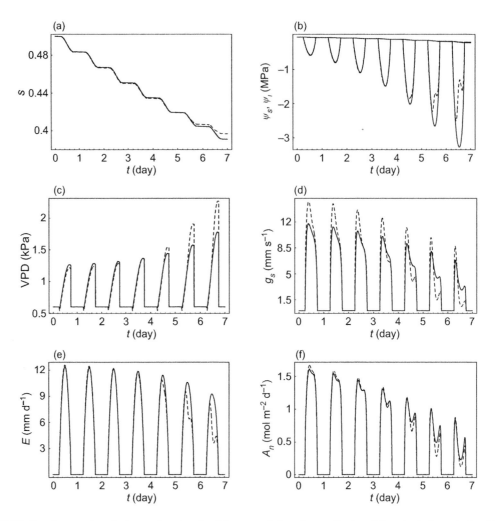

Figure 5.20 Results at an hourly timescale using Jarvis' approach (continuous line) and Leuning's formulation (dashed line) for stomatal conductance (Sec. 4.8), with an initial value of s equal to 0.45. (a) Relative soil moisture, (b) soil and leaf water potential, (c) water vapor pressure deficit, (d) stomatal conductance, (e) transpiration, and (f) net leaf carbon assimilation. RAI = 5.6, LAI = 1.4, and Z_r = 60 cm; the soil is a loam. Plant parameters as in Table 4.3. After Daly et al. (2004a).

to follow a very similar pattern to those at the hourly timescale. The behavior of the transpiration, shown in Figs. 5.22b and 5.23a, closely resembles the empirical relationships obtained from field experiments (see Fig. 5.24), providing supports for simple parameterizations in water balance models. For example, the *daily transpiration* may be approximated as a piecewise function of the relative soil moisture, s, that is constant above a certain s^* and decreases linearly to zero at the wilting point, s_w:

$$E(s) = f(s)E_{\max} = \begin{cases} 0 & 0 < s \leq s_h, \\ E_w \dfrac{s - s_h}{s_w - s_h} & s_h < s \leq s_w, \\ E_w + (E_{\max} - E_w)\dfrac{s - s_w}{s^* - s_w} & s_w < s \leq s^*, \\ E_{\max} & s^* < s \leq 1. \end{cases} \tag{5.73}$$

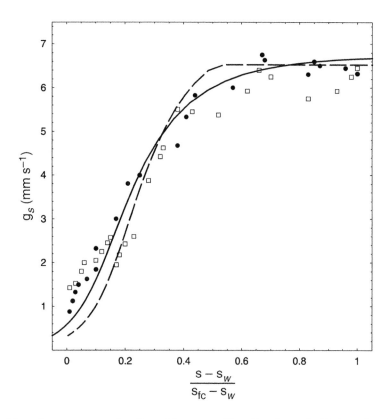

Figure 5.21 Stomatal conductance g_s as a function of the extractable soil moisture for *Nerium oleander*, measured while maintaining constant irradiance, for two different levels of vapor pressure deficit (solid circles and open squares; after Gollan et al. 1985). The model results for stomatal conductance sampled every day at a given time for Jarvis' (continuous line) and Leuning's (dashed line) formulation are given for comparison. The time of sampling was chosen to ensure the same stomatal conductances as those for the measured data under well-watered conditions. The extractable soil moisture is defined using typical values for loamy soil and C_3 plants, e.g., $s_w = 0.20$ and $s_{fc} = 0.55$. After Daly et al. (2004a).

The parameters of the transpiration function, s_w, s^*, and E_{max}, can thus be connected by temporal scaling to the plant, soil, and climate characteristics. Note that our numerical simulations assumed constant E_{max}. The daily fluctuations in E_{max} may be used to normalize the soil-moisture–evapotranspiration relationship to improve the fit of Eq. (5.73), as in Fig. 5.24. However, these fluctuations play a limited role in the soil moisture dynamics, justifying the use of a constant value of E_{max} for representative periods of time during the growing season in most applications (Daly and Porporato, 2006a; see also Sec. 7.7).

Finally, because of the links of s_w and s^* with the plant water stress, the parameters of the climate–soil–vegetation system modeled at the daily scale also include the most important feedbacks between the hydrological processes and plant conditions. This analysis justifies the model that will be used for stochastic soil moisture dynamics in Chapter 7.

A behavior similar to that of *daily transpiration* is also found for the dependence of the daily carbon assimilation on the soil moisture (Figs. 5.22c and 5.23b). This dependence may be functionally approximated as (Daly et al., 2004b).

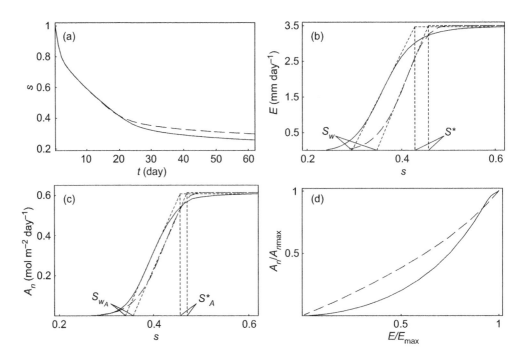

Figure 5.22 Results at a daily timescale for Jarvis' approach (solid line) and Leuning's formulation (long-dashed line) for stomatal conductance (Sec. 4.8): (a) relative soil moisture during a drying period, (b) and (c) daily transpiration and daily net assimilation, respectively, as a function of relative soil moisture (shown with short-dashed lines are the approximating piecewise functions; see the main text for details), (d) the relation between assimilation and transpiration. Loamy soil, RAI = 5.6, LAI = 1.4, and soil depth $Z_r = 60$ cm. The vegetation and soil parameters are given in Tables 6.1 and 6.2. After Daly et al. (2004b).

$$
A_n(s) = \begin{cases} 0 & s \le s_{w_A}, \\ \dfrac{s - s_{w_A}}{s_A^* - s_{w_A}} A_{\max} & s_{w_A} \le s \le s_A^*, \\ A_{\max} & s_A^* < s \le 1. \end{cases} \tag{5.74}
$$

As will be seen in the following chapters, this simple description of assimilation as a function of soil moisture at a daily timescale is key to linking photosynthesis and assimilation to a probabilistic description of soil moisture dynamics. The relation between E and A_n (Fig. 5.22d) shows a slightly nonlinear behavior of WUE, linked to the faster decrease of A_n compared with that of E at low soil moisture. This difference is less marked with Leuning's formulation because in that case E is related to A_n through g_s. Jarvis' and Leuning's approaches give almost the same s_{w_A} and s_A^*, while for Leuning's formulation the assimilation rate under stressed conditions is lower (Fig. 5.22c) and s_w and s^* are higher. This is probably an artifact in Leuning's formulation that could be resolved by making the parameter κ in Eq. (4.45) a function of soil moisture.

Figures 5.23a–c show the impact of two different the soil types on the soil moisture dynamics. As expected, the soil properties, especially the hydraulic conductivity and soil texture, control transpiration through their influence on both the soil-root conductance and soil water potential. Plants in soils with higher saturated hydraulic conductivity K_s tend to have a lower s^*, because of the lower soil resistance to water uptake and transpiration, but they reach s^* approximatively at the same time during the drying phase because of the higher leakage (Fig. 5.23c). The soil type has a negligible influence on the maximum

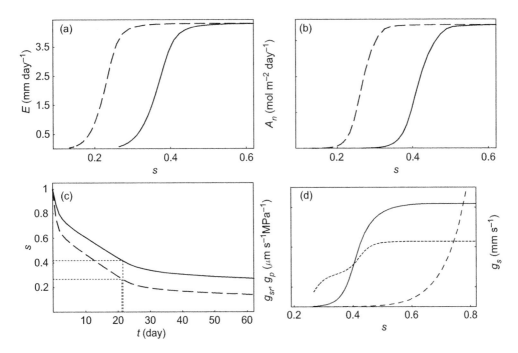

Figure 5.23 Influence of soil type on (a) mean daily transpiration, (b) mean daily assimilation, and (c) mean daily soil moisture using Jarvis' approach for stomatal conductance (Sec. 4.8): loam (solid line) and loamy sand (dashed line). The dotted lines in (c) show the times after which soil moisture reaches s^*, for the two considered soil types. (d) for comparison, the various mean daily conductances for a loam with $Z_r = 60$ cm: stomatal conductance (solid line), soil–root conductance (dashed line), and plant conductance (short-dashed line). See caption of Fig. 5.22 for other parameters. After Daly et al. (2004b).

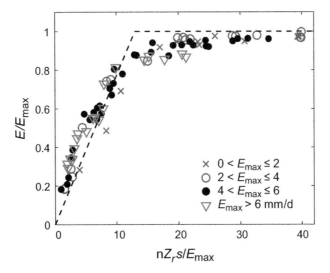

Figure 5.24 Field observations of the daily transpiration E versus the available soil water $nZ_r s$ normalized by the maximum evapotranspiration E_{max}. After Federer (1979).

transpiration and assimilation rates, which essentially depend on both the vegetation type, through the physiological parameters g_{smax} and g_{pmax}, and the atmospheric boundary layer conditions (Daly et al., 2004b).

5.5   Key Points

- Calculating the evapotranspiration requires knowledge of the evaporating-surface temperature; the latter may be obtained by the surface energy balance given by the partitioning of the incoming radiation into latent and sensible heat fluxes.
- The albedo (the fraction of reflected radiation) depends on vegetation characteristics and decreases with canopy height; it higher for dryer soils.
- Evaporation from leaf tissues occurs at negative water potentials, implying smaller driving forces of transpiration compared with a saturated surface.
- Mixed-layer models of the atmospheric boundary layer (ABL) can be written as a dynamical system for the diurnal evolution of the height, potential temperature, and specific humidity of the ABL, coupled to the land–atmosphere exchanges of water and energy driven by solar radiation and constrained by soil and free-atmosphere conditions.
- Tracking the evolution of the convective available potential energy (CAPE) and the timing of the crossing of the lifting condensation level (LCL) during ABL evolution provides useful information on the onset of cloud formation and deep convection potential.
- Upscaling to a daily timescale results in simple one-to-one relations of evapotranspiration and plant carbon assimilation to the soil moisture; such relations present a plateau in well-watered conditions and a quasi-linear decay, starting from the onset of water stress s^* to negligible values at the wilting point s_w.

5.6   Notes, including Problems and Further Reading

5.1   The solid line in Fig. 5.7 shows the Bowen ratio for conditions in which both atmosphere and surface are saturated. According to Eq. (5.21), the Bowen ratio in such conditions is the equilibrium Bowen ratio (Bo_{eq}). Plot Bo_{eq} as a function of air temperature for a wet surface with a fixed surface temperature of 293 K; find the limiting value of Bo_{eq} when $T_a = T_0$.

5.2   Write down the system of equations needed to determine transpiration as a function of soil and atmospheric conditions above the canopy for the two different types of stomatal models (Sec. 4.8). Analyze the coupling among the equations given by the main variables, ψ_l, T_l, and c_i.

5.3   Using the parameters in Table 4.3, solve numerically the system obtained from Note 5.2 and reproduce the results in Fig. 5.9.

5.4   Using the results of Note 5.3 and the necessary parameters in Tables 4.2 and 4.3, compute the plant carbon assimilation using the carbon flux and photosynthesis model of Sec. 4.7.

5.5   Write the Penman equation (5.17) using the Bowen ratio as the independent variable.

5.6   Jarvis' equation in Sec. 4.8.1 for stomatal conductance was used for calculating transpiration in Fig. 5.9. Use the physiological model in Sec. 4.8.2 to model g_s and estimate the transpiration rate.

5.7   Use the Penman–Monteith equation to explore the impact of canopy height on transpiration by considering its effect on albedo (see Fig. 5.4) and atmospheric conductance (see Sec. 2.3).

5.8   Derive the boundary-layer growth rate of Eq. (5.37) by using the ABL governing equations (5.32)–(5.34) and the linear free atmospheric profiles (5.35) and (5.36). See Porporato (2009).

5.9   Use Eq. (5.52) to prove that the potential temperature, as defined in Eq. (2.65), is constant during a dry adiabatic process.

5.10   Use Eqs. (5.52) and (5.53), the definition of potential temperature (2.65), and the ideal gas law (2.30) to derive the dry and moist adiabatic lapse rates in terms of the potential temperature. Compare the results with the schematic diagram in Fig. 5.11.

5.11   Use Eq. (5.73) to solve Eq. (5.72) assuming no leakage ($L = 0$) (see Sec. 7.2).

5.12 (CO_2 in the ABL)   Write down the balance equation for CO_2 in the ABL, assuming well-mixed conditions and a linear lapse rate in the free atmosphere, and discuss how this equation is coupled to the main ABL equations, Eqs. (5.32), (5.33), and (5.34), as well as the SPAC, carbon assimilation, and the leaf energy equations.

5.13 (NO_x and methane in the ABL)   Similarly to Note 5.12, write down the balance equation for NO_x and methane in the ABL. Assume well-mixed conditions, a linear lapse rate in the free atmosphere, and constant soil concentration during the day (see Note 9.8). Discuss how these equations are coupled to the main ABL equations and the surface conditions.

5.14 (Large-Eddy Simulations of the ABL)   Turbulent flows may be numerically modeled by means of so-called large-eddy simulations (LESs), which use a low-pass filtered approximation of the equations for turbulent flow (Navier–Stokes plus convection), in effect parameterizing the small-scale fluctuations to reduce computational costs. LES has become one of the most popular numerical tools for modeling the atmospheric circulation and boundary-layer dynamics (Deardorff et al., 1970). Interestingly, the profiles of the well-mixed ABL model introduced here in Sec. 5.3 are in good agreement with the results of LES simulations (Stevens, 2007).

5.15 (Stable Boundary Layer)   The well-mixed ABL discussed in Sec. 5.3 usually collapses after sunset in the absence of thermally generated turbulence. This leads to the formation of a stable or nocturnal boundary layer (SBL) (Stull, 1988). SBLs can also form during the day when warm air flows over a colder surface in a stably stratified atmosphere. Intermittent and sporadic turbulence in an SBL may be generated by the wind shear usually associated with strong nocturnal jets above the stable layer. SBLs are critical to several nighttime processes, including soil respiration and CAM plant transpiration (see Note 4.11); their detailed modeling remains an open area of research (Mahrt, 2014).

5.16 (Simulations with Photo3-ABL)   The Photo3 python code, originally designed for modeling C_3, C_4, and CAM photosynthesis and evapotranspiration (Hartzell et al., 2018b), has been coupled to the ABL model to analyze the impacts of vegetation on atmospheric convection (samhartz.github.io/Photo3/). Use the Photo3-ABL model to simulate LCL crossing and CAPE dynamics during the day for winter wheat, sorghum, and prickly pear with the atmospheric conditions given in Table 5.2.

5.17 (Diurnal Cloud Cycle)   Convective clouds and thunderstorms usually form in the afternoon over land (see Fig. 5.19 and Sec. 5.4.3). Global satellite observations also clearly show that clouds peak in the afternoon over regions with a strong land–atmosphere interaction (Yin and Porporato, 2017). The coupled SPAC–ABL model captures these feedbacks satisfactorily, producing realistic LCL crossing and CAPE results. However, the detailed parameterization of atmospheric convection and land–atmosphere feedbacks and the coupling to the large-scale atmospheric dynamics remain a challenge for climate prediction (Yin and Porporato, 2017, 2020).

5.18 (Topography)   Topography perturbs practically every process and variable relevant to land–atmosphere interactions, including radiation, air temperature and saturation deficit, wind and turbulence, cloudiness, precipitation, and properties of the soil and vegetation (Raupach and Finnigan, 1997). Since traditional boundary-layer scaling approaches cannot easily be applied over highly complex topography, the challenge is to find suitable extensions to be used in large-scale numerical models where the topography is not resolved (Rotach and Zardi, 2007). The review by Finnigan et al. (2020) offers an excellent starting point for more information on this important topic.

6   Stochastic Tools for Ecohydrology

This lack of determinacy is really a great stumbling block

Paul A. M. Dirac (1978)[†]

In this chapter, we introduce some basic tools to model the apparently random behavior of eco-hydrological variables driven by hydroclimatic variability. We begin by briefly reviewing the main probabilistic models for both discrete and continuous random variables, the derived-distribution approach for the distributions of functions of random variables, and the fundamental concept of stochastic processes. With the specific goal of modeling the intermittent occurrence of rainfall and the related sequences of random jumps and deterministic decays of soil moisture, we introduce the "marked" Poisson process and stochastic differential equations with random jumps. We derive the equations for the evolution of the probability density functions for such a stochastic process and obtain a general solution for the steady-state distribution. Finally, we analyze crossing properties and the mean first-passage times. These methods will be used extensively in the following chapters.

6.1   Randomness in Ecohydrology

Most quantities of interest in ecohydrology vary so erratically in time and space that they may be considered as random variables. Because of the unpredictable forcing by rainfall and other meteorological and climatic variables, ecohydrological phenomena thus require a probabilistic description. As a result, the dynamical laws for their physical, chemical, and biological behavior take the form of stochastic dynamical systems.

The origin of this randomness is manifold and has both endogenous and exogenous causes. In general, it can be traced to the presence of critical phenomena, involving aggregation processes and phase transitions as well as fluid dynamic instabilities, whereby the spatial and temporal correlation scales get very large and become capable of transferring information across scales. This causes an amplification of micro-scale variability, including that related to complicated boundary and initial conditions. This *sensitivity to initial conditions* is typical of many nonlinear dynamical systems with chaotic and turbulent behavior and originates from perturbations of phase-space saddles, homoclinic excursions, etc. (Cross and Hohenberg, 1993; Strogatz, 2001; Argyris et al., 2015).

[†]Dirac, P. A. M., (1978). "Directions in Physics." Lectures delivered during a visit to Australia and New Zealand, August–September 1975.

In addition to this intrinsic high-dimensional variability, there is often uncertainty about the detailed functioning of the system, and that by itself requires a statistical approach (Tartakovsky, 2013; Clark, 2020). To some extent, all these elements are present at once and are somewhat interlinked in each component of the soil–plant–atmosphere system. Accordingly, we refer to stochasticity as a result of the practical impossibility of precisely modeling and predicting the temporal evolution and spatial configuration of a system at all spatial and temporal scales (Katul et al., 2007). Our goal is to provide usable results based, as much as possible, on physical intuition rather than a formal treatment. We refer to the specialized literature for a more comprehensive exposition (Cox and Miller, 1965; Priestley, 1981; Van Kampen, 1992; Gardiner, 2004; Sornette, 2006; Ridolfi et al., 2011).

We have already seen instances of random behavior in the previous chapters, for example in the apparently random behavior of rainfall and soil moisture in time (e.g., Fig. 1.7), as well as in the fluctuations of turbulent quantities, where the scaling of their statistical properties allow for simple closures and simple synthetic transport laws (see Sec. 2.3). In Chapter 3, the spatial averaging of the soil aggregates at the Darcy scale led to a continuum description of porous media and to the definition of the main dynamic variables related to soil moisture and the water retention and hydraulic conductivity curves (see Sec. 3.2); similarly, the vulnerability curves in the plant xylem (see Sec. 4.4) and the branching and leaf statistics at the canopy scale allowed us to refer in effect to probabilistic laws and the main statistical properties of a 'big leaf'.

In this chapter, we model the effects of the impact of external random variability in time due to large-scale meteorology and climatology, in particular focusing on the propagation of random rainfall pulsing on soil moisture dynamics and biogeochemistry. Compared with other branches of science and engineering, where randomness is linked to symmetric, Gaussian forcing, in ecohydrology it is often related to highly intermittent and asymmetric forcing, causing apparent discontinuities (i.e., jumps) at the scale of interest in the behaviors of several variables. Figure 6.1 presents an example of measured rainfall, soil moisture, temperature, and CO_2 concentration at different depths in the soil. The figure clearly shows the random intermittency of rainfall forcing, superimposed on the seasonal variability, along with its impact in the form of jumps on the soil moisture and in turn on the soil CO_2 concentration.

6.2   Background on Probability

Random variables (RVs) are the result of processes that cannot be predicted with certainty and therefore involve chance. Random variables are described by probability distributions which define how likely it is to observe a given value of the variable. We follow the so-called empirical approach to probability and refer to Priestley (1981) for a succinct presentation of the axiomatic approach.

Discrete RVs can take values only from a discrete set, x_i ($i = 1, \ldots, N$). Ecohydrological examples of this type of variable are the number of rainy days in a month (e.g., for the month of August, $N = 31$; this should not be confused with the sample size n, which in this case would be the number of years of observation of rainy days in August), the number of plant-water stress periods in a growing season, the number of plants or species in a given area, etc. As is well known, given a sample of size n for such variables we can define the absolute frequency, $f_a(x_i)$, of the value x_i as the number of times that this value appears in the observed sample. The relative frequency is then the absolute frequency divided by the number of observations: $f(x_i) = f_a(x_i)/n$, where $0 \leq f(x_i) \leq 1$. Summing up all the frequencies relating to the values of the variable that are lower than or equal to a given x_i, one obtains, respectively, the absolute and relative cumulative frequency distributions, $F_a(x_i) = \sum_{j=1}^{i} f_a(x_j)$ and $F(x_i) = \sum_{j=1}^{i} f(x_j)$, which are

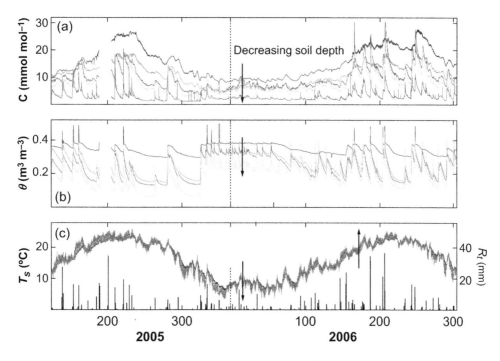

Figure 6.1 Examples of time series of (a) soil CO_2 concentration C, (b) soil volumetric water content θ, (c) soil temperature T_s and throughfall R_t for a fertilized plot at Duke Forest. The arrows indicate measures that are at shallower soil depth. After Daly et al. (2009).

increasing functions of x_i, with $F_a(x_N) = n$ and $F(x_N) = 1$. By increasing the number of outcomes of a stochastic process (i.e., the sample size n), the frequencies converge to probabilities (the law of large numbers). Thus, as $n \to \infty$, the relative frequency of x_i tends to a probability:

$$f(x_i) \to p(x_i). \tag{6.1}$$

Similarly, cumulative frequencies tend to cumulative probabilities:

$$F(x_i) \to P(x_i), \tag{6.2}$$

so that $P(x_i) = \sum_{j=1}^{i} p(x_j)$ and $P(x_N) = 1$.

For continuous RVs (e.g., wind speed, rainfall intensity, soil moisture, etc.), starting from an observed sample of n independent realizations one can subdivide the sample into k classes of equal but arbitrary size Δx. For each i-class, centered at x_i, one can then compute frequencies in the same way as was done for discrete RVs. In addition, for continuous RVs it is also useful to define the relative frequency density (Fig. 6.2), by dividing the relative frequency by the class size:

$$g(x_i) = \frac{f(x_i)}{\Delta x}. \tag{6.3}$$

Note that $g(x_i)$ is not dimensionless, but has the dimensions of the inverse of the RV. Moreover, $g(x_i)$ is always positive (but not necessarily less than 1), and has a unit-area histogram.

If the sample is large enough, the histograms of the relative frequency density and cumulative frequency indicate how the RV is distributed, i.e., which values are more likely to appear (see Fig. 6.2). As before, as $n \to \infty$, all the frequencies will tend to constant values, which however now depend on the chosen Δx. On increasing the sample size, the number of values within a given class also increases, so that it becomes

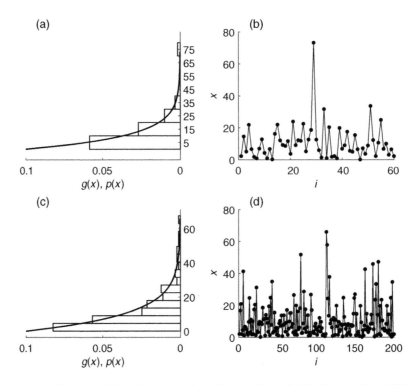

Figure 6.2 Convergence of $g(x_i)$ to $p(x)$ as the sample size is increased and the bin size is reduced. The upper panels present a numerically generated sample of $n = 60$ realizations of an exponentially distributed RV (see Sec. 6.3.5) with a mean of 10 (panel (b)), the corresponding relative frequency density, $g(x_i)$, and the theoretical probability density function of the exponential distribution (histogram and solid line in panel (a). The lower panels are similar to the upper panels, but with a sample size $n = 200$.

possible to consider simultaneously the limit $\Delta x \to 0$. Invoking again the law of large numbers, the histogram of the relative frequency density tends to a curve, a nonnegative function called the *probability density function* (PDF), $p(x)$,

$$\lim_{\substack{\Delta x \to 0 \\ n \to \infty}} g(x_i) = p(x) = \lim_{\substack{\Delta x \to 0 \\ n \to \infty}} \frac{\Delta F_i}{\Delta x} = \frac{dP(x)}{dx}. \tag{6.4}$$

Thus the cumulative probability function (CDF), $P(x)$, or non-exceedance probability, is the integral of the PDF,

$$P(x) = \int_{-\infty}^{x} p(x')dx', \tag{6.5}$$

which represents the probability of finding a value of the RV $\leq x$; moreover, $\int_{-\infty}^{+\infty} p(x)dx = 1$. These properties are illustrated in Fig. 6.3. It is important to note that $p(x)$ is not a probability, but a probability density, or probability per unit interval of x (note that the probability of a particular value x is zero, while $p(x)$ can be much larger than one!), meaning that

$$p(x)dx = \text{prob. of finding a value between } x \text{ and } x + dx. \tag{6.6}$$

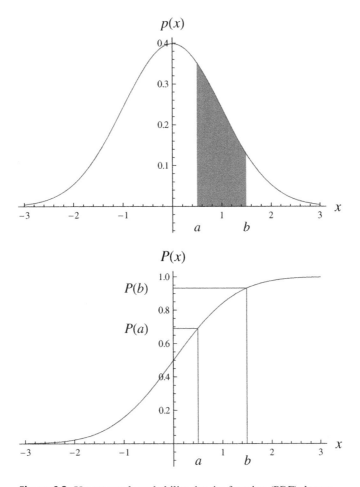

Figure 6.3 Upper panel: probability density function (PDF); lower panel: cumulative distribution function (CDF).

Thus, the probability of finding a value between a and b is $\int_a^b p(x')dx' = P(b) - P(a)$ (see the shaded area in the upper panel in the figure). The exceedance probability, i.e. the probability that the RV $\geq x$, is the complement to 1 of $P(x)$, that is $1 - P(x)$.

6.2.1 Mixed Distributions

Some RVs can take on values from both a discrete set and a continuous range. Their probability distributions are called mixed type. For a mixed-type RV that can take discrete values x_i, $i = 1, \ldots, N$ with probability p_i (which, as stated above, is not a probability) and any value within a continuous range with probability density $p_c(x)$ (note that this is not a probability), one can write

$$p_{\text{mix}} = p_c(x) + \sum_{i=1}^{N} p_i \delta(x - x_i), \tag{6.7}$$

where δ is the Dirac delta function (see Sec. 6.3.4).

Mixed distributions are not infrequent in ecohydrology. For example, daily rainfall values have a mixed distribution, since there is a finite (i.e., nonzero) probability of having a dry day, which gives rise to an "atom" of probability at zero. For evapotranspiration, whose rates are a maximum in well-watered conditions ($E = E_{\max}$ for $s > s^*$) and zero for soil moisture below the wilting point ($E = 0$ for $s \le s_w$), see Eq. (5.73), the steady-state PDF will have atoms of probability at $E = E_{\max}$ and $E = 0$, corresponding to the fraction of time spent at or above the onset of stress and at or below the wilting point, respectively. A similar consideration applies to the plant static stress and plant carbon assimilation (Secs. 8.1.1 and 8.1.3). The transient solutions of jump processes are typically also mixed distributions (see Sec. 6.7).

6.2.2 Characterization of Probability Distributions

As is well known, the shape and main properties of probability curves (PDF and CDF) may be succinctly described by special quantities, such as the mean,

$$\mu = \int_{-\infty}^{\infty} xp(x)dx, \tag{6.8}$$

which is the centroid of the distribution; the mode $\breve{\mu}$, which is the value of x corresponding to the maximum of the PDF (if the distribution has two or more relative maxima it is called bimodal or multimodal, respectively); the median $\tilde{\mu}$, which is the value with cumulative probability equal to 0.5, $P(\tilde{\mu}) = 0.5$, meaning that $\tilde{\mu}$ is exceeded in 50% of the cases. Other means, which are especially useful for asymmetric distributions, include: the geometric mean (the exponential of the mean of logarithms),

$$\mu_g = \exp\left[\int_{-\infty}^{+\infty} \ln xp(x)\,dx\right]; \tag{6.9}$$

and the harmonic mean

$$\mu_h = \left[\int_{-\infty}^{+\infty} x^{-1}p(x)dx\right]^{-1} \tag{6.10}$$

(i.e., the reciprocal of the expected value of the reciprocals of the RV). It can be shown that $\mu_h \le \mu_g \le \mu$.[1]

The (non-central) moment of order r is defined as

$$\mu_r^* = \int_{-\infty}^{\infty} x^r p(x)dx,$$

while the corresponding *central moment* is

$$\mu_r = \int_{-\infty}^{\infty} (x - \mu)^r p(x)dx.$$

---

[1] The harmonic and geometric means have numerous geophysical and environmental engineering applications, including the characterization of the large-scale effective permeability in layered porous media, heterogeneous media, and filters. The US Environment Protection Agency also recommends the use of the harmonic-mean daily streamflow as a design streamflow for the protection of human health against human exposure to suspected carcinogens (e.g., Limbrunner et al., 2000). The use of the geometric mean to estimate the most representative value of the hydraulic conductivity of given types of soils was suggested in Clapp and Hornberger (1978).

The spread of the distribution is characterized by the variance,

$$\sigma^2 = \int_{-\infty}^{\infty} (x - \mu)^2 p(x) dx,$$

(6.11)

or by the standard deviation or root mean square, σ, which has the same dimensions as the RV. The dimensionless coefficient of variation, $CV = \sigma/\mu$, is also used as a measure of dispersion.

The normalized versions of the central moments of order three and four are often used, respectively, to characterize asymmetry and peakedness (and thus also the thickness of the tails) of distributions. These are called the skewness (μ_3/σ^3), and the kurtosis or flatness factor (μ_4/σ^4) of the distribution.

Most of the previous statistics can be related to the operation of the statistical expectation $E[\cdot]$, or *ensemble average* $\langle \cdot \rangle$, of a function $f(x)$ of the RV x. Such an operation is defined as

$$E[f(x)] = \langle f(x) \rangle = \int_{-\infty}^{\infty} f(x) p(x) dx.$$

(6.12)

Thus the mean is the expected value of the RV itself, and is also indicated as $\langle x \rangle$; similarly, the rth moment may also be indicated as $\langle x^r \rangle$.

The *moment generating function* (MGF) of the RV has a special place in the theory of stochastic processes (Cox and Miller, 1965). It is defined as the (two-sided) Laplace transform of the PDF:

$$M(k) = \int_{-\infty}^{\infty} e^{kx} p(x) dx = \langle e^{kx} \rangle.$$

(6.13)

Its relation to the moments can be seen by expanding the MGF in a power series of k around the origin:

$$\langle e^{kx} \rangle = \left\langle 1 + kx + \frac{k^2}{2!} x^2 + \cdots \right\rangle = 1 + k\mu_1^* + \frac{k^2}{2!}\mu_2^* + \cdots = \sum_{r=0}^{\infty} \frac{k^r}{r!} \mu_r^*$$

(6.14)

Differentiating $M(k)$ r times with respect to k and then evaluating it at $k = 0$, we obtain the rth moment:

$$\left. \frac{d^r M}{dk^r} \right|_{k=0} = \mu_r^*.$$

(6.15)

In some problems it may be easier to deal with the MGF rather than the PDF itself, as for example when solving the evolution equations for the PDF of a stochastic process (e.g., see Sec. 6.6).

6.3 Probability Models

In this section, we present some of the main PDF models that appear as solutions of special stochastic processes of interest for ecohydrology. We will deal only with the univariate case. We begin with two related models for discrete random variables (binomial and Poisson), mostly because they appear when describing some important properties of the Poisson process, which is used extensively in the chapters to come. We then move to continuous RVs, analyzing the Gaussian or normal distribution, as well as the exponential, gamma, and Pareto distributions.

6.3.1 Binomial Distribution

The binomial distribution describes the number m of successes in a sequence of N independent zero/one experiments, each of which yields success (one) with probability ϑ and failure (zero) with probability $1 - \vartheta$ (i.e., its complementary event). A single zero/one (or success/failure) experiment is also called a Bernoulli experiment.

By definition, a discrete RV m can take on any integer value between 0 and N. To obtain its probability distribution it is useful to proceed by mathematical induction, considering first the case of $m = 2$ successes in a sequence of three experiments, $N = 3$. Because the experiments are independent, we can apply the product law to each of the Bernoulli experiments, and write for the probability of $m = 2$ in a sequence of $N = 3$ experiments

$$
\begin{aligned}
p[1,1,0] &= \vartheta\vartheta(1-\vartheta) = \vartheta^2(1-\vartheta), \\
p[1,0,1] &= \vartheta(1-\vartheta)\vartheta = \vartheta^2(1-\vartheta), \\
p[0,1,1] &= (1-\vartheta)\vartheta\vartheta = \vartheta^2(1-\vartheta).
\end{aligned}
\tag{6.16}
$$

The required total probability is their sum,

$$
p[m=2, N=3] = 3\vartheta^2(1-\vartheta).
\tag{6.17}
$$

This is the number of all possible combinations in which two successes can be found in a sequence of three trials, times the probability of such a sequence. We can now generalize this result by considering first that the number of different combinations in which we can find m object in N observations is given by the binomial coefficient

$$
\binom{N}{m} = \frac{N!}{(N-m)!m!},
$$

and then noting that the probability of having a given sequence of m events each of probability ϑ in N extractions is $\vartheta^m(1-\vartheta)^{N-m}$. As a result we obtain the binomial probability distribution:

$$
p(m) = \binom{N}{m} \vartheta^m(1-\vartheta)^{N-m},
$$

where N and ϑ are the two parameters. Figure 6.4 shows the different forms of the distribution for various N and ϑ.

It is not difficult to obtain the mean and the variance of the distribution as $N\vartheta$ and $N\vartheta(1-\vartheta)$, respectively. It is also possible to show that, for $N \to \infty$ and finite ϑ, the envelope of the binomial distribution tends to a Gaussian distribution (see Fig. 6.4), a consequence of the central limit theorem, which will be described in Sec. 6.3.3.

6.3.2 Poisson Distribution

The Poisson distribution (also called the rare-event distribution) is obtained as a limiting form of the binomial distribution when the number of observations goes to infinity ($N \to \infty$), at the same time considering a very rare event ($\vartheta \to 0$). The limit is taken in such a way that the product $N\vartheta$ (i.e., the mean) tends to a finite limit, $N\vartheta \to \alpha$, different from zero.

It can be shown (Priestley, 1981) that the resulting distribution is

$$
p(m) = \frac{\alpha^m}{m!} e^{-\alpha},
\tag{6.18}
$$

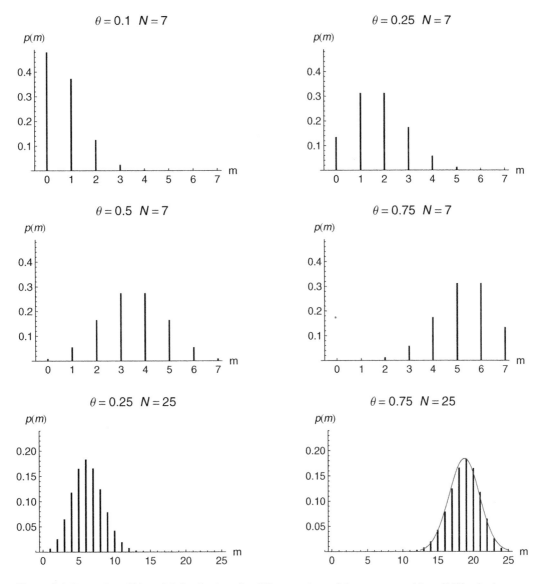

Figure 6.4 Examples of binomial distributions for different values of the parameters (the solid line in the lower right figure shows the normal PDF resulting from the central limit theorem for large N).

with parameter α, which has the noteworthy property of being equal to both the variance and the mean of the distribution. The Poisson distribution tends to a Gaussian for large values of α, as can be seen in Fig. 6.5. The Poisson distribution is related to the Poisson process, as we will see in Section 6.5.

6.3.3 Gaussian or Normal Distribution

The well-known Gaussian or normal distribution is

$$p(x) = \frac{1}{\sqrt{2\pi}\,\sigma} e^{-\frac{(x-\mu)^2}{2\sigma^2}}, \tag{6.19}$$

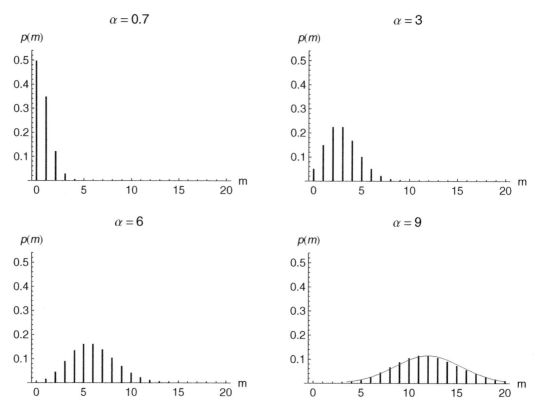

Figure 6.5 Examples of Poisson distributions for different values of the parameter α. The corresponding Gaussian distribution with same mean and variance is shown in continuous line for comparison.

for $-\infty < x < \infty$, with parameters μ (its mean) and σ (its variance). It is symmetric (zero skewness), as shown in Fig. 6.6, and has kurtosis equal to 3 (this is a reference value for all other distributions). Using the Tchebicheff inequality (Priestley, 1981), it is possible to show that about 67%, 95%, and 99% of the values are included between the intervals $\mu \pm \sigma$, $\mu \pm 2\sigma$, and $\mu \pm 3\sigma$, respectively. It is common to refer to the standard normal distribution, obtained by the linear variable transformation $u = (x - \mu)/\sigma$. Applying the derived distribution approach (Sec. 6.3.8) to Eq. (6.19) for this linear transformation yields the standard normal distribution

$$p(u) = \frac{1}{\sqrt{2\pi}} e^{-u^2/2}, \qquad (6.20)$$

which obviously has zero mean and unit variance.

The ubiquity of the Gaussian distribution is due to the very important *central limit theorem*: the distribution of the sum of n random variables, independent and identically distributed with mean μ and variance σ^2, for $n \to \infty$, tends to a Gaussian distribution with mean $n\mu$ and variance $n\sigma^2$. The theorem (see Van Kampen, 1992 for an elegant proof) is valid even under more general conditions of weak independence and

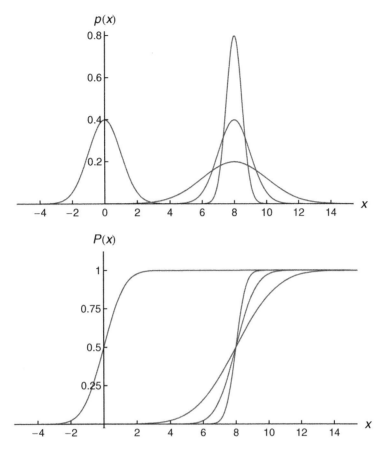

Figure 6.6 Examples of Gaussian distributions for different values of the parameters. The distribution on the left is the standard normal distribution ($\mu = 0$ and $\sigma = 1$); the other three distributions have $\mu = 8$ and $\sigma = 0.5, 1, 2$, respectively.

not necessarily identically distributed variables.[2] The Gaussian distribution has a noteworthy summability property, i.e., the sum of Gaussian RVs is still a Gaussian RV.

The CDF of the normal distribution cannot be obtained analytically but can be computed numerically. For the standard normal distribution, the CDF is related to the so-called error function (Abramowitz and Stegun, 1964)

$$P(u) = \int_{-\infty}^{u} p(u')du' = \frac{1}{2}(1 + \mathrm{Erf}(u)). \tag{6.21}$$

6.3.4 Dirac Distribution

The Dirac distribution, or Dirac delta function, is a singular function, and describes the limiting case of a distribution concentrated at a point, say x_0, for an RV which takes on the value x_0 with probability one

---

[2] The sums of RVs with distributions having infinite variance do not tend to normal distributions but follow a generalized form of the central limit theorem and tend to so-called Levy stable laws with power-law tails (Sornette, 2006).

(certainty). It is thus a PDF which is so spiky at x_0 that, although being on an infinitesimal support around x_0, when integrated its area is finite and equal to one. For example, making the variance of a Gaussian distribution tend to zero (Fig. 6.6) one obtains a Dirac distribution.

Formally, a Dirac distribution centered at $x = 0$ is defined as

$$\delta(x) = \begin{cases} 0 & \text{for} \quad x \neq 0, \\ \infty & \text{for} \quad x = 0, \end{cases} \tag{6.22}$$

plus the condition

$$\int_{-\infty}^{\infty} \delta(x) dx = 1. \tag{6.23}$$

As for all the probability distributions, its units are the inverse of those of its argument. The Dirac distribution has several interesting properties; we recall here only its symmetry $\delta(x) = \delta(-x)$, the sifting property $f(a) = \int_{-\infty}^{\infty} \delta(x - a) f(x) dx$, and the scaling property $\delta(\alpha x) = \delta(x)/|\alpha|$.[3] In the solution of stochastic processes, the Dirac delta function often appears to describe initial conditions known with certainty and when the solution is a mixed distribution. For a Dirac delta located at x_0, $\delta(x - x_0)$, the MGF is $e^{x_0 k}$.

The integral of the Dirac delta function defines another important special function, the *Heaviside step function*,

$$\Theta(x) = \int_{-\infty}^{x} \delta(x) dx = \begin{cases} 0 & x < 0, \\ 1 & x \geq 0, \end{cases} \tag{6.25}$$

describing mathematically a unit step located at $x = 0$.

6.3.5 Exponential Distribution

The exponential distribution is a simple and important distribution which appears as a solution of certain paradigmatic stochastic processes. It is also often employed as a first approximation to describe RVs that are positive definite, especially when analytical developments are required.

The exponential PDF is

$$p(x) = k e^{-kx} \tag{6.26}$$

with $x > 0$ and parameter k, and its CDF is

$$P(x) = 1 - e^{-kx}. \tag{6.27}$$

Integrating by parts the moment definition, one easily gets

$$\mu = \frac{1}{k} \quad \text{and} \quad \sigma^2 = \frac{1}{k^2}, \tag{6.28}$$

---

[3] This is a special case of the "transformation" property

$$\int_{-\infty}^{\infty} \delta(g(x)) f(x) dx = \sum_i \frac{f(x_i)}{|g'(x_i)|}, \tag{6.24}$$

where the sum extends over all the roots of $g(x)$.

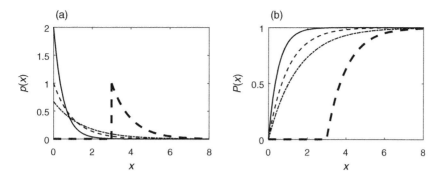

Figure 6.7 (a) PDF and (b) CDF for exponential distributions with parameters $k = 0.5, 1, 1.5$ (thin solid, dashed, and dot-dashed lines), along with a shifted version with $k = 1$ and $x_0 = 3$ (thick dashed lines).

which implies that the coefficient of variation $CV = 1$. The exponential distribution favors low values of the RV (the mode is at zero) and is positively skewed. Similarly to other distributions, it may be translated by introducing a shift parameter x_0 (e.g., Fig. 6.7) to give a bit more flexibility.

As we will see, the exponential distribution has an important relation with the Poisson process (Sec. 6.5), being the distribution of the *interarrival times* between its events (see Sec. 6.5). The memoryless property of the Poisson process derives from the invariance of the exponential distribution under "censoring" and rescaling (the exponential distribution is the only continuous distribution with this property). This is expressed as follows: given a RV x, exponentially distributed with parameter k, it is always true that, for any positive threshold a,

$$\text{Prob}[x - a > y | x > a] = \frac{\text{Prob}[x > y + a]}{\text{Prob}[x > a]} = \frac{e^{-k(y+a)}}{e^{-ka}} = e^{-ky}, \tag{6.29}$$

which follows from Eq. (6.27). In other words, taking only the values of x that are larger than a given value a (censoring) corresponds to considering a new RV y with exactly the same distribution of x.[4]

---

[4] A way to see how censoring changes the PDF of an RV is as follows. Since

$$1 - P_Y(y) = \frac{1 - P_X(y + a)}{1 - P_X(a)}, \tag{6.30}$$

the PDF of y is

$$p_Y(y) = \frac{dP_Y(y)}{dy} = \frac{d}{dy} \left[1 - \frac{1 - P_X(y + a)}{1 - P_X(a)} \right] = \frac{\frac{d}{dy} P_X(y + a)}{1 - P_X(a)}, \tag{6.31}$$

so that

$$p_Y(y) = \frac{p_X(y + a)}{1 - \int_{-\infty}^{a} p_X(x')dx'}, \tag{6.32}$$

which for an exponentially distributed y is given by

$$p_Y(y) = \frac{ke^{-k(y+a)}}{e^{-ka}}, \tag{6.33}$$

as in Eq. (6.29).

6.3.6   Gamma Distribution

The gamma distribution is another classic distribution that appears as the solution of many stochastic processes. Notably for ecohydrology, it is the steady-state solution of the minimalist stochastic processes of soil moisture with exponential jumps and decays (Sec. 7.5). Its most typical form has two parameters:

$$p(x) = \frac{b^a}{\Gamma(a)} x^{a-1} e^{-bx}, \tag{6.34}$$

where a and b are both positive and $x > 0$; a is called the shape factor, while b is the scale factor; $\Gamma(a)$ is a normalizing factor, in which $\Gamma(\cdot)$ is the gamma function defined below.

For $a = 1$ the gamma distribution becomes the exponential distribution. The gamma distribution is also related to the Poisson process (Sec. 6.5) as it gives the probability distribution of the time τ between r subsequent occurrences in a Poisson process (the Erlang distribution). To provide more flexibility, the gamma distribution is sometimes used with a shift parameter x_0, and Eq. (6.34) becomes the three-parameter gamma distribution, also called the Pearson type-III distribution.

The mean and variance of the gamma distribution are

$$\mu = \frac{\alpha}{\beta} + x_0 \quad \text{and} \quad \sigma^2 = \frac{\alpha}{\beta^2}. \tag{6.35}$$

Related to the gamma distribution is the *gamma function*, one of the special mathematical functions that appear in the solutions of linear second-order ODEs (such as the beta function, the Bessel functions, etc.; Abramowitz and Stegun, 1964). It is defined as

$$\Gamma(a) = \int_0^\infty t^{a-1} e^{-t} dt. \tag{6.36}$$

It is easy to see that $\Gamma(1) = 1$ and $\Gamma(a) = (a - 1)\Gamma(a - 1)$. The latter property allows us to interpret the gamma function as an extension of the factorial function to real numbers. In fact, applying recursively the above property to a generic real number $a = n + y$, where n is the largest integer in the factorial (thus $0 < y < 1$), gives

$$\begin{aligned}
\Gamma(a) = \Gamma(n + y) &= (n + y - 1)\Gamma(n + y - 1) \\
&= (n + y - 1)(n + y - 2)\Gamma(n + y - 2) \\
&= \dots \\
&= (n + y - 1)(n + y - 2) \cdots (y + 1)\Gamma(y + 1).
\end{aligned}$$

For $y = 0$, and recalling that $\Gamma(1) = 1$, one immediately obtains $\Gamma(n) = (n - 1)!$.

When the range of the RV is limited to values below x, the normalization constant of the *truncated gamma distribution* takes the form of an incomplete gamma function, defined as (Abramowitz and Stegun, 1964)

$$\Gamma(a, x) = \int_0^x t^{a-1} e^{-t} dt. \tag{6.37}$$

This function will be used in Chapter 7 in connection with a minimalist soil moisture model with linear losses.

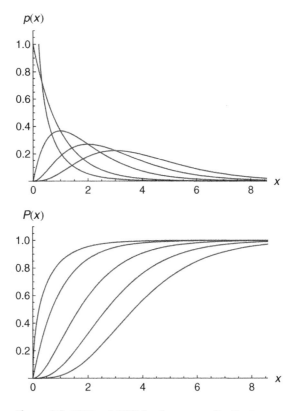

Figure 6.8 PDF and CDF for the gamma distribution with shape parameter $a = 0.5, 1, 2, 3, 4$ and scale parameter $b = 1$.

The gamma function and the *incomplete gamma function* can be computed numerically using approximate functions (Abramowitz and Stegun, 1964) or using standard numerical routines (e.g., Matlab or Mathematica).[5]

6.3.7 Pareto Distribution

The Pareto distribution can be seen as a generalization of the exponential distribution. It is a flexible distribution for positive definite variables with mode at zero and can have heavy tails (algebraic or power-law decay). Its general form with three parameters is

$$p(x) = \frac{1}{a} \left(1 - k\frac{x - x_0}{a} \right)^{1/(k-1)} \tag{6.38}$$

the parameters a, k, and x_0 are called the scale, shape, and position parameters, respectively.

For $k < 0$ the Pareto distribution is limited to $x > 0$, while for $k < 0$ it is bounded from both below and above, $x_0 < x < x_0 + a/k$. In the limit for $k \to 0$ it tends to the exponential distribution. Moreover,

---

[5] Note that Mathematica and Matlab use slightly different definitions of the incomplete gamma function.

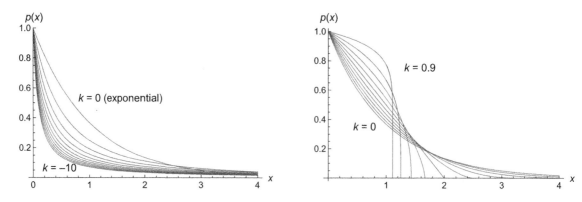

Figure 6.9 PDF and CDF (left- and right-hand panels) for the Pareto distribution for a range of values of the parameter k.

for $k \rightarrow 1$ and $x_0 = 0$ the Pareto distribution tends to the uniform distribution between 0 and 1. Its main moments are

$$\mu = x_0 + \frac{a}{1+k} \quad \text{and} \quad \sigma^2 = \frac{a^2}{(1+k^2)(1+2k)}. \tag{6.39}$$

Its variance diverges for $k < -1/2$, where it is a Levi-stable distribution (see footnote 2 regarding the central limit theorem in Sec. 6.3.3). Because of the algebraic decay of the tail, for $k < -1/3$ the skewness coefficient diverges.

6.3.8 Derived Distributions

The problem of derived distributions deals with obtaining the PDF $p_Y(y)$ of an RV y that is related to another RV x of known PDF by a one-to-one function $y = f(x)$. This is an extremely useful technique as problems of this type are often encountered.[6] Many examples in ecohydrology include the PDF for stream flow derived from the distribution of water level in a river, the distribution of daily carbon assimilation and transpiration as a function of the PDF of soil moisture (given the one-to-one relation seen in Sec. 5.4.5), and the PDF of plant water stress (Sec. 8.1).

Solving the problem of derived distributions is very easy for discrete RVs, because the probabilities remain the same in going from x_i to the corresponding y_i, i.e., $p_Y[y_i = f(x_i)] = p_X(x_i)$, and similarly for the cumulative distributions, $P_Y[y_i = f(x_i)] = P_X(x_i)$.

Things are more interesting for continuous RVs, since the PDF changes shape when going from x to $y = f(x)$. We consider explicitly the case of monotonically increasing functions. Consider a given value x^* which is mapped into $y^* = f(x^*)$. Since the probability of not exceeding y^* will be the same as that of not exceeding x^* (i.e., the fraction of values of y below y^* is the same as those of x below x^*), the non-exceedance probabilities of y^* and x^* will be the same, i.e., $\Pr[x < x^*] = \Pr[y < y^* = f(x^*)]$. This is valid for any value of x and $y = f(x)$. Therefore the CDFs of y and x will be related by $P_Y[y = f(x)] = P_X(x)$. Thus, the median of the two RVs is simply $\tilde{\mu}_y = f(\tilde{\mu}_x)$, and similarly for all the other quantiles.

---

[6] Van Kampen (1992) appropriately said: "The theory of probability is nothing but transforming variables".

Differentiating $P_Y(y)$ with respect to x using chain rule gives

$$\frac{dP_Y(y)}{dy}\frac{dy}{dx} = \frac{dP_X(x)}{dx}, \tag{6.40}$$

so that, by definition of the PDF,

$$p_Y(y) = p_X(x)\left(\frac{dy}{dx}\right)^{-1}, \tag{6.41}$$

where on the right-hand side x must be expressed as a function of y, $x = f^{-1}(y)$. Equivalently, Eq. (6.41) can be written as $p_Y(y)dy = p_X(x)dx$, which indicates that the PDF for y is distorted by the local slope of $f(x)$ in such a way as to maintain equality of the infinitesimal probabilities. We can interpret this distortion graphically as shown in detail in Fig. 6.10.

In general, for nonmonotonic functions, one can easily prove that

$$p_Y(y) = \sum_i \frac{p_X(x_i = f^{-1}(y))}{|dy/dx|_{x_i = f^{-1}(y)}}, \tag{6.42}$$

where the sum extends to all the roots of the inverse function $x_i = f^{-1}(y)$.

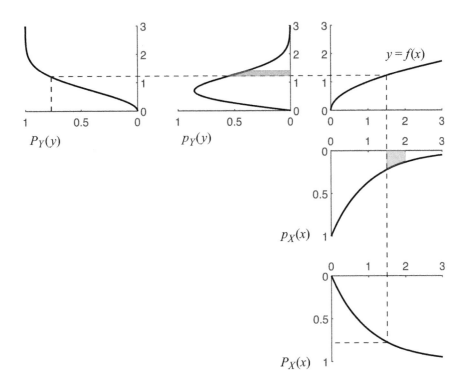

Figure 6.10 Graphical representation of the derived-distribution approach. In this example, x is exponentially distributed, $p_X(x) = e^{-x}$, and $y = f(x) = x^{1/2}$. The resulting distribution for y is a Raleigh distribution, $p_Y(y) = 2ye^{-y^2}$.

Note that when the function is flat, the transformation concentrates parts of a continuous PDF at a point, thereby producing an atom of probability at that level of y (i.e., a mixed distribution with Dirac deltas).[7]

6.3.9 Compound Distributions

In some cases, the parameters of a PDF may themselves be subject to some random variability, and it may be of interest to be able to describe how this variability modifies the behaviors of the random variable compared with the case when the parameter is a constant. For concreteness, consider a parameter β, and call the distribution of x with given value of such a parameter $p(x|\beta)$, while the PDF of the parameter is $f(\beta)$. Assuming that the values of β are independent of those of x, the overall distribution is obtained by weighting the original conditional PDF of x for a given β by the probability that that value of the parameter is actually observed,

$$p(x) = \int_{-\infty}^{\infty} p(x|\beta)f(\beta)d\beta, \tag{6.43}$$

where we have assumed β to vary between $-\infty$ and $+\infty$.

The double source of randomness present in Eq. (6.43) always increases the overall variability of x compared with the case of fixed β, and may give rise to power-law tails (see the application to the interannual rainfall variability in Sec. 7.8.2). An interesting application of compound distributions appears when one or more parameters of a stochastic process (see Sec. 6.4), which may be called internal, are in turn controlled by an external stochastic process, varying at a longer timescale. Thus one has two stochastic processes embedded into one another. If the internal stochastic process reaches equilibrium (e.g., a steady state) faster than the rate at which the external stochastic process makes the driving parameter change, one can solve the internal process with given values of the parameters and then use the compound distribution approach to get the whole solution. This type of approach was used in nonequilibrium statistical mechanics by Beck and Cohen (2003), who named it super-statistics. An application to ecohydrology and interannual rainfall variability was presented in Porporato et al. (2006).

6.4 Stochastic Processes

The theory of stochastic processes deals with systems that change randomly in time and/or space. We consider here only processes evolving in continuous time and with a continuous state space, thereby involving an RV $x(t)$ (we will limit our review to the univariate case so that x is a scalar), which evolves in time according to probabilistic laws. The assumption of continuous time often stems from mathematical convenience, although it does not necessarily mean that the physical interpretation is also valid up to infinitesimal scales. For example, rainfall and soil moisture are modeled in continuous time throughout the book but interpreted at the daily timescale; this means that below the daily timescale their physical behavior is not resolved by the model.

Assuming that a variable of interest is the result of a stochastic process means assuming that the time evolution of the random variable is completely described by the joint PDF of n values at different times (with n arbitrary):

[7] As we will see, this is the case when going from the soil moisture PDF to those of evapotranspiration assimilation and static water stress, as discussed in Sec. 8.1.

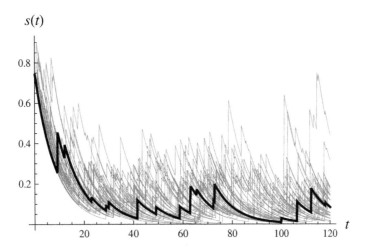

Figure 6.11 Ensemble of realizations of a stochastic process with random jumps of height, exponentially distributed and occurring at random times (according to a Poisson process), connected by exponential decays. The stochastic process may be seen as a minimalist model of soil moisture dynamics. In this example all the realizations start from a given initial condition $s_0 = 0.75$, as if originating from the end of a wet season, and evolve toward a stochastic steady state during a long homogeneous growing season.

$$p(x_1, t_1; x_2, t_2; \ldots; x_n, t_n). \tag{6.44}$$

While typically one observes only one realization of a stochastic process, it is useful – as often done in statistical physics – to refer to a hypothetical *ensemble* of realizations. Accordingly, one imagines that the probabilistic structure of a stochastic process may be obtained by analyzing (as we did for a single random variable in Sec. 6.2) an infinitely large set of identical realizations of the same stochastic process. These realizations are assumed to be independent and statistically equivalent repetitions of the same phenomenon. For each member of this ensemble, the probabilistic rules are kept the same, but the outcome differs because of the randomness of the dynamical components. From the ensemble, one can then construct distributions and all the statistics of the random variables at a given time or simultaneously at different times. Figure 6.11 illustrates this concept for a key example in ecohydrology, the evolution of a stochastic soil moisture process, $s(t)$, during a growing season. Assuming that each growing season forms a different realization of the same stochastic soil moisture dynamics (provided that the seasonal behavior remains the same and that interannual variability and autocorrelation are negligible), the ensemble of growing season realizations provides a full description of the stochastic process in Eq. (6.44).

The simplest category of stochastic processes is that where the present value of $x(t)$ is completely independent of both past and future,[8] in which case the joint PDF (6.44) can be factorized as follows:

$$p(x_1, t_1; x_2, t_2; \ldots; x_n, t_n) = p_1(x_1, t_1) p_2(x_2, t_2) \cdots p_n(x_n, t_n). \tag{6.45}$$

If the process is also time-independent, then all the one-time PDFs are the same and the process is simply a sequence of repeated Bernoulli trials, uncorrelated in time. In the continuous-time limit these processes are referred to as *white noise*, because of their flat power spectrum, which resembles that of natural (white) light.

[8] Independence of future events is typically assumed in physical systems to ensure causality.

6.4.1   Markov Processes

The next step beyond white noise is occupied by Markov processes, for which the immediate future is influenced only by the present conditions, not by the past. Thus, all the trajectories $x(t)$ having the same present value, say $x_0(t = 0)$, evolve (probabilistically) in the same way through a so-called transition probability. Formally, the Markov assumption is expressed by means of conditional probabilities as

$$p(x_1, t_1; x_2, t_2; \ldots | y_1, \tau_1; y_2, \tau_2; \ldots) = p(x_1, t_1; x_2, t_2; \ldots | y_1, \tau_1) \tag{6.46}$$

for

$$t_1 \geq t_2 \geq \cdots \geq \tau_1, \tau_2, \ldots \tag{6.47}$$

This also means that the joint probability in Eq. (6.44) can be expressed as (Gardiner, 2004)

$$p(x_1, t_1; x_2, t_2; \ldots; x_n, t_n) = p(x_1, t_1 | x_2, t_2) p(x_2, t_2 | x_3, t_3) \cdots p(x_{n-1}, t_{n-1} | x_n, t_n). \tag{6.48}$$

Although no physical process can be truly Markovian, since there is always some memory over some finite temporal interval, this assumption may in fact be a very good approximation at the proper timescale. As was said by Van Kampen (1992), "the art [...] is to make the description (approximately) Markovian." The Markovian assumption makes the problem more tractable analytically.

Limiting our general overview to continuous-time processes, the evolution equation of a Markovian stochastic process (this is the so-called Langevin approach) may be written in the form of a *stochastic differential equation* (SDE),

$$\frac{dx}{dt} = a(x, t) + b(x, t)\xi(t) + \mathcal{F}(x, t), \tag{6.49}$$

where $a(x, t)$ is a (deterministic) drift term, $b(x, t)$ is the diffusion term, $\xi(t)$ is the white Gaussian noise with zero mean and unit variance (the formal derivative of a Wiener process), and $\mathcal{F}(x, t)$ is a term inducing random jumps. When the diffusion and jump terms depend on x, the SDE is ill-defined and requires further specification of how exactly those functions are computed (e.g., using a causal prescription, with antecedent value $x(t)$, or the value at some later time during the noise effect. This leads to the Ito–Stratonovich dilemma (Van Kampen, 1992; Bartlett and Porporato, 2018)).

The previous SDE along with its noise prescription corresponds (e.g., Gardiner, 2004) to a first-order-in-time partial differential equation for the time-varying PDF $p(x, t)$, called the differential *Chapman Kolmogorov forward equation*,

$$\frac{\partial p(x, t)}{\partial t} = -\underbrace{\frac{\partial (a(x, t) p(x, t))}{\partial x}}_{\text{drift}} + \underbrace{\frac{1}{2} \frac{\partial^2 (b^2(x, t) p(x, t))}{\partial x^2}}_{\text{diffusion}}$$

$$+ \underbrace{\int_{-\infty}^{\infty} [W(x|z, t) p(z, t) - (W(z|x, t) p(x, t)] dz}_{\text{jump}}, \tag{6.50}$$

where $p(x, t)$ can be either the transition PDF, $p(x, t | x_0, 0)$, with initial condition $\delta(x - x_0)$, or the one-time PDF, $p(x, t) = \int p_0(x_0) p(x, t | x_0, 0) dx_0$, with generic initial condition $p_0(x)$ (Gardiner, 2004). The functions $W(x|z, t)$ and $W(z|x, t)$ refer to the instantaneous transition probabilities of jumping from z to x and from x to z and are related to the timing and size of the jumps (Bartlett et al., 2015; Bartlett and Porporato, 2018). It can be shown that the diffusion term (corresponding to very frequent but small Gaussian forcing) does not make the realizations of $x(t)$ discontinuous, although it does make them nondifferentiable, while the jump term causes intermittent and sudden discontinuities (Gardiner, 2004).

In ecohydrology, where the forcing is mainly due to rainfall inputs, which are intermittent in time and highly non-Gaussian (see Sec. 7.4), the resulting Chapman–Kolmogorov equations typically have diffusion terms that are negligible compared with the jump term. In the following sections, we will discuss in more detail specific cases where jumps occur as the result of a Poisson process. When jump terms are not present, the Chapman–Kolmogorov equation reduces to a nonhomogeneous advection–diffusion equation known as the Fokker–Planck equation.

6.4.2 Steady State, Ergodicity, and Seasonality

When the parameters of the drift, diffusion, and jump terms in Eq. (6.49) are time independent, their influence may balance in the long-term limit. In these cases, after an initial transient the system approaches a stochastic steady state (see Fig. 6.11), for which the statistics of $x(t)$ and $x(t + \tau)$ are the same for any τ. This means that the one-time probability is independent of time, so that

$$\lim_{t \to \infty} p(x, t) = p(x), \tag{6.51}$$

and

$$\lim_{t \to \infty} \frac{\partial p(x, t)}{\partial t} = 0. \tag{6.52}$$

Stationary stochastic processes are usually also ergodic, meaning that average (ensemble) values computed over the ensemble can be practically constructed from the subsequent x values of a single realization as long as these values are sufficiently separated in time to be statistically independent. In this case, the time average converges to the ensemble average,

$$\lim_{T \to \infty} \frac{1}{T} \int_0^T x(t)dt = \langle x \rangle. \tag{6.53}$$

The case of periodic forcing, which for example happens in ecohydrology for seasonal climates, is slightly more complicated. In such a case, in fact, the parameters in Eq. (6.49) are periodic functions of time with period T_{seas} and Eq. (6.53) obviously does not hold. However, the process statistically repeats itself periodically after the initial transient, i.e.,

$$p(x, t) = p(x, t + T_{\text{seas}}), \tag{6.54}$$

reaching a sort of periodic "steady" state. This is the case, for example, for soil moisture in seasonally dry ecosystems (Feng et al., 2012, 2015), which will be analyzed in Sec. 7.8.1.

6.5 Poisson Process

The *Poisson process*, named after Siméon-Denis Poisson (1781–1840), is one of the most important stochastic processes and its use is crucial in ecohydrology to represent the punctuated nature of rainfall at the daily level. It is the simplest and most fundamental point process used to model the occurrence of random "points" in time and space. It has been used, for example, to describe the times of radioactive emissions, the arrival times of customers at a service center, the positions of flaws in a piece of material, the telephone

Figure 6.12 Poisson process for random event occurrence and related counting process.

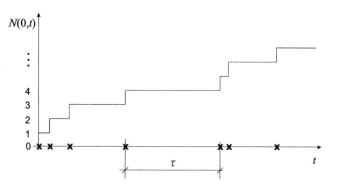

calls arriving at a switchboard, car arrivals at traffic junctions, page-view requests to a website, etc. As was mentioned previously, in ecohydrology it is a very useful model for the arrival of storm events in time. In the context of our review of continuous-time Markov processes (Sec. 6.4), Poisson processes are the simplest way to describe the random occurrence of jumps in Eqs. (6.49) and (6.50).

Specifically, a Poisson process is a continuous-time process in which events occur independently of one another with a constant mean rate (or *frequency*) of occurrence per unit time λ, which is the only parameter of the process. For example, when using the Poisson process to model rainfall events, one rainfall event occurring on average every five days corresponds to a mean rainfall rate of $\lambda = 1/5 \ \mathrm{day}^{-1}$.

To describe the Poisson process mathematically, one introduces a discrete random process $N(0, t)$, called the Poisson counting process,[9] where $N(0, t)$ is the total number of events which occur in the interval $(0, t)$; see Fig. 6.12.

The defining properties of the Poisson process are, for a positive constant λ and as $\Delta t \to 0$ (Cox and Miller, 1965):

$$\mathrm{prob}[N(t, t + \Delta t) = 0] = 1 - \lambda \Delta t + o(\Delta t), \tag{6.55}$$

$$\mathrm{prob}[N(t, t + \Delta t) = 1] = \lambda \Delta t + o(\Delta t), \tag{6.56}$$

$$\mathrm{prob}[N(t, t + \Delta t) > 1] = o(\Delta t). \tag{6.57}$$

A basic result is that for each t the random variable $N(0, t)$ has an exact Poisson distribution (6.18) with parameter $a = \lambda t$,

$$p[N(0, t) = m] = \frac{(\lambda t)^m}{m!} e^{-\lambda t}. \tag{6.58}$$

An heuristic derivation (adapted from Priestley, 1981) is as follows. Divide the time interval t into n subintervals each of length $\Delta t = t/n$. In these small intervals, the random events occur independently with probability $\vartheta = \lambda \Delta t + o(\Delta t)$, according to Eq. (6.56). This implies that the number m of occurrences in n intervals is a binomial variable with mean $n\vartheta = n\lambda \Delta t + o(\Delta t) = \lambda t + o(\Delta t)$. Letting the time interval become infinitesimal, one obtains a Poisson distribution with parameter exactly λt.

The other important result is that the interarrival times are independent[10] and exponentially distributed. To prove this, consider the PDF of the time to first occurrence $p(t)$ and the corresponding CDF $P(t)$ in

---

[9] We stress that $N(0, t)$ is a discrete RV but evolves in continuous time; the Poisson counting process is an example of a stochastic process in continuous time but with a discrete state space.

[10] Point processes with independent subsequent interarrival times are called renewal processes.

a Poisson process with rate λ. The complement of $P(t)$ is the exceedance probability, $1 - P(t)$, that is, the probability that the time of first occurrence is larger than t. It is easy to see that this is equal to the probability of having no occurrences during the period $(0, t)$. The latter is given by the Poisson distribution (6.58) with $m = 0$, i.e.,

$$1 - P(t) = e^{-\lambda t}, \tag{6.59}$$

so that by differentiation one obtains the PDF of the time of first occurrence, $p(t) = \lambda e^{-\lambda t}$. The last step is to realize that, since occurrences are independent, the time origin can be chosen arbitrarily. Therefore, it can also be chosen to coincide with an occurrence, say t_n. As a result, the distribution of times to the next arrival, $\tau = t_{n+1} - t_n$ is same as that for t obtained before,

$$p(\tau) = \lambda e^{-\lambda \tau}. \tag{6.60}$$

Note that the last consideration, about the time origin, is a result of the memoryless property of the exponential distribution in relation to the Poisson process (it is the only point process with this property). In fact, as we previously showed, when the exponential distribution is censored and the distributions are rescaled, the remaining values are still distributed exponentially: $P(s + t|s) = P(t)$ for all s and $t > 0$ (see Eq. (6.29)). This means that, if the events occur according to a Poisson process and we have already waited a time s, the waiting time to the next arrival will be the same as if we had just had an arrival at s. This reflects the fact that the future waiting time is independent of the past history of the process.[11]

6.6 Marked Poisson Process

A marked (or compound) Poisson process associates an independent random variable y extracted from a distribution $f(y)$ with each occurrence of the Poisson process. In rainfall modeling, for example, we associate with each rainfall event a random amount of rainfall, i.e., a *mark* (see Fig. 6.13). For simplicity, we will limit our exposition to positive definite marks.

We consider in particular the total accumulated up to time t, which is the sum

$$x(t) = \sum_{n=1}^{N(0,t)} y_n. \tag{6.61}$$

Clearly, this forms a stochastic jump process, with discontinuities each time a mark occurs (Fig. 6.13). The continuous RV $x(t)$ can be described probabilistically by means of its time-dependent PDF $p(x, t)$, using the properties of the Poisson process with initial conditions $p(x, t = 0) = \delta(x - x_0)$. Consider a small time interval, from t to $t + \Delta t$. According to the defining properties of the Poisson process, there are two possibilities:

$$x(t + \Delta t) = \begin{cases} x(t) & \text{no jump, with probability} & 1 - \lambda \Delta t + o(\Delta t), \\ u + y & \text{jump, with probability} & \lambda \Delta t + o(\Delta t), \end{cases} \tag{6.62}$$

---

[11] This is also known as the hitchhiker's paradox. If cars arrive according to a Poisson process, the hitchhiker who arrives at the roadside randomly in time waits, on average, the same time as the (supposedly unlucky) one who always arrives just after the arrival of the previous car. The intuitive explanation is that the hitchhikers' probability of arriving during a long interval is greater than that of arriving during a short interval.

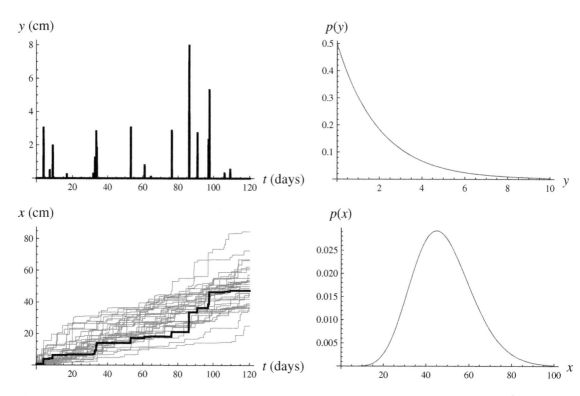

Figure 6.13 Growing-season rainfall as a realization of a marked Poisson process, with rate $\lambda = 0.2$ day^{-1} and exponential marks with parameter $\gamma = 0.5$ cm^{-1}, and accumulated rainfall, $x(t)$, along with their corresponding PDFs (see Eq. (6.68)).

where u is the state before the jump, and the jump y, assumed to be positive definite, has a probability distribution $f(y)$. The probability that the process takes a value in $(x, x + dx)$ at time $t + \Delta t$ can therefore be expressed as (Cox and Miller, 1965, p. 239)

$$p(x, t + \Delta t)dx = (1 - \lambda\Delta t)\,p(x, t)dx + \lambda\Delta t \int_0^x p(u, t)f(x - u)dudx + o(\Delta t), \qquad (6.63)$$

which is the sum of the probability of finding x at time t without any jump since the previous measurement plus the probability of arriving at x with a jump (this, in turn, is equal to the product of the probability of jumping times the probability of being in $[u, u + du]$ before the jump times the probability of jumping by an amount $x - u$ to reach x from u, summed over all possible values of u below x). A diagram indicating the derivation of the above terms is provided in Fig. 6.14. Note that, since Δt can be made infinitesimally small, we can assume the presence of only one jump in that interval, according to the properties of the Poisson process.

Dividing by dx, subtracting $p(x, t)$ from both sides, dividing by Δt, and taking the limit $\Delta t \to 0$, one gets

$$\frac{\partial}{\partial t}p(x, t) = -\lambda p(x, t) + \lambda \int_0^x p(u, t)f(x - u)du, \qquad (6.64)$$

which is the forward Chapman–Kolmogorov equation for the process $x(t)$.

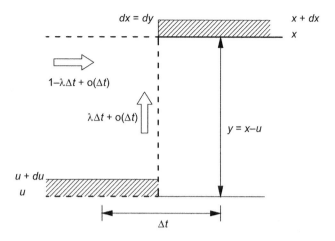

Figure 6.14 Diagram for the derivation of Eq. (6.63).

By taking the *Laplace transform* (Andrews, 1986) of Eq. (6.64),

$$p^*(k, t) = p^*(k, 0)e^{-\lambda(1-f^*(k))t}, \qquad (6.65)$$

where $p^*(k, t)$ and $f^*(k)$ are the Laplace transforms of $p(x)$ and $f(y)$, respectively ((see Eq. (6.13) for a link to the moment generating function)). For an initial condition concentrated at $x = x_0$ at time $t = 0$, which corresponds to $p(x, 0) = \delta(x - x_0)$, one has

$$p^*(k, t) = e^{-x_0 k}e^{-\lambda t}e^{\lambda f^*(k)t}, \qquad (6.66)$$

from which, using the properties of the MGF, one easily obtains expressions for the mean and the variance of the process as a function of time (see Eq. (6.13)).[12]

A specific solution of (6.66) can be obtained for the case of exponential jumps with parameter γ (recall that γ is the inverse of the mean jump size), $f(y) = \gamma e^{-\gamma y}$, for which $f^*(k) = \frac{\gamma}{\gamma - k}$,

$$p^*(k, t) = e^{-x_0 k}e^{-\lambda t}e^{\lambda \gamma t/(\gamma - k)t}. \qquad (6.67)$$

To anti-transform this equation, it is useful first to note that $e^{-\lambda t}$ can be treated as a constant and then to use the property that $e^{-x_0 k}g^*(k)$ is anti-transformed as $\Theta(x - x_0)g(x - x_0)$, where $\Theta(\cdot)$ is the Heaviside function. Then the Laplace transform tables (or Mathematica) provide the anti-transform for the remaining part, $g^*(k) \to e^{-\gamma(x)}\sqrt{\frac{\lambda \gamma t}{x}}I_1(2\sqrt{\lambda \gamma(x)t})$, so that Eq. (6.67) can be anti-transformed as

$$p(x, t) = e^{-\lambda t}\left[e^{-\gamma(x-x_0)}\sqrt{\frac{\lambda \gamma t}{x - x_0}}I_1(2\sqrt{\lambda \gamma(x - x_0)t}) + \delta(x - x_0)\right], \qquad (6.68)$$

for $x \geq x_0$, where $I_\nu(\cdot)$ is the modified Bessel function of the first kind of order ν (Abramowitz and Stegun, 1964), which for large times tends to a Gaussian, as expected from the central limit theorem (Sec. 6.3.3). The above solution is plotted in the bottom right panel of Fig. 6.13.

---

[12] Note that for the Laplace transform, which has a minus sign at the exponent, $\mathcal{L}\{f\}(k) = \int_0^\infty f(x)e^{-kx}dx = M(-k)$.

6.7   Master Equation for Jump Processes

In the previous section the variable x, the cumulated total of the inputs by a marked Poisson process, remained constant during interarrival periods. We now generalize the analysis by considering a one-dimensional dynamical system where the variable $x(t)$ evolves deterministically during the interarrival periods of a Poisson process driven by a differential term $\rho(x)$, that is,

$$\frac{dx}{dt} = -\rho(x), \tag{6.69}$$

and with jumps of a random amount y distributed as $f(y)$ each time a Poisson event occurs. The case of the previous section is recovered by simply putting $\rho(x) = 0$ in Eq. (6.69). In ecohydrology, the relative soil moisture dynamics, to be analyzed in detail in Chapter 7, provides an important example of a dynamical system driven by marked Poisson noise. We will limit our analysis to cases where $\rho(x) \geq 0$ and where there are positive definite jumps, although generalizations are not difficult.

The overall evolution of the system with deterministic decays and random jumps may be formally described as a Langevin-type SDE (see Eq. (6.49)), valid at all times,

$$\frac{dx}{dt} = -\rho(x) + \mathcal{F}_{\lambda, f(y)}(t), \tag{6.70}$$

where $\mathcal{F}_{\lambda, f(y)}(t)$ represents the random forcing in time given by the formal derivative (e.g., its time integral must give the jumps' cumulative totals as in Eq. (6.61)) for a marked Poisson process,

$$\mathcal{F}_{\lambda, f(y)}(t) = \sum_{n=1}^{N(0,t)} y_n \delta(t - t_n). \tag{6.71}$$

As before, for exponentially distributed jumps with parameter γ, $f(y) = \gamma \exp(-\gamma y)$, $\mathcal{F}_{\lambda, f(y)}(t)$ can be simplified to $\mathcal{F}_{\lambda, \gamma}(t)$.

The evolution equation for the PDF of x, $p(x, t)$, corresponding to the previous SDE, can be constructed as in Sec. 6.6, with some extra attention because of the presence of the drift term. In fact, as is apparent from Fig. 6.15, the drift term tends to distort the trajectories during their deterministic decay, with the

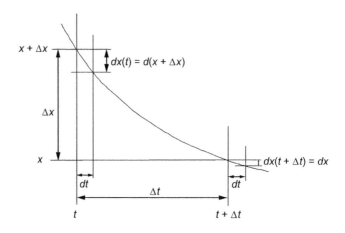

Figure 6.15 Diagram for the derivation of Eq. (6.77).

result that the infinitesimal intervals in x are different at different times. We will use the notation $d(x + \Delta x)$ for $dx(t)$, and dx for $dx(t + \Delta t)$.

Consider again a small time interval from t to $t + \Delta t$, where now, because of the drift, one has

$$
x(t + \Delta t) = \begin{cases} x(t) - \Delta x & \text{with probability} \quad 1 - \lambda \Delta t + o(\Delta t), \\[2mm] u - \Delta x' + y - \Delta x'' & \text{with probability} \quad \lambda \Delta t + o(\Delta t), \end{cases} \tag{6.72}
$$

where $\Delta x = \int_t^{t+\Delta t} \rho[x(\tau)]d\tau = \rho[x(t)]\Delta t + o(\Delta t)$ is the decrement due to deterministic decay corresponding to the solution of (6.69), $\Delta x'$ and $\Delta x''$ are similar decrements just before and after the jump, and as in the previous section y is the jump with PDF $f(y)$ and u is the state before the jump.

The probability that the system is in the interval $(x, x + dx)$ at time $t + \Delta t$, arriving there without a jump, is equal to the probability that the system does not jump times the probability that it is in the interval $(x, x + d(x + \Delta x)$ at time t, that is,

$$
(1 - \lambda \Delta t)p(x + \Delta x, t)d(x + \Delta x) + o(\Delta t), \tag{6.73}
$$

while the probability that the system arrives in $(x, x + dx)$ at time $t + \Delta t$ with a jump is

$$
\lambda \Delta t \int_0^x p(u + \Delta x', t)dx(t)f(x - u + \Delta x'')du + o(\Delta t). \tag{6.74}
$$

Noting that in Eq. (6.73) the differential of x at time t can be simplified as

$$
d(x + \Delta x) = dx + \frac{\partial \rho(x)\Delta t}{\partial x}dx + o(\Delta t) \tag{6.75}
$$

and that within the integral in Eq. (6.74) one can similarly expand and neglect terms $o(\Delta t)$, one can then write the overall probability that the system arrives in $(x, x + dx)$ at time $t + \Delta t$ as

$$
p(x, t + \Delta t)dx = (1 - \lambda \Delta t)\left[p(x, t) + \rho(x)\Delta t \frac{\partial}{\partial x}p(x, t) \right]\left[1 + \frac{\partial}{\partial x}\rho(x)\Delta t \right]dx
$$
$$
+ \lambda \Delta t\, dx \int_0^x p(u, t)f(x - u)du. \tag{6.76}
$$

Finally, dividing by Δx, subtracting $p(x, t)$ from both sides, dividing by Δt, and taking the limit as $\Delta t \to 0$, one arrives at the *master equation* or Chapman–Kolmogorov forward equation for x:

$$
\frac{\partial}{\partial t}p(x, t) = \frac{\partial}{\partial x}\left[\rho(x)p(x, t) \right] - \lambda p(x, t) + \lambda \int_0^x p(u, t)f(x - u)du. \tag{6.77}
$$

The various terms on the right-hand side of the integro-differential equation (6.77) represent the contributions to $p(x, t)$ of the different mechanisms acting on the variable x. The first term is related to the gain of probability due to the deterministic decay (drift) caused by $\rho(x)$, the second is the loss of probability due to possible jumps with frequency λ which cause the process to leave the given trajectory at the level x at time t, and the last term is the positive contribution to the probability due to jumps to level x starting from lower values. The master or Chapman–Kolmogorov forward equation (6.77) is valid for general choices of the loss function $\rho(x)$ and positive-jump distribution $f(y)$. For exponentially distributed jumps with parameter γ, the master equation becomes

$$
\frac{\partial}{\partial t}p(x, t) = \frac{\partial}{\partial x}\left[\rho(x)p(x, t) \right] - \lambda p(x, t) + \lambda \gamma \int_0^x e^{-\gamma(x-u)}p(u, t)du. \tag{6.78}
$$

The master equation can be expressed as the conservation of the number of trajectories at a point x in time (Daly and Porporato, 2010; Bartlett et al., 2015; Bartlett and Porporato, 2018):

$$\frac{\partial}{\partial t} p(x,t) = -\frac{\partial J(x,t)}{\partial x}, \tag{6.79}$$

where $J(x,t)$ is the *probability current*:

$$J(x,t) = -\left[\rho(x)p(x,t)\right] + \lambda \int_0^x \left(\int_{x-z}^{+\infty} f(y)dy\right) p(z,t)dz. \tag{6.80}$$

For exponentially distributed jumps with parameter γ the probability current is

$$J(x,t) = -\left[\rho(x)p(x,t)\right] + \lambda \int_0^x e^{-\gamma(x-u)}p(u,t)du. \tag{6.81}$$

6.7.1 Steady-State PDF

A general solution for jump processes under stochastic steady-state conditions (see Sec. 6.4.2) is possible for exponentially distributed jumps. Taking the limit as $t \to \infty$ in Eq. (6.78) and using Eq. (6.52), one obtains

$$\frac{d}{dx}\left[\rho(x)p(x)\right] - \lambda p(x) + \lambda \int_0^x p(u)\gamma e^{-\gamma(x-u)}du = 0. \tag{6.82}$$

Multiplying by $e^{\gamma x}$ and differentiating with respect to x, one obtains (Cox and Isham, 1986; Rodríguez-Iturbe et al., 1999a)

$$\frac{d^2}{dx^2}\left[\rho(x)p(x)\right] + \frac{d}{dx}\left[\gamma\rho(x)p(x) - \lambda p(x)\right] = 0, \tag{6.83}$$

which can be immediately integrated once to obtain

$$\frac{d}{dx}\left[\rho(x)p(x)\right] + \gamma\rho(x)p(x) - \lambda p(x) = \text{const.} \tag{6.84}$$

To evaluate the integration constant, we first evaluate Eq. (6.82) at $x = 0$:

$$\frac{d}{dx}\left[\rho(x)p(x)\right]\Big|_{x=0} - \lambda p(0) = 0, \tag{6.85}$$

and then evaluate Eq. (6.84) at $x = 0$:

$$\frac{d}{dx}\left[\rho(x)p(x)\right]\Big|_{x=0} + \gamma\rho(0)p(0) - \lambda p(0) = \text{const.} \tag{6.86}$$

In the case where $\rho(0) = 0$ (e.g. no water loss for a dry soil), the integration constant becomes 0. As a result, the general form of the steady-state solution is easily obtained from Eq. (6.84). Multiplying and dividing the last term on the left-hand side of Eq. (6.84) by $\rho(x)$, and setting the constant to zero, the general solution can be obtained as follows:

$$\frac{d}{dx}f + \left[\gamma - \frac{\lambda}{\rho(x)}\right]f = 0, \tag{6.87}$$

where $f = \rho(x)p(x)$. Separating the variable f and integrating, one obtains an indefinite integral:

$$\ln[\rho(x)p(x)] = -\gamma x + \lambda \int^x \frac{du}{\rho(u)} + C', \tag{6.88}$$

so that, finally, the steady-state PDF becomes

$$p(x) = \frac{C}{\rho(x)} \exp\left(-\gamma x + \lambda \int^x \frac{du}{\rho(u)}\right), \tag{6.89}$$

where $C = e^{C'}$ is normalization constant such that

$$\int_0^\infty p(x)dx = 1. \tag{6.90}$$

In the case of a piecewise loss function $\rho(x)$, the limits of the integral in the exponential term of Eq. (6.89) must be chosen to ensure the continuity of $p(x)$ at the points of discontinuity of the loss function (Cox and Isham, 1986; Rodríguez-Iturbe et al., 1999b).

6.7.2 Effect of an Upper Bound

In ecohydrology, the dynamics of soil moisture $x(t)$, which has positive jumps and positive loss function $\rho(x)$, cannot exceed saturation. It is thus important to discuss the effects of an upper bound, say x_b, on the evolution of $p(x, t)$, and especially on its steady-state forms. The bound at x_b does not produce an atom of probability at x at that level, since the process decays immediately from that state. As a result, at steady state the PDF must be rescaled to have area one over the bounded domain, but otherwise it retains the same mathematical form. Thus for the steady-state PDF, all the complications brought by the presence of the bound are contained in the integration constant. Calling $p_b(x)$ the steady-state PDF of the bounded process, we thus have

$$\int_0^{x_b} p_b(x)dx = C_b \int_0^{x_b} p(x)dx = 1, \tag{6.91}$$

so that $C_b = 1/\int_0^{x_b} p(x)dx$. It follows that

$$p_b(x) = C_b p(x) = C_b \frac{C}{\rho(x)} \exp\left(-\gamma x + \lambda \int^x \frac{du}{\rho(u)}\right). \tag{6.92}$$

This is similar to the problem of the censoring of random variables (Sec. 6.3.5). The explanation for this lies in the Markovian nature of the process (Daly and Porporato, 2010). If excursions of the process above x_b are impossible, the process will spend more time in states $x \leq x_b$ than would be the case otherwise, but the relative proportions of the times in those states will be unchanged. Imagine two processes with, and without, the restriction that x is bounded to values less than one. In the latter case, trajectories of the soil moisture process will jump above the level $x = x_b$ and, eventually, drift down across this level once more. In the former case, these excursions are effectively excised, as the process jumps only to $x = x_b$ and then immediately begins its downward decay. The trajectories below x_b in the two processes are indistinguishable.

6.8 Crossing Properties

The frequency and duration of excursions of a stochastic process above or below a specific threshold provide a description of the process that is complementary to that given by its PDF. In ecohydrology, the time spent between specific levels of soil moisture or plant water potential is directly related to the physiological dynamics of plants or to the fluxes in the soil water balance (e.g., wilting and onset of stress, percolation, carbon assimilation, etc.). Hence, knowledge of these quantities is of crucial importance.

Typically, only the mean statistics of threshold or level crossing can be deduced analytically. In this section, we provide a brief review of the main analytical expressions for stationary processes. We refer to Chapter 3 of Rodríguez-Iturbe and Porporato (2004) and the references therein for a more detailed exposition of the subject.

6.8.1 Level Crossings

We begin with level-crossing statistics (see Fig. 6.16). We refer in particular to the mean duration in which the process $x(t)$ is below or above a threshold ξ, and the frequency and mean number of such intervals during a period of time T_{seas} (e.g., a growing season).

For stationary processes, the mean *frequency of crossing* of a level ξ is given by very general formula of Rice (Vanmarcke, 2010),

$$\nu(\xi) = \int_{-\infty}^{+\infty} |\dot{x}| p(\xi, \dot{x}) d\dot{x} = \langle |\dot{x}| \,|\xi\rangle p(\xi), \tag{6.93}$$

where $p(\xi, \dot{x})$ is the joint PDF of ξ and $\dot{x} = dx/dt$. Thus, the mean frequency of crossing is related to the mean of the absolute value of the slope $\dot{x}$ at the crossing level ξ (also see Brill et al., 2008). In stationary conditions, the number of downcrossings is equal to that of upcrossings,

$$\nu^{\downarrow}(\xi) = \nu^{\uparrow}(\xi) = \frac{\nu_{tot}(\xi)}{2}, \tag{6.94}$$

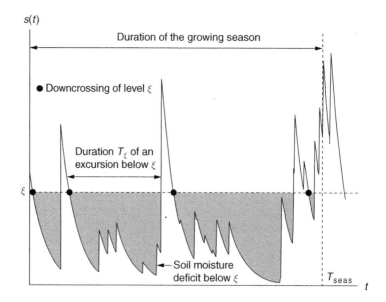

Figure 6.16 Crossing properties of a threshold ξ. After Ridolfi et al. (2000).

where $v^\downarrow(\xi)$ and $v^\uparrow(\xi)$ are the frequencies of downcrossing and upcrossing, respectively. These frequencies can also be found by separating the positive and negative contribution in Eq. (6.93):

$$v^\downarrow(\xi) = -\int_{-\infty}^{0} \dot{x}p(\xi, \dot{x})d\dot{x} = -\langle \dot{x}^- | \xi \rangle p(\xi) \tag{6.95}$$

and

$$v^\uparrow(\xi) = \int_{0}^{\infty} \dot{x}p(\xi, \dot{x})d\dot{x} = \langle \dot{x}^+ | \xi \rangle p(\xi). \tag{6.96}$$

For processes with positive jumps and negative deterministic decays, such as those discussed in the previous section, the slope of downcrossing at a given level is a deterministic quantity equal to $\rho(\xi)$, so that Eq. (6.95) immediately gives

$$v^\downarrow(\xi) = \langle \dot{x}^- | \xi \rangle p(\xi) = \rho(\xi)p(\xi), \tag{6.97}$$

An intuitive derivation of the previous result is as follows (Porporato et al., 2001). Given the process $x(t)$, with instantaneous upward jumps followed by continuous decays with rate $\rho(x)$, the time spent by the process between the levels ξ and $\xi + d\xi$ is zero during an upcrossing and is equal to $d\xi/\rho(\xi)$ during a downcrossing. Thus, since the total fraction of time spent by the process between ξ and $\xi + d\xi$ is, by definition, $p(\xi)\,d\xi$, the frequency of downcrossings (or upcrossings) can be simply obtained as the ratio of such a fraction of time and the time spent during a single downcrossing, i.e.,

$$v^\downarrow(\xi) = \frac{p(\xi)d\xi}{d\xi/\rho(\xi)}, \tag{6.98}$$

which gives Eq. (6.97). The mean number of upcrossings (or downcrossings) during a period of length T_{seas} is then readily obtained as

$$\overline{n}_\xi^\downarrow(\xi) = v_\xi^\downarrow T_{\text{seas}} = \rho(\xi)p(\xi)T_{\text{seas}}. \tag{6.99}$$

The *mean duration* of an excursion below the soil moisture level ξ, $\overline{T}_\xi^\downarrow$, can be found by combining the steady-state PDF and the frequency of downcrossing. The product of $v_\xi^\downarrow$ and $\overline{T}_\xi^\downarrow$ gives the fraction of time that the trajectory spends below ξ, which in steady-state conditions is equal to the cumulative probability distribution evaluated in ξ, $P(\xi) = \int_0^\xi p(x)dx$. The mean time between a downcrossing and the subsequent upcrossing can thus be calculated as

$$\overline{T}_\xi^\downarrow = \frac{P(\xi)}{v_\xi^\downarrow}, \tag{6.100}$$

and, substituting Eq. (6.97) into Eq. (6.100),

$$\overline{T}_\xi^\downarrow = \frac{P(\xi)}{\rho(\xi)p(\xi)}. \tag{6.101}$$

Similarly, the average time spent above ξ is

$$\overline{T}_\xi^\uparrow = \frac{1 - P(\xi)}{v_\xi^\downarrow} = \frac{1}{v^\downarrow(\xi)} - \overline{T}_\xi^\downarrow. \tag{6.102}$$

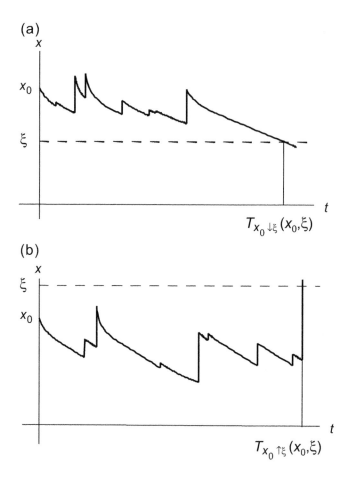

Figure 6.17 Definition of the first passage times. After Laio et al. (2001a).

6.8.2 Mean First Passage Times

The first passage time (MFPT) to reach a given level ξ starting from an initial point x_0 is often also important in ecohydrology. We distinguish between the two cases in which x_0 is either above or below the final level ξ (Fig. 6.17), and use the notation $\overline{T}_{x_0\uparrow\xi}(x_0,\xi)$ and $\overline{T}_{x_0\downarrow\xi}(x_0,\xi)$, respectively, to refer to their mean values. We state here only the main results and refer to Laio et al. (2001c) and Rodríguez-Iturbe and Porporato (2004) for details on the derivations.

It can be shown that, for an initial point x_0 above ξ,

$$\overline{T}_{x_0\downarrow\xi}(x_0,\xi) = T^{\text{det}}_{x_0\downarrow\xi}(x_0,\xi) + \lambda \int_\xi^{x_0} \frac{\overline{T}^{\uparrow}_u(u)}{\rho(u)} du \qquad (6.103)$$

where $T^{\text{det}}_{x_0\downarrow\xi}(x_0,\xi) = -\int_{x_0}^\xi du/\rho(u)$.

For the time to reach an upper threshold starting from a point x_0 below it,

$$\overline{T}_{x_0\uparrow\xi}(x_0,\xi) = \overline{T}^{\downarrow}_\xi(\xi) - \lambda \int_{x_0}^\xi \frac{\overline{T}^{\downarrow}_u(u)}{\rho(u)} du - T^{\text{det}}_{\xi\downarrow x_0}(x_0,\xi), \qquad (6.104)$$

where $T^{\text{det}}_{\xi\downarrow x_0}(x_0,\xi) = -\int_\xi^{x_0} du/\rho(u)$.

In the case of exponentially distributed jumps with parameter γ, it is also possible to rewrite the previous expressions as

$$\overline{T}_{x_0 \downarrow \xi}(x_0, \xi) = \overline{T}_{x_0}^{\uparrow}(x_0) + \gamma \int_{\xi}^{x_0} \overline{T}_u^{\uparrow} du \tag{6.105}$$

and

$$\overline{T}_{x_0 \uparrow \xi}(x_0, \xi) = \overline{T}_{x_0}^{\downarrow}(x_0) + \gamma \int_{x_0}^{\xi} \overline{T}_u^{\downarrow}(u) du. \tag{6.106}$$

6.9 Key Points

- Unpredictable hydroclimatic fluctuations, especially rainfall, propagate through the soil–plant system, giving rise to a random-like variability of ecohydrological quantities.
- Nonlinear and intermittent components of ecohydrological dynamics often produce non-Gaussian random variables with asymmetric distributions and processes with sudden jumps or other discontinuities.
- It is both mathematically and physically convenient to represent ecohydrological dynamics as Markovian processes in continuous time, i.e., as stochastic differential equations; the evolution equation of the related probability density functions (PDFs) is given by the so-called Chapman–Kolmogorov (or master) equation.
- The Poisson process is fundamental to modeling the punctuated random nature of rainfall-event occurrence, which is reflected in its only parameter, λ, the rainfall frequency. The process is characterized by an exponential distribution of the interarrival times, with mean $1/\lambda$, and by a Poisson distribution of the number of occurrences in a given time interval T, with mean λT.
- The Chapman–Kolmogorov equation for a random variable with jumps and decays (such as soil moisture) is an integro-differential equation that cannot be solved in general; however, its steady-state solution can be obtained for any loss function $\rho(x)$ and exponentially distributed jumps; see Eq. (6.89). For this solution at steady state, the effect of a bound is simply reflected in a renormalization of the truncated PDF.
- For jump-decay processes, the mean frequency of crossing a threshold and the mean time below it can be computed exactly as a function of the steady-state PDF and the loss function; see Eqs. (6.97) and (6.101).

6.10 Notes, including Problems and Further Reading

6.1 Prove Eq. (6.15) with the definition of the MGF in Eq. (6.13).

6.2 Derive the MGF of the exponential distribution and use it to derive its first moments. Show that the skewness of the exponential distributions is equal to 2 and the kurtosis is equal to 9.

6.3 Assume that rainfall depths are exponentially distributed and that interception eliminates all events which have depth lower than a given amount, Δ. What is the distribution of throughfall events that reach the ground?

6.4 Equation (6.29) is a special case of the censoring of RVs or truncation of distributions. Discuss what happens in case of a non-exponential distribution.

6.5 Reproduce the plots of the gamma distribution of Fig. 6.8. Use numerical routines for the gamma function to normalize the PDF. Plot $\Gamma(\alpha)$ as a function of α, noting that for α integer it is equal to $(\alpha - 1)!$. Write the cumulative probability function for the gamma PDF in terms of the incomplete gamma function (6.37).

6.6 Using l'Hôpital's rule show that the Pareto distribution becomes an exponential in the limit $k \to 0$. Obtain the CDF for the generalized Pareto distribution.

6.7 Derive the formula corresponding to Eq. (6.41) for monotonically decreasing functions.

6.8 Derive analytically the case represented in Fig. 6.10, where x is exponentially distributed and $f(x) = \sqrt{x}$.

6.9 Assume a linear function $y = ax + b$: How does the PDF of y differ from that of x? How do the mean and variance of y relate to those of x?

6.10 Use the derived distribution approach (Sec. 6.3.8) to show that the square of a Gaussian RV is exponentially distributed.

6.11 Show that, when the parameter k of the exponential distributions (6.26) is itself an exponentially distributed RV with parameter β, the compound distribution is a Pareto PDF. Link the parameters of the Pareto distribution to the original parameters of the exponential PDFs.

6.12 Assuming that rainfall events occur according to a Poisson process with rate $\lambda = 0.3 \, \mathrm{day}^{-1}$, compute the probability of having three rain events in a period of 30 days.

6.13 Show that the distribution of times between r events in a Poisson process follows the Erlang distribution,

$$p(\tau) = \frac{\lambda^r}{(r-1)!} \tau^{r-1} e^{-\lambda \tau}$$

for $\tau > 0$ (a special form of gamma distribution). Hint: this is a sum of independent and identically distributed RVs, which requires a convolution integral of Eq. (6.60).

6.14 (Rainfall Superstatistics)   The mean of the exponential interarrival time distribution between storms in a growing season can be well approximated by a gamma distribution (Porporato et al., 2006). Show that the resulting compound distribution is the Pareto distribution

$$p(\tau) = \frac{a_\tau b_\tau^{a_\tau}}{(\tau + b_\tau)^{a_\tau + 1}}. \tag{6.107}$$

What are the shape and scale factors? What is the new mean and the variance of the interarrival times? Compare these values with those obtained with the same interannual mean but no variability.

6.15 (Poisson Process: Further Properties)   The Poisson process has a number of other important properties (Cox and Miller, 1965; Ross, 2014). For example, the composition, censoring, and splitting of Poisson processes will still give a Poisson process. Moreover, the Poisson process can be generalized to space–time, and multidimensional domains, as well as to time-dependent (or nonhomogeneous) conditions in which λ is a function of time (Cox and Miller, 1965; Ross, 2014), or it can even be made state-dependent (Porporato and D'Odorico, 2004; Daly and Porporato, 2006, 2007).

6.16   Assuming that, in a growing season of 180 days, rainfall occurs as a Poisson process with rate $\lambda = 0.2$ day^{-1} and with exponential marks with mean $\alpha = 1.5$ cm, plot a realization of the process and compare the simulated and theoretical means, variances, and distributions of total rainfall during the growing season.

6.17   Obtain the mean and variance as a function of time for the cumulated process (6.61) for a generic jump distribution.

6.18   Prove that the Laplace antitransform (backward transform) of $e^{-x_0 k} g^*(k)$ is $\Theta(x - x_0) g(x - x_0)$, where $\Theta(\cdot)$ is the Heaviside function and $g(x)$ is the antitransform of $g^*(k)$.

6.19   How would the master equation (6.64) change for two-sided jumps (e.g., jumps that can be both negative and positive)?

6.20 (Fokker–Planck Equation)   Show that the master equation (6.64) becomes an advection–diffusion equation in the limit to large times (or frequent and small marks), with a Gaussian solution. This is expected from the central limit theorem (CLT) (Sec. 6.3.3). Hint: expand in Taylor series the jump distribution around its mean inside the integral term.

6.21   The transient solution starting from an initial condition x_0 at time $t = 0$ is always of mixed type, with an atom of probability on the deterministic trajectory from x_0 corresponding to the probability that the system does not make any jump up to time t. Write formally this part of the solution in terms of the Dirac delta function.

6.22 (Piecewise Loss Function)   In the case of a piecewise loss function $\rho(x)$, the limits of the integral in the exponential term of Eq. (6.88) must be chosen to ensure the continuity of $p(x)$ at the points of discontinuity of the loss function (e.g., Cox and Isham, 1986; Rodríguez-Iturbe et al., 1999a). Obtain explicitly the solution for the bounded system $x < 1$ with linear losses between $x = 0$ and $x = x^*$ and constant loss for $x^* < x < 1$.

6.23 Obtain explicitly the steady-state PDF (6.88), along with the normalization constant, for the case of a linear loss function $\rho(x) = \eta x$, for an unbounded process and for a process bounded at $x = 1$.

6.24 Derive the frequency and mean time of downcrossing for the steady-state linear-loss jump process $\rho(x) = \eta x$, and plot it as a function of the level of x.

7 Stochastic Soil Moisture Dynamics

By this formulation we have traded off some fidelity in physical behavior for the ability to include in the derived output statistics an explicit representation of both the essential system dynamics and the input statistics.

Eagleson (1978)[†]

In this chapter, we analyze probabilistically the soil water balance. We begin by considering each term of the moisture stochastic differential equation, describing the sequence of sudden jumps caused by rainfall infiltration (i.e., rainfall minus canopy interception and runoff) followed by the deterministic decays due to evapotranspiration and percolation. The steady-state PDF for the soil moisture and the mean soil water balance are discussed as a function of the main soil, climate, and vegetation parameters. A synthetic representation of the hydrologic partitioning is obtained by combining a minimalist version of the stochastic soil moisture model with the dimensionless framework of Budyko. Finally, we explore the role of other forms of hydroclimatic variability in addition to rainfall pulsing, namely the role of daily fluctuations in potential evapotranspiration, the origin of preferential dry and wet states due to soil moisture feedbacks on rainfall frequency, and the effects of seasonal and interannual rainfall fluctuations.

7.1 Soil Moisture Balance Equation

In Chapters 3–5 we described the temporal evolution of the main ecohydrological processes within the soil–plant–atmosphere continuum, in the absence of rainfall variability. Starting from a finer timescale description, where the fastest processes resolved are on the order of tens of minutes, we obtained a representation of the water, energy, and carbon fluxes at the daily level (see Sec. 5.4.5), expressed as a function of mean daily soil moisture and atmospheric characteristics. In Chapter 6 we introduced the probabilistic tools necessary to account for the fact that the forcing terms, rainfall in particular, are highly unpredictable at longer timescales. In this chapter, we finally combine these concepts to arrive at a stochastic description of soil moisture dynamics.

By analyzing the probabilistic dynamics of soil moisture and the related soil–plant water, energy, and carbon fluxes, we can obtain a meaningful description of the long-term ecohydrological dynamics, explicitly accounting for the uncertainty in the hydroclimatic forcing. The models developed here will be used

---

[†] Eagleson, P. S. (1978). "Climate, soil, and vegetation: 1. Introduction to water balance dynamics." *Water Resources Research* 14.5, pp. 705–712.

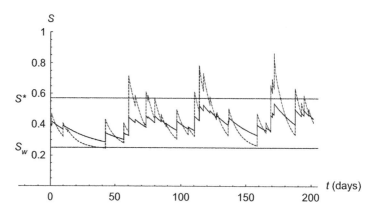

Figure 7.1 Example of soil moisture traces for the same rainfall sequence ($\alpha = 0.15$ cm; $\lambda = 0.2$ day^{-1}) in a loamy soil with the vegetation typical of semi-arid environments (see the caption of Fig. 7.8 for other parameters). The solid line refers to $Z_r = 90$ cm and the dashed line to $Z_r = 30$ cm. After Porporato et al. (2001).

in the following chapters to analyze several ecohydrological processes, including plant water stress, plant physiology, soil nutrient cycle, and ecosystem management.

The main focus of this chapter is the soil moisture balance, Eq. (1.5), recalled here for convenience in conditions of no irrigation,

$$nZ_r\frac{ds}{dt} = R - C_i - Q - E - L. \tag{7.1}$$

where s is the vertically averaged relative soil moisture level, n is the soil porosity, R is the rainfall rate, C_i is the canopy interception, Q is the runoff, E is the evapotranspiration, and L is the percolation or leakage. This equation is obtained by integrating Richards' equations over the rooting zone of depth Z_r (see Sec. 3.3.3 and Eq. (3.31)) and will be interpreted as a stochastic differential equation by considering the rainfall input as a marked Poisson process (Sec. 6.6).

An example of simulated soil moisture time series is given in Fig. 7.1, where the jumps are associated with the instant input of rainfall while the decay is related to the evaporation and leakage losses. The various terms of Eq. (7.1) are discussed in detail in the following sections and then brought together for a stochastic description of the soil moisture balance, following the presentation of Rodríguez-Iturbe and Porporato (2004).

7.2   Dry-down and Soil Water Losses

Canopy interception and infiltration take place during and immediately after a rainfall event. Therefore, at the daily level, evapotranspiration and percolation (i.e., the last two terms in Eq. (7.1)) can be considered to be the only losses that remain in the soil moisture equation between storms during soil moisture dry-downs.

7.2.1   Evapotranspiration

In Sec. 5.4 we analyzed the coupled dynamics of soil, plant, and atmospheric boundary layer models to obtain the upscaled dynamics of the water and energy fluxes at the daily timescale. We found that the evapotranspiration rate is primarily controlled by the soil moisture, s, and the potential evapotranspiration, E_{max} (see Sec. 5.4.5). This behavior was approximated by a piecewise function, Eq. (5.73), reported here for convenience (see Fig. 7.2),

$$E(s) = \begin{cases} 0 & 0 < s \leq s_h, \\ E_w \dfrac{s - s_h}{s_w - s_h} & s_h < s \leq s_w, \\ E_w + (E_{max} - E_w)\dfrac{s - s_w}{s^* - s_w} & s_w < s \leq s^*, \\ E_{max} & s^* < s \leq 1, \end{cases} \tag{7.2}$$

where s^* is the onset of plant water stress, s_w is the wilting point, and s_h is the hygroscopic point (see e.g. Table 3.1); E_w and E_{max} are the evaporation at s_w and s^*.

The characteristic soil moisture levels in Eq. (7.2) correspond to the water-potential levels (see Fig. 3.3), which are important for plant physiology (Chapter 4). For simplicity, we assign a fixed value to E_w (0.01 cm/day); this is expected to depend on the rates of soil evaporation, linked to soil and atmospheric properties, as well as to the cuticular transpiration, which continues even if the stomata are fully closed. Thanks to the relatively small losses involved at low soil moisture levels, both E_w and s_h usually need not be known with great precision, compared to that required for other parameters.

The potential evapotranspiration E_{max} may be obtained by using the soil–plant–atmosphere continuum model (Sec. 5.4; see especially Eqs. (5.19) and (5.20)), resulting in approximations of the Priestley–Taylor type, which assume the daily potential evaporation to be proportional to the equilibrium evaporation,

$$\rho_w \lambda_w E_{max} = \alpha_{PT} \frac{\Delta}{\Delta + \gamma^*} R_n, \tag{7.3}$$

where ρ_w is the water density, λ_w is the latent heat of vaporization of water, γ^* is the psychrometric constant (see Eq. (5.16)), R_n is the net radiation, Δ is the temperature-dependent slope of the saturation vapor pressure curve, and α_{PT} is the Priestley–Taylor coefficient, which has a typical value of 1.26 but tends to be larger in dry and cold regions. A linkage between α_{PT} and the dryness index can be used to provide an estimate of E_{max} (Yin et al., 2019).

7.2.2 Percolation and Capillary Rise

As discussed in Sec. 3.3.3, according to Richards' equation the water flux at the lower boundary of the root zone is $K(s)(1 - \partial \Psi / \partial z)$. For gravity-driven percolation, this loss of water following the hydraulic conductivity $K(s)$. Here, for reasons of mathematical tractability, the hydraulic conductivity is assumed to decay exponentially from a value equal to the saturated hydraulic conductivity K_s at $s = 1$ to a value of zero at field capacity s_{fc}:

$$L(s) = K(s) = \frac{K_s}{e^{\beta(1 - s_{fc})} - 1} \left[e^{\beta(s - s_{fc})} - 1 \right], \qquad s_{fc} < s \leq 1, \tag{7.4}$$

where β is a coefficient which is used to fit the above expression to the more familiar power-law form (see Sec. 3.2.4). The value of β depends on the type of soil, varying from $\simeq 12$ for sand to $\simeq 26$ for clay (see Table 3.1). The strong nonlinear behavior of these losses as a function of soil moisture content is evident for $s_{fc} \leq s \leq 1$. We recall from Sec. 3.2.4 that the field capacity s_{fc}, the soil moisture content below which drainage by gravity practically ceases to occur, is operationally defined here as the value of soil moisture at which the hydraulic conductivity becomes 0.05 cm/day.

For shallow groundwater, the capillary flux may play an important role in vegetation dynamics and soil water balance. Analytical solutions or approximations for this flux have been found in the literature with specific expressions for the hydraulic conductivity and the soil water retention curve (e.g., Eagleson, 1978b; Salvucci, 1993). These expressions can be used in the stochastic soil water balance model to account for groundwater-table effects on root-zone soil moisture (Vervoort and Zee, 2008; Ridolfi et al., 2008; Laio et al., 2009; Tamea et al., 2009).

7.2.3   Soil-Drying Process

Upon normalization with respect to the active soil depth, the complete form of the losses described in Secs. 7.2.1 and 7.2.2 for a dry-down process can be expressed as

$$\rho(s) = \frac{\chi(s)}{nZ_r} = \frac{E(s) + L(s)}{nZ_r} = \begin{cases} 0 & 0 < s \le s_h, \\ \eta_w \dfrac{s - s_h}{s_w - s_h} & s_h < s \le s_w, \\ \eta_w + (\eta - \eta_w)\dfrac{s - s_w}{s^* - s_w} & s_w < s \le s^*, \\ \eta & s^* < s \le s_{fc}, \\ \eta + m\left[e^{\beta(s - s_{fc})} - 1\right] & s_{fc} < s \le 1, \end{cases} \tag{7.5}$$

where $\rho(s)$ stands for the normalized loss function, to which we will refer hereafter, and

$$\eta_w = \frac{E_w}{nZ_r}, \tag{7.6}$$

$$\eta = \frac{E_{max}}{nZ_r}, \tag{7.7}$$

$$m = \frac{K_s}{nZ_r\left[e^{\beta(1 - s_{fc})} - 1\right]}. \tag{7.8}$$

Figure 7.2 shows the loss function $\chi(s) = E(s) + L(s)$ for typical semi-arid ecosystem. With this notation, the soil moisture dry-down can be expressed as

$$\frac{ds}{dt} = -\rho(s), \tag{7.9}$$

the general solution of which is obtained implicitly by the separation of variables as

$$\int \frac{du}{\rho(u)} = t - t_0. \tag{7.10}$$

Using the full expression for the loss function in Eq. (7.5), the analytical expression for the soil moisture decay from an initial condition $s_0 \ge s_{fc}$ in the absence of rainfall events is

$$s(t) = \begin{cases} s_0 - \dfrac{1}{\beta}\ln\left\{\dfrac{\left[\eta - m + me^{\beta(s_0 - s_{fc})}\right]e^{\beta(\eta - m)t} - me^{\beta(s_0 - s_{fc})}}{\eta - m}\right\} & 0 \le t < t_{s_{fc}}, \\ s_{fc} - \eta(t - t_{s_{fc}}) & t_{s_{fc}} \le t < t_{s^*}, \\ s_w + (s^* - s_w)\left[\dfrac{\eta}{\eta - \eta_w}e^{-\frac{\eta - \eta_w}{s^* - s_w}(t - t_{s^*})} - \dfrac{\eta_w}{\eta - \eta_w}\right] & t_{s^*} \le t < t_{s_w}, \\ s_h + (s_w - s_h)e^{-\frac{\eta_w}{s_w - s_h}(t - t_{s_w})} & t_{s_w} \le t < \infty, \end{cases} \tag{7.11}$$

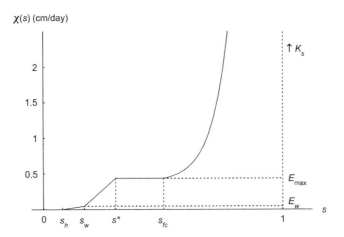

Figure 7.2 Soil water losses (evapotranspiration and leakage), $\chi(s)$, as a function of relative soil moisture for typical climate, soil, and vegetation characteristics in semi-arid ecosystems. After Laio et al. (2001c).

where

$$t_{s_{\mathrm{fc}}} = \frac{1}{\beta(m - \eta)} \left\{ \beta(s_{\mathrm{fc}} - s_0) + \ln\left[\frac{\eta - m + m e^{\beta(s_0 - s_{\mathrm{fc}})}}{\eta}\right] \right\},$$

$$t_{s^*} = \frac{s_{\mathrm{fc}} - s^*}{\eta} + t_{s_{\mathrm{fc}}}, \tag{7.12}$$

$$t_{s_w} = \frac{s^* - s_w}{\eta - \eta_w} \ln\left(\frac{\eta}{\eta_w}\right) + t_{s^*},$$

represent the times to evolve, in the absence of rainfall, from s_0 to s_{fc}, s^*, and s_w respectively. Notice that, since the moisture decays exponentially toward s_h, the process will only be at $s = s_h$ if it starts at this value.

Figures 7.3a and 7.3b show examples of Eq. (7.11) starting from saturated conditions. One can see the remarkable control exerted by both the soil texture and the active soil depth. Under the same climatic conditions and with the same vegetation type (i.e., the same $E_{\max}$, Ψ_{s,s_w}, and Ψ_{s,s^*}) the time to reach the wilting point in the absence of precipitation can vary from around 20 days for a shallow loamy sand, up to well beyond 60 days for a deep loamy soil. It is also interesting to notice the different effects of soil texture on the water availability for vegetation. On the one hand, because of its lower hydraulic conductivity a finer soil has less water loss from percolation and thus longer time to reach the wilting point; on the other hand, the higher values of s_h, s_w, and s^* lead to an increase in the amount of residual water which is not extractable by plants for transpiration.

The dry-down solutions just presented can be linked to a two-stage sequence of evapotranspiration losses, such as that described by Brutsaert and Chen (1995, 1996) for a short-grass steppe using data sampled at a daily timescale (see Note 3.4). According to their analysis, during the first stage of drying after abundant rainfall, soil water evaporates both from the soil surface and from the vegetation, at a rate which is governed by the available energy supply. This phase corresponds in our model to the range of soil moisture $\{s^*, 1\}$ or, according to Eq. (7.11), to the time interval $0 < t \le t_{s^*}$. Then, after the soil moisture content has decreased below a critical level, a transitional stage sets in, during which the soil moisture is the controlling factor. This agrees with the assumptions of the model presented in this chapter, where

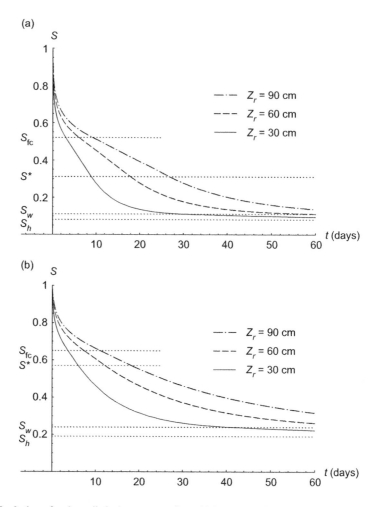

Figure 7.3 Plots of solutions for the soil-drying process for a (a) loamy sand and (b) loam, with different values of active soil depth (30, 60, and 90 cm). The parameters used were $E_{max} = 0.45$ cm/day, and $E_w = 0.01$ cm/day. The values of s_h, s_w, s^*, s_{fc}, n, β, and K_s are reported in Table 3.1. After Laio et al. (2001c).

soil moisture controls transpiration through stomatal closure between s_w and s^*, or that is, in the interval $t_{s^*} < t \leq t_{s_w}$. Brutsaert and Chen (1995) referred to this as a desorption phase (see Sec. 3.4.1). Finally, there may be a third-stage drying when the soil moisture decreases below the plant wilting level and drying takes place mainly as evaporation from the soil surface or plant cuticle ($s_h < s \leq s_w$, or $t_{s_w} < t \leq \infty$).

7.3   Daily Timescale Description of Rainfall and Infiltration

7.3.1   Rainfall

When precipitation is considered at the daily timescale, rainfall events (storms), which typically last a few hours, correspond to pulses that can be assumed to be concentrated at an instant in time, so that the temporal structure within each rain event may be ignored. This can be seen in Fig. 7.4a, which shows

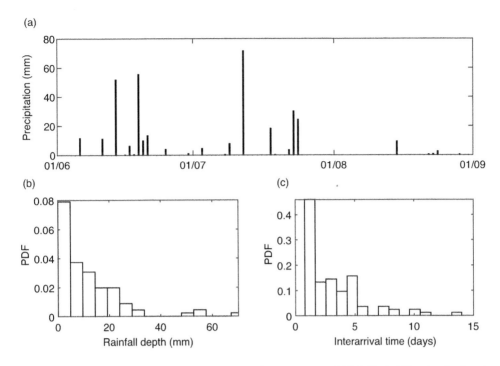

Figure 7.4 (a) Time series of rainfall in Princeton, NJ, USA in the summer of 2019. Probability density function of (b) daily rainfall depth and (c) inter-arrival time of rainfall in Princeton in the summer of 2016–2020. Data obtained from the National Climatic Data Center.

summertime rainfall events in Princeton, NJ. One also sees that the rainfall time series is highly unpredictable, a fact which calls for a probabilistic description. The marked Poisson process, introduced in Sec. 6.6, is a very good candidate for this, where both the occurrence and amount of rainfall are considered to be stochastic. Accordingly, we idealize the occurrence of rainfall as a series of point events in continuous time due to a Poisson process of rate λ, each carrying a random amount of rainfall extracted from a given distribution. Notice that, because of its memoryless property, the Poisson process has no clustering or correlation between rainfall events; simulating rainfall with memory components requires more sophisticated models (e.g., Richardson, 1981).

With these assumptions, the distribution of the interarrival time of rainfall events τ is exponential with mean $1/\lambda$ (e.g., Cox and Miller, 1965),

$$f_T(\tau) = \lambda e^{-\lambda \tau}, \quad \text{for } \tau \geq 0, \tag{7.13}$$

while the depth of the rainfall in an event is assumed to be an independent random variable h, described by an exponential probability density function:

$$f_H(h) = \frac{1}{\alpha} e^{-h/\alpha}, \quad \text{for } h \geq 0, \tag{7.14}$$

where α is the mean depth of a rainfall event. Since the model is interpreted at the daily timescale, α may be estimated as the mean daily rainfall for the days when precipitation occurs. In the following, we will often refer to the value of the mean rainfall depth normalized by the active soil depth, i.e.,

$$\frac{1}{\gamma} = \frac{\alpha}{nZ_r}. \tag{7.15}$$

Both a Poisson process for rainfall events and an exponential distribution of rainfall depth are of common use in simplified models of rainfall at the daily timescale. The assumptions are well supported by daily rainfall data (see Fig. 7.4b and c) and, at the same time, they allow analytical tractability (Eagleson, 1978a,b; Benjamin and Cornell, 2014).

The average rainfall rate can be expressed as

$$\langle R \rangle = \alpha \lambda, \tag{7.16}$$

and the total average rainfall in a season is $\mathcal{R}_{\text{seas}} = \langle R \rangle T_{\text{seas}} = \alpha \lambda T_{\text{seas}}$, in which T_{seas} is the length of the season. Notice that the rainfall rate $R(t)$ in Eq. (7.1) can be linked to the probability distributions (7.13) and (7.14), if one expresses the marked Poisson process as a temporal sequence, i.e.,

$$R(t) = \sum_i h_i \delta(t - t_i) \tag{7.17}$$

where $\delta(\cdot)$ is the Dirac delta function (see Sec. 6.3.4), $\{h_i, i = 1, 2, 3, \ldots\}$ is the sequence of random rainfall depths with distribution given by Eq. (7.14), and $\{\tau_i = t_i - t_{i-1}, i = 1, 2, 3, \ldots\}$ is the interarrival time sequence of a stationary Poisson process with rate λ.

For the present, the values of α and λ are assumed to be time-invariant quantities, representative of a typical growing season. In cases where there are significative changes in the parameters throughout the season, it may be useful to consider separately the early and late periods of the growing season, each with its own set of parameters (Feng et al., 2012). A nonhomogeneous Poisson process may be used to model the details of the rainfall seasonality and/or interannual variability (see Sec. 7.8). When the contribution of local soil moisture feedback on rainfall is important, the marked Poisson process can be treated as a state-dependent process (see Sec. 7.9).

7.3.2 Canopy Interception

Some rainfall is intercepted by the aerial part of vegetation, while the remaining rainfall reaches the soil directly as *throughfall* (e.g., Fig. 1.6). Especially in arid and semi-arid climates, where rainfall events are generally short and evaporation demand is high, a sizeable fraction of the intercepted rainfall is lost directly through evaporation, while the rest reaches the soil as stem flow (e.g., Scholes and Archer, 1997; Waring and Running, 2010; Dingman, 2015). Because of its strong dependence on the type of plant and more specifically on its leaf area index, the amount lost by interception can indeed be quite different for, say, trees and grasses. As an example, well-vegetated trees in South African savannas can intercept up to over 0.2 cm of rainfall per storm event (Scholes and Walker, 1993), whereas grasses usually intercept much less. At small spatial scales, interception is also known to alter the spatial distribution of the water reaching the soil (e.g., through stem flow and crown shading; Scholes and Archer, 1997).

The precise mechanisms of interception are quite complicated to model and depend on vegetation type as well as on the intensity and duration of the rainfall. However, in the context of a parsimonious model and focusing on daily and longer timescales, the throughfall process may be treated with relatively simple approximations. From empirical observations (see Fig. 7.5), canopy interception tends to operate as a threshold process on small rainfall depths and then progressively reducing the larger depth. Accordingly, the throughfall may be assumed to be a linear function of rainfall with slope k_{int} and intercept Δ:

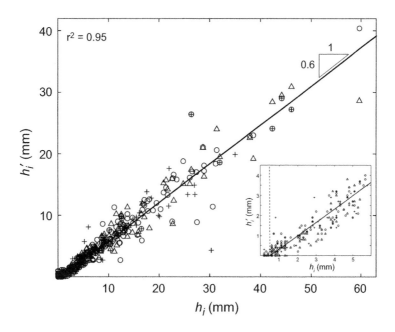

Figure 7.5 Observed relationship between incident rainfall, h, and throughfall, h', during winter (+) and during growing seasons (◦ and △). The inset is a zoom-in of the relationship, showing the threshold below which the incident rainfall is entirely intercepted by the foliage. After Daly et al. (2008).

$$h' = \begin{cases} k_{\text{int}}(h - \Delta) & h > \Delta, \\ 0 & h \leq \Delta. \end{cases} \tag{7.18}$$

This implies fixing a threshold for the rainfall depth, Δ, below which no water reaches the ground. If the depth of a given rainfall event, h, is higher than Δ then the actual rainfall depth reaching the soil, h', is assumed to be $k_{\text{int}}(h - \Delta)$. The amount of water intercepted can be related to the type of vegetation by simply using different values for Δ and k_{int}. The effect of fluctuations in wind and air temperature on interception losses are assumed to be negligible compared with the role of the differences in canopy coverage. This model of interception is illustrated in Fig 7.6a. The previous simple scheme provides a representation of the process which is in good agreement with the experimental evidence. As shown in Fig. 7.6b, the percentage of rainfall intercepted as a function of total rainfall reproduces quite well the values found in field experiments (e.g., Lai and Katul, 2000).

From a mathematical viewpoint the consideration of a slope and a threshold for the rainfall Poisson process does not complicate its analytical tractability. The rainfall process is in fact transformed into a new marked Poisson process called a censored and rescaled process (because of the rescaling properties of the exponential distribution; Sec. 6.3), where the frequency of throughfall events is now

$$\lambda' = \lambda \int_{\Delta}^{\infty} f_H(h)dh = \lambda e^{-\Delta/\alpha}, \tag{7.19}$$

the mean depth α' is reduced,

$$\alpha' = k_{\text{int}}\alpha, \tag{7.20}$$

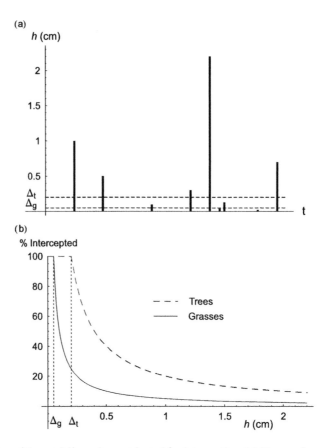

Figure 7.6 Representation of the modeling scheme adopted for interception. (a) Temporal sequence of rainfall events (h is the rainfall depth) along with the thresholds of interception, $\Delta_t = 0.2$ cm, $\Delta_g = 0.05$ cm, typical for trees and grasses in some savannas. (b) Percentage of intercepted rainfall as a function of the total rainfall per event. After Laio et al. (2001c).

and the normalized mean depth becomes

$$\frac{1}{\gamma'} = \frac{\alpha'}{nZ_r}. \tag{7.21}$$

Thus, one can simply write the throughfall sequence as

$$R(t) - C_i(t) = \sum_i h_i' \delta(t - t_i'), \tag{7.22}$$

where $\{\tau_i' = t_i' - t_{i-1}', i = 1, 2, 3, \ldots\}$ is the interarrival time sequence of a stationary Poisson process with frequency λ', and $\{h_i', i = 1, 2, 3, \ldots\}$ is a sequence of random throughfall depths from an exponential distribution with mean α'.

7.3.3   Infiltration and Runoff

Once throughfall reaches the ground it can be partitioned into infiltration and runoff (Sec. 3.4). In hydrology, infiltration in turn is often categorized into two types of mechanisms: the "infiltration excess runoff" or "Hortonian runoff" (Horton, 1933), when throughfall rate exceeds the infiltration capacity and the excess part is converted into surface runoff, and the "saturation excess runoff" or "saturation from below" (Dunne, 1978), whereby the throughfall rate is lower than the surface infiltration capacity, so that overland flow starts only when the soil layer becomes saturated. In vegetated soil and in the absence of widespread soil crust, the saturation excess runoff tends to be the dominant mechanism at the plot scale.

For the saturation excess runoff, infiltration depends as a threshold-like process (e.g., Fig. 3.14) on the antecedent soil moisture content. A very simple way to model this, while keeping analytical tractability, is to assume that, for a throughfall event of depth h' with antecedent soil moisture s, the cumulative infiltration in this event is

$$\mathcal{I} = \begin{cases} h' & h' \leq nZ_r(1-s), \\ nZ_r(1-s) & h' > nZ_r(1-s). \end{cases} \tag{7.23}$$

Guswa et al. (2002) and Rigby and Porporato (2006) addressed the validity of such hypotheses by comparing models of different degrees of complexity, where rainfall is either assumed to be an instantaneous event (Milly, 1993; Rodríguez-Iturbe et al., 1999b) or a rectangular event with specified intensity and duration (Eagleson, 1978a; Kim et al., 1996). More complex forms of threshold-type runoff were considered, for example, in Bartlett et al. (2016a); see also Note 3.12.

Because of Eq. (7.23), when rainfall is treated as a marked Poisson process, infiltration becomes a stochastic, state-dependent process, whose magnitude and temporal occurrence are controlled by the soil moisture dynamics. The probability distribution of the infiltration component may be written in terms of the exponential rainfall-depth distribution (7.14) and the soil moisture state s. Referring to its dimensionless counterpart, $y = \mathcal{I}/(nZ_r)$, one can write

$$f_Y(y, s) = \gamma e^{-\gamma y} + \delta(y - 1 + s) \int_{1-s}^{\infty} \gamma e^{-\gamma u} du, \quad \text{for } 0 \leq y \leq 1 - s, \tag{7.24}$$

where γ is defined in Eq. (7.15). Note that γ and λ should be replaced by γ' and λ' when there is canopy interception (see Sec. 7.3.2). Equation (7.24) is thus the probability distribution of having a jump in soil moisture equal to y, starting from a level s. The atom of probability at $1 - s$ represents the probability that a storm produces saturation when the soil has moisture s (Fig. 7.7). This sets the upper bound of the process at $s = 1$, making the soil moisture balance evolution (e.g., Eq. (7.1)) a bounded shot-noise process. Although the bounded character of the process complicates its mathematical analysis, it will be seen that the threshold process of infiltration preserves the Markovian nature of the process and still allows for a complete analytical solution in the case of steady-state conditions.

Similarly to Eq. (7.22), the infiltration from rainfall can be written as

$$I(t) = R(t) - C_i(t) - Q(s(t), t) = nZ_r \sum_i y_i \delta(t - t'_i), \tag{7.25}$$

where $\{y_i, i = 1, 2, 3, \ldots\}$ is a sequence of random infiltrations events having the distribution given by Eq. (7.24). With the assumption of saturation excess runoff (Eq. (7.23)), the probabilistic occurrence of runoff is related to the crossing properties of the level $s = 1$ (see the details in Sec. 6.8). We note that the runoff process by itself is not Markovian and that the temporal distribution of its time of occurrence is not exponential. It is also worth noticing that runoff production may be interpreted in a different but equivalent manner by considering an alternative process where runoff occurs from deterministic losses which take place at an infinite rate above $s = 1$. In this interpretation, which may help us to develop intuition in

Figure 7.7 The probability density function describing infiltration from rainfall y. The asterisk represents the atom of probability corresponding to soil saturation. After Rodríguez-Iturbe et al. (1999a).

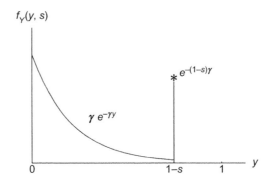

some mathematical derivations, the state variable s is not required to be formally bounded at $s = 1$, but the instantaneous decay above $s = 1$ makes the process completely equivalent to the bounded process for $s < 1$.

7.4   Probabilistic Soil Moisture Dynamics

When substituting the loss function (7.5) and the infiltration equation (7.25) into the soil moisture balance equation (7.1), one obtains a stochastic differential equation (SDE) with drift and jump terms (see Eq. (6.70)):

$$\frac{ds}{dt} = \underbrace{-\rho(s)}_{\text{drift}} + \underbrace{\mathcal{F}_{\lambda,\gamma}(s, t)}_{\text{jumps}}, \tag{7.26}$$

where the drift term expresses the water loss due to evaporation and leakage, and the jump term describes the infiltration. This probabilistic soil moisture model was originally proposed by Rodríguez-Iturbe et al. (1999b) and extended by Laio et al. (2001c). Note that the parameters λ and γ should be replaced by λ' and γ' when there is canopy interception.

As discussed in the previous chapter (see Eq. (6.77)), this SDE can be related to the partial differential equation for the time-varying PDF of soil moisture, the master equation,

$$\frac{\partial}{\partial t}p(s, t) = \frac{\partial}{\partial s}\left[\rho(s)p(s, t)\right] - \lambda p(s, t) + \lambda\gamma \int_0^s e^{-\gamma(s-u)}p(u, t)du. \tag{7.27}$$

This equation synthesizes the effects of climate, soil, and vegetation on the soil moisture dynamics. Unfortunately, because of the soil moisture dependence of the loss function and the presence of the bound at $s = 1$, it cannot be solved in closed form for general conditions.

7.4.1   Steady-State PDF for the Complete Model

The steady-state analysis can be done following Sec. 6.7. The general solution to Eq. (7.27) for $\lim_{t\to\infty} \partial p/\partial t = 0$ was obtained as

$$p(s) = \frac{C}{\rho(s)} \exp\left(-\gamma s + \lambda \int_s \frac{du}{\rho(u)}\right). \tag{7.28}$$

With the losses $\rho(s)$ specified by the expression in Eq. (7.5), the previous solution for the soil moisture PDF becomes

$$
p(s) = \begin{cases}
\dfrac{\mathcal{C}}{\eta_w}\left(\dfrac{s - s_h}{s_w - s_h}\right)^{\frac{\lambda(s_w - s_h)}{\eta_w} - 1} e^{-\gamma s} & s_h < s \leq s_w, \\[2ex]
\dfrac{\mathcal{C}}{\eta_w}\left[1 + \left(\dfrac{\eta}{\eta_w} - 1\right)\left(\dfrac{s - s_w}{s^* - s_w}\right)\right]^{\frac{\lambda(s^* - s_w)}{\eta - \eta_w} - 1} e^{-\gamma s} & s_w < s \leq s^*, \\[2ex]
\dfrac{\mathcal{C}}{\eta}e^{-\gamma s + \frac{\lambda}{\eta}(s - s^*)}\left(\dfrac{\eta}{\eta_w}\right)^{\lambda \frac{s^* - s_w}{\eta - \eta_w}} & s^* < s \leq s_{fc}, \\[2ex]
\dfrac{\mathcal{C}}{\eta}e^{-(\beta + \gamma)s + \beta s_{fc}}\left(\dfrac{\eta e^{\beta s}}{(\eta - m)e^{\beta s_{fc}} + m e^{\beta s}}\right)^{\frac{\lambda}{\beta(\eta - m)} + 1} \\[2ex]
\times \left(\dfrac{\eta}{\eta_w}\right)^{\lambda \frac{s^* - s_w}{\eta - \eta_w}} e^{\frac{\lambda}{\eta}(s_{fc} - s^*)} & s_{fc} < s \leq 1.
\end{cases}
\tag{7.29}
$$

An expression for the constant $\mathcal{C}$ can be obtained analytically, but it is not reported here because it is quite involved due both to the piecewise form of the losses and to the presence of the bound. Similarly, the expressions for the mean and variance of the soil moisture are also rather cumbersome and so are not reported here (see Rodríguez-Iturbe et al., 1999b).

Figure 7.8 shows some examples of the PDFs derived from Eq. (7.29). The two different types of soil, already specified in Fig. 7.3, are loamy sand and loam, each with two different values of active soil depth. These are chosen in order to emphasize the role of soil texture in the soil moisture dynamics. The parameters related to vegetation are kept fixed and correspond to those used in Table 3.1. The role of climate is studied only in relation to changes in the frequency of storm events, λ, keeping fixed the mean rainfall depth α and the maximum evapotranspiration rate $E_{\max}$. A coarser soil texture corresponds to a consistent shift of the PDF toward drier conditions, which in the most extreme case can reach a difference of 0.2 in the location of the mode. The shape of the PDF also undergoes marked changes, the broadest PDFs occurring for shallower soils.

Figure 7.1 shows a superposition of two traces of soil moisture with the same rainfall realization, for the case of loam with two different active soil depths (the parameters are those used for the broken line PDFs in Figs. 7.8c and d). In both cases, the levels of soil moisture s^* and s_w are the same and so approximately is also the mean soil moisture. The continuous line, corresponding to the deeper soil, is almost always between s^* and s_w, and is different from the shallower soil where the trace jumps and decays between low and high soil moisture levels. This demonstrates the importance of such contrasting soil moisture dynamics for different types of plants and the possible implications for the development of different strategies of adaptation to water stress.

General experimental support for the theoretical PDFs presented earlier has been reported in the literature. Salvucci (2001) showed that for a site in Illinois the soil moisture losses agreed with the form of Eq. (7.5) as reported in Fig. 7.9a. With estimates of the frequency of occurrence and mean depth of daily rainfall, λ and α, the relative soil moisture PDFs computed from Eq. (7.29) matched the observed frequency distributions (Fig. 7.9b).

In the analysis of Salvucci (2001), the steady-state assumption required subdividing the growing season into an early and later part. In fact, the validity of the steady-state solution requires approximately homogeneous rainfall and negligible influence of the initial conditions. This happens where rainfall is mostly

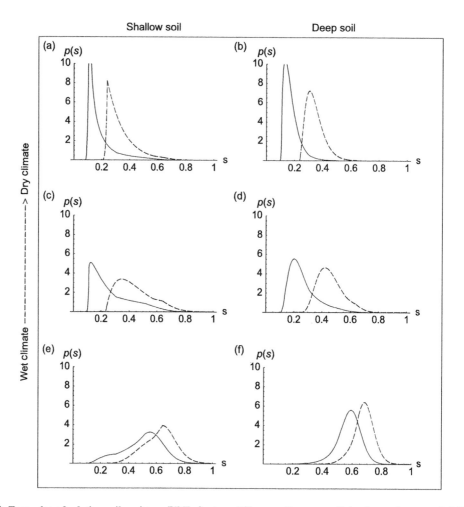

Figure 7.8 Examples of relative soil moisture PDFs for two different soil types, soil depths, and mean rainfall rates. The solid lines refer to loamy sand and the dashed lines to loam (see Table 3.1 for the values of the soil parameters). The left column corresponds to a soil depth $Z_r = 30$ cm and the right column to $Z_r = 90$ cm. The top, center, and bottom graphs have mean rainfall rates λ of 0.1, 0.2, and 0.5 day^{-1} respectively. The parameters common to all graphs are $\alpha = 1.5$ cm, $\Delta = 0$ cm, $k_{\text{int}} = 1$, $E_w = 0.01$ cm/day, and $E_{\text{max}} = 0.45$ cm/day. After Laio et al. (2001c).

concentrated in a warm growing season and winter is usually temperate and dry. Examples of these conditions are the savannas of South Africa (Scholes and Walker, 1993), the shrublands in Southern Texas, at least for most of the growing season (Archer et al., 1988; Scholes and Archer, 1997), and the shortgrass steppe in Colorado (Sala et al., 1992). Transient soil moisture dynamics and climatic seasonality are important in other semi-arid environments, especially at the beginning of the growing season; these are discussed in Sec. 7.8.1.

7.4.2 Mean Water Balance

The components of the mean water balance, obtained as ensemble averages, are also essential to our understanding of the interaction between climate, soil, and vegetation. For Eq. (7.1), the long-term average is simply given by conditions of steady state for the ensemble averages (see Sec. 6.4.2):

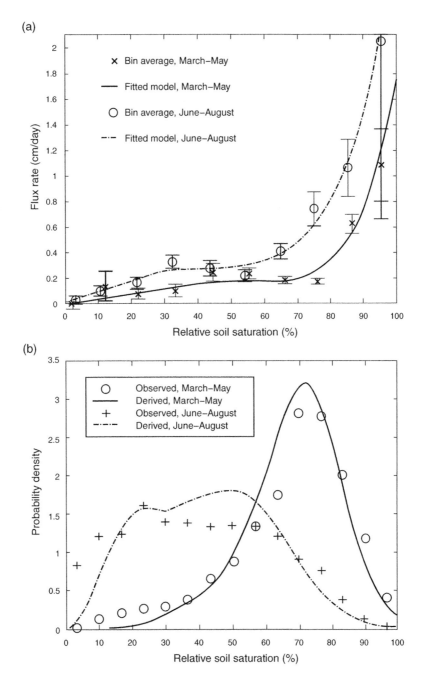

Figure 7.9 (a) Estimation of water loss from conditional mean precipitation for a site in Illinois. (b) Observed and derived probability distributions for the relative soil saturation. After Salvucci (2001).

$$\langle I \rangle = \langle R \rangle - \langle C_i \rangle - \langle Q \rangle = \langle L \rangle + \langle E \rangle = \langle L \rangle + \langle E_s \rangle + \langle E_{ns} \rangle, \tag{7.30}$$

where $\langle I \rangle$, $\langle R \rangle$, $\langle C_i \rangle$, $\langle Q \rangle$, $\langle L \rangle$, $\langle E \rangle$ stand for the mean rates of infiltration, rainfall, interception, runoff, leakage, and evapotranspiration, respectively. The last equality further partitions evapotranspiration into that under stressed ($s < s^*$) and unstressed ($s \geq s^*$) conditions. An extension to larger scales in relation to the Budyko curve is further discussed in Sec. 7.6.

The mean rainfall intensity is the product of the mean rainfall depth and the mean rainfall frequency,

$$\langle R \rangle = \alpha \lambda, \tag{7.31}$$

and the mean throughfall intensity can be expressed as

$$\langle R \rangle - \langle C_i \rangle = \alpha' \lambda'. \tag{7.32}$$

The difference between the two is the mean interception rate,

$$\langle C_i \rangle = \alpha \lambda - \alpha' \lambda'. \tag{7.33}$$

The mean rate of water loss from the soil can be calculated as the expectation of the loss function (see Eq. (6.12)):

$$
\begin{aligned}
\langle E_s \rangle + \langle E_{\mathrm{ns}} \rangle + \langle L \rangle &= n Z_r \int_{s_h}^{1} \rho(s) p(s) ds \\
&= \int_{s_h}^{s^*} E(s) p(s) ds + \int_{s^*}^{1} E(s) p(s) ds + \int_{s_{\mathrm{fc}}}^{1} L(s) p(s) ds.
\end{aligned}
\tag{7.34}
$$

Analytical expressions for the terms in Eq. (7.34) can be obtained by integrating Eq. (6.87) at the given soil moisture range. Specifically, the total mean loss can be expressed as

$$\langle E \rangle + \langle L \rangle = \alpha' \lambda' - \alpha' \left(\eta + \frac{K_s}{n Z_r} \right) p(1). \tag{7.35}$$

Finally, Eqs. (7.30) and (7.35) allow us to write the mean runoff as

$$\langle Q \rangle = \alpha' \left(\eta + \frac{K_s}{n Z_r} \right) p(1). \tag{7.36}$$

Figure 7.10 presents examples of the behavior of the various components of the water balance, normalized by the mean rainfall rate, for some specific rainfall, soil, and vegetation characteristics. The influence of the frequency of rainfall events, λ, is shown in Fig. 7.10a for the case of a shallow loam soil. Since the amount of interception changes proportionally to the rainfall rate, it is not surprising that the fraction of water intercepted remains constant when normalized by the total rainfall, $\alpha \lambda$. The percentage of runoff increases almost linearly. More interesting is the interplay between leakage and the two components of evapotranspiration. The fraction of water transpired under stressed conditions rapidly decreases as λ goes from 0.1 to about 0.4, while the evapotranspiration under unstressed conditions evolves in a much more gentle manner. This last aspect has interesting implications for vegetation productivity, as will be discussed in Chapter 8. It is clear that in semi-arid conditions most of the water that actually reaches the soil is lost by evapotranspiration (in particular transpiration), a result in agreement with many field observations (Sarmiento, 1984; Eagleson and Segarra, 1985; Sala et al., 1992; Scholes and Walker, 1993).

Figure 7.10b shows the role of the active soil depth in the water balance. For relatively shallow soils, there is a strongly nonlinear dependence on soil depth of all the components of the water balance (with the obvious exception of the interception, which is constant because the rainfall is constant). For example, changing from $nZ_r = 5$ cm to $nZ_r = 20$ cm, the amount of water transpired is practically double in this particular case.

Figure 7.10c shows the impact on the water balance when the frequency and amount of rainfall are varied while keeping constant the total amount of rainfall in a growing season. The result is interesting, because of the existence of two opposite mechanisms regulating the water balance. On one hand, runoff production, for a given mean rainfall input, strongly depends on the ratio between of the soil depth and the mean depth of rainfall events. The rapid decrease of runoff is thus somewhat analogous to that in the first part of Fig. 7.10b, where a similar behavior is produced by an increase in soil depth. On the other hand, interception increases almost linearly with λ. The interplay between these two mechanisms determines a maximum of both leakage and evapotranspiration at moderate values of λ (of course, the position of the maxima changes according to the parameters used). This is particularly important from the vegetation point of view, since the mean transpiration rate is linked to the productivity of ecosystems (e.g., Kramer and Boyer, 1995). The role of not only the amount but also the timing of rainfall in soil moisture dynamics (Noy-Meir, 1973), is made clear by the existence of an optimum for transpiration and productivity which is directly related to the climate–soil–vegetation characteristics. The particular position of this maximum in the parameter space is governed by the interplay of all the mechanisms acting in the soil water balance, namely the intensity and amount of rainfall, interception, the active soil depth, and the nonlinear losses due to evapotranspiration and leakage.

7.5 Minimalist Models of Soil Moisture Dynamics

While the nonlinear loss function discussed in Sec. 7.4 offers a detailed description of ecohydrological processes at the plot scale, a more parsimonious description with fewer parameters may be desirable for larger-scale analysis of the soil–plant water balance. To this purpose, here we discuss two minimalist models, from Milly (1993) and from Porporato et al. (2004), with simple loss functions. In both models, the soil moisture is assumed to fluctuate only between the wilting point s_w and s_1, a calibration parameter whose value is between field capacity and saturation. Accordingly, one can introduce a new variable

$$x = \frac{s - s_w}{s_1 - s_w}, \tag{7.37}$$

called the *effective relative soil moisture*, and define the available water storage as $w_0 = (s_1 - s_w)nZ_r$. The corresponding water balance equation normalized from Eq. (7.1) is

$$w_0 \frac{dx(t)}{dt} = R(t) - E(x(t), t) - \mathrm{LQ}(x(t), t), \tag{7.38}$$

where the interception has been neglected (or included in the censored process of precipitation) and LQ stands for the sum of L and Q.

In the model of Milly (1993), the water loss rate is assumed to be a constant, $E_{\max}$, between s_w and the level s_1. The loss becomes zero at s_w and infinity at s_1, where leakage and runoff losses take place

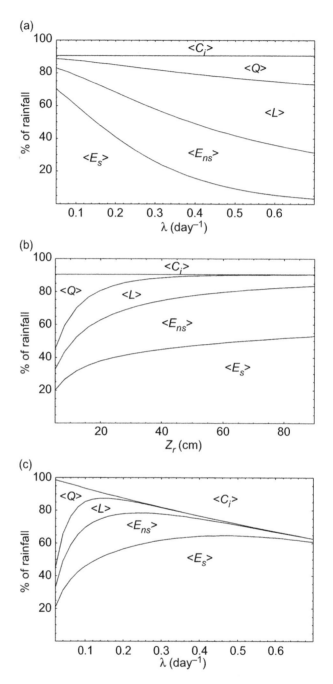

Figure 7.10 Components of the water balance normalized by the total rainfall. (a) Water balance as a function of the mean rainfall rate λ, for a shallow loamy soil ($Z_r = 30$ cm, $\alpha = 2$ cm). (b) Water balance as a function of the soil depth Z_r, for a loamy sand ($\alpha = 2$ cm, $\lambda = 0.2$ day^{-1}). (c) Water balance for a loamy sand as a function of the frequency of rainfall events for a constant mean total rainfall during a growing season, 60 cm. Other common parameters are $E_w = 0.01$ cm/d, $E_{max} = 0.45$ cm/day, $\Delta = 0.2$ cm, and $k_{int} = 1$ (see Table 3.1 for the soil parameters). After Laio et al. (2001c).

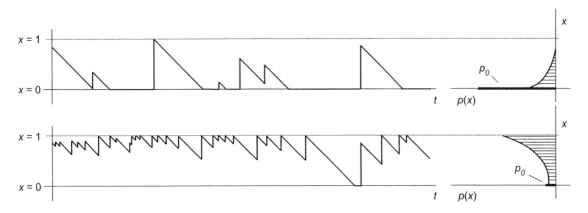

Figure 7.11 Examples of traces and steady-state PDFs for Milly's model. From Rodríguez-Iturbe and Porporato (2004).

instantaneously. Since the losses approach s_w (i.e., $x = 0$) discontinuously, the soil moisture PDF is a mixed distribution with an atom of probability at $x = 0$ (see Fig. 7.11). The normalized steady-state solutions can be obtained from the master equation (7.27) as (Milly, 1993)

$$p(x) = \frac{\lambda}{\eta} p_0 e^{(\lambda/\eta - \gamma)x}, \tag{7.39}$$

where $\eta = E_{\max}/w_0$, $\gamma = w_0/\alpha$ (see Eqs. (7.7) and (7.15)), and p_0 is the atom of probability at $x = 0$,

$$p_0 = \frac{\lambda/\eta - \gamma}{\lambda/\eta \, e^{\lambda/\eta - \gamma} - \gamma}. \tag{7.40}$$

The continuous part (i.e., Eq. (7.39)) is thus an exponential function (see Figs. 7.11 and 7.12a) which becomes uniform for $\lambda/\eta = 1$ (i.e., $D_I = 1$).

Another minimalist model was proposed by Porporato et al. (2004), assuming linearly increasing evapotranspiration losses with soil moisture. In this case the soil moisture decays exponentially, so that x approaches zero only asymptotically. As a consequence, the corresponding steady-state PDF has no atom at zero and reads

$$p(x) = \frac{C}{\eta} e^{-\gamma x} x^{\lambda/\eta - 1}, \tag{7.41}$$

where

$$C = \frac{\eta \, \gamma^{\lambda/\eta}}{\Gamma\left(\frac{\lambda}{\eta}\right) - \Gamma\left(\frac{\lambda}{\eta}, \gamma\right)}, \tag{7.42}$$

where $\Gamma(\cdot)$ and $\Gamma(\cdot, \cdot)$ are the gamma function and incomplete gamma function[1] (see Eqs. (6.36) and (6.37) in Sec. 6.3). As shown in Fig. 7.12, the PDF of this new minimalist model shows richer behaviors than the corresponding model with constant losses.

The behaviors of the PDF of the effective relative soil moisture as a function of the governing parameters may be used for a general classification of soil moisture regimes. Note that these nondimensional parameters can be derived from the Π theorem (see Sec. 7.6.1 and Note 7.13). Accordingly, the boundaries between different shapes of the PDF may be used to define an "arid" regime (soil moisture PDFs

---

[1] In Matlab, calculating $\Gamma(a, b)$ by (1-gammainc(b,a))*gamma(a), instead of igamma(a,b), is more efficient.

(a)   (b)

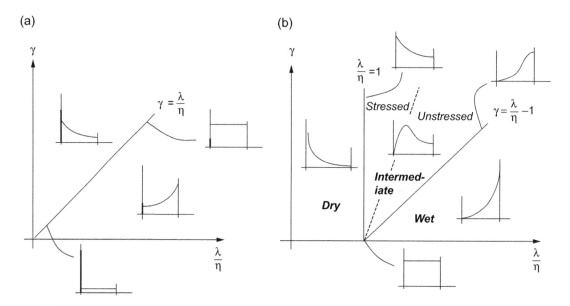

Figure 7.12 Different forms for the relative soil moisture steady-state PDFs for the minimalist model with (a) constant losses and (b) losses linearly increasing with soil moisture. The dashed line in panel (b), $\gamma = 1/x^* \ (\lambda/\eta - 1)$, is the locus of points where the mode of the soil moisture PDF is equal to the threshold x^* which marks the onset of plant water stress. After Porporato et al. (2004).

with zero mode), an "intermediate" regime (corresponding to PDFs with a central maximum), and a "wet" regime (with the mode at saturation), as indicated in Fig. 7.12b. A further distinction within the intermediate regime can be made on the basis of plant response to the soil moisture dynamics. Using the effective relative soil moisture value (x^*) as a threshold marking the onset of plant water stress (x^* is typically on the order of 0.3–0.4), along the dashed line of slope $1/x^*$ in Fig. 7.12b the mode of the effective relative soil moisture PDF is equal to x^* and thus where plants are more likely to be at the boundary between stressed and unstressed conditions. Accordingly, this line may be used to divide water-stressed (or semi-arid) types of water balance on the left side from unstressed types on the right side (Fig. 7.12b).

7.6   Mean Water Balance and Budyko's Curve

The long-term water balance discussed in Sec. 7.4.2 can be revisited for the case of the minimalist models, using dimensional analysis (Sec. 7.6.1) to provide a physical interpretation of the Budyko curve for hydrologic partitioning (Sec. 7.6.2).

7.6.1   Dimensional Analysis of Hydrologic Partitioning

Let us consider long-term rainfall partitioning and assume steady-state conditions, so that the time derivative in Eq. (7.38) vanishes. The long-term averages can then be written as

$$\langle R \rangle = \langle E \rangle + \langle LQ \rangle, \tag{7.43}$$

where the angle brackets indicate a long-term average. This partitioning equation can be applied at the plot scale as well as to an entire watershed or small river basin; in the latter case the $\langle LQ \rangle$ term should be interpreted as the average streamflow.

We want to express the mean evapotranspiration as a function of the main governing quantities. For a start, it is reasonable to assume that this flux depends on the mean rainfall, which in turn can be written as the product of the mean depth α of storm events ($[\alpha] = L$) and their frequency of occurrence λ ($[\lambda] = T^{-1}$),

$$\langle R \rangle = \alpha \lambda. \tag{7.44}$$

We can further assume that the mean evapotranspiration is also a function of the maximum evapotranspiration rate, E_{max}, and of the mean effective soil moisture, $\langle x \rangle$. The latter, according to Eq. (7.38), is related to w_0, $\langle R \rangle$, and $\langle LQ \rangle$, which in turn go back to $\langle R \rangle$ and $\langle E \rangle$ in Eq. (7.43). With this list of quantities, a general equation for the annual average evapotranspiration can be written as follows (Porporato et al., 2004; Feng et al., 2013; Daly et al., 2019b):

$$\langle E \rangle = f(\alpha, \lambda, E_{max}, w_0), \tag{7.45}$$

where the dimensions of the various terms are:[2]

$$
\begin{aligned}
&\text{mean depth of storms, } \alpha &&[\alpha] = L, \\
&\text{frequency of storm occurrence, } \lambda &&[\lambda] = T^{-1}, \\
&\text{max. transpiration rate, } E_{max} &&[E_{max}] = LT^{-1}, \\
&\text{soil water storage capacity, } w_0 &&[w_0] = L.
\end{aligned}
$$

From a dimensional point of view (see Sec. 2.2), the system (7.45) is characterized by only two primary dimensions, a length (e.g., cm) and time (e.g., days), so that $k = 2$. It is easily verified that, selecting α and λ as the dimensionally independent quantities, the Π theorem (see Sec. 2.2.2) gives

$$\Pi = \phi(\Pi_1, \Pi_2), \tag{7.46}$$

where

$$\Pi = \frac{\langle E \rangle}{\alpha \lambda} = \frac{\langle E \rangle}{\langle R \rangle}, \tag{7.47}$$

$$\Pi_1 = \frac{E_{max}}{\alpha \lambda} = D_I, \quad \text{dryness index,} \tag{7.48}$$

$$\Pi_2 = \frac{w_0}{\alpha} = \gamma, \qquad \text{storage index.} \tag{7.49}$$

The relationship (7.46) provided by the Π theorem gives a powerful synthesis of hydrologic partitioning, with the dryness index and the storage index playing a fundamental role. It is also possible to choose different pairs of variables in Eq. (7.45) as the dimensionally independent quantities, which allows us to link hydrologic partitioning to the humidity index (i.e., the inverse of the dryness index) and the storage index (see Note 7.13). While the operation is mathematically equivalent to Eq. (7.46), the different specific results could be more suitable for analyzing water balance in regions where the dryness index is small or where the role of the storage index is prominent (Daly et al., 2019a,b).

---

[2] Note that by limiting the list to such quantities, we are implying that other variables play a secondary role in the long-term water balance, compared with those listed in Eq. (7.45). This does not mean that such quantities will not be important in determining other ecohydrological processes or in controlling the hydrologic partitioning at a finer spatial and/or temporal scale.

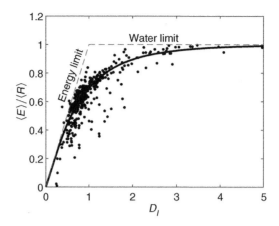

Figure 7.13 Budyko's curve, representing the hydrologic partitioning between evapotranspiration and streamflow as a function of the dryness index, Eq. (7.50). The dots are the observations from 438 watersheds in the United States recorded by the international Model Parameter Estimation Experiment (MOPEX) and the dashed lines show the water and energy limits.

7.6.2   Physical Interpretation of Budyko's Curve

For an entire watershed or river basin, one can assume that the term $\langle LQ \rangle$ represents the long-term discharge at the outlet. The classical analysis by Budyko, using data from different watersheds, showed that the dryness index is the dominant term in Eq. (7.46), while the storage index plays a less important role. Budyko (1974) used the expression

$$\frac{\langle E \rangle}{\langle R \rangle} = [D_I(1 - \exp(-D_I))\tanh(1/D_I)]^{1/2} \tag{7.50}$$

to interpolate between the limiting behaviors of extremely wet conditions $\langle E \rangle \sim E_{\max}$ for $D_I \to 0$ (i.e., energy limitation, when the evaporation rate is limited by the available energy and cannot exceed the rate $E_{\max}$ set by the atmospheric energy demand), and the dry limit $\langle E \rangle \sim \langle R \rangle$ for $D_I \to \infty$ (i.e., water limitation, when the potential evaporation is very high but the actual evapotranspiration can at most reach the rainfall rate). These limits are indicated as dashed lines in Fig. 7.13, along with Budyko's original curve (i.e., Eq. (7.50)). Several subsequent studies have confirmed this result, at the same time also showing some scatter and deviations from this general behavior and its underlying assumptions (Zhang et al., 2004; Daly et al., 2019b).

We can derive a corresponding formula using the minimalist models in Sec. 7.5. For the model of Milly (1993), where evapotranspiration is constant for $x > 0$, the long-term mean evapotranspiration is

$$\langle E \rangle = E_{\max}(1 - p_0), \tag{7.51}$$

where p_0 is the atom of probability at $x = 0$. Dividing the equation by the mean rainfall intensity and substituting the expression for p_0 from Eq. (7.40), one finds an expression for the fraction of rainfall partitioned into evapotranspiration:

$$\frac{\langle E \rangle}{\langle R \rangle} = D_I\,(1 - p_0) = \frac{e^{\gamma\left(1 - D_I^{-1}\right)} - 1}{e^{\gamma\left(1 - D_I^{-1}\right)} - D_I^{-1}}. \tag{7.52}$$

When expressed as a function of the dryness index and the storage index as in Eq. (7.52), it shows the Budyko-type curve in Fig. 7.14.

An alternative formula for Budyko's curve can be obtained from the minimalist model of Porporato et al. (2004). From Eq. (7.41), one can find the long-term mean of the effective soil moisture as

$$\langle x \rangle = \int_0^1 s p(x')\, dx' = \frac{\lambda}{\gamma \, \eta} - \frac{\gamma^{\lambda/\eta - 1} e^{-\gamma}}{\Gamma\left(\frac{\lambda}{\eta}\right) - \Gamma\left(\frac{\lambda}{\eta}, \gamma\right)}, \tag{7.53}$$

Since E is a linear function of x, the long-term mean evapotranspiration can be obtained as

$$\langle E \rangle = E_{\max} \langle x \rangle. \tag{7.54}$$

Following the same approach as in Eq. (7.52), one obtains (Daly et al., 2019b)

$$\frac{\langle E \rangle}{\langle R \rangle} = D_I \langle x \rangle = 1 - D_I \frac{\gamma^{\gamma/D_I - 1} e^{-\gamma}}{\Gamma(\gamma/D_I) - \Gamma(\gamma/D_I, \gamma)}, \tag{7.55}$$

which is again a function of the dryness index, D_I, and the storage term, γ.

Figure 7.14 compares the Budyko curves from Milly (1993) and Porporato et al. (2004) for different values of γ. Interestingly, the results based on the stochastic water balance model are qualitatively analogous to the original empirical curve fitted from observations in various watersheds. Particularly, the curve from Porporato et al. (2004) with $\gamma \sim 5.5$ reproduces very well Budyko's curve (see the dotted lines in Fig. 7.14). Therefore, using typical values of the parameters (e.g., the average rainfall depth per event $\alpha = 1.5$ cm, the relative soil moisture at the wilting point $s_w = 0.2$, the relative soil moisture threshold for deep infiltration and runoff $s_1 = 0.85$, and the porosity $n = 0.4$), we find that Budyko's curve corresponds to an average active soil depth of approximately 30–35 cm. Interestingly, this is the average depth within which most roots typically exist (Jackson et al., 1997; Schenk and Jackson, 2002b). The model interpretation also makes clear the possible climate effects on Budyko's curve. As an example, depending on the degree to which evapotranspiration, rainfall regime and plant characteristics are affected by climate change, alterations in the mean depth of rainfall per event and in the rooting depth will imply a vertical

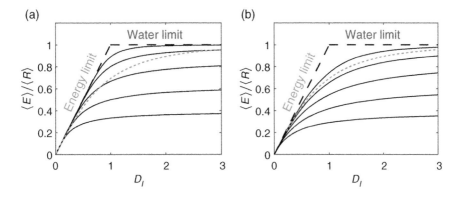

Figure 7.14 For comparison, the two minimalist models. The fraction of precipitation that is evapotranspired, as a function of the dryness index for various values of the dimensionless soil water holding capacity γ. The values of γ are, from bottom to top, 0.5, 1, 2, 4, 8. (a) Minimalist model with constant losses; (b) minimalist model with losses linearly increasing with soil moisture. The dotted lines show Budyko's original curve as in Eq. (7.50) and the dashed lines are the energy and water limits.

shift in the diagram. In contrast, a shift along the x-axis will entail changes in the potential transpiration and the total amount of rainfall (Porporato et al., 2004).

7.7   Role of Fluctuations in Potential Evapotranspiration

Up to this point, precipitation has been the only stochastic component in the soil water balance model (Eq. (7.1)). While rainfall is the main driver of soil moisture variability, the daily evapotranspiration under well-watered condition (i.e., the potential evapotranspiration E_{max}) shows a considerable day-to-day variability (e.g., Fig. 7.15a). In this section, we follow the work of Daly and Porporato (2006a) to explore the impact of fluctuations on evapotranspiration using the stochastic water balance model. The analysis shows that the effects of fluctuation of E_{max} on the soil water balance are much reduced compared to those of rainfall.

Potential evapotranspiration can be decomposed into a constant, $\bar{E}_{max}$, and fluctuations, $E'_{max}(t)$. The latter can be modeled as an Ornstein–Uhlenbeck (OU) process to account for its autocorrelation, which decays to zero in an exponential manner (Gardiner, 2004),

$$\frac{dE'_{max}}{dt} = -kE'_{max} + a\xi(t), \tag{7.56}$$

where $\xi(t)$ is a Gaussian white noise, with $\langle \xi(t) \rangle = 0$ and $\langle \xi(t)\xi(u) \rangle = \delta(t - u)$ (e.g., the formal time derivative of a Wiener process).

Equation (7.56) is a realistic description of the E'_{max} fluctuations, but its temporal correlation makes it a colored Gaussian noise, which is challenging to deal with analytically. Invoking the so-called adiabatic elimination of fast variables (Gardiner, 2004), the colored noise of Eq. (7.56) can be replaced by an equivalent idealized Gaussian white noise with zero memory and adjusted noise intensity. Accordingly, the potential evapotranspiration can be written as (Daly and Porporato, 2006a)

$$E_{max} = \bar{E}_{max} + E'_{max} = \bar{E}_{max} + b\xi(t), \tag{7.57}$$

where $b = a/k$ (Gardiner, 2004). The accumulated E_{max} over the time t has mean $\bar{E}_{max}t$ and variance $b^2 t$, whereas the accumulated precipitation has mean $\alpha\lambda t$ and variance $2\lambda\alpha^2 t$ (see the PDF of the accumulated rainfall in Eq. (6.68)). The ratio of these two variances is close to unity for the case of Duke Forest (see Figs. 7.15 a and c), showing that the fluctuations in the potential evapotranspiration are of similar intensity to those of the rainfall.

To consider how the rainfall and evapotranspiration variability propagate through the daily soil moisture dynamics, Daly and Porporato (2006a) treated evapotranspiration as a linear function of soil moisture, similarly to Sec. 7.5, but now with E_{max} as a stochastic process given by Eq. (7.57). After incorporating the fluctuations in E_{max}, the soil water balance equation (7.38) becomes

$$\frac{dx}{dt} = \mathcal{F}_{\lambda,\gamma}(x(t), t) - [\eta x(t) + \beta x(t) \diamond \xi(t)], \tag{7.58}$$

where x is the effective relative soil moisture (see Eq. (7.37)), $\mathcal{F}_{\lambda,\gamma}$ is the jump term (see Eq. (7.26)) with $\gamma = w_0/\alpha$, $\eta x(t)$ is the deterministic component of evapotranspiration ($\eta = \bar{E}_{max}/w_0$) and $\beta x(t) \diamond \xi(t)$ is the stochastic forcing related to evapotranspiration, where $\beta = b/w_0$. Equation (7.58) is a nonlinear (because of the bound at 1) stochastic differential equation driven by two different forms of noise: a multiplicative Gaussian noise and a state-dependent Poisson noise. The *multiplicative noise* $x(t) \diamond \xi(t)$ must be interpreted according to Stratonovich (e.g., Van Kampen, 1981), because the multiplicative Gaussian white noise is obtained as a limit of a colored noise (Daly and Porporato, 2006a).

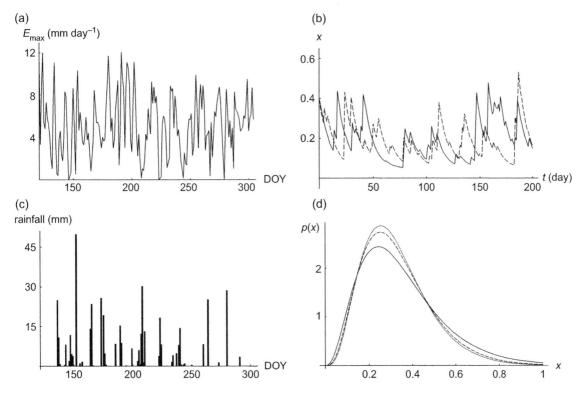

Figure 7.15 Time series of (a) potential evapotranspiration and (c) precipitation recorded during the 2001 growing season at the Duke Forest site, North Carolina, USA. (b) Time series of modeled soil moisture with two different values of noise intensity ($b = 2.5$, dashed line, and $b = 5$ mm day$^{-1/2}$, solid line). (d) The corresponding PDFs (the dashed line represents the PDF with $b = 0$ mm day$^{-1/2}$). Other parameters are $\lambda = 0.2$ day^{-1}, $\alpha = 4.5$ mm, $n = 0.373$, $Z_r = 300$ mm, $s_w = 0.047$, $s_1 = 0.6$, and $\overline{E}_{max} = 3.7$ mm day^{-1}. After Daly and Porporato (2006a).

The forward Chapman–Kolmogorov equation when both Poisson and Gaussian noises are present was given in Eq. (6.50). For this specific case, it becomes (Czernik et al., 1997; Daly and Porporato, 2006a,b)

$$\frac{\partial}{\partial t}p(x, t) = \frac{\partial}{\partial x}\left[\left(\eta x - \frac{1}{2}\beta^2 x\right)p(x, t)\right] - \lambda p(x, t)$$
$$+ \lambda\gamma \int_0^x e^{-\gamma(x-u)}p(u, t)du + \frac{1}{2}\beta^2\frac{\partial^2}{\partial x^2}\left[x^2 p(x, t)\right]. \tag{7.59}$$

It is interesting to note that the fluctuations in evapotranspiration give rise to the state-dependent diffusion (last term in Eq. (7.59)), while for $b = 0$ the problem reduces to the minimalist model (see Sec. 7.5).

The steady-state solutions can be obtained as (Daly and Porporato, 2006a)

$$p(x) = \mathcal{C}e^{-\gamma x}x^{\kappa_1 - 1}L_{-\kappa_1 - 1}^{\kappa_1 - \kappa_2}(\gamma x), \tag{7.60}$$

where $\mathcal{C}$ is a constant of normalization, $L_n^m(\cdot)$ is the generalized Laguerre polynomial (Abramowitz and Stegun, 1964), and

$$\kappa_{1,2} = \frac{-\eta \pm \sqrt{\eta^2 + 2\lambda\beta^2}}{\beta^2}. \tag{7.61}$$

For $b = 0$ in the case of only Poisson noise, Eq. (7.60) reduces to a truncated gamma distribution (see Sec. 7.5).

Figures 7.15b and d show two realizations for two different values of the noise, with for comparison the corresponding PDFs for the case with no evapotranspiration fluctuations ($b = 0$). While the effect of b is hardly visible in the time series, it is evident that increases in b tend to move the mode of the PDF toward lower values of x, as is typical in stochastic processes with multiplicative noise (e.g., Schenzle and Brand, 1979). Deeper soils tend to attenuate the effect of b on x, since nZ_r largely controls the relaxation time of the system while different climatic conditions, i.e., different values of λ and γ, induce slight changes in the PDF of x. The most important conclusion, however, is that evapotranspiration fluctuations do not qualitatively alter the probabilistic properties of soil moisture dynamics. This implies that the simplified stochastic soil moisture models at a daily timescale provide a quite realistic description of the main soil moisture probabilistic properties even if they only account for rainfall variability.

7.8   Seasonal and Interannual Variations of Climate and Soil Moisture

In many regions and for timescales beyond a few months, seasonal and inter-annual variations in rainfall may become important for the soil water balance. Here, we focus on the impact of this variability on the stochastic soil water balance, leaving to Chapter 8 the implications for plants and ecosystems. We first follow the work of Laio et al. (2002) and Feng et al. (2015) to analyze the impacts of a seasonal climate on the time-dependent PDF and mean soil moisture. We then follow D'Odorico et al. (2000) and Porporato et al. (2006) to analyze the impacts of interannual climate variability on the ecohydrological processes.

7.8.1   Seasonality

Starting from the stochastic water balance equation (7.38), we consider climate seasonality by assuming that the normalized potential evapotranspiration $k(t) = E_{\max}(t)/w_0$, average rainfall depth $\alpha(t)$, and rainfall frequency $\lambda(t)$ are periodic functions. For simplicity, Feng et al. (2015) adopted sinusoidal forms to describe these variables:

$$v_t = \mu_v + A_v \sin (\omega t + \phi_v) \tag{7.62}$$

where time dependence is denoted by the subscript t, v_t is a stand-in variable for α_t, λ_t, or k_t, μ_v is the annual mean, A_v is the amplitude of seasonal variation, ϕ_v is the phase, and the period ω is set to a year. Note that the forms of these seasonal variations do not influence the following analytical development (e.g., the stepwise functions for climate seasonality in Feng et al., 2012). With climate seasonality, Eq. (7.38) describes a time-dependent marked Poisson process that is nonhomogeneous in time with time-dependent rate parameters α_t, λ_t, and k_t.

The equation for the time-dependent mean soil moisture is obtained by multiplying the master equation (7.27) by s and integrating over the whole range of s (Laio et al., 2002; Feng et al., 2012, 2015):

$$\frac{d\langle x_t \rangle}{dt} = \underbrace{\frac{\lambda_t}{\gamma_t}}_{\text{rainfall}} - \underbrace{\int_0^1 \frac{E(u)}{w_0} p_t(u) \mathrm{d}u}_{\text{evapotranspiration}} - \underbrace{\frac{\lambda_t}{\gamma_t} \int_0^1 e^{-\gamma(t)(1-u)} p_t(u) \mathrm{d}u}_{\text{leakage and runoff}}, \tag{7.63}$$

where $\gamma_t = w_0/\alpha_t$, $p_t(x)$ is the time-dependent PDF of the soil moisture, and $\langle \cdot \rangle$ denotes an ensemble average. Following the minimalist approach of Sec. 7.5 with a linear loss function, the evaporation term can be further simplified as follows:

$$\int_0^1 \frac{E(u)}{w_0} p_t(u) du = \frac{E_{\max}(t)}{w_0} \int_0^1 u p_t(u) du = k_t \langle x_t \rangle. \tag{7.64}$$

Equation (7.63) is often referred to as a macroscopic equation for the stochastic process (Van Kampen, 1992). From a physical viewpoint, the macroscopic equation (multiplied by w_0) shows that the change in the mean soil moisture is adjusted by the mean rainfall rate, mean evapotranspiration rate, and mean leakage/runoff rate.

The major difficulty for the solution of the macroscopic equation derives from the presence of the last two terms, whose evaluation would require knowledge of $p(x, t)$. Feng et al. (2015) considered three different approximations. The first, referred to as the quasi-steady-state approximation, assumes that the inhomogeneous process instantaneously reaches a steady state at every point in time, which is the minimalist model discussed in Sec. 7.5. The instantaneous PDF of the soil moisture can be expressed as

$$p_{t,\text{ss}}(x) = \frac{b_t^{a_t} x^{a_t-1} e^{-b_t x}}{\Gamma(a_t) - \Gamma(a_t, b_t)} \tag{7.65}$$

where the subscript ss refers to the steady state and the shape and rate parameters $a_t = \lambda_t/k_t$ and $b_t = \gamma_t$ are "frozen" as constants and applied to the parallel homogeneous stochastic process until it has reached a steady state, represented by $p_{t,\text{ss}}(x)$. Examples of this approximation are shown in Fig. 7.16 at different times of the year for a typical seasonal dry climate. Substituting $p_{t,\text{ss}}(x)$ in place of $p_t(x)$ in Eq. (7.63) results in a closed ordinary differential equation for $\langle x_t \rangle$:

$$\frac{d \langle x_t \rangle}{dt} = \frac{\lambda_t}{\gamma_t} - k_t \langle x_t \rangle - \frac{k_t}{\gamma_t} \frac{\gamma_t^{\lambda_t/k_t}}{\Gamma(\lambda_t/k_t) - \Gamma(\lambda_t/k_t, \gamma_t)} e^{-\gamma_t}. \tag{7.66}$$

The approximation of $p_t(x)$ by $p_{t,\text{ss}}(x)$ is most appropriate when the soil moisture can quickly adjust to changes in its environment, which is the case when the soil water storage is small. Figure 7.16 also shows that the difference between the true steady-state soil moisture and $p_{t,\text{ss}}(x)$ diminishes during persistently wet or dry periods. During the drydown and wetting-up periods, the quasi-steady-state approximation predicts soil moisture values that are, respectively, lower and higher than the true steady-state values because it does not capture the effect of soil moisture transfer over time. Thus, this treatment of the leakage or percolation term is expected to work well at small γ_t values and in conditions where the environmental parameters do not change quickly.

The second approximation, referred to as the negligible fluctuations approximation, assumes that the soil moisture PDF itself is concentrated entirely on its mean value, represented by a Dirac delta function $p_t(x) \approx \delta(x - \langle x_t \rangle)$, are shown by the spikes at $\langle x_t \rangle$, for various values of t, in Fig. 7.16. The substitution of this Dirac delta function into Eq. (7.63) yields

$$\frac{d \langle x_t \rangle}{dt} = \frac{\lambda_t}{\gamma_t} - k_t \langle x_t \rangle - \frac{\lambda_t}{\gamma_t} e^{-\gamma_t(1 - \langle x_t \rangle)}. \tag{7.67}$$

The bias of this approximation rises from the difference between $\langle e^{-\gamma(1-x_t)} \rangle$ and $e^{-\gamma(1-\langle x_t \rangle)}$, which decreases as the soil moisture PDF becomes naturally concentrated around its mean; this happens as the

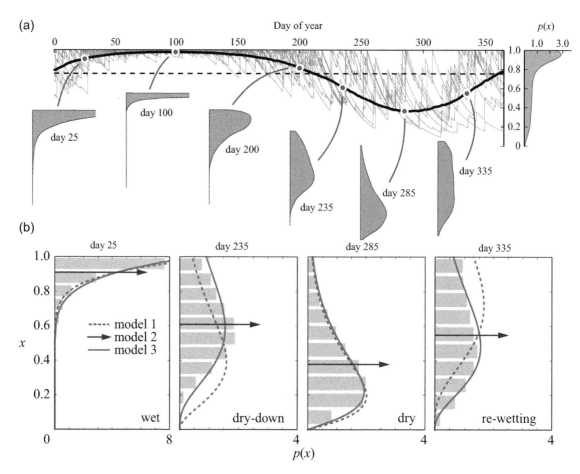

Figure 7.16 (a) Evolution of the soil moisture trajectories and PDFs over a year. The jagged gray lines represent single realizations of the stochastic soil moisture. The thick black line is their ensemble average, and the dashed line is its time average over a year. The PDF in the right-hand panel was compiled for values over the entire year, whereas the PDFs along the trajectories correspond to a particular day of the year. (b) For comparison, the three approximations of the soil moisture PDFs, at different times of the annual cycle. The histograms were compiled from 2000 stochastic simulations. The quasi-steady-state and the self-consistent truncated gamma approximations (models 1 and 3) give PDFs as truncated gamma distributions. In contrast, the negligible fluctuations approximation (model 2) gives PDFs as Dirac delta functions (thick arrows). After Feng et al. (2015).

true value of soil moisture approaches its upper or lower bounds (for example, as it does during the wet period in Fig. 7.16).

The last model, referred to as the self-consistent truncated-gamma approximation, combines the advantages of the first two models but is analytically more complex. It approximates the instantaneous soil moisture PDF by a truncated gamma distribution, as in the first model, Eq. (7.65), but with parameters consistent with the evolution of the mean soil moisture. This is done by setting $b_t = \gamma_t$ and obtaining a_t from an implicit function,

$$\int_0^1 p_{t,\text{ss}}(x; a_t, \gamma_t)x\,dt = \frac{a_t}{\gamma_t} - \frac{\gamma_t^{a_t-1}}{\Gamma(a_t) - \Gamma(a_t, \gamma_t)}e^{-\gamma_t} = \langle x_t \rangle. \tag{7.68}$$

Substituting the PDF of $p_{t,\mathrm{ss}}(x; a_t, \gamma_t)$ into Eq. (7.63) yields

$$\frac{\mathrm{d}\,\langle x_t\rangle}{\mathrm{d}t} = \frac{\lambda_t}{\gamma_t} - k_t\,\langle x_t\rangle - \frac{k_t}{\gamma_t}\frac{\gamma_t^{a_t}}{\Gamma(a_t) - \Gamma(a_t, \gamma_t)}e^{-\gamma_t}, \qquad (7.69)$$

where a_t is given implicitly in Eq. (7.68). This forms a closed set of ordinary differential equations, which can be solved using simple numerical solvers. The difference between this approximation and the quasi-steady-state approximation is that the leakage/runoff term is no longer predetermined by environmental conditions but rather needs to be solved implicitly using current values of $\langle x\rangle$. This modification produces considerable improvements in the approximation to the true values (see Fig. 7.16).

Using the previous approximations for the soil moisture PDF under seasonally varying climates, we can analyze the seasonality in the mean hydrologic partitioning, and this can be visualized in seasonal Budyko-type curves. Figures 7.17a and b show the instantaneous partitioning, e.g. $\langle E_t\rangle/\langle R_t\rangle$ versus $D_{I,t}$ ($= \langle E_{\max,t}\rangle/\langle R_t\rangle$) as a function of time during the year. The results demonstrate a hysteretic behavior whereby the same climate condition may result in different hydrological responses owing to the effect of transient

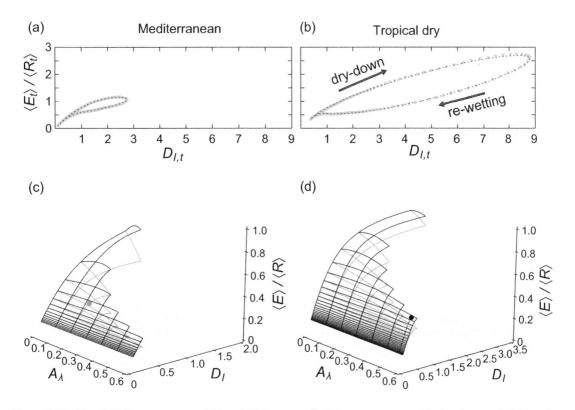

Figure 7.17 (a) and (b) Instantaneous and (c) and (d) long-term Budyko-type water partitioning from typical (a) and (c) Mediterranean and (b) and (d) tropical dry climates. In (a) and (b), the instantaneous dryness index, $D_{I,t}$, is defined as $\langle E_{\max,t}\rangle/\langle R_t\rangle$; in (c) and (d), the rainfall is respectively in phase and out of phase with the potential evapotranspiration. Other common parameters, if not given in the figure, are $(\mu_\lambda, A_\lambda) = (0.3, 0.2)$, $(\mu_k, A_k) = (0.03, 0.02)$, $\phi_k - \phi_\lambda = 180°$, for the Mediterranean climate; $(\mu_\lambda, A_\lambda) = (0.6, 0.58)$, $(\mu_k, A_k) = (0.06, 0.02)$, $\phi_k - \phi_\lambda = 0°$ for the tropical dry climate. The gray curves in (a) and (b) are from numerical simulations and the dots are from the self-truncated gamma approximation; the results in (c) and (d) are from the quasi-steady-state approximation. After Feng et al. (2015).

soil water storage. This can be seen in the "loop" of the transient Budyko curves. For any given dryness index $D_{I,t}$, the corresponding $\langle E_t \rangle / \langle R_t \rangle$ value can fall within two domains – one along the upper "drydown" trajectory and one along the lower "rewetting" trajectory. The reason why more evapotranspiration can occur on average during the drydown trajectory is that stored soil moisture can be carried over in greater amounts from a wet period than from a dry period, accentuating the role of soil water storage.

The annually averaged Budyko curves calculated from seasonal climatic inputs are shown in Figs. 7.17c and d as a function of seasonal rainfall amplitudes, with the rainfall signal both in-phase (black) and out-of-phase (gray) with the potential evapotranspiration, with phase differences of $0°$ and $180°$, respectively. It can be seen that the annually averaged evapotranspiration ratio is lower for an out-of-phase climate than for an in-phase climate, regardless of the seasonal rainfall amplitude. In general, increasing the seasonal rainfall amplitude decreases the annual evapotranspiration ratio. Both the instantaneous and long-term Budyko-type water partitioning suggest that soil water storage tends to transfer rainfall in the wet season for use during the dry periods and so reduces the impact of the unsynchronized rainfall and evaporation demand; this highlights the interactive roles of soil, plant, and climate on hydrological processes. The implications for plant productivity and ecosystems will be further discussed in Chapter 8.

7.8.2 Interannual Variability

The changes in the hydro-climatic regime from one growing season to another are generally controlled by large-scale circulation patterns that are well known for their interannual and interdecadal variability (Greenland et al., 2003). This is especially true for the rainfall, where random interarrival times and intensities (e.g., Figs. 7.18a and b) tend to exhibit marked interannual variability (Figs. 7.18c, e, and g). To understand this interannual variability of rainfall, Porporato et al. (2006) analyzed the statistics of rainfall as a marked Poisson process, using the framework of compound distributions or "superstatistics" (see Sec. 6.3.9). For an individual growing season, the interarrival time for rainfall τ and the rainfall amount h were found to fit exponential distributions,

$$p_\tau(\tau|\lambda) = \lambda \exp(-\lambda\tau), \tag{7.70}$$

$$p_h(h|\alpha) = (1/\alpha) \exp(-h/\alpha), \tag{7.71}$$

where the means of the interarrival times and the rainfall amounts are $1/\lambda$ and α. However, when applied to all years combined, the model underestimated the rainfall variability, on the basis of a comparison of the measured total precipitation with the total precipitation modeled including only daily fluctuations (Fig. 7.18h). The strong interannual variability in the mean rainfall frequency and event depth (λ and α, respectively) is even more evident in Figs. 7.18c and e.

Formally, the interannual variability can be defined as the changes in the year-by-year statistics that cannot simply be explained as the result of different realizations of the same stochastic process describing intra-annual fluctuations. As shown by D'Odorico et al. (2000), interannual fluctuations in rainfall depth and frequency tend to be independent, and a good model for such interannual fluctuations (Figs. 7.18d and f) is a two-parameter gamma distribution,

$$g_x(x) = \frac{b_x^{a_x}}{\Gamma(a_x)} x^{a_x-1} e^{-b_x x}, \tag{7.72}$$

where x stands for either α or λ, b_x is the scale parameter, and a_x is the shape parameter of the distribution.

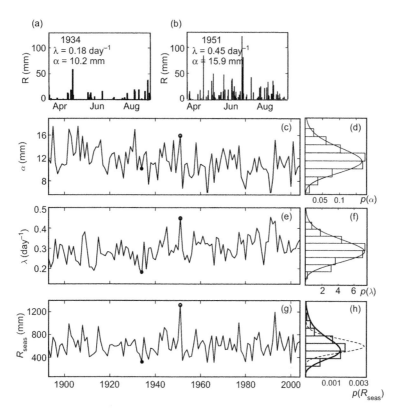

Figure 7.18 Growing season (April to September) rainfall regime at Manhattan, Kansas. Daily precipitation during (a) a very dry and (b) a very wet growing season. (c) Time series of the mean depth of rainfall events, α; (e) mean rate of storm arrival, λ; (g) total rainfall during the growing season, $\mathcal{R}_{\text{seas}}$. The frequency distribution and fitted two-parameter gamma distributions of (d) α and (f) λ. (h) The frequency distribution of the seasonal total precipitation, $\mathcal{R}_{\text{seas}}$, and corresponding theoretical model (solid line); the PDF obtained assuming no interannual variability is also plotted for comparison (dashed line). After Porporato et al. (2006).

The overall distributions of the rainfall depth and the interval between rainfalls can be obtained analytically by combining Eqs. (7.70) and (7.71) with (7.72) using Eq. (6.43):

$$p_\tau(\tau) = \frac{a_\lambda b_\lambda^{a_\lambda}}{(\tau + b_\lambda)^{a_\lambda+1}}, \tag{7.73}$$

$$p_h(h) = \frac{2b_\alpha^{(a_\alpha+1)/2}}{\Gamma(a_\alpha)} h^{(a_\alpha-1)/2} K_{1-a_\alpha}[2\sqrt{b_\alpha h}]. \tag{7.74}$$

Assuming the independence of α and λ, one can find the joint distribution of these two quantities, $g_{\alpha,\lambda}(\alpha, \lambda) = g_\alpha(\alpha)g_\lambda(\lambda)$, which along with the PDF for the seasonal total rainfall expressed in Eq. (6.68) can be used to derive the distribution of the long-term seasonal total rainfall.

The tails of these PDFs are similar to those of stretched exponential distributions (Sornette, 2006) and, therefore, they are intermediate between a power-law (e.g., scaling) behavior and an exponential behavior. Interannual variability increases the frequency of extreme rainfall events with respect to the model that includes only daily variability. The PDFs from Eqs. (7.73) and (7.74) well describe the observed

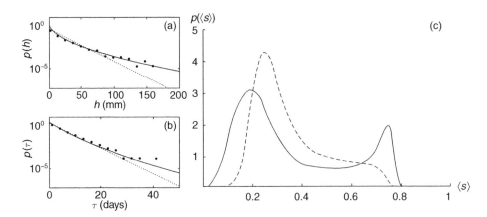

Figure 7.19 (a) PDFs of rainfall depths and (b) interarrival times obtained as superstatistics of daily and interannual variability (continuous lines); for comparison, the exponential PDFs corresponding to no interannual variability (dotted lines), and the observed frequency distributions at Manhattan, Kansas (solid symbols). After Porporato et al. (2006). (c) Probability density function of the average soil moisture during the growing season. The rainfall is characterized by averages of 12.4 mm per storm and 0.21 storms per day, with coefficients of variation $CV(\alpha) = 0.45$, $CV(\lambda) = 0.23$ (solid curve); $CV(\alpha) = 0.22$, $CV(\lambda) = 0.11$ (dashed curve). After D'Odorico et al. (2000).

rainfall patterns, while the exponential distributions without consideration of interannual variability are less suitable to characterize the extreme rainfall depths of all the years combined (Fig. 7.19a and b).

The interannual rainfall variability may have important consequences for the interannual variability of soil moisture. D'Odorico et al. (2000) used a Monte Carlo procedure to numerically estimate the probability distribution of the mean soil moisture $\langle s \rangle$ resulting from the random interannual fluctuations of α and λ. Shown in Fig. 7.19c are the distributions of $\langle s \rangle$ for a given climate, soil, and vegetation with different hypotheses on the coefficients of variation of α and λ. For high values of the coefficient of variation of any or both of these parameters, one observes the emergence of a bimodal behavior driven by the variability of the climatic parameters. This behavior suggests that the system tends to switch between two states, one characterized by high average soil moisture and the other characterized by low average soil moisture. This implies that ecosystems tend to remain in states with far from average conditions. The bimodal behavior disappears when the fluctuations become weaker (dashed line in Fig. 7.19c), clearly showing the importance of the strength of the fluctuations. These patterns can lead to differences in the estimates of extreme events in terms of the intensity of storms and the duration of droughts and their impacts on water resource availability and ecosystem function; this will be discussed further in Chapter 8.

7.9   Soil Moisture Feedback on Convective Rainfall

As we saw in Sec. 5.4.3, the dynamics of convective precipitation is in part affected by the soil moisture, which, through the partitioning of surface heat fluxes, controls the growth of the atmospheric boundary layer and the initiation and intensity of atmospheric moist convection. Moreover, the surface soil moisture may also significantly contribute to the total rainfall input through large-scale precipitation recycling (e.g., Rodríguez-Iturbe et al., 1991; Eltahir and Bras, 1996). Especially in continental regions during the warm seasons, these land–atmosphere interactions often result in a close relationship between the amount of

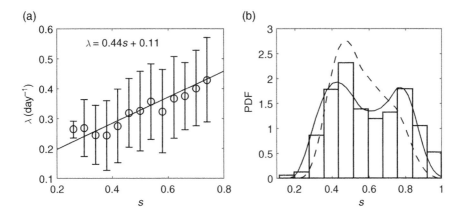

Figure 7.20 (a) Relationship between the relative soil moisture, s, and the rainfall frequency λ in Illinois, calculated for different soil moisture classes (i.e., in intervals of width 0.04). The solid line represents a linear fit ($r^2 = 0.87$). (b) The histograms show the soil moisture distribution estimated from the Illinois data. The figure shows also the probability distribution of the regional soil moisture calculated by the stochastic soil water balance model with (solid line) $\lambda = as + b$ and (dashed line) constant λ. After D'Odorico and Porporato (2004).

rainfall and the antecedent soil moisture (Findell and Eltahir, 1997; Yin et al., 2015). As a result, the rainfall input can no longer be treated as an external forcing, as we have done so far, but becomes in part an internal component of the dynamics.

D'Odorico and Porporato (2004) explored this interaction by analyzing the feedback between soil moisture and rainfall frequency. They used data from the state of Illinois, where local recycling and soil-moisture–rainfall feedbacks had already been analyzed (Findell and Eltahir, 1997; Koster et al., 2004). For each day in the warm seasons, the rainfall frequency in the subsequent 21 days was calculated and related to the regional average soil moisture. Figure 7.20a shows a plot of the average storm frequency, λ, as a function of the statewide average soil moisture for a station centrally located in the region of interest. The results show the existence of a relatively strong dependence between the frequency of rainstorm occurrences and antecedent soil moisture conditions (Fig. 7.20a), whereas almost no correlation exists between soil water content and subsequent storm depths. An enhancement of the likelihood of rainfall occurrence, rather than an increase in rainfall amount, is consistent with other results in the literature (Emanuel and Bister, 1996; Zeng et al., 2000; Findell and Eltahir, 2003; Konings et al., 2010).

From a modeling point of view, one can assume that average rainfall frequency is a linear function of soil moisture,

$$\lambda = as + b, \tag{7.75}$$

where coefficient a may be linked to the strength of land-atmosphere interaction, while coefficient b is related to rainfall events that are not affected by the regional soil moisture. One may associate a with the convective rainfall triggered by the initial soil moisture, while b may be thought to represent the component of rainfall less sensitive to local effects, such as the frontal rainfall (Yin et al., 2014).

The master equation (7.27) with the inclusion of a state-dependent frequency as in Eq. (7.75) can be written as (Porporato and D'Odorico, 2004)

$$\frac{\partial p(s,t)}{\partial t} = \frac{\partial}{\partial s}[p(s,t)\rho(s)] - \lambda(s)p(s,t) + \int_0^s \lambda(z)p(z,t)f_Y(s-z;z)dz, \tag{7.76}$$

where $p(s, t)$ is the time-dependent PDF of s and f_Y is the probability of normalized infiltration (see Eq. (7.24)). A solution to this master equation is possible under steady-state conditions (Porporato and D'Odorico, 2004):

$$p(s) = \frac{C}{\rho(s)} \exp\left(-\gamma s + \int_s [\lambda(u)/\rho(u)]du\right) \tag{7.77}$$

where C is a normalization constant. Interestingly, the mathematical form of Eq. (7.77) is the same as that of the homogeneous case (see Eq. (7.28) in Sec. 7.4.1). The resulting soil moisture distribution from Eq. (7.77) with $\lambda(s)$ in Fig. 7.20a, presented in Fig. 7.20b (solid line), is consistent with the empirical distribution estimated from the observation data (the histogram). The well-defined bimodal distribution for the soil moisture indicates a relatively high likelihood for the summer soil moisture dynamics to be either in a dry state or in a wet state. The bimodality disappears when λ is assumed to be constant, i.e., independent of soil moisture (Fig. 7.20b, dashed line). Thus, a state-dependence in the timing of the forcing of the terrestrial water balance can generate bimodality in the probability distribution of the soil water content, suggesting that the positive feedback between soil moisture and the timing of precipitation may lead to the observed emergence of two preferential states in the summer soil moisture dynamics. In contrast with the bimodality due to interannual rainfall variability (see Fig. 7.19c), here the biomodality corresponds to the genuine presence of dry and wet preferential states; this is clearly shown in Fig. 7.20b.

A simple explicit solution for which the appearance of preferential states can be analyzed in detail can be obtained by extending the minimalist model of Milly presented in Sec. 7.5 (i.e., $\rho(s) = \beta = $ const.) with Eq. (7.75). In this case, the steady-state distribution is a mixed one (Cox and Miller, 1965; Laio et al., 2001a), with an atom of probability at zero:

$$p(s) = C\left[\frac{1}{\beta} \exp(-[\gamma - (a/\beta)]s + (b/2\beta)s^2) + \frac{\delta(s)}{a}\right] \tag{7.78}$$

with

$$C = \left[\frac{1}{a} + \sqrt{\frac{\pi}{2b\beta}} e^{-\eta^2}[\mathrm{Erfi}(\eta + b) - \mathrm{Erfi}(\eta)]\right]^{-1}, \tag{7.79}$$

where $\eta = (a - \beta\gamma)/\sqrt{2b\beta}$ and $\mathrm{Erfi}(\cdot)$ is the imaginary error function (Abramowitz and Stegun, 1964). Figure 7.21 shows a plot of the PDF of s, where the continuous part can be bimodal with modes at $s = 0$ and $s = 1$. The existence of these preferential states is observed in the soil moisture time series (Fig. 7.21). It is interesting to compare the similar bimodal behavior of the soil moisture distributions from this minimalist model with the behavior with the full loss function (Fig. 7.20).

The initial transient of the mean trajectory plays an important role in the development of noise-induced phase transitions (Van den Broeck et al., 1997). In the present case, the so-called macroscopic equation for the temporal evolution of the mean $\langle s_t \rangle$ can be derived from Eq. (7.76) as

$$\frac{d\langle s_t \rangle}{dt} = \frac{\langle \lambda(s) \rangle}{\gamma} - \langle \rho(s) \rangle - \frac{1}{\gamma} \int_0^1 \lambda(z)p(z, t)e^{-\gamma(1-z)}dz \tag{7.80}$$

which has the same format as Eq. (7.63) but expresses λ as a function of s to account for the impact of the land–atmosphere interaction. Assuming $p(s, t = 0) = \delta(s - s_0)$ as the initial condition, the slope of the mean trajectory starting from s_0 is found to be

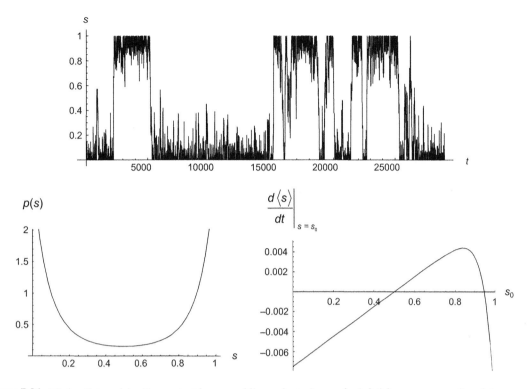

Figure 7.21 Minimalist model with constant losses and linear dependence of rainfall frequency on soil moisture ($a = 0.1$, $b = 0.3$, $\beta = 0.0125$, $\gamma = 20$). Upper panel: Example of time series for rainfall. Lower panel: (left) Continuous part of the steady-state PDF from Eq. (7.78) and (right) initial slope of the mean trajectory, Eq. (7.81), as a function of s_0. After Porporato and D'Odorico (2004).

$$\left. \frac{d\langle s_t \rangle}{dt} \right|_{s=s_0} = -\rho\,(s_0) + \frac{\lambda\,(s_0)}{\gamma} \left[1 - e^{-\gamma(1-s_0)} \right]. \tag{7.81}$$

This is plotted in Fig. 7.21, where the destabilizing action of noise in some parts of the soil moisture domain is clearly evident.

In the presence of seasonal variations, the soil-moisture–rainfall feedback is typically stronger in summer than in winter (Yin et al., 2014). This adds to the seasonal climatic forcing discussed in Sec. 7.8.1 and further enhances the seasonality in the soil moisture dynamics. Yin et al. (2014) addressed this problem by considering a and b in Eq. (7.75) as time-dependent coefficients, where $a(t)$ is expected to be larger in warm seasons, corresponding to the stronger land–atmosphere interaction. The resulting time-dependent soil moisture PDF shows that the soil moisture is unimodal in winter and spring but becomes consistently bimodal in summer owing to the increased soil-moisture–rainfall feedback strength, and returns back to unimodal in the late fall when the feedback fades away (Yin et al., 2014). These results not only show that land–atmosphere feedbacks may sustain and enhance the effect of initial moisture anomalies occurring at the beginning of the warm season but also that, because of such feedbacks, the summer soil moisture dynamics may evolve toward either a dry or a wet state in which the system may remain locked for the rest of the warm season. These preferential states of soil moisture may induce important fluctuations in the interarrival times of rainstorms (Daly and Porporato, 2007) and in the average duration of the periods of water stress, thus posing a significant challenge to vegetation.

7.10 Key Points

- Soil moisture dynamics is modeled as a stochastic differential equation with instantaneous jumps due to infiltration; the latter is a marked Poisson process censored by canopy interception, which reduces both its frequency and amount, and by a state-dependent threshold representing the infiltration excess runoff and its dependence on the antecedent soil moisture conditions.
- The loss of soil moisture is well approximated by a nonlinear function of soil moisture featuring a linear increases between the wilting point (s_w) and the onset of stress (s^*), a subsequent constant plateau up to field capacity (s_{fc}), and a rapid increase following the hydraulic conductivity until saturation.
- For homogeneous seasons, the steady-state probability density function (PDF) of the soil moisture can be used to determine the likelihood of having soil moisture conditions between certain critical thresholds, including those of plant water stress and percolation, as well as to compute mean loss rates, which define the average hydrologic partitioning, as a function of climate, soil, and vegetation parameters.
- The Π theorem allows us to write the hydrologic partitioning as a function of two dimensionless groups: the dryness index (D_I) and the storage index (γ); the shape of Budyko's curve is explained by the stochasticity of the soil moisture dynamics and the presence of a threshold due to the finite storage capacity of the soil.
- Due to the soil moisture modulation of evapotranspiration, the daily fluctuations of potential evapotranspiration have a minor impact on the soil moisture PDF and the soil water balance, compared with the rainfall fluctuations.
- The seasonality of rainfall and evapotranspiration create hysteresis loops in the curves of the hydrologic partitioning related to the seasonal changes in soil storage capacity; the interannual rainfall variability can be accounted for by random fluctuations in the frequency and amount of rainfall, and it results in increased variance of the mean annual rainfall and possible bimodality of the long-term mean soil moisture PDF.
- A dependence of rainfall frequency on soil moisture due to land–atmosphere feedbacks has the potential to induce bimodality in the soil moisture PDF as a result of persisting dry and wet conditions.

7.11 Notes, including Problems and Further Reading

7.1 (Constant vs. Stochastic Rainfall) To appreciate the importance of rainfall stochasticity and intermittency, consider the soil moisture balance equation (7.1) with constant rainfall. Discuss qualitatively the differences compared with the corresponding stochastic differential equation. Consider the differences in the water balance, and elaborate on the consequences for the related ecohydrological processes.

7.2 (Periodic vs. Stochastic Rainfall) Following a similar conceptual reasoning to that in the previous question, compare the stochastic soil moisture balance equation (7.1) forced by rainfall as a marked Poisson process with frequency λ and mean depth α with an equation forced by periodic rainfall events, consisting of regularly spaced rainfall occurrences with the same frequency λ and depth α. Discuss the differences in the water balance, and elaborate on the consequences for the related ecohydrological processes.

7.3 For rainfall as a marked Poisson process with exponential jumps, what is the probability that the total rainfall is less than the mean for a growing season of 180 days? Hint: obtain the CDF of the integrated rainfall process by numerical integration of (6.68).

7.4 Plot the PDF of cumulated rainfall for two cases with the same average rainfall rate but different rainfall frequencies for a growing season of 180 days. What happens when the frequency of rainfall becomes very large and the depth small?

7.5 Prove that the exponential distribution with mean α under censoring and rescaling (Eq. (7.18)) is still an exponential distribution with reduced mean of $\alpha' = k_{\text{int}}\alpha$ (Eq. (7.20)).

7.6 Find the analytical expression for the percentage of intercepted rainfall for given amount of rainfall, using the framework developed in Sec. 7.3.2.

7.7 Figure 7.8 shows some examples of soil moisture PDFs for two different types of soil. Extend these examples by including other types of soil (e.g., sand, sandy loam, and clay) using Table 3.1.

7.8 Find analytical expressions for $\langle E_s \rangle$, $\langle E_{ns} \rangle$, and $\langle L \rangle$ by integrating Eq. (7.34). Verify these results by summing them up and comparing the result with the total mean loss in Eq. (7.35).

7.9 (Time-Dependent Soil Moisture) The time-dependent mean soil moisture can be found by multiplying the master equation (7.27) by s and integrating over the whole range of s (see Laio et al., 2002). Obtain the resulting time-dependent mean soil moisture equation (the macroscopic equation) for linear losses and no runoff (for unbounded soil moisture). For sinusoidal rainfall and constant evapotranspiration (ET), plot the solution for different rooting depths. Extend the analysis to sinusoidal ET in-phase and out-of-phase with rainfall.

7.10 The long-term mean runoff in Eq. (7.36) is derived from the water balance (see Sec. 7.4.2). Show that it can also be obtained as the product of the mean rainfall depth and the frequency of upcrossings at the saturation level of soil moisture (see Sec. 6.8).

7.11 Find analytical expressions for the variances of the effective relative soil moisture from the two minimalist models under the steady-state conditions in Sec. 7.5.

7.12 (Minimalist Model with Constant Loss) The minimalist model from Milly (1993) describes the *Takacs process* with an upper bound, which can be useful for modeling the queuing and storage problems. In such a system, the state variable can be interpreted as the total time it would take to serve all customers in an office at time t (Cox and Miller, 1965). In Chapter 10 it will be related to micro-irrigation and to the water stored in a rainwater-harvesting cistern.

7.13 (Dimensional Analysis of Hydrologic Partitioning) Show that besides (a, λ) it is possible to choose $(E_{\max}, \lambda)$, $(E_{\max}, \alpha)$, $(E_{\max}, w_0)$, or (w_0, λ) as dimensionally independent quantities in Eq. (7.45). Note that the Π theorem gives equations of the following types (Daly et al., 2019b):

$$\frac{\langle E \rangle}{E_{\max}} = \phi_1 \left(\frac{\alpha \lambda}{E_{\max}}, \frac{\lambda w_0}{E_{\max}} \right) = \phi_1 \left(H_I, \frac{\lambda}{\eta} \right), \tag{7.82}$$

$$\frac{\langle E \rangle}{E_{\max}} = \phi_2 \left(\frac{\alpha \lambda}{E_{\max}}, \frac{w_0}{\alpha} \right) = \phi_2 \left(H_I, \gamma \right), \tag{7.83}$$

$$\frac{\langle E \rangle}{E_{\max}} = \phi_3 \left(\frac{\lambda w_0}{E_{\max}}, \frac{w_0}{\alpha} \right) = \phi_3 \left(\frac{\lambda}{\eta}, \gamma \right), \tag{7.84}$$

$$\frac{\langle E \rangle}{\lambda w_0} = \phi_4 \left(\frac{E_{\max}}{\lambda w_0}, \frac{\alpha}{w_0} \right) = \phi_4 \left(\frac{\eta}{\lambda}, \gamma^{-1} \right) \tag{7.85}$$

$$\frac{\langle E \rangle}{\alpha \lambda} = \phi_5 \left(\frac{E_{\max}}{\alpha \lambda}, \frac{w_0}{\alpha} \right) = \phi_5(D_I, \gamma) \tag{7.86}$$

where $H_I = 1/D_I$ is the humidity index. The resulting hydrological spaces provide different, if related, scaling laws that allow us to focus on the role of specific combinations of parameters and emphasize different hydrological conditions (Daly et al., 2019a, b).

8   From Plant Water Stress to Ecosystem Structure

The cactus thrives in the desert while the fern thrives in the wetland. The fool will try to plant them in the same flowerbox.

Nazarian, V. (2010)[†]

In this chapter we model the plant physiological effects of low soil moisture described in Chapter 4 and couple them with the stochastic models of soil moisture developed in Chapter 7 to explore some strategies developed by vegetation to cope with water stress. Using the crossing properties of soil moisture dynamics we define a measure of plant water stress, which is used along with the stochastic properties of the soil water balance and plant assimilation to analyze different ecosystems in semi-arid regions. We focus on tree–grass coexistence along the Kalahari precipitation transect, the role of precipitation variability on shrub encroachment in Southern Texas, and the so-called inverse texture effect in the Colorado steppe. Finally we discuss different strategies of water use by plants in conditions of stochastic variability and the role of soil moisture fluctuations on plant biodiversity.

8.1   Probabilistic Description of Plant Water Stress and Carbon Assimilation

In water-controlled ecosystems, water demand by plants is generally higher than water availability, leading to plant water stress. To cope efficiently with water stress, plants have developed different strategies, which become more sophisticated the more intense and unpredictable the water deficit is. Many species combine a number of complementary measures to develop such strategies, the most extreme of which include permanent forms of adaptation, such as changes in resource allocation, specialized root growth (e.g., cacti build a dense network of roots to capture light rainfall events, while some desert shrubs – the so-called *phreatophytes* – develop deep roots to tap the water table when present), specialized photosynthetic pathways (e.g., the CAM pathway), short and intense life cycles during favorable periods, dormancy, drought deciduousness, specialized metabolism and leaf structure to reduce water losses (high cuticular resistance, protection and changes in dimension and density of the stomata), etc. (Jones, 1992; Larcher, 1995). Our focus is on the hydrologic role of such strategies and our approach will be simplified to make use of the analytical tools previously developed.

    With this goal in mind, on the basis of how plant status and the related physiological response are impacted by reductions in soil moisture levels (Sec. 4.9), in this section we formalize a simple quantitative

---

[†]Nazarian, V. (2010). *The Perpetual Calendar of Inspiration*. Highgate Center, Vermont.

description of plant water stress for a given soil moisture level (Sec. 8.1.1). This description is then coupled with the crossing properties of soil moisture, discussed in Sec. 6.8, to analyze plant water stress in a dynamic context, including information on the duration and frequency of water stress during the growing season (Sec. 8.1.2). Finally, these concepts are extended to consider the impact of water stress on the dynamics of carbon assimilation (Sec. 8.1.3).

8.1.1   Static Stress

As we saw in Chapter 4, the range of stomatal closure spans a wide range of plant responses to water stress, from the *onset of stomatal closure* at s^* to an almost complete halting of plant functions when the stomata are fully closed, at the *wilting point* s_w. From these considerations, following Porporato et al. (2001), one can assume a "static stress" function of the soil moisture, $\zeta(s)$, which is zero when the soil moisture is above the level of incipient reduction of transpiration, s^*, and which reaches a maximum value of one when the soil moisture is at the level of complete stomatal closure (wilting). For soil moisture values between s_w and s^*, the effects described in Fig. 4.14 clearly suggest a nonlinear increase in plant water stress with soil moisture deficit. Figure 8.1a, adapted from Bradford and Hsiao (1982), may be used as a reference sketch for the modeling of ζ, whose form can thus be expressed as (Porporato et al., 2001)

$$\zeta(s) = \begin{cases} 0 & s > s^*, \\ \left(\dfrac{s^* - s}{s^* - s_w}\right)^q & s_w \leq s \leq s^*, \\ 1 & s < s_w. \end{cases} \tag{8.1}$$

Typical values for s^* and s_w and their relation to the soil and plant characteristics have been discussed in Secs. 3.2, 5.4.5, and 7.2.1. The parameter q is a measure of the nonlinearity of the effect of soil moisture deficit on plant conditions (Fig. 8.1). The nonlinearity implied by Eq. (8.1) is justified by the sequence of effects and the increase in intensity of each effect with soil moisture deficit (see Fig. 4.14). As in the case of s_w and s^*, the value of q can also vary with plant species and, to a smaller extent, with the soil type.

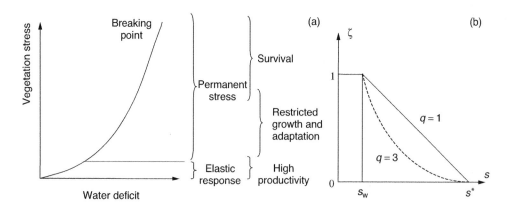

Figure 8.1 Relationship between soil water content and plant water stress: (a) qualitative representation of the effect of water deficit on vegetation; redrawn from Bradford and Hsiao (1982). (b) static water stress ζ versus soil moisture, s, for two values of the nonlinearity parameter q; see Eq. (8.1). After Porporato et al. (2001).

The simple relationship between the static water stress $\zeta(t)$ and soil moisture content s allows one to invert Eq. (8.1) and obtain the probability density function $f_Z(\zeta)$ as a derived distribution of the steady-state PDF for the soil moisture (Sec. 6.3.8). Owing to the particular form of the stress function, ζ has an atom of probability F_Z at both 0 and 1:

$$F_Z(0) = 1 - P(s^*), \tag{8.2}$$

$$F_Z(1) = P(s_w), \tag{8.3}$$

where $P(s^*)$ and $P(s_w)$ are the values of the soil moisture CDFs calculated for s^* and s_w. Including the continuous part of the soil moisture in Eq. (7.29), one can obtain the distribution for the static water stress:

$$f_Z(\zeta) = \delta(\zeta)\big[1 - P(s^*)\big] + \delta(\zeta - 1)P(s_w)$$
$$+ \frac{C_\zeta}{\eta_w}\left[\left(1 - \frac{\eta}{\eta_w}\right)\zeta^{1/q} + \frac{\eta}{\eta_w}\right]^{(\lambda(s^*-s_w)/\eta-\eta_w)-1} e^{\gamma\left[(s^*-s_w)\zeta^{1/q}-s^*\right]}, \tag{8.4}$$

where $\delta(\cdot)$ is the Dirac delta function (see Sec. 6.3.4), and C_ζ is a constant of integration.

Figure 8.2 shows some examples of PDFs for the static water stress for different types of climate and active soil depth. Except for the case of wet climates, shallow-rooted plants tend to experience higher stress conditions than deep-rooted species (i.e., greater values of $F_Z(1)$ and modes of the continuous part at higher ζ). This behavior is accounted for by the temporal dynamics of soil moisture, examples of which are shown in Fig. 8.3. In deeper soils (continuous line), the trace of soil moisture tends to remain almost always between s_w and s^*, so that in this example vegetation does not experience very high stress conditions. On the contrary, for shallow-rooted plants, the trace of stress frequently reaches high values, but also remains close to zero over longer periods. In contrast, for wet climates the mean is above s^* and the trajectory of soil moisture for deep soils hardly ever crosses s^*, thus reducing the level of water stress with respect to that of shallow-rooted plants (Figs. 8.2e and f).

The mean value of water stress given that the plant is under stress is more meaningful for our purposes. To obtain the latter, denoted by $\bar{\zeta}'$, only the part of the PDF corresponding to ζ above zero should be considered, i.e.,

$$\bar{\zeta}' = \frac{1}{1 - F_Z(0)} \int_{0^+}^1 f_Z(\zeta)\zeta d\zeta = \frac{1}{P(s^*)} \int_0^1 f_Z(\zeta)\zeta d\zeta = \frac{\bar{\zeta}}{P(s^*)}, \tag{8.5}$$

where the integration interval automatically excludes the atom of probability at $\zeta = 0$.

The sensitivity of $\bar{\zeta}'$ to different rainfall conditions is shown in Fig. 8.4a: the relationship between the mean conditional stress and λ is strongly nonlinear. The difference between $q = 1$ and $q = 3$ remains approximately constant for all values of λ except for the lowest, when s is almost always below s_w and differences in q have a smaller effect (see Fig. 8.1b). The value of $\bar{\zeta}'$ remains recognizably different from 0 for very wet climates also, since $\bar{\zeta}'$ by definition does not take into account the probability of occurrence of water stress, which for wet climates is very low. The influence of the soil depth, Z_r, is shown in Fig. 8.4b for moderate values of the average intensity and frequency of rainfall. As can be seen in Fig. 8.3, for this climate the water stress decreases with soil depth, both for $q = 1$ and, more rapidly, for $q = 3$.

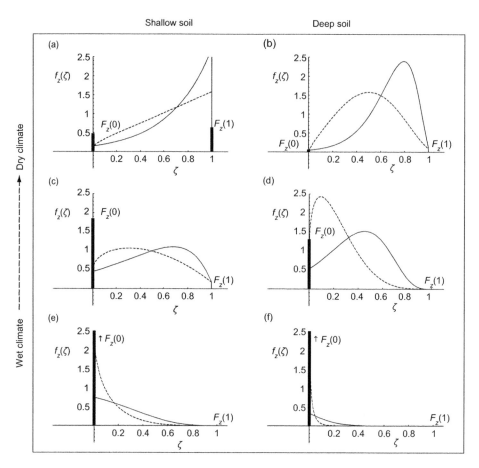

Figure 8.2 Examples of PDFs of static water stress ζ for different active soil depths and frequency of rainfall events λ (the atoms of probability at $\zeta = 0$ and $\zeta = 1$ have a different scale from $f_Z(\zeta)$). The solid lines refer to $q = 1$ and the dotted lines to $q = 3$. The other climate and soil parameters are the same as those in Fig. 7.8. After Porporato et al. (2001).

8.1.2 Dynamic Water Stress

To extend the previously defined static stress with the mean duration and frequency of the water stress, Porporato et al. (2001) developed a dynamic water stress, $\overline{\theta}$, defined as follows:

$$\overline{\theta} = \begin{cases} \left(\dfrac{\overline{\zeta'}\,\overline{T}_{s*}}{k\,T_{\text{seas}}} \right)^{\overline{n}_{s*}^{-r}} & \text{if } \overline{\zeta'}\,\overline{T}_{s*} < kT_{\text{seas}}, \\ 1 & \text{otherwise.} \end{cases} \tag{8.6}$$

In this definition, the average static water stress, $\overline{\zeta'}$ in Eq. (8.5), takes into account the mean intensity of the water deficit but contains no information on its duration and frequency. If we assume a linear dependence between the time under stress and the intensity of the dynamic stress, the mean amount of plant water stress during a period of stress becomes a function of the product $\overline{\zeta'}\,\overline{T}_{s*}$. The actual plant water stress, however, cannot increase with $\overline{\zeta'}\,\overline{T}_{s*}$ indefinitely, since there must be a point, corresponding

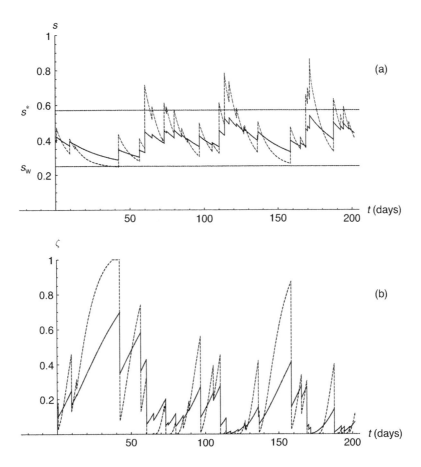

Figure 8.3 Example of traces of soil moisture (a) and static water stress (b) for the same rainfall sequence ($\alpha = 0.15$ cm; $\lambda = 0.2$ day^{-1}) in a loamy soil with vegetation typical of semi-arid environments (see the caption of Fig. 7.8 for other parameters). The solid lines refer to $Z_r = 90$ cm and the dashed lines to $Z_r = 30$ cm. After Porporato et al. (2001).

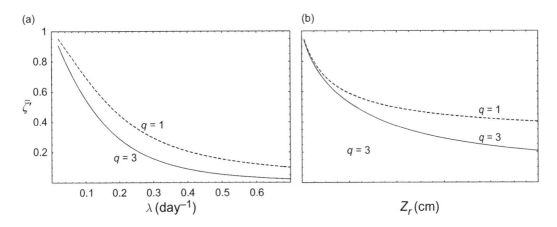

Figure 8.4 Sensitivity of the mean static stress $\bar{\zeta}'$ with respect to: (a) rainfall frequency λ, and (b) active soil depth Z_r. Parameters for this figure are $\alpha = 1.5$ cm, $T_{seas} = 200$ days, $Z_r = 60$ cm in (a); $\lambda = 0.2$ day^{-1} in (b); see the caption of Fig. 7.8 for other parameters. After Porporato et al. (2001).

to the onset of permanent damage, from whereon the stress is at its maximum level. The plant physiology literature frequently makes reference to this upper value of the water stress (e.g., Levitt, 1980; Bradford and Hsiao, 1982). The parameter k in Eq. (8.6) allows us to fix the value of such a threshold: permanent damage appears when $\overline{\zeta'}\,\overline{T}_{s*} > kT_{\text{seas}}$, with k representing an index of plant resistance to water stress. The quantity k may also be interpreted as the average static stress $\overline{\zeta'}$ that a plant can experience without suffering permanent damage, when the duration of the period of stress is the whole growing season.

In addition, the effect of multiple periods of stress on a plant's status is contained between the following two extreme conditions: (1) the plant completely recovers after any period of water stress; (2) the plant does not recover and each new occurrence of a period of water deficit contributes to increase the existing stress of the plant. In the first case, the frequency of the periods of water stress has no importance, and a good measure of the total dynamic stress during the growing season is simply $\overline{\zeta'}\,\overline{T}_{s*}/(kT_{\text{seas}})$. In the latter case, assuming simply an additive accumulation of stress throughout time or, equivalently, a linear relationship between the number of crossings and the dynamic stress, the total water stress during the growing season is $\overline{n}_{s*}\overline{\zeta'}\,\overline{T}_{s*}/(kT_{\text{seas}})$. With such a formulation and Eqs. (6.99), (6.101), and (8.5), the dynamic water stress would be reduced to

$$\overline{n}_{s*}\frac{\overline{\zeta'}\,\overline{T}_{s*}}{kT_{\text{seas}}} = \frac{\overline{\zeta}}{k}, \tag{8.7}$$

so that the information on the duration and frequency of the water stress cancels out, resulting again in a static formulation of the water stress. An intermediate condition that includes this effect employs a decreasing function of $\overline{n}_{s*}$ as the exponent of $\overline{\zeta'}\,\overline{T}_{s*}/(kT_{\text{seas}})$. Porporato et al. (2001) found that the particular choice $r = 1/2$ tempers satisfactorily the importance of very high values of $\overline{n}_{s*}$.

Figure 8.5 shows how the value of the mean dynamic water stress $\overline{\theta}$ decreases with an increase in the storm frequency λ (the mean depth of rainfall per storm is fixed). As rainfall events become more frequent, the soil water becomes more abundant and the water stress tends to disappear. The behavior of the curves is interesting: a rapid decay of the dynamic stress for low values of λ is first followed by a

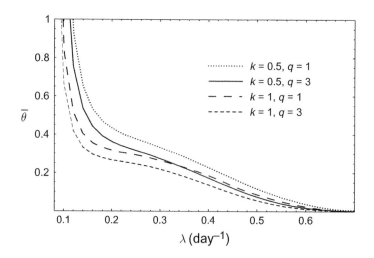

Figure 8.5 Mean dynamic water stress $\overline{\theta}$ versus frequency of rainfall events λ for four different choices of the parameters k and q ($T_{\text{seas}} = 200$ days; $\alpha = 1.5$ cm, and $Z_r = 60$ cm; see the caption of Fig. 8.2 for the soil and vegetation parameters. After Porporato et al. (2001).

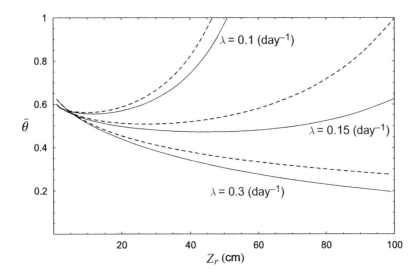

Figure 8.6 Mean dynamic water stress $\overline{\theta}$ versus active soil depth Z_r, for three different values of λ and two values of the parameter q ($q = 1$, dashed lines, $q = 3$, solid lines). $T_{\text{seas}} = 200$ days, $k = 0.5$, and $\alpha = 1.5$ cm; see the caption of Fig. 8.2 for the soil and vegetation parameters. After Porporato et al. (2001).

plateau for intermediate frequencies of rainfall events and then by a slow decay to zero stress for very wet climates. Notice also from Fig. 8.5 that the behavior of $\overline{\theta}$ is relatively robust to changes in the parameters k and q.

Figure 8.6 shows the influence on $\overline{\theta}$ of the effective soil depth (or rooting depth) Z_r. The response of plants to Z_r is very different for dry and wet climates. When the climate is wet, deeper-rooted species seem more fitted than shallow-rooted species, while the opposite is true for very dry climates. The reason for this is that in dry climates the soil layers near the surface are the layers generally wetted by weak storm events, so that it becomes fundamental for a plant to have mostly superficial roots in order to cope with rapidly occurring evaporation losses. When the climate is wet, the surface soil layers are frequently not much wetter than for drier climates, due because the storage limit imposed by the field capacity and the soil evaporation occurs mainly in those layers (which explains why the curves for $\overline{\theta}$ are so close to one another for shallow soil depths). Nevertheless a sizeable amount of water infiltrates to the deeper layers, so that the presence of deep roots becomes very important. Notice that this contrasting behavior causes the curve for an intermediate climate to develop a broad minimum-stress condition for Z_r close to 30 cm.

Finally, the interplay between the timing and amount of rainfall can be studied, considering as a variable the storm frequency λ, while keeping fixed the total precipitation per growing season, $\mathcal{R}_{\text{seas}} = \lambda \alpha T_{\text{seas}}$. Figure 8.7 shows that, except for very wet climates, there is an optimal partition between the timing and amount of rainfall (corresponding to the minimum of the dynamic water stress) which provides the best condition for vegetation. More frequent and lighter rainfall (i.e., toward the right-hand side of the diagram) tends to save the water loss from leakage and runoff and to keep a relatively invariant soil water content, leading to lower dynamic water stress in a wet climate where the soil moisture often stays above s^* but resulting in a higher dynamic water stress in a dry climate where the soil moisture constantly stays below s^*.

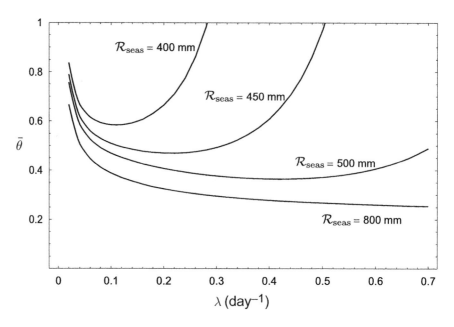

Figure 8.7 Impact of timing and amount of rainfall on dynamic water stress. Each curve corresponds to a constant total rainfall during the growing season, $\mathcal{R}_{seas} = T_{seas}\alpha\lambda$. $T_{seas} = 200$ days, $q = 2$, $k = 0.5$, and $Z_r = 60$ cm; see the caption of Fig. 8.2 for other soil and vegetation parameters. After Porporato et al. (2001).

8.1.3   Plant Assimilation and Productivity

Reasoning analogous to the previous sections can be followed to link the statistics of soil moisture to those of plant carbon assimilation. As a starting point, recall that the relationship between daily carbon assimilation and soil moisture was modeled as a piecewise function in Eq. (5.74), reported here for convenience:

$$A_n(s) = \begin{cases} 0, & s \leq s_{w_A}, \\ \dfrac{s - s_{w_A}}{s_A^* - s_{w_A}} A_{max}, & s_{w_A} \leq s \leq s_A^*, \\ A_{max}, & s_A^* < s \leq 1. \end{cases} \tag{8.8}$$

This expression can be used along with the distribution of soil moisture (see Sec. 6.3.8) to derive the probabilistic structure of the daily assimilation (Daly et al., 2004b). Given the similarity between Eqs. (8.1) and (8.8), it can be expected that the distribution of A_n has atoms of probability both at 0 and 1 and a continuous part in between:

$$\begin{aligned} P_{A_0} &= \int_0^{s_{w_A}} p(s)ds = P(s_{w_A}), \\ P_{A_{st}} &= \int_{s_{w_A}}^{s_A^*} p(s)ds = P(s_A^*) - P(s_{w_A}), \\ P_{A_{max}} &= \int_{s_A^*}^1 p(s)ds = 1 - P(s_A^*), \end{aligned} \tag{8.9}$$

where $P(s_{w_A})$ and $P(s_A^*)$ are the values of the soil moisture CDFs calculated for s_{w_A} and s_A^*. These three probabilities represent the fractions of time in which the soil moisture is too low for a plant to perform photosynthesis, in which the assimilation takes place in stressed conditions, and in which the plant achieves maximum assimilation, respectively.

The net carbon assimilation at the leaf level can be considered as an index of plant productivity. Similarly to the mean static water stress in Eq. (8.5), one can obtain the mean assimilation during a growing season as

$$\langle A_n \rangle = \int_0^{A_{\max}} A_n p(A_n) dA_n. \tag{8.10}$$

Porporato et al. (2004) used this metric in conjunction with the minimalistic soil moisture model of Sec. 7.5 to assess the effects of rainfall regime on ecosystem productivity (Fig. 8.8). They compared the theoretical results with the findings of a four-year manipulative experiment (Knapp et al., 2002), in which the response of a native grassland to increased rainfall variability was investigated by artificially reducing storm frequency and increasing rainfall quantity per storm, while keeping the total annual rainfall unchanged. The theoretical mean carbon assimilation was found to reproduce well the $\sim 20\%$ decrease in measured net assimilation for the altered rainfall pattern, going from a mean net assimilation of 23 μmol m^{-2}s^{-1} in natural conditions to ~ 18.4 μmol m^{-2}s^{-1} when the total rainfall was the same but concentrated in fewer

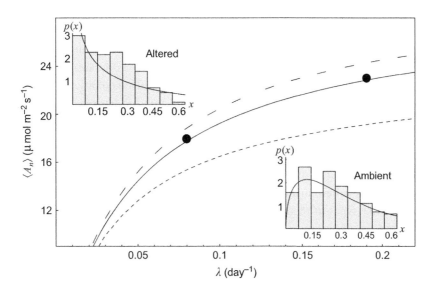

Figure 8.8 Mean daily carbon assimilation rate as a function of the frequency of rainfall events for constant total amount of precipitation during a growing season. The lines are the theoretical curves derived from the soil moisture PDF, while the two large black points are field data published by Knapp et al. (2002), who compared the response of a mesic grassland to the ambient rainfall pattern (the point on the right, $\lambda = 0.19$ day^{-1}) and the response to an artificially increased rainfall variability (the point on the left, $\lambda = 0.08$ day^{-1}). The solid line is for a mean total rainfall during the growing season of 500 mm, the long-dashed line is for 700 mm, and the dotted line is for 300 mm. The two insets show the observed and theoretical soil moisture PDFs for the ambient and altered conditions. Other parameters are: $n = 0.38$, $Z_r = 30$ cm, $E_{\max} = 0.6$ cm day^{-1}, $A_{\max} = 28$ μmol m^{-2}s^{-1}, $s_w = 0.12$, $s^* = 0.30$, $s_1 = 0.65$. After Porporato et al. (2004).

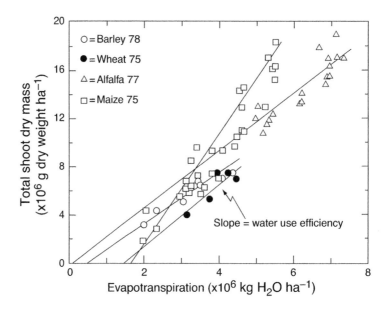

Figure 8.9 Production of above-ground shoot dry matter at various levels of water use in several crops near Logan, Utah. The slope of the linear relation is the water use efficiency, which corresponds to 2.11 g of dry matter per kg of water for barley, 2.50 for wheat, 2.36 for alfalfa, and 4.49 for maize. Notice that maize is a C_4 plant and the others are C_3 plants. After Kramer and Boyer (1995).

events. As shown by the effective relative soil moisture PDFs, the dramatic shift in the rainfall frequency changes the grassland water balance from an intermediate to a dry one (see Fig. 8.8).

Plant productivity is closely related to A_n and is often linked empirically to transpiration. For crop species, Kramer and Boyer (1995) reported a linear relationship between the total shoot dry mass and the amount of water transpired (see Fig. 8.9). The greater slope of this relationship (which is precisely the water use efficiency) for C_4 species implies higher productivity than for C_3 species (see Note 4.10). For natural ecosystems, any relationship between productivity and transpiration is likely to be more complicated than for agricultural crops due to external disturbances and competition among species. Nevertheless, it is reasonable to believe that similar relationships also hold for example for competition between trees and grass. These types of relationships also deviate from linearity when extended over wider ranges of moisture (and thus transpiration) conditions, as was the case for the crop-yield data shown in Fig. 1.11.

8.1.4 Optimal Plant Conditions

It can be seen in Fig. 7.10c that for a given total rainfall during the growing season there is an intermediate value of the storm frequency λ that produces a maximum of evapotranspiration and possibly optimal conditions for plant growth. Along the same lines, Fig. 8.7 shows that for constant seasonal rainfall, the interplay between λ and α produces an optimal condition in the dynamic water stress, again for intermediate values of λ.

Optimal conditions are also found in the relation between rainfall regime and rooting depth, shown in Fig. 8.10a, where the vegetation dynamic water stress is calculated for different values of Z_r and λ, keeping fixed the total rainfall during the growing season at $\mathcal{R}_{seas} = 50$ cm. The dynamic stress is equal

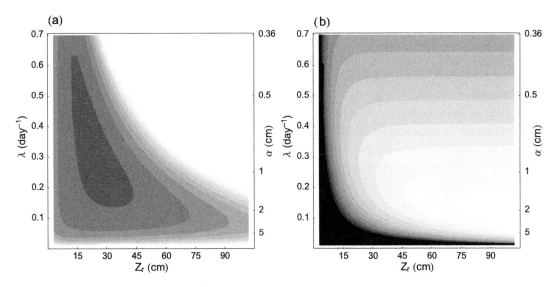

Figure 8.10 Optimal vegetation conditions in terms of plant water stress and total evapotranspiration. (a) Mean dynamic water stress and (b) total evapotranspiration during the growing season versus frequency of rainfall events, mean depth of events, and active soil depth, keeping fixed the total amount of rainfall per growing season. In (a), $\overline{\theta} = 1$ in the white part of the diagram, $\overline{\theta} < 0.5$ in the darker area of the diagram, and the fixed interval between two adjacent contour lines is 0.05; in (b), evapotranspiration is more than 92% of incoming rainfall in the white part of the diagram and less than 50% in the black area, with a distance of 3% between adjacent contour lines. $T_{\text{seas}} = 200$ days, $q = 2$, and $k = 0.5$. Canopy interception is included here with a value of $\Delta = 0.1$ cm. After Porporato et al. (2001).

to 1 in a large region near the upper right corner of the diagram, where the values of α are low and the soil depth is large. With a low α the percentage of vegetation-effective rainfall decreases because interception becomes more important. In such conditions plants that are unable to develop a shallow rooting system end up suffering permanent damage, because in deeper soils the average level of s is consistently too low to sustain effective transpiration. The dynamic stress $\overline{\theta}$ is also high for very low values of λ, mainly because in this case the rainfall events become too rare and plant survival becomes very difficult. For the specific plant parameters and the particular total rainfall considered in this example, the area of best fitness (minimum dynamic stress) covers a wide range of values of λ but tends to be limited to medium to small rooting depths (the dark gray area in Fig. 8.10a). Of course, the region of optimal conditions for a given amount of rainfall may be different when the transpiration characteristics of the plants or the soil properties are changed. Therefore, even when dealing with only a single plant resource (e.g., soil moisture), multiple optimal conditions for vegetation may be a possible way for the coexistence of very diverse species and the maximization of species diversity. The investigation of such aspects could help clarify how hydrological processes drive and control many aspects of the biological richness of ecosystems (see Sec. 8.7).

A minimum in the plant water stress does not alone suffice to define the optimal conditions for a given type of vegetation. A more comprehensive measure of the favorableness of an environment, besides the plant water stress, should also take into account the effective plant productivity and reproduction capacity. A preliminary indication of this can be obtained by comparing the optimum fitness region of Fig. 8.10a with analogous regions of maximum evapotranspiration, which is a good surrogate for plant productivity as discussed in Sec. 8.1.3. Figure 8.10b shows the total evapotranspiration as a function of Z_r and λ, keeping the total amount of rainfall fixed to $\mathcal{R}_{\text{seas}} = 50$ cm. The unfavorable conditions for vegetation are now found near the left bottom corner of the diagram, where evaporation is low because of an excessive production of runoff and leakage due either to a large amount of rainfall per event or to a small rooting depth. Very high values of λ are also unfavorable, because with very light rainfall events canopy interception becomes increasingly more important. The maximum evapotranspiration is attained in this case for deeper soils with values of α of about 2 cm (the white area in Fig. 8.10b).

In water-controlled ecosystems, optimal plant conditions are likely to be subordinated to a compromise between low water stress and high productivity, which is best accomplished through some specific combinations of climate, soil, and vegetation parameters. For the particular example of Fig. 8.10, in the case of very frequent but light rainfall events (e.g., $\lambda > 0.3$) the controlling factor is the plant water stress. Thus shallow-rooted species are preferred both because of their better exploitation of the incoming water (the amount of water transpired is approximately constant except for very low values of Z_r) and because of less severe conditions of water deficit. In the case of more intense and infrequent rainfall events (e.g., $\lambda \leq 0.3$), the preferable range of Z_r shifts toward deeper values, where one finds both high transpiration (e.g., productivity) and low plant water stress.

8.2   Tree–Grass Coexistence in the Kalahari Precipitation Gradient

The Kalahari precipitation gradient in Southern Africa (see Fig. 8.11) offers an excellent case study for investigating how the differences in water balance and plant water stress between trees and grasses generate varying preferences for vegetation types along the transect, with deeper-rooted trees favored in the more mesic regions of the northern part and grasses favored in the drier zones of the southern part. The soils of the Kalahari are relatively homogenous, mainly made up of sandy sediments. The distribution of vegetation essentially follows the gradient in precipitation and soil moisture. The vegetation types form a relatively orderly progression of increasing woody plant cover and height with rainfall gradient, in which fine-leafed savanna gives way to broad-leafed savanna woodlands. The analysis in Porporato et al. (2003b) shows that the decrease in the mean rainfall amount, when moving from north to south, is mostly a consequence of a reduction in the rate of storm arrivals (see Fig. 8.11).

Assuming a spatially uniform mean rainfall depth per event, $\alpha = 10$ mm, Porporato et al. (2003b) varied the mean rate of event arrivals λ in the range between 0.5 and 0.1 day^{-1} when going from north to south along the transect. The hygroscopic point and the field capacity were assumed to be $s_h = 0.04$ and $s_{\text{fc}} = 0.35$, respectively, on account of the uniform sandy soil with low content of soil organic matter, while the porosity was set equal to 0.42. Trees tend to have higher s_w and lower s^* than grasses, which can be explained by the reduced drought resistance and high water-use efficiency of C_4 grasses. On the other hand, grasses in these regions often have maximum transpiration rates that are of the order of 10% higher than trees in well-watered conditions. Mean typical values for trees were assumed to be $s_w = 0.065$, $s^* = 0.12$, and $E_{\text{max}} = 0.45$ cm/day, while for grasses $s_w = 0.05$, $s^* = 0.17$, and $E_{\text{max}} = 0.50$ cm/day. Although all the species are expected to root throughout the upper one-meter of soil, grass roots are typically concentrated closer to the soil surface, while the density of tree roots is more uniform throughout the profile. On account of this, the parameter Z_r (representing the effective rooting depth) was taken as $Z_r = 100$ cm for tress and $Z_r = 40$ cm for grasses (Porporato et al., 2003b). Since interception can be quite important where the canopy cover is denser, the interception parameters (see Sec. 7.3.2) used were $\Delta_t = 2$ mm for tress and $\Delta_g = 1$ mm for grasses.

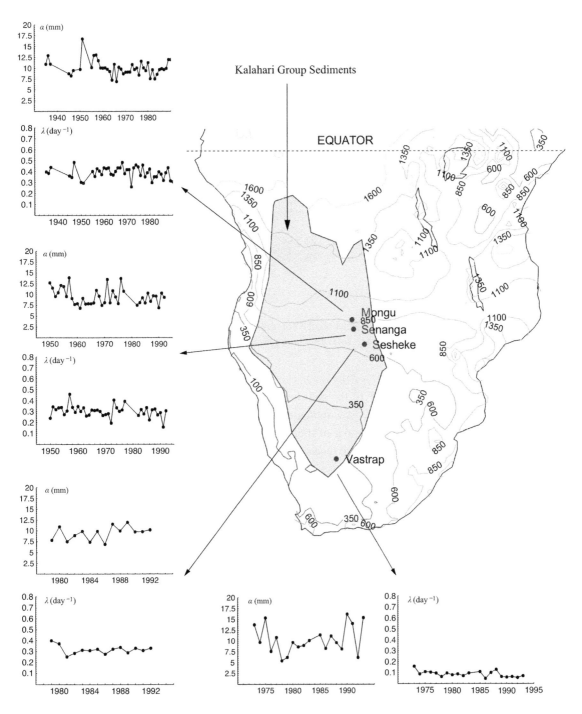

Figure 8.11 Mean annual rainfall (mm) calculated over Southern Africa during 1901–2018 using grid data from Climatic Research Unit high-resolution gridded datasets, version 4.04. The insets show the interannual variability of α and λ estimated for each growing season from the available historic records in Mongu, Sesheke, Senanga, and Vastrap. The shaded region in the main figure refers to the relatively homogenous soils of the Kalahari group sediments. Modified from Porporato et al. (2003b).

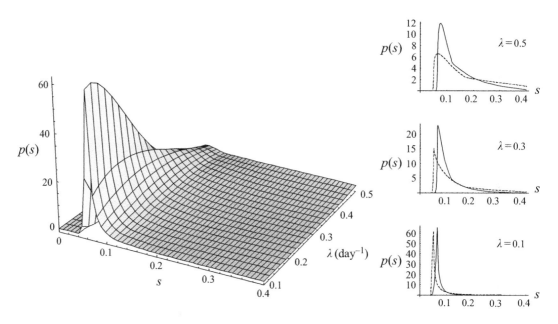

Figure 8.12 Soil moisture PDF as a function of λ for mean rainfall rates that are typical of the northern, central, and southern Kalahari, respectively (trees, solid gray line; grasses, dashed line). After Porporato et al. (2003b).

Figure 8.12 shows the PDF of the soil moisture for trees as a function of λ over the range of values encountered along the transect. As λ increases, the shift of the PDF towards higher soil moisture values is evident, progressively moving out of the wilting region and into the region of unstressed conditions. The variance of the distribution also increases with rainfall, mostly because in very arid climates the PDF is bounded from below by the wilting and the hygroscopic points. The soil moisture for grasses shows a similar behavior but with higher variance, reaching extreme high or low levels more frequently than the counterpart for trees. This fact was discussed earlier in the framework of the general dependence of the soil moisture PDF on Z_r (Sec. 7.4.1).

The role of the rooting depth on overall plant conditions, as well as that of the competition between plant transpiration and the other soil water losses, becomes more evident when considering the intensity and the temporal statistics of the plant stress, as discussed in Sec. 8.1.2. Figure 8.13 shows the dynamical water stress computed as a function of rainfall frequency for the mean parameter values representative of trees and grasses. As expected, the general behavior is one of progressive increase of plant stress in going from wet to dry climates. The plateau for intermediate values of λ is a consequence of the interplay between the frequency of periods of water stress, which attains its maximum in such a zone, and the duration of the stress periods, which increases with decreasing λ (Fig. 8.5). The dependence of the dynamic stress on λ is more marked for trees than for grasses. In particular, for very low rainfall amounts, grasses can lower the water stress more than trees can. This is partly due to their lower wilting point, which reduces the effect of water deficit on the plant. More importantly, grasses benefit from their shallow rooting depth, which allows increased access to light rainfall events and reduces the occurrence of long periods of water stress.

The point of equal stress, which could be interpreted as identifying a general region of *tree–grass coexistence*, is located near $\lambda = 0.2$ day^{-1}, which corresponds to a total rainfall of approximately 420 mm for

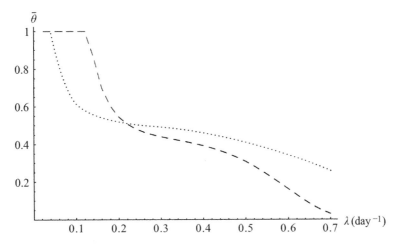

Figure 8.13 Behavior of the dynamical water stress as a function of the mean rate of arrival of rainfall events for trees (dashed line) and grasses (dotted line). $\alpha = 1$ cm, $T_{\text{seas}} = 210$ days; see Sec. 8.2 for the values of the other parameters. After Porporato et al. (2003b).

the seven-month period of the wet season (October to April). The fact that the slopes of the two curves near the crossing point are fairly shallow may contribute to explaining the existence of a wide region suitable for tree–grass coexistence at average rainfall rates. The pronounced interannual variability of both rainfall parameters might further enhance the possibility of coexistence by randomly driving the ecosystem from an increase in grasses during dry years to tree encroachment during wet years. Such a mechanism is similar to the one found to facilitate tree–grass coexistence in the savannas of southern Texas, as discussed in Sec. 8.4. In addition to the temporal fluctuations in soil moisture, spatial heterogeneities and plant competition help to give rise to complex spatial patterns of tree clusters in savannas (see Notes 8.9 and 8.10).

8.3 The Inverse Soil-Texture Effect

Soil texture influences the vegetation structure in water-controlled ecosystems through its impact on the soil water availability (e.g., Noy-Meir, 1973). The effects of soil structure are not always intuitive, as they change simultaneously the holding capacity through the retention curve (and therefore the soil water potential/soil moisture relation) and the water flow properties (e.g., the unsaturated hydraulic conductivity), as discussed in Sec. 3.2. In general, compared with sandy soils, clayish soils tend to store more water but in conditions such that the water is more tightly held within the pore spaces and thus less available to plants and soil microbial life. Which of these two opposite effects ends up being more favorable in reducing plant water stress depends on the rainfall regime, and in particular on the frequency and intensity of rainfall. Assessing such conditions requires consideration of the full stochastic dynamics of soil moisture and plant water stress.

Here we analyze what Noy-Meir (1973) called the *inverse soil texture effect*, whereby in arid and semi-arid ecosystems, the climate and soil texture may interact to give rise to different patterns of soil water availability with the result that the same plant can occur at lower rainfall conditions on coarse soils and at higher rainfall on fine soils. We follow Laio et al. (2001b), who analyzed this effect on the dominant species of the shortgrass steppe in north-central Colorado, the C_4 perennial bunchgrass blue grama *Bouteloua gracilis*.

Although the predominant soil type is sandy loam, more than 95% of the area has a sand content greater than 35%, while approximately 70% of the area has a sand content greater than 50% (Burke et al., 1999). To account for the effect of soil texture, Laio et al. (2001b) assigned soil properties to all the textural classes of the USDA soil texture triangle (see Fig. 3.2). The soil physical parameters such as the porosity n, the saturated hydraulic conductivity K_s, the matric potential at saturation $\overline{\Psi}_s$, and the pore size distribution index b were obtained using the regressions in Cosby et al. (1984), which relate these quantities to soil texture. *Bouteloua gracilis* is characterized by a soil water potential at wilting, Ψ_{s,s_w}, of about -4 MPa (Lauenroth et al., 1987). The onset of water stress was estimated to occur at $\Psi_{s,s^*} = -0.1$ MPa (Sala et al., 1981) and the soil water potential at the hygroscopic point was taken as $\Psi_{s,s_h} = -10$ MPa. The evaporation at wilting point was estimated as $E_w = 0.01$ cm day^{-1} and the daily evapotranspiration rate under well-watered conditions was estimated at $E_{max} = 3.7$ mm day^{-1} (Lauenroth and Sims, 1976). The rooting depth was taken to be $Z_r = 30$ cm, since about 75% of the root biomass in the area has been reported to be in the top 30 cm soil layer (Leetham and Milchunas, 1985; Liang et al., 1989). The semi-arid shortgrass steppe has an annual rainfall from 107 to 588 mm with mean 321 mm and standard deviation 88 mm. Due to the large range of precipitation, Laio et al. (2001b) considered two extreme conditions: the relatively dry year of 1974 had mean depth of rainfall events $\alpha_{dry} = 5.76$ mm and mean frequency of storm arrivals $\lambda_{dry} = 0.17$ day^{-1}, and the relatively wet year of 1957 had $\alpha_{wet} = 6.74$ mm and $\lambda_{wet} = 0.28$ day^{-1}.

The mean dynamic water stress, $\bar{\theta}$, is analyzed in Fig. 8.14, which shows the USDA soil texture triangle for *Bouteloua gracilis* during the dry and wet years. Figure 8.14a shows values of $\bar{\theta}$ varying from 0.58 (for sand; black in the figure) to 0.95 (for silty clay/silty loam; white in the figure), indicating the large sensitivity to soil texture in the overall condition of *Bouteloua gracilis* under a relatively dry climate. Under such a climate, this C_4 grass does better in a coarse soil than in a fine soil. In contrast, Fig. 8.14b

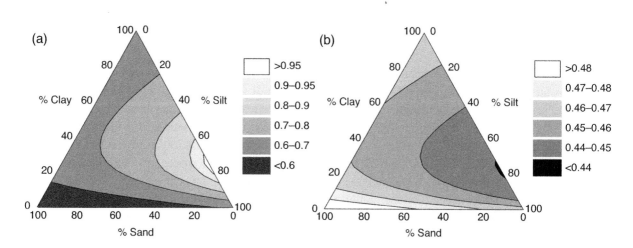

Figure 8.14 Dynamic water stress and soil texture triangles for *Bouteloua gracilis* (a) under a relatively dry climate and (b) a relatively wet climate. $T_{seas} = 183$ days, $q = 3$, and $k = 0.5$. See the main text for other parameter values. After Laio et al. (2001b).

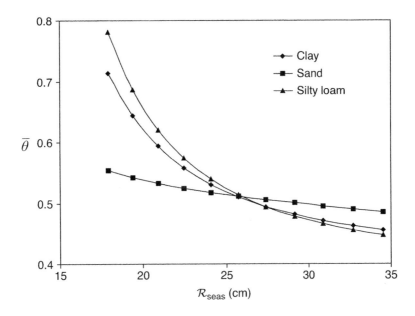

Figure 8.15 Dynamic water stress $\bar{\theta}$ for *Bouteloua gracilis* at the Central Plains Experimental Range (CPER) as a function of the total incoming rainfall during the growing season, $\mathcal{R}_{\text{seas}}$. The parameters α and λ vary linearly with $\mathcal{R}_{\text{seas}}$. $T_{\text{seas}} = 183$ days, $k = 0.5$, and $q = 3$. See the main text for other parameter values. After Laio et al. (2001b).

shows $\bar{\theta}$ varying from 0.44 (for silty clay/silty loam; black in the figure) to 0.48 (for sand; white in the figure), indicating reduced sensitivity to soil texture in the overall condition of *Bouteloua gracilis* under a relatively wet climate. In addition, for these wetter conditions it performs better in a fine soil than in a coarser one.

The above results indicate a preference of *Bouteloua gracilis* for silty loam soils during wet periods (Fig. 8.14b), in agreement with Lauenroth et al. (1994), who listed soil texture as a key factor in the recruitment of *Bouteloua gracilis*, concluding that for dry to intermediate climates the region of minimum water stress is the "sand region", while for wetter climates the most favorable soil texture region is the "silty loam region". This supports the inverse texture effect described by Noy-Meir (1973): 'The same vegetation can occur at lower rainfall on coarse soils than it does on fine ones. The balance point between the advantage of coarser texture and its disadvantage occurs somewhere between 300 and 500 mm rainfall.'

To investigate this further, Laio et al. (2001b) computed the mean dynamic stress for *Bouteloua gracilis* on three soil types, namely sand, clay, and silty loam, for continuously varying values of the total growing-season rainfall $\mathcal{R}_{\text{seas}}$ (Fig. 8.15). The values of α and λ were linearly increased from those corresponding to the relatively dry case ($\mathcal{R}_{\text{seas}} = 179$ mm) to those of the relatively wet year ($\mathcal{R}_{\text{seas}} = 345$ mm). Figure 8.15 shows how the preferential soil type for this grass differs as a function of $\mathcal{R}_{\text{seas}}$. For a relatively dry year, as Fig. 8.14a indicated, *Bouteloua gracilis* is fitter in sand than in silty loam or clay. Its better fitness in coarse soils than in fine ones is true for $\mathcal{R}_{\text{seas}}$ up to approximately 260 mm. As the total growing season rainfall increases above that value, *Bouteloua gracilis* undergoes a lower mean dynamic stress in fine soils than in coarse soils. Taking into account that the rainfall in the total growing season (April–September) is approximately 70% of the total annual rainfall for this area (Lauenroth et al., 1978), the point at which coarse soils become more favorable than fine soils for *Bouteloua gracilis*, or vice versa, occurs at an annual rainfall of approximately 370 mm, which is in the range of values indicated by Noy-Meir (1973).

8.4   Impact of Interannual Variability on Plants: Tree and Shrub Encroachment

In this section, we investigate the impact of interannual variations of rainfall on grassland-to-shrubland conversion in the Rio Grande plains of southern Texas (Archer et al., 1988). These dynamics give rise to vegetation patterns that are in continuous evolution, as shown in Fig. 8.16. Following Laio et al. (2001b), the focus here is to investigate how different rooting depths and plant physiological characteristics may lead to different responses when there are interannual changes in the frequency and amount of rainfall during the growing season.

Figure 8.17 shows the fluctuations in annual rainfall experienced by the region from 1931 to 1985. One observes large and persistent deviations from the long-term mean rainfall over the region. The period 1942–1960 was characterized by a severe drought, while the period 1961–1982 was wet in terms of total amount as well as in number of years with rainfall higher than average. Archer et al. (1988) documented changes in the woody plant cover in the region during those periods. The total woody plant coverage increased from 8% in 1960 to 36% in 1983. In 1941 the woody plant coverage was 13%. The values of α and λ for the statistical description of rainfall during the growing season (150 days, from May through September) were obtained from the rain station at Alice, Texas (see Table 8.1).

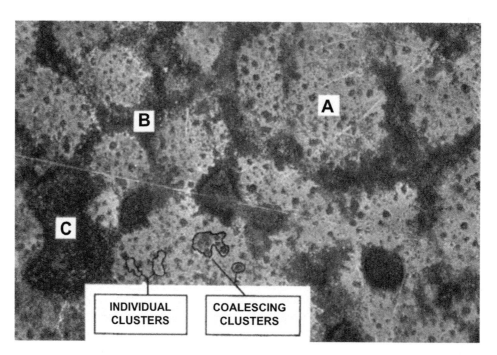

Figure 8.16 Aerial view of the savanna of La Copita (Southern Texas, USA). The two-phase patterns of discrete clumps of woody vegetation are scattered throughout a grassy matrix (A). Bordering the two-phase portions of the landscape are monophasic woodlands associated with more mesic drainages (B) and low-centers polygons of playas (C). The clusters, organized about mesquite (*Prosopis glandulosa*), represent chrono sequences whose species compositions at latter stages of development are similar to that of the closed canopy woodlands in region B. The largest clusters in the two-phase zone represent a mosaic of coalesced clusters. After Archer et al. (1988).

Table 8.1 Common parameters for both trees and grasses in the Texas Agriculture Experiment Station at La Copita.

α_{wet} (cm)	α_{dry} (cm)	λ_{wet} (day^{-1})	λ_{dry} (day^{-1})	n —	K_s (cm day^{-1})	s_{fc} —	s_h —	b —
1.42	1.34	0.20	0.17	0.43	82.2	0.56	0.14	4.9

Table 8.2 Specific parameters for trees and grasses in the Texas Agriculture Experiment Station at La Copita.

	Ψ_{s,s_w} (MPa)	s_w —	$\Psi_{s,s*}$ (MPa)	s^* —	E_{max} (mm day^{-1})	Δ (cm)	Z_r (cm)
P. glandulosa (C$_3$)	-3.2	0.18	-0.12	0.35	4.42	0.2	100
P. setaceum (C$_4$)	-4.5	0.17	-0.09	0.37	4.76	0.1	40

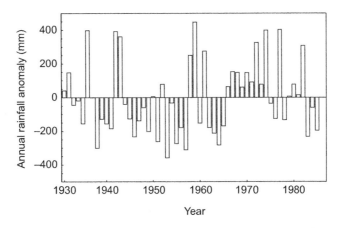

Figure 8.17 Fluctuations around the mean annual rainfall at Alice (Texas) during the period 1931–1985. After Rodríguez-Iturbe et al. (1999a).

The soil is generally described as an A horizon of fine sandy loam and the vegetation is characterized by tree–grass coexistence with a population of woody plants, consisting mostly of *Prosopis glandulosa* (Honey mesquite), coexisting with C$_4$ grasses, mainly *Paspalum setaceum* (Scifres and Koerth, 1987; Archer et al., 1988). The parameters for this ecosystem are reported in Tables 8.1 and 8.2.

With these climate, soil, and vegetation parameters, we are ready to analyze the corresponding hydrological process and its impacts on the ecosystem patterns, using the stochastic water balance model developed in Chapter 7 and the water stress metrics proposed in Sec. 8.1. Figure 8.18 gives a comparison between the normalized loss function, $\rho(s)$, for sites with *Prosopis glandulosa* or with *Paspalum setaceum*. Despite the presence of the same type of soil, the two loss functions look quite different, both because of the differences in the plant physiological characteristics and because of the different active soil depths. In

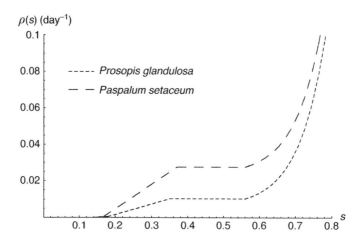

Figure 8.18 Loss functions for the two species considered at La Copita (Texas) (Eq. (7.5)). The soil is a loamy sand with $n = 0.43$, $K_s = 82.2$ cm day^{-1}, $b = 4.9$, $s_h = 0.14$, and $s_{fc} = 0.56$. *Prosopis glandulosa* is characterized by $E_{max} = 0.442$ cm day^{-1}, $s^* = 0.35$, $s_w = 0.18$, $Z_r = 100$ cm, and $E_w = 0.02$ cm day^{-1}. *Paspalum setaceum* has $E_{max} = 0.476$ cm day^{-1}, $s^* = 0.37$, $s_w = 0.167$, $Z_r = 40$ cm, and $E_w = 0.013$ cm day^{-1}. After Laio et al. (2001b).

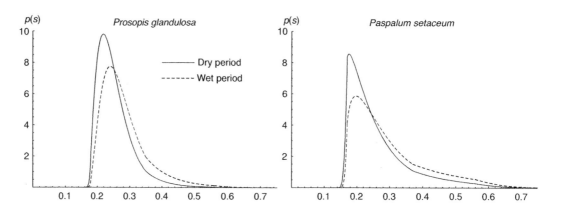

Figure 8.19 Soil moisture probability density functions for *Prosopis glandulosa* and *Paspalum setaceum* at La Copita (Texas). For the dry period, $\alpha_{dry} = 1.342$ cm, $\lambda_{dry} = 0.166$ day^{-1}; for the wet period, $\alpha_{wet} = 1.417$ cm and $\lambda_{wet} = 0.202$ day^{-1}. The canopy interception is $\Delta_t = 0.2$ cm for *Prosopis glandulosa*, $\Delta_g = 0.1$ cm for *Paspalum setaceum*. See Fig. 8.18 for the other parameter values. After Laio et al. (2001b).

particular, the normalization of the loss function by nZ_r makes the effective losses in shallow-rooted sites considerably higher than in deeper-rooted sites.

The resulting soil moisture PDFs for the two species in the cases of wet and dry climates are presented in Fig. 8.19. The effect of the different climates is relevant, but quite similar for the two functional vegetation types. Regarding the differences between trees and grasses, it appears that grasses tend to occur more frequently at low soil moisture levels. At the same time, however, they also have a fatter right tail, owing to the more frequent soil moisture excursions at relatively high soil moisture levels. The results are consistent with Fig. 8.3, where the soil moisture trace for deeper soils was found to be much less fluctuating than that for shallow soils. The transpiration regime also contributes to enhance the discrepancies in soil moisture dynamics through the values of E_{max} and s^*. As a result, the soil moisture in grassy soils decays more quickly and grasses experience wilting more frequently. Their shallower rooting zone, however, is brought to favorable soil moisture conditions even by very small rainfall events. On the other hand, the greater storage capacity of sites dominated by *Prosopis glandulosa* together with their more parsimonious transpiration regime prevent trees from being too frequently under high stress. As a consequence, mesquite hardly ever experiences high levels of soil moisture, and is most often under conditions of water stress, although these are only occasionally extreme.

The previous differences in soil moisture dynamics are also related to the water use efficiency (i.e., the total dry matter produced by plants per unit of water used) of trees and grasses. Mesquite is a C_3 plant with very low transpiration efficiency (e.g., Nilsen et al., 1983; Wan and Sosebee, 1991), while C_4 grasses, like *Paspalum setaceum*, have high water use efficiency and a very quick response to rainfall events (e.g., Kemp and Williams III, 1980; Sala et al., 1982). In other words, *Prosopis glandulosa* is an extensive soil water user while *Paspalum setaceum* can be classified as an intensive soil water user. Such physiological characteristics are perfectly suited to the soil water dynamics they contribute to produce: a low water use efficiency for deep-rooted trees subject to a slower soil moisture dynamics, and a prompt and efficient response to a rapidly varying soil water dynamics for grasses. All the above represents an example of opposite strategies of adaptation to water stress (e.g., Grime et al., 1979). This diversification in the use of the soil water resource may be among the possible mechanisms for tree–grass coexistence in savannas (e.g., Scholes and Archer, 1997).

The values of the dynamic stress, shown in Fig. 8.20, are relatively close for different species under the same climatic characteristics, although *Paspalum setaceum* has a lower dynamic stress during the dry

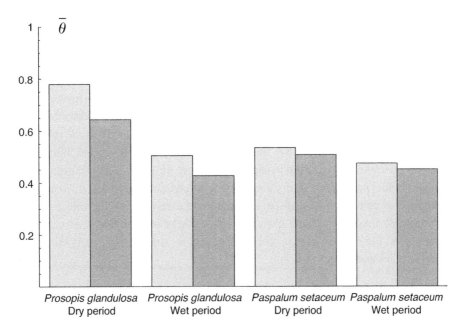

Figure 8.20 Dynamic water stress $\bar{\theta}$ of the two species at La Copita (Texas) corresponding to dry and wet growing season periods (light gray, $q = 1$; dark gray, $q = 3$). $T_{seas} = 150$ days, $k = 0.5$. See Figs. 8.18 and 8.19 for other parameters. After Laio et al. (2001b).

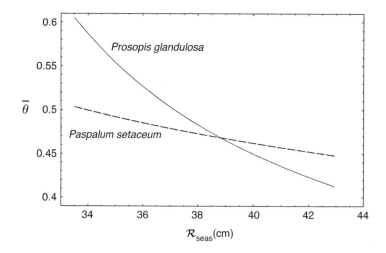

Figure 8.21 Variations of the dynamic water stress $\overline{\theta}$ for the two species at La Copita (Texas) as a function of the total incoming rainfall during the growing season, $\mathcal{R}_{\text{seas}}$. The parameters α and λ vary linearly with $\mathcal{R}_{\text{seas}}$, increasing from their dry values to those of wet conditions. $T_{\text{seas}} = 150$ days, $k = 0.5$, and $q = 3$. See Figs. 8.18 and 8.19 for other parameters. After Laio et al. (2001b).

period. This is very important for the issue of tree–grass coexistence, because it shows that even with very different dynamics of soil moisture the two species are similarly suited for this particular environment. A second observation concerns the notable difference in the sensitivity to climatic changes between trees and grasses. The dynamic water stress of *Prosopis glandulosa* changes considerably for the two different climatic conditions, while that of *Paspalum setaceum* is much less sensitive to climate fluctuations. As the rainfall increases, there is an inversion in the relative condition of the two species: during the dry period grasses are in better condition, but when the rainfall is more abundant the environment becomes more favorable for trees. This is evident in Fig. 8.21, where the dynamic water stress is computed for trees and grasses as function of the amount of rainfall per growing season. The computation was made by linearly increasing the values of α and λ from their dry values to those for wet conditions. Interestingly, for total rainfall above 39 cm trees become less stressed than grasses and the historical long-term average rainfall during the growing season is precisely around that value. Thus the rainfall variability acts as an external forcing that randomly drives the system towards different vegetation conditions: dry periods drive the ecosystem towards a reduction in canopy coverage, while wet periods favor tree encroachment. Owing to the interannual persistence of dry and wet conditions, the ecosystem fluctuates over the years between tree-domination and grass-domination. It is important to remark that the values of canopy coverage depend on the response time of trees and grasses to favorable or unfavorable conditions as well as on the actual productivity of the species and that coexistence can be stabilized by competition and interaction among species (Scholes and Archer, 1997).

8.5   Impact of Seasonality on Plants: Intensive and Extensive Water Users

In Sec. 7.8.1, we investigated soil moisture dynamics with different seasonal patterns of climate forcing. A simplified time-dependent soil moisture PDF provided a good approximation for the mean soil moisture and the related seasonal dry-down and rewetting (see Figs. 7.16 and 7.17 and Feng et al., 2015). Even

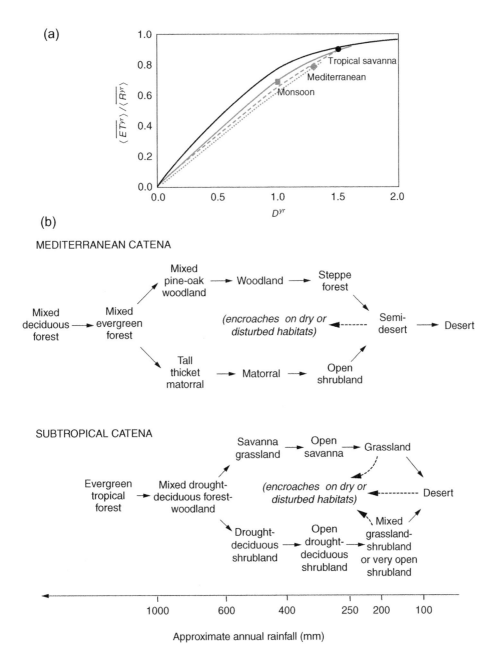

Figure 8.22 (a) Effects of climate seasonality on Budyko-type long-term water balance. E_{max} in the first half year is equal to (gray solid line), twice (gray dashed line), or three times (gray dotted line) its value in the second half of the year. After Feng et al. (2012). (b) Different vegetation catenae, from an arboreal temperate formation to a contracted desert, as examples of specific ecosystem structures resulting from different seasonal patterns of water availability. As modified by Porporato and Rodríguez-Iturbe (2002) after Shmida and Burgess (1988).

with the same total annual precipitation and average evapotranspiration, these seasonal patterns of soil water availability give rise to very different vegetation structures. Regions with fewer seasonal variations (e.g., a tropical climate) tend to have higher evaporation-to-rainfall rates. Figure 8.22a shows the effects of seasonality on the mean annual partitioning as summarized by Budyko's curve (Sec. 7.6.2). Because the rainfall and potential evapotranspiration are out of phase in Mediterranean climates, the evapotranspiration ratio decreases as rainfall accumulates in the soil during the cold wet season without being used by plants, enhancing losses to runoff and deep percolation.

The differences in ecosystem structure induced by seasonality are schematically illustrated in Fig. 8.22b, which reports the typical vegetation catenae (from an arboreal temperate formation to a contracted desert) for Mediterranean and subtropical climates: in the region between 250 and 400 mm of annual rainfall, for instance, one may find steppe forests and open shrublands in Mediterranean climates or open savannas and drought deciduous shrublands in ecosystems with subtropical climates.

In ecosystems with typically dry growing seasons following a wet winter season (e.g., Mediterranean climates), some plants rely on the dependable winter recharge, which is stored deep in the soil, as opposed to others that, not being able to tap such a resource, quickly respond to the intermittent and uncertain rainfall during the hot growing season. These kinds of plants are called extensive and intensive users, respectively. Intensive users develop a dense network of shallow roots to absorb moisture originating from rainfall during the growing season before it evaporates. Typical examples are shallow-rooted grasses with a C_4 photosynthetic pathway and a fine-tuned system of stomatal control that allows a rapid response to intermittent rainfall. In contrast, extensive users are well adapted to low temperatures and have root systems that penetrate larger and deeper volumes of soils and C_3 photosynthetic pathways. They extract water from both shallow and deep soil layers and are favored by winter rains that infiltrate deep into the soil (Burgess, 1995).

To analyze quantitatively these strategies, we follow the analysis of Rodríguez-Iturbe et al. (2001) and concentrate on the difference between the extensive and intensive use of soil moisture by plants that, respectively, adopt C_3 and C_4 photosynthetic pathways and different rooting depths. The inset in Fig. 8.23 shows typical daily soil moisture traces for extensive and intensive users during the growing season. As previously discussed in Sec. 6.8, the effects of the winter moisture storage, namely the initial soil moisture condition at the start of the growing season, is much more important for deep-rooted plants whose transpiration dynamics and larger soil reservoir lead to a smoother soil moisture evolution. Intensive users respond quickly even to light and brief rainfall events, leading to a succession of water deficit periods substantially different from that experienced by extensive users.

In the presence of an initial transient period, the mean time $\overline{T}_{s^*}(s_0)$ to reach the threshold of water stress s^* from an arbitrary soil water content s_0 at the start of the growing season becomes crucial in the vegetation's strategy to cope with stochastic water availability. To account for the possible water storage at the start of the growing season, Rodríguez-Iturbe et al. (2001) extended the dynamic water stress, $\overline{\theta}$ (see Sec. 8.1.2), assuming that the water stress is zero at the beginning of the growing season, until $s(t)$ reaches s^*, and is then equal to $\overline{\theta}$, i.e.,

$$\overline{\theta}' = \frac{T_{\text{seas}} - \overline{T}_{s^*}(s_0)}{T_{\text{seas}}} \, \overline{\theta}, \tag{8.11}$$

where T_{seas} is the duration of the growing season.[1] Despite its simplicity, this new normalized water stress provides an effective synthesis of the action of the soil water balance on plant conditions in ecosystems when the transient at the beginning of the growing season is important.

---

[1]   When the steady-state conditions are very dry and the mean time to reach steady state from s^* is very long, the lower threshold s^* in Eq. (8.11) could be replaced by a more representative level, such as the mean or the mode of the steady-state distribution.

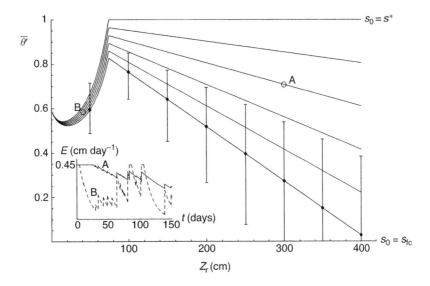

Figure 8.23 Mean total water stress $\overline{\theta}'$ during the dry season as a function of the average rooting depth Z_r for the case of the eastern Amazonian forest. From top to bottom, the curves refer to values of s_0 linearly increasing from s^* to s_{fc}. The climate parameters during the dry season are $\alpha = 1.7$ cm and $\lambda = 0.1$ day^{-1}, the soil is a clay-loam, $E_{max} = 0.45$ cm/day and $s^* = 0.5$ (Nepstad et al., 1994). The error bars on the bottom curve show that the effect on the total water stress of variations of E_{max} and s^* in a range of 20% around their mean values does not change the pattern of favorable conditions for extensive or intensive water users. The insert shows the dramatic difference in the transpiration patterns of an extensive water user (A) and an intensive water user (B) for such climate and soil conditions. After Rodríguez-Iturbe et al. (2001).

The evergreen forests of eastern Amazonia are subject to prolonged dry seasons (typically five months from July to November) during which they are able to maintain evapotranspiration by taking up water from depths of 6 to 8 meters or more (Nepstad et al., 1994). Deforestation transforms such forests into pastures with much shallower roots, which are able to withstand the dry season by responding more efficiently to the scarce precipitation (intensive users). Using the extended definition of stress to account for the high soil moisture at the start of the dry season (Eq. (8.11)), Rodríguez-Iturbe et al. (2001) showed that the pastures may be a low-stress alternative to deep-rooted forests. As it appears from Fig. 8.23, because of the scarcity of precipitation during the dry season, the rooting depth required to sustain a strategy of extensive soil water use needs to be larger than 2 m. Moreover the interannual coefficient of variation (CV ≈ 0.25), characteristic of the total rainfall during the wet season, imposes a large degree of uncertainty on the soil moisture storage s_0 at the start of the dry season. The likely occurrence of s_0 values smaller than the soil field capacity, s_{fc}, makes roots considerably deeper than 4 m a necessity in order for extensive users to be competitive.

8.6   Water-Use Behaviors: Isohydric and Anisohydric Plants

Besides adopting intensive or extensive water use strategies to minimize water stress, plants may also take osmoregulatory measures to provide optimal water use in conditions of stochastic water availability. In Sec. 4.9.2, we have described the role of osmotic adjustment on plant physiology, but we did not discuss its links to stomatal behavior and the related transpiration rates, e.g., through the functions $f_{\psi_l}(\psi_l)$ in Eqs. (4.26) and (4.40). Here, we analyze in more depth the role of different strategies related to how the transpiration changes with plant water status and in turn to the onset of plant water stress at s^*.

In general, the behavior of plant stomatal controls on water loss and carbon gain as a function of water stress varies from species to species and is conventionally categorized as either *isohydric* or *anisohydric* (Berger-Landefeldt, 1936). Isohydric plants close their stomata under water stress conditions to keep a relatively constant leaf water potential, which avoids hydraulic failure but may lead to carbon starvation during long droughts (e.g., McDowell et al., 2008; Martínez-Vilalta et al., 2014). Anisohydric plants allow a drop-off in leaf water potential in response to decreasing soil moisture, which prevents carbon starvation during short-term droughts but increases the risk of xylem cavitation (e.g, West et al., 2008). The delayed stomatal response to water stress is made possible by osmoregulatory measures (i.e., osmotic adjustment), which help maintain plant turgor and continue performing photosynthesis under water stress conditions (see Sec. 4.9.2).

Several metrics have been formulated from plant water potentials or physiological traits to identify plants as either isohydric or anisohydric (Feng et al., 2019). These metrics link the leaf water potential to soil moisture states so that models can adjust the soil-moisture–transpiration relation according to different degrees of isohydricity, as will be discussed later. Since these metrics do not always vary consistently under changing environmental conditions (e.g., vapor pressure deficit and temperature) in relation to the plants' stomatal regulation (Hochberg et al., 2018), formulations based on the SPAC model (see Sec. 5.4) are more suitable for diagnosing these plant dynamics (Feng et al., 2019; Novick et al., 2019).

From the hydrological point of view, the most important consequence of osmotic adjustment in anisohydric plants is to facilitate water uptake at low soil moisture levels, to avoid dehydration and keep performing photosynthesis. However, as the soil water reserve is very limited in such conditions, the osmotic adjustment in anisohydric plants is somewhat in contrast with the requirement of reducing soil water use (as a parsimonious strategy of water use would suggest). This, in addition to the existence of other soil water losses and the unavoidable cost of osmotic adjustment, clearly indicates that an efficient transpiration strategy may be realized only by taking into account the actual environmental variability of the soil water balance. Moreover, the unpredictability of future rainfall forces the plants to seek a sort of stochastic optimization of such clashing exigencies: a parsimonious water use that could imply low productivity and competitiveness versus an unnecessary and costly consumption of soil water that could end up in intense water stress. According to Jones (1992), such two extremes correspond to so-called *pessimistic* and *optimistic water use*. The balance between these two strategies depends on the intrinsic resistance and productivity potential under intense stress of each species (or plant), as well as on the existing environmental conditions.

It is assumed here, following Rodríguez-Iturbe and Porporato (2004), that plants with anisohydric traits adjust their transpiration function by changing the point of incipient stomatal closure, s^*, and that osmotic adjustment results in a reduction of s^* (Kumagai and Porporato, 2012). During a drydown process (Fig. 8.24), soil moisture is depleted faster, with lower values of s^*, but at the same time the appearance of static stress is delayed (Sec. 8.1). With high values of s^*, plants save soil water (provided no other losses are present) but also start experiencing water stress earlier than plants that perform osmotic adjustment. Obviously, osmotic adjustment cannot prevent the appearance of stress, and the situation is inverted after some time. The general behavior shown in Fig. 8.24b is maintained also when different types of nonlinearities are used in the definition of the static stress.

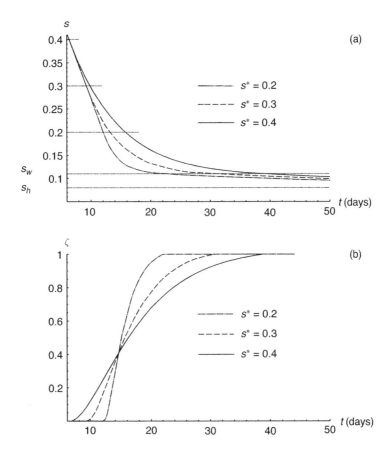

Figure 8.24 Temporal evolution of (a) soil moisture and (b) instantaneous static water stress (with $q = 2$) as a function of s^*. Other parameters: $\alpha = 10$ mm, $\lambda = 0.3$ day^{-1}, $E_{max} = 4.5$ mm day^{-1}, $E_w = 0.2$ mm day^{-1}, $s_h = 0.08$, $s_w = 0.11$, $Z_r = 60$ cm, $\Delta = 0.2$ cm. The soil is a loam (see Table 3.1). After Rodríguez-Iturbe and Porporato (2004).

The actual advantage of reducing s^* crucially depends on the stochastic aspects of precipitation, which control the mean duration of the stress periods as well as their mean frequency and intensity. In arid regions, plants with lower s^* may experience higher static water stress most of the time (see Fig. 8.24b), but they can easily recover after only a light rainfall event. The combination of the stress intensity and duration is synthesized by the dynamic water stress $\overline{\theta}$ (Sec. 8.1.2). When varying the value of s^*, there is a minimum of dynamical water stress that corresponds to the most advantageous level of osmotic adjustment, which in turn provides the optimum point for stomatal control to begin, to minimize water stress (see Fig. 8.25).

Figure 8.25 also shows how the optimal transpiration function depends on the environmental conditions. The drier the climate and the longer the interstorm period, the lower is the value of s^* required to reduce the average global stress level. One should notice, however, that, in the case of very arid conditions, the physiological cost of osmotic adjustment in anisohydric plants may be unbearable if not associated with other remedies. Moreover, since the dynamic water stress does not take into account either the relationship between transpiration and productivity or the cost of osmotic adjustment at very low soil moisture values, the actual advantage of osmotic adjustment may be overestimated by the present analysis. The exact

Figure 8.25 Dynamic water stress as a function of s^* for different values of the mean rate of storm arrival (the other parameters are the same as in Fig. 8.24). After Rodríguez-Iturbe and Porporato (2004).

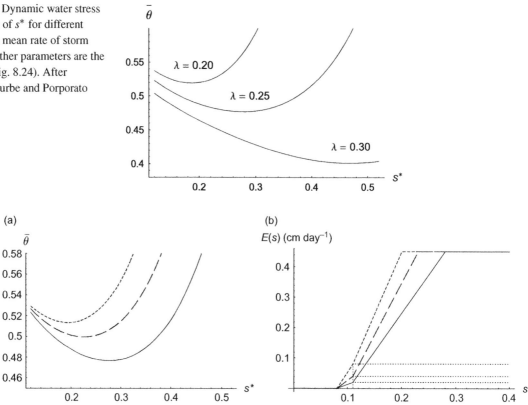

Figure 8.26 Simulation of the effect of competition for water from neighboring plants as well as of increased soil evaporation (E_w) on the optimal value of s^*. (a) Minimization of the dynamical water stress and (b) the corresponding transpiration function. Solid line, $E_w = 0.02$ cm day^{-1}; dashed line, $E_w = 0.05$ cm day^{-1}; dotted line: $E_w = 0.08$ cm day^{-1}. After Rodríguez-Iturbe and Porporato (2004).

location of the minimum is also sensitive to the specific value of the nonlinear exponent q used to define the static water stress.

The sensitivity of the optimal transpiration function to changes in the other losses in the water balance provides interesting insights. By increasing the value of E_w, one may account for the effects of an increase in both the soil evaporation and the competition for moisture from neighboring plants. The behavior of the dynamical water stress as a function of s^* for different values of E_w is shown in Fig. 8.26. Owing to the more intense competition for water from the other losses, the optimal values of s^* must be reduced in order to ensure a more efficient (and aggressive) water use strategy. This agrees with the results by Cowan (1986, Fig. 5.13), who found steeper curves of optimal daily rate of assimilation (which is itself closely related to transpiration) in the case of higher values of water deficit due to increased water losses by neighboring plants.

This framework was applied by Rodríguez-Iturbe and Porporato (2004) to the Nylsvley savanna in southern Africa (see Sec. 8.2), where woody species (*Burkea africana, Ochna pulchra*) and herbaceous species (*Eragostris pallens, Digitaria eriantha*) are coexisting. Figure 8.27a gives a comparison between the field measurements of s^* and the theoretical values of s^* obtained by minimizing the dynamic water stress. The values of s^* for grasses are in very good agreement with the field measurements from Scholes

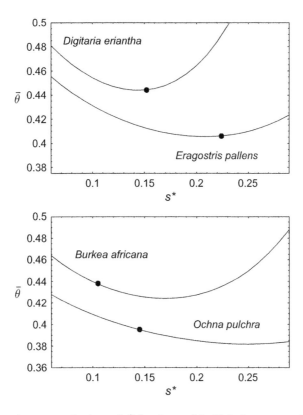

Figure 8.27 For comparison, the measured values of s^* for plants of the Nylsvley savanna (Sec. 8.2) and those computed by minimizing the dynamical water stress. After Rodríguez-Iturbe and Porporato (2004).

and Walker (1993); the theoretical values for trees, however, are slightly overestimated. Competition for water by neighboring plants (see Fig. 8.26) may be one factor responsible for the difference (i.e., the value of E_w used in Fig. 8.27 is too low to account for competition).

Finally, anisohydric plants tend to prosper in moist environments, where there is little risk of hydraulic failure. Kumagai and Porporato (2012) considered the tropical rainforest of Borneo. Using Eqs. (8.8) and (8.10), along with different transpiration functions for isohydric and anisohydric plants to measure the corresponding mean assimilation rates, they analyzed the differences in mean assimilation rate between anisohydric and isohydric plants, $\langle A_n \rangle_a - \langle A_n \rangle_i$. The results, shown in Fig. 8.28, reveal that the productivity of anisohydric plants is higher than that of isohydric plants for any rainfall regime under very moist conditions (e.g., mean rainfall > 1100 mm) and that the productivity of isohydric plants surpasses that of anisohydric plants below a mean annual rainfall of approximately 800 mm.

8.7 Soil Moisture Controls on Plant Biodiversity

Hydrologic fluctuations at different spatial and temporal scales have shaped the richness of plant species across the world, with the diversification of plant species facilitated by various water use behaviors, strategies of competition and colonization, and drought tolerance mechanisms. Ecosystems with more species

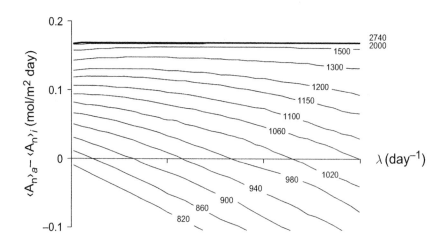

Figure 8.28 The differences between the mean assimilation rates computed for anisohydric ($\langle A_n \rangle_a$) and for isohydric plants ($\langle A_n \rangle_i$) as a function of the frequency of rainfall events (λ) for various values of annual total rainfall (represented by the number (in mm) on or beside each line). After Kumagai and Porporato (2012).

may experience smaller declines in productivity during drought, showing greater resilience to changing climate (Anderegg et al., 2018).

Resource availability and variability interact with other abiotic factors to create *ecological niches* favoring plant biodiversity (Silvertown, 2004; Porporato and Rodríguez-Iturbe, 2013). Here we follow Bonetti et al. (2017), who analyzed the species richness in tropical South America using niche theory (the theory of ecological configurations or niches), with emphasis on soil water (see Note 8.13 on biodiversity theories). The model assumes that the number of possible niches increases with the amount of resources available. Since the available resources also fluctuate in space and time, the overall number of configurations is also assumed to depend on resource variability. The latter can be quantified using the statistical distribution of the resource, which is measured in information theory through the exponential of the Shannon entropy (Cover, 1999).

Focusing on soil moisture as a limiting resource, a higher number of ecological configurations or niches can be achieved by either increasing the soil moisture abundance without changing its distribution or through a more dispersed PDF that maintains constant the total water available to the plant. Specifically, assuming that soil moisture is the primary resource driving plant diversity (Hawkins et al., 2003; Engelbrecht et al., 2007; Silvertown et al., 2015), Bonetti et al. (2017) combined, in a multiplicative way, the soil moisture abundance, represented by the mean soil moisture at a given time ($\langle x_t \rangle$, see Secs. 7.5 and 7.8.1) and its variability, represented by the exponential of the Shannon entropy of the PDF of the soil moisture state, $\exp[H_x(p(x_t))]$, giving the number of species as

$$N = \mathcal{N} e^{-I_R} \chi \frac{1}{T} \int_0^T \langle x_t \rangle^a e^{H_x(p(x_t))} dt, \qquad (8.12)$$

where $\mathcal{N}$ is a scaling factor, representing a reference value of the species richness, T is a representative one-year period, a is used to express the nonlinear dependence of species richness on soil moisture availability (e.g., as a result of waterlogged conditions), I_R is the incoming radiation, and finally the coefficient χ accounts for the observed reduction in species richness with elevation; based on the data in tropical South

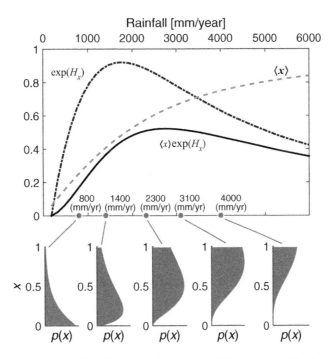

Figure 8.29 Conceptual framework of soil moisture abundance and variability. Mean effective soil moisture $\langle x \rangle$ (dashed line), its relative Shannon entropy (dotted line), and their product (solid line) as a function of total rainfall for a minimalist stochastic soil moisture model with typical parameters (see Sec. 7.5). Note that an intermediate rainfall regime (e.g. 2300 mm/yr in this figure) allows a more spread-out PDF and thus can accommodate more ecological niches, while in very dry or very wet conditions the soil moisture PDF is maximally peaked. After Bonetti et al. (2017).

America, the latter is simply assumed to vary linearly from 0 (at 4500 meters above sea level) to 1 (at zero meters above sea level).

The Shannon entropy measures the resource variability:

$$H_x = -\int_0^1 p\left(x'\right) \log_2 \left(p\left(x'\right)\right) dx', \tag{8.13}$$

where $p(x)$ is the probability density function for the effective soil moisture (Sec. 7.5). This implies that the more peaked the PDF of the soil moisture (i.e. for very dry or wet conditions), the lower the Shannon entropy and, thus, the number of niches. Conversely, the more spread out the PDF of the soil moisture (i.e., for intermediate rainfall regimes), the higher is H_x and the number of possible ecological configurations (see Fig. 8.29). The mean and the PDF of soil moisture are obtained by solving the minimalist stochastic soil water balance model taking into account climate seasonality (see Secs. 7.5 and 7.8.1).

Bonetti et al. (2017) applied the biodiversity model to tropical South America, using species richness and elevational data from the Al Gentry Dataset (Phillips and Miller, 2002), climate data from the Climate Research Unit (CRU, Harris et al., 2014), and soil plant parameters from the ISLSCP II dataset (Kleidon et al., 2011). They first related the species richness data (Fig. 8.30a) to monthly rainfall and potential evapotranspiration, latitude, and elevation. As shown in Fig. 8.30b, a generally positive correlation was found between species richness and mean annual rainfall, with possible saturation at high rainfall values.

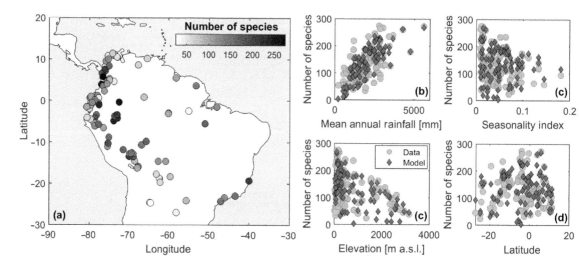

Figure 8.30 (a) Geographic location of the species richness transects: the gray scale bar represents the number of species. On the right, the modeled (black diamonds) and measured (gray circles) relationships between the number of species and (b) mean annual rainfall, (c) seasonality index, (d) elevation, in meters above sea level, and (e) latitude. After Bonetti et al. (2017).

Figure 8.30c shows an interesting pattern with seasonality, whereby a decrease in maximum species richness with seasonality occurs concomitantly with an increase in the minimum species richness. The negative correlation between seasonality and species richness was also noted by Givnish (e.g., 1999), while the positive correlation between the minimum number of species and seasonality suggests that some variability in rainfall distribution over the year could contribute to increasing the number of available niches that are able to accommodate species, thus leading to higher species richness. Figure 8.30d shows a decrease of species richness with altitude (Phillips and Miller, 2002), while Fig. 8.30e shows a slight decrease of species number with latitude in the southern hemisphere.

The model was compared with measured species richness data by considering all terms in Eq. (8.12), followed by quantifying the role of different environmental constraints in explaining the variation in species richness; this was done by repeating the previous analysis while setting each of the environmental variables constant in turn. The results show that the model simultaneously captures the main species richness trends, while the remaining error can be probably attributed to other factors (e.g. topographic features, nutrient availability) which are not explicitly captured in the model. Unlike the elevation gradient (its omission would reduce the accuracy of the model by 80%), the energy-related variables (i.e., the E_{max} seasonality and latitudinal gradients) play only a secondary role. However, excluding from the model the effects of changes in soil moisture abundance, seasonal variability, and intra-seasonal variability, showed reductions of the model accuracy of about 50%, 48%, and 64%, respectively.

Bonetti et al. (2017) also separated species richness into three functional categories corresponding to different water-use strategies and analyzed their trends with respect to rainfall (Fig. 8.31): two types of ground-rooted plants (trees and shrubs or lianas) and canopy species (hemiepiphytes). When considered separately, the behavior of trees and shrubs resembles that of the entire community, with an increase in

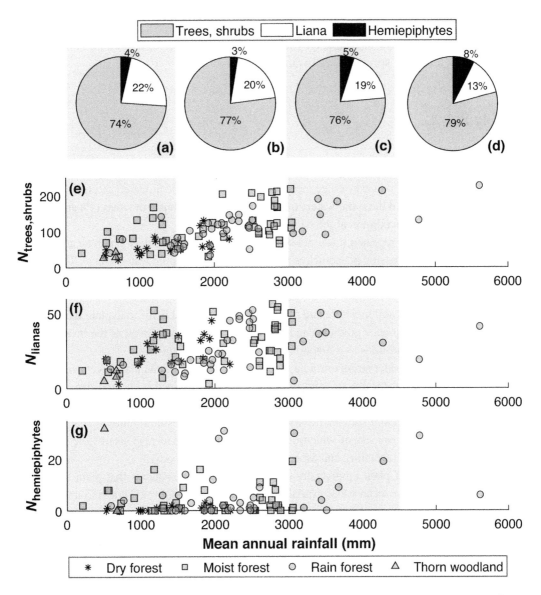

Figure 8.31 Species richness composition: mean percentages of trees and shrubs, lianas, and hemiepiphytes for mean annual rainfall between 0–1500 mm (a), 1500–3000 mm (b), 3000–4500 mm (c), and 4500-6000 mm (d). The numbers of trees and shrubs, lianas, and hemiepiphytes as a function of mean annual rainfall (e)–(g). After Bonetti et al. (2017).

species richness with rainfall and a possible slope reduction at high rainfall values. On the other hand, the mean percentage of lianas decreases from 22% to 13% with increasing mean annual rainfall (Fig. 8.31), owing at least in part to oxygen limitations. A very interesting and separate behavior is displayed by the hemiepiphytes (Fig. 8.31g), which represent a higher proportion of the community with increasing rainfall. This may be explained by considering that such species strongly rely on atmospheric humidity and direct precipitation inputs (Gotsch et al., 2015). For these plants, atmospheric moisture becomes a separate resource creating additional hydrological niches for canopy species. Even if this modeling framework does not directly account for possible water vapor and dew uptake by aerial plants, Bonetti et al. (2017) argued that this effect is implicit in the increase in the total number of species with mean soil moisture, since the increase in the number of hemiepiphytic species compensates for a possible diminished increase in the number of terrestrial species in well-watered conditions.

8.8   Key Points

- The reduction in plant water potential as soils dry gives rise to a sequence of increasingly more serious physiological consequences for plants. This can be modeled as a nonlinear function of soil moisture, called the static water stress, starting from the onset of stress (s^*) and then progressively growing to a maximum at the wilting point (s_w).
- The impact of repeated instances of stress during a growing season can be accounted for by considering the crossing properties of stress levels (the dynamic water stress). The resulting measure can be used to assess long-term plant conditions as a function of soil, vegetation, and climate parameters.
- The stochastic soil moisture and plant water stress model helps to explain the tree–grass response to rainfall-frequency gradients and interannual variability, as well as the inverse texture effect on vegetation conditions as a function of hydroclimatic conditions.
- To minimize water stress during a recurrent dry season, plants may develop either deep roots and an extensive water-use strategy or shallow roots combined with intensive water use, while the intermediate case corresponds to higher plant water stress.
- Adjustments of stomatal sensitivity to water stress (s^*) result in more or less aggressive water-use strategies, the success of which depends on the combined response of the soil–plant system to hydroclimatic variability and especially to rainfall frequency.
- An analysis of plant biodiversity data in South America shows that plant biodiversity depends not only on the mean water availability but also on the soil moisture variability (measure by the Shannon entropy of the soil moisture PDF), which creates niches for different plants to coexist at different moisture conditions.

8.9   Notes, including Problems and Further Reading

8.1 (Crop Stress) Consider a rainfed cultivation of common wheat (*Triticum aestivum*), with root-zone depth of 30 cm, rainfall frequency of 0.2 day^{-1}, mean rainfall depth of 15 mm, $E_w = 0.01$ cm day^{-1}, $E_{max} = 0.45$ cm day^{-1}, and loamy sand soil ($n = 0.43$, $\beta = 13.8$, $K_s = 80$ cm day^{-1}, $s_h = 0.14$, $s_w = 0.18$, $s^* = 0.46$, $s_{fc} = 0.56$; see Table 3.1). Estimate the frequency with which the soil moisture goes below the stress level for a homogeneous growing season (i.e., in a statistical steady state).

8.2 (Rainfall Regime and Rooting Depth) Using the parameters in the previous problem, draw a contour plot of dynamic water stress for given constant long-term rainfall rate of 700 mm yr^{-1} as a function of rainfall frequency, λ and rooting depth, Z_r.

8.3 Use the soil–plant parameters in Note 8.1 and compute the dynamic water stress for sandy and clay soil (see Table 3.1), with α = 10 mm, as a function of rainfall frequency λ. Calculate the value of λ at which the two dynamic water-stress curves intersect.

8.4 For the case of linear losses (see Sec. 7.5), and assuming constant water-use efficiency (WUE = $\langle A_n \rangle / \langle E \rangle$), compute the reduction in mean assimilation for a reduction in rainfall frequency while keeping the same mean rainfall depth per event.

8.5 (Optimal Rooting Depth) In Note 4.3, we briefly reviewed various models of root architecture and functioning. The discussion in the present chapter makes it clear that root-growth strategies must take into account the stochastic variability of water stress and carbon assimilation. The optimization approach of Guswa (2008, 2010) addresses affectively the impact of water availability on rooting depth.

The model considers the tradeoff between the extra carbon cost of deeper roots and the greater access to additional soil moisture for photosynthesis and carbon uptake. The optimal root depth related to water availability is found by equating the marginal carbon cost of adding deeper roots to the marginal benefit of carbon gain due to water supply from deep soil (Guswa, 2008):

$$\frac{dC(Z_r)}{dZ_r} = \text{WUE} f_{\text{seas}} \frac{d\langle E \rangle}{dZ_r}, \tag{8.14}$$

where Z_r is the rooting depth, $C(Z_r)$ is the carbon cost of roots per unit area of ground surface as a function of the rooting depth, WUE is the water-use efficiency, f_{seas} is the fraction of the growing season in one year, and $\langle E \rangle$ is the average rate of transpiration during the growing season.

Guswa (2010) used the stochastic water balance framework in Sec. 7.4 and the Budyko curve in Sec. 7.6.2 to obtain analytical expressions for $d\langle E \rangle / dZ_r$, which allow us to link the optimal rooting depth to the dryness index. Consistently with field observations (Schenk and Jackson, 2002a), the results show a tendency to develop deeper roots when the dryness index is close to unity. This implies that deep roots in wet regions are less necessary as water is regularly available near the surface, while deep roots in dry regions are not useful as the deep soil water is usually very low. There are of course exceptions to these results, related to the presence of a deep water table (phreatophytes may develop very deep tap roots to access this moisture) and deep percolation due to macroporosity and preferential flows (see Note 3.11).

8.6 Using Eq. (8.14) and assuming $1/(f_{\text{seas}}\text{WUE})dC/dZ_r = 0.0004$ day^{-1}, calculate the optimal depths for *Bouteloua gracilis* for clay, sand, and silty loam soils, with other parameters reported in Sec. 8.3.

8.7 (Hierarchical Competition–Colonization Model) A simple framework for tree–grass coexistence driven by rainfall variability is provided by the hierarchical competition–colonization model (Fernandez-Illescas and Rodríguez-Iturbe, 2003), which describes the dynamics of species of different ranks (Tilman, 1994). In case of two species, the fractions of sites occupied by the superior (p_1) and inferior (p_2) competitors can be expressed as

$$\frac{dp_1}{dt} = c_1 p_1 (1 - p_1) - m_1 p_1, \tag{8.15}$$

$$\frac{dp_2}{dt} = c_2 p_2 (1 - p_1 - p_2) - c_1 p_1 p_2 - m_2 p_2, \tag{8.16}$$

where as m_1 and c_1 are the mortality and colonization rates of the superior competitor, whose evolution is unaffected by the inferior competitor, which can only colonize empty sites and has parameters m_2 and c_2. The colonization rate may be modeled as a function of dynamic water stress as

$$c = c_0(1 - \overline{\theta}). \tag{8.17}$$

Consider the two species *Prosopis glandulosa* and *Paspalum setaceum*, either of which is the superior competitor if it has lower dynamic water stress. For *Prosopis glandulosa*, c_0 is 0.5 yr^{-1} if it is the superior competitor and 7.6 yr^{-1} if it is the inferior competitor. For *Paspalum setaceum*, c_0 is 0.6 yr^{-1} if it is the superior competitor and 7.9 yr^{-1} if it is the inferior competitor. The mortality rates for both species are set at 0.1 yr^{-1} (Fernandez-Illescas and Rodríguez-Iturbe, 2003). Use the soil and climate parameters in Sec. 8.4 and Eqs. (8.15) and (8.16) to simulate the fractions of sites occupied by both species in dry and wet years, as defined in Table 8.1.

8.8 (Evolutionarily Stable Strategies) In this chapter, we have emphasized simple strategies of water use by plants in conditions of stochastic water availability. The general problem of plant strategies and competition for water and other resources can be framed within the context of game theories and the related ecological principles, such as evolutionarily stable strategies (ESS). The paper by Zea-Cabrera et al. (2006) provides an interesting analysis of ESS for water use. Bridging these theories to stochastic formulations of water stress is an interesting open problem.

8.9 (Vegetation Pattern Formation) The formation of spatial patterns in semi-arid regions has become a popular topic in ecohydrology after some early work (Lefever and Lejeune, 1997; Klausmeier, 1999). We refer to Cross and Hohenberg (1993) for an introduction to models and mechanisms of pattern formation and to Borgogno et al. (2009), Meron (2015), and Gandhi et al. (2019) for an overview of the vast literature related to this topic.

Typically, these models describe the joint evolution of soil moisture and biomass in time and space, where in the absence of spatial variability a nonlinear coupling gives rise to a transition (a pitchfork bifurcation) from a stable bare-soil state at low soil moisture values to a condition with both vegetation and bare soil as stable solutions. This bistability scenario (Sec. 2.4.1 and Note 2.13) is usually linked to the tendency of vegetation to act as an ecosystem engineer, thus facilitating its own survival, e.g., by favoring infiltration and soil moisture uptake or somehow reducing evapotranspiration losses. The literature is replete with a variety of formulations of different degree of complexity, some using additional variables (e.g., representing surface water, nutrients, etc.), and involving disparate mechanisms of spatial interaction, ranging from seed dispersal to soil moisture and surface water lateral transport and plant competition for water). The important implications of biogeochemistry, fauna (e.g., Tarnita et al., 2017), and geomorphology (Istanbulluoglu and Bras, 2005; Dietrich and Perron, 2006), as well as the complications brought about by the vast range of spatial scales and timescales involved, contribute to make it a challenging open field of research.

8.10 (Savannas as Critical States) When analyzed in terms of average behavior at large scales, both the tree–grass coexistence in savannas and the vegetation survival in conditions that approach desertification present characteristics of an abrupt transition. In several instances the structure of the spatial patterns displays clusters with long-range correlation and power-law scaling laws typical of random fractals. In contrast with the pattern formation models mentioned in the previous note, the patterns here are reminiscent of those appearing at critical points in phase transitions, typically analyzed with the methods and models of statistical mechanics. We refer to Scanlon et al. (2007b) for an analysis of the clustering of vegetation using the Ising spin model.

8.11 (Forest Dynamics and Perfect Plasticity Approximation) In this chapter, we used various metrics related to water stress, assimilation, and productivity to explain the competition between and co-existence of different plant species. More specific models such as SORTIE can be used to analyze forest dynamics by simulating the dynamics of each individual plant and its interaction with the surrounding throughout its life cycle (Mitchell, 1975). However, the complex structures of these models limit their use in understanding the carbon and ecological dynamics of our terrestrial system. Pacala et al. (1996) identified the plastic growth of tree crowns as the key modeling ingredient of canopy growth. With a perfect-plasticity approximation, the authors scaled up from individual plants to the forest and developed a simple model to predict the dynamics of forest biomass, stand structure, and ecological succession (Strigul et al., 2008). Such a simple and analytically tractable model can accurately predict the basal area dynamics and ecological succession over various regions in the USA and has been implemented in the latest version of Earth System Models (Purves et al., 2008; Dunne et al., 2020).

8.12 (Drought-Induced Plant Mortality) Under extreme climate conditions, drought-induced critical and permanent damage (see Sec. 4.9) increases the risk of plant mortality, which has been widely observed with significant implication for global water and carbon cycling (Allen et al., 2010). Several drought-induced mortality mechanisms have been identified, including hydraulic failure, carbon starvation, and biotic attack. Plants may die when water or carbon or both are depleted below the levels necessary to support physiological processes. Further, a water and/or carbon deficit may increase susceptibility to insect or pathogen outbreaks, or alter plant adaptation, seed production, and germination, which can also cause widespread mortality (McDowell et al., 2013). Parolari et al. (2014) introduced a forest mortality model which is built upon the simplified soil–plant–atmosphere continuum model (Sec. 5.4) in the stochastic soil water balance framework (Sec. 7.4) with different plant water-use strategies (Sec. 8.6) to assess the likelihood of reaching carbon starvation and hydraulic failure. Under this framework and using climate model outputs, Liu et al. (2017) predicted a higher global forest mortality risk due to changes in precipitation and air temperature, which would be largely alleviated by increases in atmospheric specific humidity and CO_2 concentration. Schwantes et al. (2018) further investigated the impacts of landscape differences in aspect, topography, and soils. When spatially distributed soil depths, topographic wetness index, and heat load index are incorporated into the stochastic soil water balance framework, the model was able to describe the spatial patterns of drought-induced canopy loss across a watershed in central Texas (see Fig. 8.32).

8.13 (Niche and Neutral Theories of Biodiversity) Understanding the determinants of species assembly and biodiversity is a central theme of theoretical ecology (Chesson, 2000; Chave et al., 2002; Silvertown, 2004); its complexity represents a formidable challenge for modern statistical mechanics (Phillips and Quake, 2006; Goldenfeld and Woese, 2011; Azaele et al., 2016). The various existing frameworks

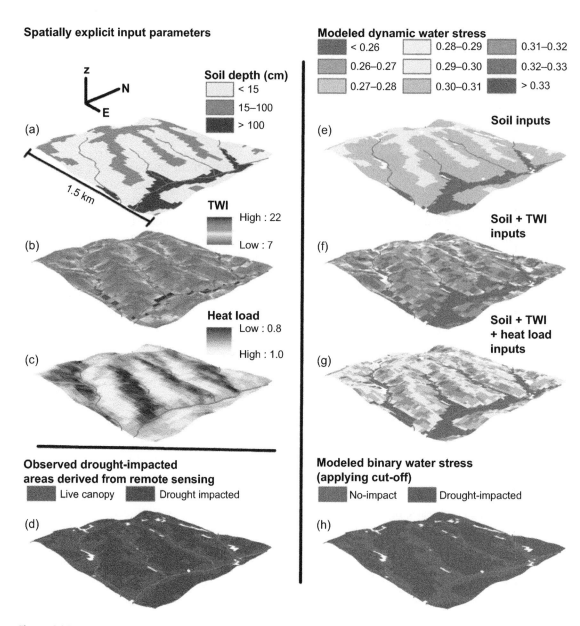

Figure 8.32 Simulations for a watershed in central Texas: (a) soil depth, (b) topographic wetness index (TWI), (c) heat load index, (d) remotely sensed drought-impacted area, and mean dynamic water stress for scenarios including (e) heterogeneous soil inputs, (f) lateral redistribution of water using TWI, (g) spatially variable potential evapotranspiration using a heat load index, and (h) a binary dynamic water stress with cutoff value of 0.28. After Schwantes et al. (2018).

that seek to explain the origin of observed biodiversity patterns can be classified into two broad categories – that of classical niche theory and the more recently updated neutral theory, which has received intense attention since being revisited by Bell (2000), Hubbell (2001), and Volkov et al. (2003). While classical studies on community assembly focus on species-level interactions and differences in their environmental niches (Tilman, 1982; Chesson, 2000), neutral theory adopts a null model in which species are taken to be indistinguishable from each other under all environmental conditions, and community assembly is dominated instead by the probabilistic birth, death, speciation, and immigration occurring at the individual level. Many authors now call for a broader theoretical framework which reconciles neutral and non-neutral processes (Rosindell et al., 2012; Matthews and Whittaker, 2014), both of which are thought to operate in tandem in natural communities (Chave, 2004; Leibold and McPeek, 2006; Adler et al., 2007).

9   Soil Carbon and Nitrogen Cycles

零落成泥碾作尘，只有香如故。

Her petals may be ground into mud, but her fragrance will endure.

Lu You

In this chapter, we turn to the study of the cycles of carbon and nitrogen in soils. In the first part we review the soil carbon and nitrogen cycles and the impact of soil moisture on their dynamics, followed by an analysis of simple models of organic matter decomposition. The stochastic soil moisture model of Chapter 7 is then coupled to a system of equations describing the temporal dynamics of soil carbon and nitrogen. This detailed model provides a basis for investigation of hydrologic control of the various components of the soil nitrogen cycle. An application of the model to the Nylsvley savanna in South Africa explores this interesting and well-studied ecosystem. We close with a discussion of how soil microbial life and global carbon patterns respond to aridity. The final notes outline some topics of interest for further study, including the biogenic emission of N-oxides, the impact of iron-redox fluctuation on decomposition, and methane emissions from wet soils.

9.1   Background

The dynamics of soil carbon and nitrogen are extremely important for the life and growth of vegetation and have impacts on climate and several soil processes. Hydrological fluctuations play an important role in modulating the speed of these cycles, from the microbial decomposition of organic matter and the related soil respiration and nitrogen mineralization to nitrogen uptake by plants and leaching to groundwater and streams. By enhancing some processes and quenching others, the patterns of soil moisture regulate the sequence of fluxes between different components and determine the dynamics of the other state variables of the system. Plants are often both water and nutrient limited, so it is difficult to determine the extent to which net primary production is controlled by water or nutrient availability. As noticed by Pastor et al. (1984), the nitrogen cycle needs to be explained through the water balance; not less marked is how the soil water balance controls the carbon cycle.

Here we will concentrate mostly on the direct influence of soil moisture on the soil carbon and nitrogen cycles, without considering vegetation growth. In general, however, the interactions between hydrology, soil organic matter (SOM), and nutrient cycles are extremely broad. Figure 9.1 indicates a few of the feedbacks between hydrological processes and soil biogeochemistry. We will focus on soil moisture controls on carbon and nitrogen cycles, with special attention to semi-arid systems.

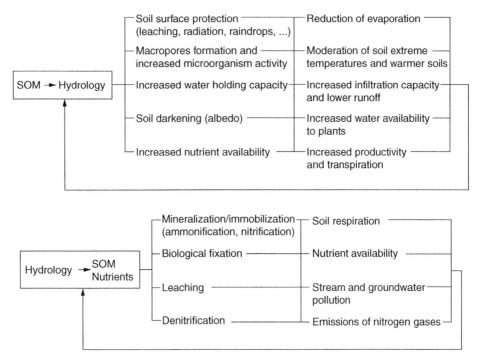

Figure 9.1 Interactions between soil organic matter (SOM) and hydrology. After Rodríguez-Iturbe and Porporato (2004).

9.1.1 Soil Organic Matter and Carbon Cycle

Soil organic matter (SOM) is a complex and varied mixture of organic substances, which may be conceptually subdivided into three main components: plant residues, microorganism biomass, and humus (also known as stabilized organic matter). Figure 9.2 shows the carbon cycle within the soil–plant–atmosphere system. Globally, the soil contains more carbon than the vegetation and atmospheric compartments (or pools) combined. Table 9.1 gives an indication of the amount and distribution of carbon in various types of natural soils of interest here. The amount of organic matter in soils varies widely and usually increases with humidity and, to a smaller extent, with temperature (also see the global distribution of carbon in Fig. 9.23). Aridisols (dry soils) are generally lowest in organic matter (mineral surface soils contain mere traces), while some humid vegetated A horizons (histosols) arrive at carbon levels as high as 20%–30% (see Sec. 3.1.1). The reason is that, in poorly drained soils with high productivity, the small amount of aeration inhibits organic matter decomposition and promotes the accumulation of SOM. For this reason, bogs and marshes are very important carbon pools in the terrestrial carbon cycle.

Plants take carbon from the atmosphere through photosynthesis; part of it is used by the plants as a source of energy and then directly released by respiration, while the other part is assimilated by vegetation and later transferred as plant litter to the soil, where it becomes part of SOM. Soil moisture has an important long-term influence on the amount and quality of litter, especially on its carbon-to-nitrogen (C/N) ratio, which in turn affects the rates of SOM decomposition.

SOM *decomposition*, or *mineralization*, involving an enzymatic oxidation, produces mineral compounds (e.g., ammonium) and carbon dioxide (CO_2), which is then returned to the atmosphere via *soil respiration*. While part of the carbon is lost as soil respiration and the simpler compounds are metabolized

Table 9.1 Typical content of SOM in the uppermost meter of various soils (see Sec. 3.1). Histosols are formed from materials high in organic matter and are typical of bogs and marshes; aridisols are soils of dry climates; mollisols are soils quite rich in organic matter formed under grasslands; spodosols are soils with alluvial accumulation of organic matter formed under forests in humid, temperate climates; alfisols form under forests or savannas in climates with slight-to-pronounced moisture deficit. Modified from Brady and Weil (1996).

Soil order	Organic carbon (kg/m^2)	SOM (kg/m^2)	Organic nitrogen (kg/m^2)
Histosols	205	350	10.2
Aridisols	< 3	< 5	< 0.3
Mollisols	13	22	1.1
Spodosols	15	26	1.2
Alfisols	7	12	0.6

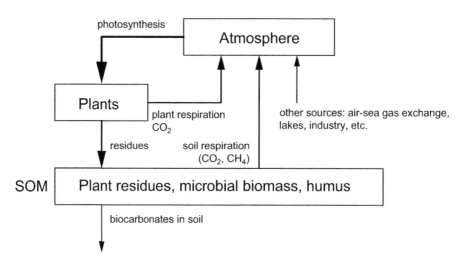

Figure 9.2 A simplified representation of the soil–plant–atmosphere carbon cycle. The thickness of the arrows and the sizes of the rectangles indicate the relative importance of the fluxes or pools. Adapted from Porporato et al. (2003a).

by soil microbes, the most complex compounds are not metabolized but, along with other compounds polymerized by soil microbes, are combined to form so-called *humic substances*, including both resistant humus and less-resistant nonhumic substances. Humic substances are very stable and contribute to maintaining high organic levels in soils and protecting the associated essential nutrients against mineralization and loss from the soil. Such protection may be further helped by some clay and other inorganic components that combine with humic materials. The so-called half-time (i.e., the time required to destroy half the amount of a substance) of humic substances varies from decades to centuries; because of their stability, humic substances comprise up 60%–80% of the SOM. An example of the turnover of litter and soil organic fractions in a grassland soil is shown in Fig. 9.3.

In soils of mature natural ecosystems, the release of carbon as CO_2 is approximately balanced in the long term by the input of plant residues, while at shorter timescales (e.g., seasonal-to-interannual) the

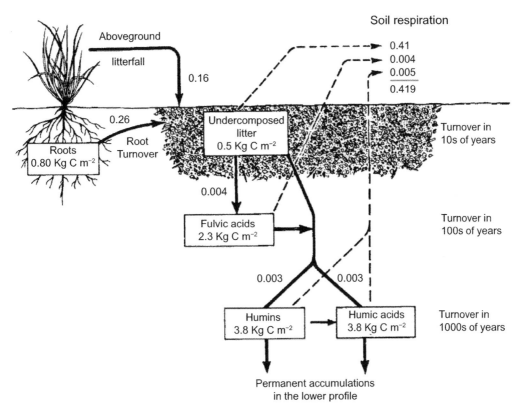

Figure 9.3 Turnover of litter and soil organic fractions in a grassland soil. The flux estimates are in kg C m^{-2} yr^{-1}. After Schlesinger (1997).

carbon content is subject to fluctuations induced by climatic and hydrologic variability, so that the entire soil carbon cycle is quite sensitive to external disturbances (see Sec. 9.3.4).

The decomposition process is related to *immobilization* in a complex manner and is regulated by the SOM carbon-to-nitrogen ratio and by the environmental conditions. It is usually modeled as first-order kinetics (i.e., with decomposition proportional to the amount of substance to be decomposed and to the amount of existing decomposing bacteria). When plant residues are added to the soil and the conditions are favorable, the bacterial colonies grow fast, but as soon as the decomposable SOM is reduced, they starve and die as easily. The decomposition of dead microbial cells is associated with the release of simple products, such as nitrate and sulphates. Different bacterial colonies exist, and each specializes in a given part of the decomposition process. Such bacteria are very sensitive to environmental conditions and, in particular, to the soil water potential (e.g., Fenchel et al., 1998). This in great part determines the close relationship between mineralization and soil moisture that will be discussed below.

Since most of the organic residues are deposited and incorporated at the surface, SOM tends to accumulate in the upper layers of soils. In general, under similar climatic conditions the total SOM is higher and vertically more uniform in soils under grasslands than under forests. This is so because a relatively high proportion of plant residues in grasslands consists of root matter, which decomposes more slowly and contributes more effectively to soil humus than does forest litter. Other ecosystems with deeper-rooted plants, such as the Nylsvley savanna (see Sec. 9.3.4), also tend to have a uniform distribution of SOM and mineral nitrogen over the whole root layer (except perhaps for the most superficial layer). This reduces the importance of the vertical dimension in the modeling of the carbon and nitrogen dynamics and allows us to use the same vertical domain (i.e., the active soil depth or rooting depth Z_r) as that used for soil moisture dynamics and the water balance (see Chapter 7). This simplification will be employed in the mathematical development of the model (see Sec. 9.3).

9.1.2   Soil Nitrogen Cycle

Although it is an essential nutrient for plants, soil nitrogen is mostly in the form of organic compounds that protect it from loss but leave it largely unavailable to vegetation. Plants almost always uptake mineral nitrogen only in the form of *ammonium* (NH_4^+) and *nitrate* (NO_3^-), which are made available through SOM decomposition. For this reason, the nitrogen cycle is intimately linked to that of carbon. The greatest amount of nitrogen in terrestrial ecosystems is in the soil, which contains 10-20 times as much nitrogen as does the living vegetation. Soil organic matter typically contains about 5% nitrogen, while inorganic (i.e., mineral) nitrogen is usually less than 1% to 2% of the total nitrogen in the soil. The atmosphere, which is composed of 78% nitrogen in form of dinitrogen (N_2, being quite inert, is not usable by most plants and animals), is a practically limitless reservoir of this element.

Figure 9.4 provides a schematic representation of the main components of the nitrogen cycle in soils. The existence of an internal cycle, which involves only soil and plants through nitrogen uptake and

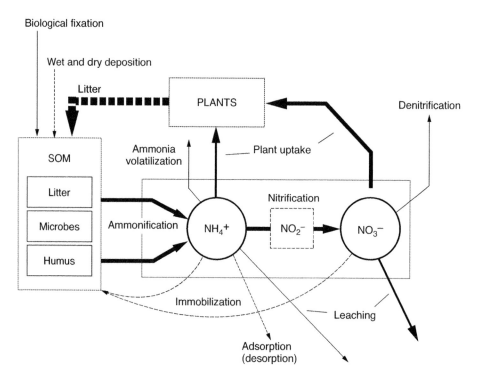

Figure 9.4 Schematic representation of the soil nitrogen cycle (the thickness of the arrows indicates the relative importance of the various fluxes in the cycle; the solid lines refer to processes for which the impact of soil moisture is relevant). After Porporato et al. (2003a).

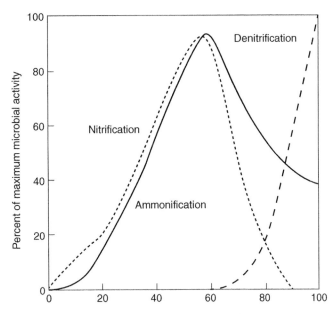

Figure 9.5 Rate of microbial activity related to the various phases of nitrogen transformation as a function of soil water content; after Brady and Weil (1996).

production of SOM and dominates the nitrogen turnover at the daily-to-seasonal timescales, is clear. The other external fluxes, such as wet and dry deposition and the biological fixation, become important only in the long-term balance and will be neglected in this analysis.

For the nitrogen cycle, decomposition of SOM produces ammonium and nitrate (ammonification and nitrification); approximately 1.5%–3.5% of the organic nitrogen of a soil mineralizes annually, depending on environmental factors, such as pH, temperature, and soil moisture. The influence of soil moisture on mineralization is due mainly to the balance between aeration, which diminishes with soil moisture and favorable humid conditions for microbial biomass (Fig. 9.5). At high soil moisture levels, anoxic conditions prevent bacteria from performing the aerobic oxidation necessary for decomposition. The reason for the reduction in the decomposition rate at low soil moisture levels rate is twofold (Stark and Firestone, 1995). Dry soils not only slow down the diffusion rate of the substrate but also reduce hydration and enzyme activity, thus reducing microbial activity (see Eq. (9.8)). In water-controlled ecosystems, temperature is usually less important than soil moisture, especially at the daily timescale. The most favorable conditions for mineralization are found around 20–30 °C; mineralization practically ceases outside the temperature range 5–50 °C and also tends to decline in acid soils.

When the conditions are favorable, nitrification is quite rapid. In hot and dry environments, sudden water availability can cause a flush of soil nitrate production, which may greatly influence the growth patterns of natural vegetation (see Fig. 9.19 and Note 9.4). For this reason, under warm and wet conditions, nitrate is the predominant form of nitrogen in most soils. Release of ammonia gas (NH_3) may have a certain importance (ammonia volatilization), especially in drying and hot sandy soils with ammonium accumulation in the soil top layers. The presence of nitrite (NO_2^-), however, is always negligible as its transformation to NO_3^- is practically immediate. This is important, since NO_2^- is quite toxic.

As already mentioned, the mineralization rate also depends on the composition of plant residues and, in particular, on their C/N ratio. The growth of microbial colonies takes place with a fixed proportion of carbon and nitrogen so that their C/N ratio remains practically constant. For example, if the microbial biomass has a C/N ratio of 8 then, for each part of nitrogen metabolized, 24 parts of carbon are needed of which eight are metabolized and 16 are respired as CO_2 (Brady and Weil, 1996). As a consequence, if the nitrogen content of the organic matter being decomposed is high (i.e., C/N < 24), mineralization proceeds unrestricted and mineral components in excess are released into the soil. In contrast, when the litter is nitrogen poor (i.e., C/N > 24), microbes can use some of the mineral nitrogen through the process of immobilization. If mineral nitrogen is not available, then mineralization may be halted. The modeling of this delicate balance is discussed later in Secs. 9.2.3 and 9.3.2. The overall dependence of mineralization rate on environmental conditions (soil moisture, temperature, pH, etc.) and litter composition (e.g., the C/N ratio) is summarized in Fig. 9.6.

Nitrate is easily soluble in water and, although this facilitates its uptake by plants, it also makes it prone to losses by leaching at high soil moisture levels. In contrast, the positive charge of ammonium ions attracts them to the negatively charged surfaces of clays and humus, thus partially protecting them from leaching. Although held in an exchangeable form, this may be a problem for plant uptake since the rate of release of the fixed ammonium is often too slow to fulfill plant needs. At high soil moisture levels the process of denitrification may take place (see Fig. 9.5), releasing greenhouse nitrogen gases (see Note 9.8).

Plant nitrogen requirements are met either passively by the soil solution during the transpiration process or actively through a diffusive flux driven by concentration gradients produced by the plant itself (e.g., Russell, 1931; Engels and Marschner, 1995; Larcher, 1995). As this second mechanism is energy intensive, active uptake seems to take place only when the nitrogen demand by a plant is higher than the passive supply by transpiration. If both mechanisms are insufficient to meet such a demand, the plant is under a condition of nitrogen deficit. On the other hand, if the concentration in the soil solution is high, the passive uptake rate may exceed the actual plant demand. Plants appear to have little control on passive uptake and nitrogen excess may even result in toxicity (Brady and Weil, 1996).

9.2   Reduced-Order Models of Soil Organic Matter Decomposition

The decomposition of SOM is an interesting process controlled by the abundance and quality of the SOM substrate as well as by the soil microbial activity, which in turn is modulated by the soil moisture. Disentangling these effects is best achieved by starting from simplified models, in which the essential interactions are brought to the forefront. The relatively simple mathematical results obtained here help shed light on the complex patterns observed in field experiments and guide the interpretation of results from more complex models.

9.2.1   SOM Dynamics as a Predator–Prey System

Following Manzoni and Porporato (2007), we consider a simplified soil system where a single pool of carbon substrate (C_s) interacts with the microbial biomass (C_b) (Fig. 9.7). The SOM carbon fraction (SOM-C) is assumed to be a passive substrate, while the microbial pool includes all the biotic components of the

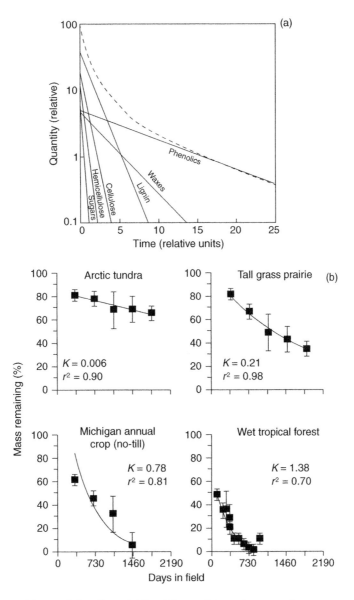

Figure 9.6 Time courses of decay of plant litter. (a) Breakdown of the various organic components of litter; after Chapman (1976) and Larcher (1995). (b) Mass loss from litter bags containing the same type of wheat litter placed in four different ecosystems; after Sala et al. (1976).

soil system. The variables are expressed in terms of the mass of carbon per unit volume of soil; the system is assumed to be open, with a plant residue input (ADD) and an output of mineralized carbon to the atmosphere. The carbon balances for the substrate and the microbe pools can thus be expressed as

$$\frac{dC_s(t)}{dt} = \text{ADD} - \text{DEC} + \text{BD}, \tag{9.1}$$

$$\frac{dC_b(t)}{dt} = (1 - r)\text{DEC} - \text{BD}. \tag{9.2}$$

Figure 9.7 Schematic representation of carbon flow in soil. After Manzoni and Porporato (2007).

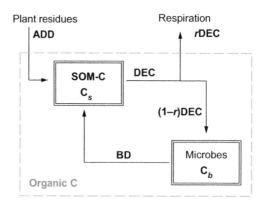

where DEC is the flux of decomposed carbon without nitrogen limitation (see Sec. 9.3.2). The heterotrophic respiration is rDEC, and $1 - r$ is the carbon utilization efficiency. This model essentially describes a general loss–win interaction known as the predator–prey relationship (Strogatz, 2001); see Sec. 2.4.2.

For simplicity, the microbial decay, BD, is modeled using first-order kinetics as in most biogeochemical models (Parton et al., 1987):

$$\text{BD} = k_b C_b. \tag{9.3}$$

The quantity DEC represents the carbon output due to microbial decomposition, which is one of the most critical issues in soil C and N models (Manzoni and Porporato, 2009). Decomposition can be regarded as an enzyme-catalyzed reaction, where the C flux from the organic substrate degradation depends on both the substrate and the available enzyme concentration (e.g., Schimel and Weintraub, 2003). While complex models may take into account the feedbacks from microbial density and co-metabolism effects, other models are often related to the Michaelis–Menten equation (MM, see Note 4.9) (e.g., Panikov and Sizova, 1996),

$$\text{DEC} = k_s C_b \frac{C_s}{K_m + C_s}, \tag{9.4}$$

where K_m is the Michaelis constant, and k_s is a rate constant. If $C_s \ll K_M$, Eq. (9.4) converts to a multiplicative model;

$$\text{DEC} = k_s C_b C_s. \tag{9.5}$$

If C_b is assumed to change slowly, Eq. (9.5) converts to a linear model with respect to C_s:

$$\text{DEC} = k_s C_s. \tag{9.6}$$

As demonstrated later by linear stability analysis, the choice of these models has a strong impact on the ability to describe complex dynamics (Manzoni and Porporato, 2007). A more comprehensive review of different types of decomposition models is given in Manzoni and Porporato (2009).

The decomposition rate k_s in Eqs. (9.4)–(9.6) can be modeled as a multiplicative function to address the effects of soil moisture and temperature:

$$k_s = k_s^* f_s(s) f_T(T), \tag{9.7}$$

where k_s^* is the potential decomposition rate, while $f_s(s)$ and $f_T(T)$ are reduction factors accounting for soil moisture s and soil temperature T, respectively. The temperature effect can be simply modeled as a quadratic function of T, normalized in order that $0 \leq f_T(T) \leq 1$ (Ratkowsky et al., 1982). Microbial activity depends on s in a strongly nonlinear way (see Fig. 9.5): low substrate diffusivity and cellular water potential are limiting factors under dry soil conditions, while a low oxygen content for saturated soil can also reduces microbial activity. A piecewise function can be used to model $f_s(s)$ (Cabon et al., 1991; Porporato et al., 2003a):

$$f_s(s) = \begin{cases} 0 & s \leq s_B, \\ \dfrac{s - s_B}{s_{\mathrm{fc}} - s_B} & s_B < s \leq s_{\mathrm{fc}}, \\ s_{\mathrm{fc}}/s & s > s_{\mathrm{fc}}, \end{cases} \tag{9.8}$$

where s_{fc} is the soil field capacity and s_B the microbial biomass stress point, below which (in analogy with the plant wilting point) biological activity is completely halted. Microbial activity in soils is severely inhibited when soil moisture decreases during dry periods, as recognized as early as the 1920s (Greaves and Carter, 1920). Two complementary explanations are often invoked: (i) physiological water stress ensues during drying, thereby limiting the metabolism of at least some organisms, with clear effects at the community level, and (ii) solute transport is reduced during drying, leaving microbes under substrate-limited conditions at a time when they need additional resources to face water stress. The compound effects of these abiotic and biological factors contribute to the strong declines in microbial activity in drying soil (Manzoni et al., 2012; Manzoni and Katul, 2014; Manzoni et al., 2019).

9.2.2 Nonlinearities in the Soil Carbon Cycle

The predator–prey model in Sec. 9.2.1 can be used to simulate the soil carbon cycle, and particularly how the decomposition models influence the system's behavior. We first analyze the time series of microbial biomass using different decomposition models (see Fig. 9.8), and then study Eqs. (9.1) and (9.2) as a dynamical system (Sec. 2.4.2), in the idealized case of constant soil moisture, temperature, and plant residue input in the SOM pool, to assess how climatic factors control their dynamic behavior.

The performance of these decomposition models are compared in Fig. 9.8. The nonlinear decay function used by Zelenev et al. (2000) introduces a further feedback between biological activity and its substrate, resulting in enhanced oscillation amplitudes and a good fit of the measured oscillations (the dashed line corresponds to a correlation coefficient $\rho_c = 0.66$). The multiplicative model (Eq. (9.5)) is able to simulate the oscillatory behavior, but the resulting trajectories appear to be overdamped (solid line, $\rho_c = 0.43$). In contrast, the linear model (Eq. (9.6)) captures only the general trend of microbial biomass; it cannot describe its fluctuations (dotted line, $\rho_c = 0.23$). Thus, nonlinearity seems to be necessary to explain these oscillations.

It is also interesting to explore the role of parameter values on a model's dynamic behavior around equilibrium in the idealized case of constant soil moisture, temperature, and plant residue input in the SOM pool. Recall the linear stability analysis in Sec. 2.4 and Table 2.4: real negative eigenvalues imply exponential convergence to a stable equilibrium, while complex values indicate convergence through damped oscillations. The equilibrium values and their eigenvalues depend on both model type and parameter values (see Fig. 9.9a). The dynamic behavior of the linear model (Eq. (9.6)) does not qualitatively change around the equilibrium, and the steady state is always a stable node since both eigenvalues are real and

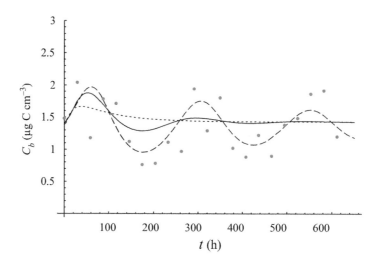

Figure 9.8 For comparison, different model simulations and also measured microbial biomass concentrations. All the models share the same initial conditions and external input (ADD). Only the decomposition and microbial decay functions change between models. The dots are data from Semenov et al. (1999); the dashed line refers to Zelenev et al. (2000); the solid lines are from multiplicative model simulations (Eq. (9.5)); the dotted lines are from linear model simulations (Eq. (9.6)). After Manzoni and Porporato (2007).

negative. When either Eq. (9.4) or Eq. (9.5) is used for decomposition modeling, a richer dynamic behavior is observed because changes in the rate constant, k_s, can dramatically affect the eigenvalues and result in qualitative changes in the equilibrium.

Owing to the nonlinear character of Eq. (9.8), in moving from low soil moisture to values close to saturation, while keeping the temperature constant (the vertical dot-dashed line in Fig. 9.9b), two dynamic changes occur. For soil moisture below the microbe stress point ($s \leq s_B$), biological activity is not sustained and the equilibrium is characterized by zero biomass. When the soil moisture is increased above s_B, microbial activity is low but sufficient to sustain decomposition, resulting in a stable focus with damped oscillations to equilibrium (Re(λ) < 0, see Table 2.4). Approaching field capacity, the first bifurcation takes place and, in the region where decomposition is more efficient, the equilibrium point becomes a stable node (λ < 0, see Table 2.4). The situation is inverted at high soil moisture values, where the damped oscillations result in the reappearance of a stable focus. Figure 9.9c illustrates the effects of climatic factors on the nonlinear system governed by Eq. (9.4). Interestingly, in this case the system undergoes two changes in stability, not only when the soil moisture is increased (as in Fig. 9.9a) but also when temperature is increased, resulting in an even more complex dynamic pattern as a function of the climatic parameters.

Both the data–model comparison (Fig. 9.8) and the linear stability analysis (Fig. 9.9) show that linear models are less suited to describe the biomass–substrate interaction, and lack the oscillating behavior observed on small space–time scales under certain climate conditions. The multiplicative model appears more flexible and represents a useful compromise between the linear model and more complex nonlinear formulations. The type of damped oscillations produced by the simplified predator–prey model for SOM decomposition model (see Sec. 9.2.1) are also found under certain conditions in more complete models of the C and N dynamics at different timescales (see Sec. 9.3).

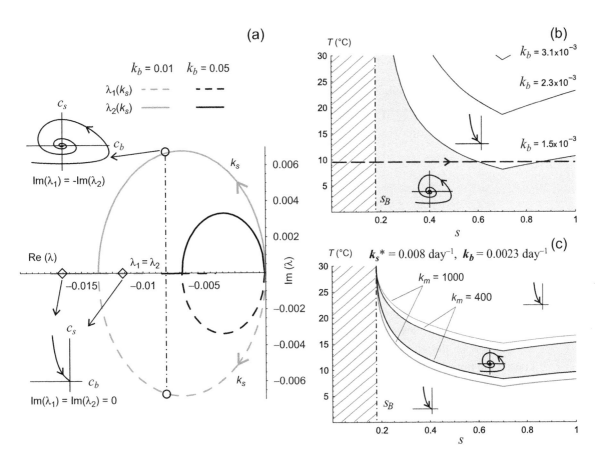

Figure 9.9 Stability analysis of the system of equations (9.1) and (9.2). (a) The real and imaginary parts of the two eigenvalues λ_1 and λ_2 of the system with nonlinear decomposition of Eq. (9.5). These eigenvalues can be used to identify the system stability following Table 2.4. (b) and (c) Bifurcation diagrams as a function of soil moisture s and temperature T for (b) the multiplicative decomposition model (Eq. (9.5)) and (c) the Michaelis–Menten decomposition model (Eq. (9.4)). The shaded and blank areas are regions of parameter space corresponding to a stable focus and a stable node, respectively (see Sec. 2.4.2), while the hatched areas are not associated with any equilibrium points. After Manzoni and Porporato (2007).

9.2.3 Stoichiometry of Litter Decomposition

Besides the abiotic and biotic factors that control decomposition processes (see Sec. 9.2.1), nutrient availability (e.g., of nitrogen, phosphorus) also has a significant impact on microbial activity. Decomposers sequester carbon and nutrients in approximately fixed proportions, thus imposing a strict constraint on the decomposition processes and the resulting evolution of SOM chemical composition, soil respiration, and nutrient-release patterns. Here, we follow Manzoni et al. (2008) to derive a general equations for the nutrient-releasing pattern of plant residue in order to explain global observations of litter decomposition.

The model is constructed to follow a so-called *litter cohort*, representing a litter-bag experiment (Fig. 9.10). In these experiments, some bags that allow the free exchange of water and nutrients with the rest of the soil are filled with plant residues of known composition. The bags are buried in the soil

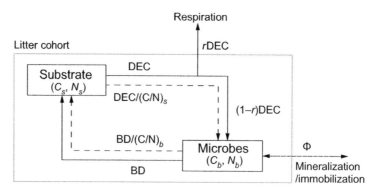

Figure 9.10 Scheme of the carbon and nitrogen cycles of a litter cohort. The solid arrows indicate the carbon fluxes and the long-dashed arrows indicate the nitrogen fluxes.

and their content is left to decompose. Several bags are placed simultaneously under similar environmental conditions so that some of them can be recovered at different times to track the composition changes and reconstruct the release patterns of carbon, nitrogen, and other nutrients. The decomposition process depends on the coupling between the nitrogen and the carbon cycle, because of the stoichiometric constraints imposed by microbial growth. The main carbon and nitrogen fluxes are shown in Fig. 9.10.

The total carbon and nitrogen balance within the litter cohort ($C_{tot} = C_s + C_b$, $N_{tot} = N_s + N_b$) can be expressed as

$$\frac{dC_{tot}}{dt} = -r\text{DEC}, \tag{9.9}$$

$$\frac{dN_{tot}}{dt} = -\Phi. \tag{9.10}$$

Plant residues of given composition are added at the initial time $t = 0$, after which the litter cohort start to lose carbon at a rate specified by the decomposition function DEC weighted by the respiration rate r. Other possible inputs or outputs of carbon and nitrogen (e.g., due to atmospheric deposition, advection, diffusion, or the transfer of biomass by root growth, microbes and soil fauna) are considered negligible.

The litter decomposition patterns are essentially controlled by microbes, which have to assimilate carbon and nitrogen in fixed proportion. This type of constraint is referred to as the "decomposer stoichiometric requirement". To formulate this constraint explicitly, we focus on the carbon and nitrogen fluxes to the microbes (see Fig. 9.10):

$$\frac{dN_b}{dt} = \frac{\text{DEC}}{(\text{C/N})_s} - \Phi - \frac{\text{BD}}{(\text{C/N})_b}, \tag{9.11}$$

$$\frac{dC_b}{dt} = (1 - r)\text{DEC} - \text{BD}, \tag{9.12}$$

where $(\text{C/N})_s$ and $(\text{C/N})_b$ are the C/N ratios for the substrate and microbial biomass. Dividing Eq. (9.11) by (9.12) and setting $dC_b/dN_b = (\text{C/N})_b$ (i.e., the stoichiometric requirement) yields

$$\Phi = \text{DEC}\left[\frac{1}{(\text{C/N})_s} - \frac{1 - r}{(\text{C/N})_b}\right]. \tag{9.13}$$

Given that the substrate accounts for the majority of the litter cohort mass, we may approximate $(C/N)_s$ by $(C/N)_{tot}$ and rewrite Eq. (9.13) as

$$\Phi = \text{DEC} \left[\frac{N_{tot}}{C_{tot}} - \frac{1-r}{(C/N)_b} \right]. \tag{9.14}$$

Combining Eqs. (9.9), (9.10), and (9.14) eliminates the direct time dependency of C_{tot} and N_{tot}, producing an ordinary differential equation in which time enters only implicitly and r and $(C/N)_b$ are parameters,

$$\frac{dN_{tot}}{dC_{tot}} = \frac{\Phi}{r\text{DEC}} = \frac{N_{tot}}{rC_{tot}} - \frac{1-r}{r(C/N)_b}. \tag{9.15}$$

Solving this differential equation by separation of variables yields (Manzoni et al., 2008, 2010)

$$n(c) = c\frac{(C/N)_0}{(C/N)_b} + \left[1 - \frac{(C/N)_0}{(C/N)_b} \right] c^{1/r}, \tag{9.16}$$

where $n = N_{tot}(t)/N_{tot}(0)$ and $c = C_{tot}(t)/C_{tot}(0)$ are the resulting fractions of remaining carbon and nitrogen after adding the plant residue, and $(C/N)_0$ is the initial C/N ratio of the substrate. Equation (9.16) describes the dynamic relationship between carbon and nitrogen, the so-called nutrient-releasing curves of the decomposition process in the litter cohort. One can also derive similar equations for the stoichiometric balance of other nutrients such as phosphorus (Manzoni et al., 2010; Manzoni, 2017).

These nutrient-release patterns have been observed in different climatic regions across diverse terrestrial ecosystems over the world (Fig. 9.11). During decomposition, the fractions of remaining N and lost C are illustrated by the curves going from left to right, moving at a speed dictated by biogeochemical and environmental conditions embedded in the decomposition function (which here is left unspecified). All the curves show slower N loss than C loss, meaning that N tends to accumulate, and the N/C ratio of the litter increases throughout decomposition. Where the curves increase with respect to the initial condition, not only is N retained more efficiently than C, but net immobilization occurs. At the point on each curve where n is a maximum, immobilization ends and net mineralization begins. Conversely, if the curve decreases monotonically then there is no initial net immobilization, as in Figs. 9.11a and b. The maximum of the N-release curve thus corresponds to the litter critical C/N ratio; its value can found by setting Eq. (9.15) to zero,

$$(C/N)_{CR} = (C/N)_b/(1-r), \tag{9.17}$$

a function of the decomposer characteristics. When $(C/N)_{CR} > (C/N)_0$, net release occurs from the beginning of decomposition. Conversely, if $(C/N)_{CR}$ is low compared with the initial $(C/N)_0$, large amounts of mineral N have to be immobilized to increase the litter's N concentration to its critical value.

The N-release patterns of decomposing litter appear to be regulated by the initial chemical composition of the litter and the stoichiometric requirements of the decomposers. Because the decomposer C:N ratio is relatively constant, this pattern suggests that decomposer communities are able to adapt to low-nitrogen substrates by decreasing their carbon use efficiency (i.e., increasing r). Such a pattern has been observed in aquatic environments and at other trophic levels and appears to be a general response of decomposers in nutrient-poor conditions (Mattson Jr., 1980; Del Giorgio and Cole, 1998; Elser et al., 2000; Manzoni et al., 2008). Decreasing efficiency results in higher heterotrophic respiration per unit mass of litter humified or unit nutrient released, confirming that the soil carbon cycle is closely linked to its nutrient states.

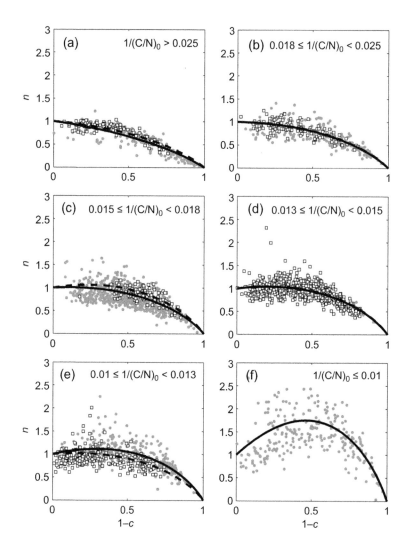

Figure 9.11 Observed and modeled fractions of the initial nitrogen, $n = N_{tot}(t)/N_{tot}(0)$, as a function of the decomposed fractions of initial carbon, $1 - c = 1 - C_{tot}(t)/C_{tot}(0)$, for leaf litter with values of $1/(C/N)_0$ decreasing from panels (a) to (f). The data and analytical N-release curves are represented by dots and solid lines for the Long-term Intersite Fine Litter Decomposition Experiment (LIDET) datasets, and are represented by squares and dashed lines for the Canadian Intersite litter Decomposition Experiment (CIDET) datasets. After Manzoni et al. (2008).

9.3   Coupled Model of the Carbon and Nitrogen Cycles

Several processes in the soil carbon and nitrogen cycles were neglected in the previous sections on reduced-order models, while environmental conditions were either kept constant, as in the predator–prey system for SOM decomposition (Sec. 9.2.1), or only entered implicitly into the solution of the stoichiometric model, where the time variable entered only as a parameter (Sec. 9.2.3). A more complete model, proposed by Porporato et al. (2003a), for the analysis of the soil carbon and nitrogen cycles and their relationship with stochastic soil moisture dynamics, is discussed here. The highly intertwined carbon and

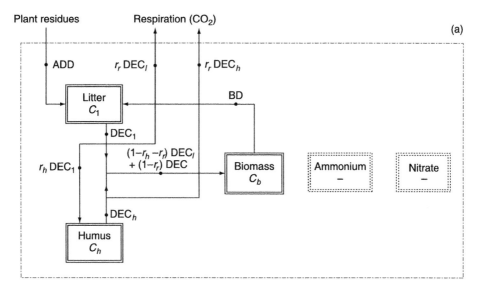

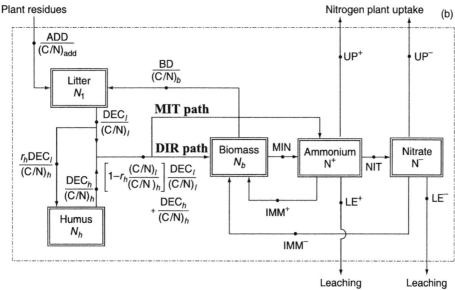

Figure 9.12 Schematic representation of the main components of the model. (a) Soil carbon cycle; (b) soil nitrogen cycle. Modified from Porporato et al. (2003a).

nitrogen cycles described in Sec. 9.1 are modeled by employing five separate pools for each main component of the system (Fig. 9.12). The soil organic matter is divided into three pools, representing litter, humus (or stabilized organic matter), and microbial biomass, respectively, while the inorganic nitrogen in the soil is divided into ammonium (NH_4^+) and nitrate (NO_3^-). For the sake of model simplicity, no distinction among the different bacterial colonies is made: all of them are included in a single biomass pool. Similarly, because of the high rate of nitrification, the presence of nitrite (NO_2^-) is usually very low and is neglected here.

Only the input of the added litter and the losses due to soil respiration, leaching, and plant uptake are considered among the external fluxes to the soil system, neglecting in this first analysis other fluxes that are less important at the daily-to-seasonal timescale, such as ammonium adsorption and desorption by clay colloids, the volatilization or absorption of ammonium, and the nitrogen input due to wet and dry deposition, biological fixation, and denitrification.

Since differences of some orders of magnitude are to be expected between the decomposition rate of the faster (proteins) and that of the slower (lignin) organic components (Brady and Weil, 1996), two or more pools of organic matter have been considered (e.g., Jenkinson, 1990; Hansen et al., 1995). The separate consideration of litter and humus addresses this aspect since, in general, litter compounds have a faster decomposition rate than humic ones. Although a range of components is present in the SOM, the litter and humus pools are assumed to be characterized by unique values of the C/N ratio and of the resistance to microbial decomposition, representing weighted averages of the various components (Porporato et al., 2003a).

As we saw in Sec. 9.1 as well as in the analysis of the stoichiometric model in Sec. 9.2.3, the C/N ratios of the pools containing organic matter are very important in the dynamics of the entire carbon and nitrogen cycles. Both the humus and the biomass C/N ratio remain approximately constant in time, while the litter C/N ratio may be highly variable. Typically, the humus C/N ratio, $(C/N)_h$, is on the order of 10–12, that of the biomass, $(C/N)_b$, is usually around 8–12 (depending on the type of microbial community), and that of the litter, $(C/N)_l$, ranges from 20 to over 50. The constancy of $(C/N)_b$ and the variability of $(C/N)_l$ play an essential role in controlling the rates of decomposition, mineralization, and immobilization.

All the components of the soil carbon and nitrogen cycles considered in the model of Porporato et al. (2003a) are represented in Fig. 9.12. Eight state variables in terms of the mass per unit volume of soil (e.g., grams of carbon or nitrogen per m^3 of soil) are needed to characterize the system (see Table 9.2). The temporal dynamics of such variables is controlled by a system of as many coupled differential equations are required to describe the balance of carbon and nitrogen in the various pools. All the equations represent balances of fluxes of nitrogen or carbon in terms of mass per unit volume per unit time (e.g., g m^{-3} day^{-1}). Since many of the fluxes are heavily dependent on soil moisture content, the system is coupled to the soil moisture evolution, Eq. (1.5). In this way, the main hydrologic control on the carbon and nitrogen cycles is explicitly considered.

Table 9.2 State variables for the coupled carbon and nitrogen model.

Variable	Description
C_l	Carbon concentration in the litter pool
C_h	Carbon concentration in the humus pool
C_b	Carbon concentration in the biomass pool
N_l	Organic nitrogen concentration in the litter pool
N_h	Organic nitrogen concentration in the humus pool
N_b	Organic nitrogen concentration in the biomass pool
N^+	Ammonium concentration in the soil
N^-	Nitrate concentration in the soil

9.3.1   Soil Organic Matter

The Fast Pool

According to the fluxes depicted in Fig. 9.12, the carbon balance equation for a fast decomposing litter pool can be written as

$$\frac{dC_l}{dt} = \text{ADD} + \text{BD} - \text{DEC}_l. \tag{9.18}$$

The term ADD is the external input into the system, representing the rate at which carbon in plant residues is added to the soil and made available to the attack of the microbial colonies. Its temporal variability depends on the evolution of vegetation biomass. The term BD represents the rate at which carbon returns to the litter pool owing to the death of microbial biomass (see Eq. (9.3)).

The quantity DEC_l (and also DEC_h in Sec. 9.3.1) represents the carbon output due to microbial decomposition, which is one of the most critical issues in soil C and N models (Manzoni and Porporato, 2009). As summarized in Sec. 9.2.1, multiple decomposition models have been used to simulate this decomposition process. Here, we use the multiplicative model (Eq. (9.5)) with consideration of nitrogen limitation,

$$\text{DEC}_l = \varphi f_s(s) k_l C_b C_l. \tag{9.19}$$

The coefficient φ is a non-dimensional factor accounting for a possible reduction in the decomposition rate when the litter is very poor in nitrogen and immobilization is not sufficient to integrate the nitrogen required by bacteria. We will return later to this point. The factor $f_s(s)$, first introduced in Sec. 9.2.1, describes the soil moisture effects on decomposition. The value of the constant k_l defines the rate of decomposition for the litter pool as a weighted average of the decomposition rates of the different organic compounds in the plant residues. Its average value is usually much higher than the corresponding value for humus. The rate of decomposition depends also on the microbial biomass concentration, C_b (e.g., Hansen et al., 1995), whose evolution will be described later. The relationship between the concentration of microbial biomass and its activity is assumed to be linear.

The nitrogen balance in the litter pool is similar to Eq. (9.18), with each term divided by the C/N ratio of its respective pool, i.e.,

$$\frac{dN_l}{dt} = \frac{\text{ADD}}{(\text{C/N})_{\text{add}}} + \frac{\text{BD}}{(\text{C/N})_b} - \frac{\text{DEC}_l}{(\text{C/N})_l}, \tag{9.20}$$

where $(\text{C/N})_{\text{add}}$, the C/N ratio of added plant residues, ranges from 10 in legumes and young green leaves to more than 200 in sawdust (Brady and Weil, 1996). This large variability may produce pronounced changes in the C/N ratio of the litter pool, which has a very important role in regulating decomposition, immobilization, and mineralization.

The Slow Pool

The more recalcitrant part of organic matter, the humus or slow pool, contributes to maintaining high organic levels in soil, preventing rapid nutrient loss from the soil via leaching or denitrification (Brady and Weil, 1996). The decomposition in the slow pool is constrained by physical processes that "protect" organic matter in aggregates and by adsorption to clay, making it less accessible to extracellular enzymes and microbes (Schimel and Schaeffer, 2012; Manzoni et al., 2019). According to Fig. 9.12, the balance equation for carbon in the slow pool is

$$\frac{dC_h}{dt} = r_h \text{DEC}_l - \text{DEC}_h, \tag{9.21}$$

where the only input is represented by the fraction r_h of the decomposed litter undergoing humification (see Fig. 9.12). The coefficient r_h, which is sometimes referred to as the "isohumic coefficient" (Russell, 1931), is in the range of 0.15–0.35 (Brady and Weil, 1996), depending on the composition of the plant residues.

The output due to humus decomposition is modeled in the same way as litter decomposition (see Eq. (9.19)),

$$\text{DEC}_h = \varphi f_d(s) k_h C_b C_h, \tag{9.22}$$

where the value of the constant k_h, encompassing the various components of the humus pool, is much smaller than the corresponding value for the litter pool, k_l, because of the greater resistance to microbial attack of humic substances.

The nitrogen balance equation may be simply obtained by dividing Eq. (9.21) by $(\text{C/N})_h$, i.e.,

$$\frac{dN_h}{dt} = r_h \frac{\text{DEC}_l}{(\text{C/N})_h} - \frac{\text{DEC}_h}{(\text{C/N})_h}. \tag{9.23}$$

As noted by Porporato et al. (2003a), this relation implies an assumption that the products of the humification process from litter have the same characteristics, and thus also the same C/N ratio, as the soil humus. As a consequence, the value of $(\text{C/N})_h$ remains constant in time, making Eq. (9.23) redundant. Moreover, the fraction r_h cannot exceed $(\text{C/N})_h/(\text{C/N})_l$, as follows from the obvious condition that the fraction of nitrogen entering the humus pool from litter decomposition cannot exceed the total nitrogen flux decomposed from the litter, i.e., $r_h \text{DEC}_l/(\text{C/N})_h \leq \text{DEC}_l/(\text{C/N})_l$.

The Biomass Pool

The carbon balance in the biomass pool is given by

$$\frac{dC_b}{dt} = (1 - r_h - r_r)\text{DEC}_l + (1 - r_r)\text{DEC}_h - \text{BD}. \tag{9.24}$$

The input is represented by the fraction of organic matter incorporated by the microorganisms from litter and humus decomposition (see Fig. 9.12). The constant r_r ($0 \leq r_r \leq 1 - r_h$) defines the fraction of decomposed organic carbon that goes into respiration (CO_2 production), usually estimated in the interval 0.6–0.8 (Brady and Weil, 1996). The only output is BD, already defined in the context of Eq. (9.18).

The balance of the nitrogen component in the biomass is associated with the *mineralization–immobilization pathways* (Garnier et al., 2001; Manzoni and Porporato, 2009). Two different schemes have been developed to model these complex pathways: the *mineralization–immobilization turnover* (MIT) scheme (Jansson, 1958) and the *direct assimilation* (DIR) scheme (Molina et al., 1983). The MIT scheme maintains that all the organic nitrogen is mineralized to ammonium prior to assimilation by the soil microbial biomass, whereas the DIR scheme assumes that the exocellular enzymes hydrolyze the organic compounds to their basic amino acids, which are then directly assimilated by the microbes (see Fig. 9.12b). At the macroscopic level, a combination of the DIR and MIT schemes may be observed

(Hadas et al., 1992); this is referred to as the *parallel* (PAR) scheme (Barraclough, 1997). The nitrogen balance under the PAR scheme can be expressed as

$$\frac{dN_b}{dt} = \eta \left[1 - r_h \frac{(C/N)_l}{(C/N)_h} \right] \frac{DEC_l}{(C/N)_l} + \eta \frac{DEC_h}{(C/N)_h} - \frac{BD}{(C/N)_b} - \Phi. \tag{9.25}$$

The first two terms on the right-hand side represent the incoming nitrogen from decomposition and do not contain r_r, because the respiration process involves only the carbon component. The partition coefficient η in this PAR scheme defines which fraction of the decomposed organic N enters the microbial biomass directly, following the DIR scheme, while the fraction $1 - \eta$ is mineralized through the MIT pathway. The third term of Eq. (9.25) is the output of nitrogen due to microbial death, while the fourth term, Φ, takes into account the nitrogen contribution due to either net mineralization or net immobilization, as already seen in the stoichiometric model in Sec. 9.2.3; these processes will now be discussed in more detail.

9.3.2 Mineralization and Immobilization Rates

The term Φ in Eq. (9.25) is the net nitrogen flux, which attains positive or negative values in relation to the difference between the rate of gross mineralization and the total rate of immobilization of ammonium, IMM^+, and nitrate, IMM^-, i.e.,

$$\Phi = MIN - IMM, \tag{9.26}$$

where

$$IMM = IMM^+ + IMM^-. \tag{9.27}$$

Since it is the net fluxes among the various pools that are important to the nitrogen balance, only the net amounts of mineralization and immobilization need to be modeled. This can be done as if they were mutually exclusive processes. Thus, when $\Phi > 0$, we will assume that

$$\begin{cases} MIN = \Phi, \\ IMM = 0, \end{cases} \tag{9.28}$$

while, when $\Phi < 0$,

$$\begin{cases} MIN = 0, \\ IMM = -\Phi. \end{cases} \tag{9.29}$$

The switch between the two states is regulated by microbial stoichiometry, which requires a relatively constant $(C/N)_b$. Accordingly, when the average C/N ratio of the biomass is lower than the value required by the microbial biomass, decomposition results in a surplus of nitrogen, which is not incorporated by the bacteria, and net mineralization takes place. If, instead, the decomposing organic matter is nitrogen poor, bacteria try to meet their nitrogen requirement by increasing the immobilization rate from ammonium and nitrate, thus impoverishing the mineral nitrogen pools. If the nitrogen supply from immobilization is not enough to ensure a constant C/N ratio for the biomass, the decomposition rates may change correspondingly to account for the imbalance in C and N availability. This has been modeled by several schemes, including N inhibition and C overflow (Manzoni and Porporato, 2009). Here we follow Porporato et al. (2003a) and introduce the *N-inhibition hypothesis*. It assumes that the rates of decompositions are reduced below their potential values by means of the parameter φ (e.g., see Eqs. (9.19) and (9.22)), as follows.

The condition of a constant C/N ratio for the biomass pool is implemented analytically by setting $dC_b/dN_b = (C/N)_b$. Similarly to the results in Eq. (9.13) from the two-pool model, the combination of Eqs. (9.24) and (9.25), along with the expressions for DEC_l and DEC_h in Eqs. (9.19) and (9.22), yields

$$\Phi = \varphi f_d(s) C_b \left\{ k_h C_h \left[\frac{\eta}{(C/N)_h} - \frac{1 - r_r}{(C/N)_b} \right] \right.$$
$$\left. + k_l C_l \left[\frac{\eta}{(C/N)_l} - \frac{\eta r_h}{(C/N)_h} - \frac{1 - r_h - r_r}{(C/N)_b} \right] \right\}. \tag{9.30}$$

Notice that Eq. (9.30), and not (9.25), is the real dynamical equation to be associated with Eq. (9.24) for biomass evolution. Equation (9.30), in fact, makes Eq. (9.25) redundant, as the biomass C/N ratio is set constant.

When the term in braces in Eq. (9.30) is positive, net mineralization takes place, while no net immobilization occurs, as is indicated by Eq. (9.28). In such conditions, humus and litter decomposition proceed unrestricted and the parameter φ is equal to 1.

In the opposite case, when the term in braces in Eq. (9.30) is negative, net mineralization is halted and immobilization sets in (Eq. (9.29)). The latter is partitioned proportionally between ammonium and nitrate on the basis of their concentrations and according to two suitable coefficients, k_i^+ and k_i^-, i.e.,

$$\left\{ \begin{array}{l} IMM^+ = \dfrac{k_i^+ N^+}{k_i^+ N^+ + k_i^- N^-} \ IMM, \\[3mm] IMM^- = \dfrac{k_i^- N^-}{k_i^+ N^+ + k_i^- N^-} \ IMM. \end{array} \right. \tag{9.31}$$

The rate of immobilization may be limited by environmental factors, biomass concentration, and especially by insufficient mineral nitrogen. For this reason, we assume the existence of an upper bound for the rate of immobilization, i.e.,

$$IMM \le IMM_{max}, \tag{9.32}$$

which, in view of Eq. (9.30), becomes

$$-\varphi f_d(s) C_b \left\{ k_h C_h \left[\frac{\eta}{(C/N)_h} - \frac{1 - r_r}{(C/N)_b} \right] \right.$$
$$\left. + k_l C_l \left[\frac{\eta}{(C/N)_l} - \frac{\eta r_h}{(C/N)_h} - \frac{1 - r_h - r_r}{(C/N)_b} \right] \right\} \le IMM_{max}. \tag{9.33}$$

Since immobilization is the conversion of inorganic nitrogen ions into organic forms operated by microorganisms, which incorporate mineral ions to synthesize cellular components (Brady and Weil, 1996, p. 405), it is assumed that it depends on the concentration of the microbial biomass and on the soil moisture, similarly to the decomposition process,

$$IMM_{max} = (k_i^+ N^+ + k_i^- N^-) f_s(s) C_b. \tag{9.34}$$

Two possible regimes thus exist for immobilization. In the first, immobilization is unrestricted, as the immobilization rate is lower than this maximum value, and the coefficient φ is equal to 1. This means that the bacteria can meet their nitrogen requirement and decompose the organic matter at a potential rate. The second regime occurs when the requirement of mineral nitrogen to be immobilized becomes higher than the maximum possible rate and thus the decomposition rate needs to be reduced. In the model, this is accomplished by reducing φ to a value lower than 1. By imposing the equality in Eq. (9.33), i.e., IMM $=$ IMM$_{max}$, one obtains the value of φ by which the decomposition rates must be reduced (Eqs. (9.19) and (9.22), respectively),

$$\varphi = -\frac{k_i^+ N^+ + k_i^- N^-}{k_h C_h \left[\dfrac{\eta}{(C/N)_h} - \dfrac{1 - r_r}{(C/N)_b}\right] + k_l C_l \left[\dfrac{\eta}{(C/N)_l} - \dfrac{\eta r_h}{(C/N)_h} - \dfrac{1 - r_h - r_r}{(C/N)_b}\right]}. \tag{9.35}$$

Equations (9.30) and (9.35) regulate the entire dynamics of decomposition, mineralization, and immobilization. Their functioning is illustrated by the numerical examples in Fig. 9.13. In the first example (the top rows of numbers in the figure) the litter C/N ratio is rather low (20) so that the decomposed organic matter from litter and humus (node A in Fig. 9.13) is nitrogen rich, i.e., it has an overall C/N ratio lower than 8. As a consequence, a fraction of the nitrogen must go into net mineralization ($\Phi > 0$), and the inequality in Eq. (9.33) is satisfied by $\varphi = 1$. In the second example $(C/N)_l = 30$, so that the C/N ratio in A is higher than 8. Immobilization of some mineral nitrogen is thus necessary to maintain a C/N ratio equal to 8. The rate of net immobilization is determined by Eq. (9.30). The inequality in (9.33) is satisfied because the resulting immobilization rate is lower than the maximum potential rate (IMM$_{max} = 1$), and φ is equal to 1. The third case presents an example of nitrogen-poor litter, $(C/N)_l = 40$, where the required net immobilization rate (equal to 1.5) is greater than IMM$_{max} = 1$. This means that the inequality in (9.33) is not satisfied and the decomposition rate must be reduced by imposing $\varphi = 1/1.5 = 0.67$, as is obtained from Eq. (9.35). With this value of φ, the C/N ratio of the fluxes into the biomass pool is equal to 8, and the immobilization rate is equal to IMM$_{max}$.

9.3.3 The Mineral Nitrogen

The balance of ammonium and nitrate in the soil is modeled as (see Fig. 9.12)

$$\frac{dN^+}{dt} = \text{MIN} - \text{IMM}^+ - \text{NIT} - \text{LE}^+ - \text{UP}^+$$
$$+ (1 - \eta)\left[1 - r_h\frac{(C/N)_l}{(C/N)_h}\right]\frac{\text{DEC}_l}{(C/N)_l} + (1 - \eta)\frac{\text{DEC}_h}{(C/N)_h}, \tag{9.36}$$

where the last two terms account for the nitrogen flux through the MIT pathway (see also Eq. (9.25)) and

$$\frac{dN^-}{dt} = \text{NIT} - \text{IMM}^- - \text{LE}^- - \text{UP}^-, \tag{9.37}$$

in which the rates of mineralization and immobilization have already been described. The nitrification rate can be modeled using first-order kinetics, i.e.,

$$\text{NIT} = f_n(s)k_n C_b N^+, \tag{9.38}$$

where C_b expresses the dependence of nitrification on microbial activity. The constant k_n defines the rate of nitrification, while $f_n(s)$ (nondimensional) accounts for the soil moisture effects on nitrification. As seen in Fig. 9.5, the optimum conditions for nitrification are very similar to those for decomposition, with the difference that the nitrification tends to zero at soil saturation (Linn and Doran, 1984; Skopp et al., 1990).

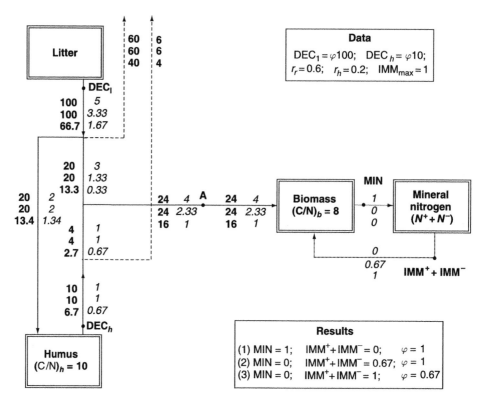

Figure 9.13 Numerical examples of decomposition, mineralization, and immobilization rates with the DIR scheme ($\eta = 1$). The three examples have the common features reported in the upper right corner of the figure, while they differ in their values of the C/N ratio of the litter pool, which is 20 in the first example (top rows of numbers), 30 in the second example (central rows), and 40 in the third examples (lowest rows). The boldface numbers refer to carbon fluxes and the italic numbers to nitrogen fluxes. The resulting mineralization and immobilization rates, along with the corresponding values of φ, are reported in the lower right corner of the figure. After Porporato et al. (2003a).

This behavior is reproduced by a linear increase up to field capacity followed by a linear decrease to soil saturation,

$$f_n(s) = \begin{cases} s/s_{\text{fc}} & s \le s_{\text{fc}}, \\ (1-s)/(1-s_{\text{fc}}) & s > s_{\text{fc}}. \end{cases} \tag{9.39}$$

For both ammonia and nitrate, leaching occurs when the nitrogen in the soil solution percolates below the root zone. It is thus simply proportional to the leakage term $L(s)$ modeled in Sec. 7.4,

$$\text{LE}^{\pm} = a^{\pm} \frac{L(s)}{snZ_r} N^{\pm}, \tag{9.40}$$

where the superscript plus and minus refer to ammonium and nitrate, respectively, and where $L(s)$ is divided by the volume of water per unit area, snZ_r, so that the term $L(s)/(snZ_r)$ assumes the dimension of the inverse of time. The nondimensional coefficients $a^{\pm}$, $0 \le a^{\pm} \le 1$, are the fractions of dissolved ammonium and nitrate, respectively, and are related to the corresponding solubility coefficients. Since nitrate is a mobile ion, a^- can be taken as equal to one, while a^+ is much lower because a large fraction of ammonium is absorbed by the soil matrix. This is why the leaching of ammonium is seldom important. At very low soil moisture levels, when liquid water becomes a highly disconnected system, the coefficients $a^{\pm}$ may be reduced by evaporation and become soil moisture dependent, $a^{\pm} = a^{\pm}(s)$. Although this is certainly not important for leaching, it may modify the water uptake at low soil moisture levels. The terms UP^{+} and UP^{-} in Eqs. (9.36) and (9.37) are the rates of ammonium and nitrate plant uptake, which are described below.

Nitrogen Uptake by Plants

The process of nitrogen uptake by plants is complex and not easy to model. Plants uptake their nutrients from the soil solution principally via the transpiration stream (*passive uptake*); they can also resort to forms of uptake which require more complicated physiological processes and are typically energetically expensive (*active uptake*). For both ammonium and nitrate, passive uptake and active uptake can be regarded as additive processes (Russell, 1931), i.e.,

$$UP^{\pm} = UP_p^{\pm} + UP_a^{\pm}. \tag{9.41}$$

The passive uptake can be assumed to be proportional to the transpiration rate, $T(s)$, and to the nitrogen concentration in the soil solution, i.e.,

$$UP_p^{\pm} = a^{\pm} \frac{T(s)}{snZ_r} N^{\pm}, \tag{9.42}$$

where, as already pointed out in the case of leaching, $T(s)$ must be divided by snZ_r for dimensional reasons and $a^{\pm}$ represent the fractions of dissolved inorganic nitrogen. As noted above, $a^{\pm}$ may be reduced at very low soil moisture levels, $a^{\pm} = a^{\pm}(s)$, but very little information seems to be available about this regard feature. Given the behavior of the transpiration rate (linearly increasing up to s^* and then constant with s), the passive uptake has the form shown in light gray in Fig. 9.14. The reason for the decrease above s^* is the dilution of the soil solution.

The active uptake mechanism is closely associated with the plant's metabolic processes (Russell, 1931; Engels and Marschner, 1995; Larcher, 1995). It is accomplished by establishing a concentration gradient between the root surface and the soil, which triggers a diffusion flux of the nitrogen ions. The intensity of the flux is limited by the gradient itself, which, in part, is controlled by the plant on account of its nitrogen demand and rate of passive uptake and by the diffusion coefficient. Accordingly, we assume that the plant tries to compensate the deficit with the active mechanism of uptake only if the passive uptake is lower than a given plant demand, $DEM^{\pm}$. When the diffusion of nitrogen ions into the soil is limiting, the active uptake is assumed to be proportional to the nitrogen concentration in the soil through a suitable diffusion coefficient; otherwise, the active uptake is simply the difference between the demand and the passive component (i.e., the total uptake satisfies the plant demand). Three possible cases can be assumed to occur in the representation of the active component (Porporato et al., 2003a):

$$UP_a^{\pm} = \begin{cases} 0 & DEM^{\pm} - UP_p^{\pm} \leq 0, \\ DEM^{\pm} - UP_p^{\pm} & 0 < DEM^{\pm} - UP_p^{\pm} \leq k_u N^{\pm}, \\ k_u N^{\pm} & DEM^{\pm} - UP_p^{\pm} > k_u N^{\pm}. \end{cases} \tag{9.43}$$

The term $k_u N^{\pm}$ expresses the dependence of the diffusive flux on the gradient of concentration between the root surface and the bulk of the soil, which at a first approximation is taken to be proportional to the nitrogen concentration in solution (i.e., the concentration within the roots is supposed to be nearly zero). The parameter k_u (day^{-1}) expresses the dependence of the diffusion process on soil moisture and can be modeled as

$$k_u = \frac{a^{\pm}}{snZ_r} Fs^d, \tag{9.44}$$

where the factor $a^{\pm}/(snZ_r)$ transforms the concentration in the soil into a concentration in solution, the term F is a rescaled diffusion coefficient (Russell, 1931), and d expresses the nonlinear dependence of the diffusion process on soil moisture. If one considers that the diffusion coefficient is often related to the product of the soil moisture and a tortuosity factor (which in turn has a quadratic dependence on soil moisture), a typical value of d is around 3. Porporato et al. (2003a) did not relate the coefficient F to the real diffusion coefficients for ammonium and nitrate, but merely chose F in such a way that the active contribution to nitrogen uptake was in the range 50%–80% of the total uptake at high values of soil moisture (Engels and Marschner, 1995, and references therein), as shown in Fig. 9.14.

When the nitrogen concentration is low compared with the plant demand (Fig. 9.14a), the active uptake is limited by diffusion and plants are under nitrogen deficit; on the other hand, when the nitrogen concentration is high (Fig. 9.14c), plant requirements are usually met, either by passive uptake alone at low soil moisture values (i.e., when the nitrogen in solution is more concentrated) or by both active and passive uptakes at higher soil moisture values. As already noted, the passive uptake may overtake the plant demand and, if this situation lasts for some time, the excessive nitrogen uptake may have harmful effects and decrease the plant's growth rate (Russell, 1931). At low soil moisture, the role of the "solubility coefficient" $a^{\pm}$ may also become important: in the case of a nonlinear dependence on soil moisture, a reduction in the passive uptake at low soil moisture appears (the dashed line in Figs. 9.14a–c).

It should be clear that DEM is an average value which is representative of the typical nitrogen requirement of a given species. Since it is difficult to distinguish between ammonium and nitrate, Porporato et al. (2003a) considered an overall nitrogen demand and then, according to the type of plant, split it proportionally into DEM$^+$ and DEM$^-$. When an ecosystem has reached stable conditions, the amount of uptaken nitrogen on annual timescales tends to balance the nitrogen returned to the soil in the form of plant litter (Russell, 1931; Larcher, 1995). It is thus reasonable to assume that the plant nitrogen demand is of the same order of magnitude (or eventually a little bit higher) as the average rate of nitrogen added with the litter. Such a linkage confers to both the rate of added litter and the plant nitrogen demand an essential regulative control of the rate of nitrogen recycling in the ecosystem, in which the physiological characteristics of plants and environmental conditions play an important role.

The present model of nitrogen uptake is in part a demand-driven one (e.g., Hansen et al., 1995), in agreement with the fact that the dependence of nitrogen uptake on nitrogen concentration is described by a Michaelis–Menten kinetics, corrected with a linear increase at high nitrogen concentration (e.g., Haynes, 1986; Engels and Marschner, 1995). Such a behavior corresponds to an increasing rate of uptake up to a saturation concentration, followed by a further increase at very high nitrogen concentrations. Figure 9.15 shows the total nitrogen uptake in our model as a function of the nitrogen concentration. The plateau where the uptake remains constant is found at DEM and its width depends on the soil moisture value. The linear increase that follows at high N concentrations, due to the passive uptake, presents a lower slope than the first part of the curve. Lower soil moisture values give a similar behavior, but with a narrower plateau.

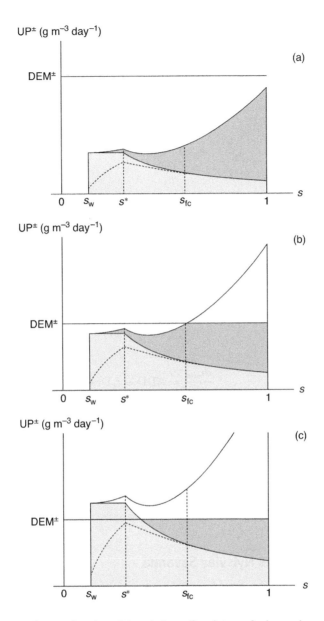

Figure 9.14 Plant nitrogen uptake as a function of the relative soil moisture s for increasing concentrations of mineral nitrogen from (a) to (c). The light gray regions represent the passive component of the uptake and the dark gray regions represent the active component. The dashed curves refer to the nonlinear reduction in the solubility coefficients $a^{\pm}$ with soil moisture, which lowers the passive uptake. After Porporato et al. (2003a).

Summary of the Model

The evolution of carbon and nitrogen in the soil is described by Eqs. (9.18), (9.20), (9.21), (9.24), (9.36), and (9.37), along with the conditions (9.28), (9.29), and (9.33) defining net mineralization and immobilization. The system is coupled with the soil moisture evolution equation, which accounts for the hydrologic forcing.

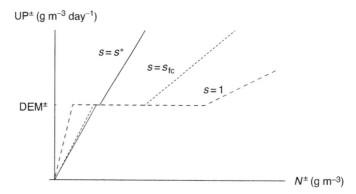

Figure 9.15 Nitrogen plant uptake as a function of nitrogen concentration, for constant values of soil moisture. After Porporato et al. (2003a).

By summing up the three equations for the carbon and the five equations for the nitrogen, we obtain corresponding differential equations for the total carbon and nitrogen in the system, $C_{tot} = C_l + C_h + C_b$ and $N_{tot} = N_l + N_h + N_b + N^+ + N^-$, respectively. In particular, from the sum of Eqs. (9.18), (9.21), and (9.24) one obtains

$$\frac{dC_{tot}}{dt} = \text{ADD} - r_r\text{DEC}_l - r_r\text{DEC}_h \tag{9.45}$$

and, from the sum of Eqs. (9.20), (9.23), (9.25), (9.36), and (9.37),

$$\frac{dN_{tot}}{dt} = \frac{\text{ADD}}{(\text{C/N})_{add}} - \text{LE}^+ - \text{UP}^+ - \text{LE}^- - \text{UP}^-. \tag{9.46}$$

The terms in Eqs. (9.45) and (9.46) represent the global gains and losses of the soil system. In this way, the continuity of nitrogen and carbon in the soil system is ensured, closely following the scheme of Fig. 9.12.

9.3.4   An Application to the Nylsvley Savanna

In this section, we follow D'Odorico et al. (2003) to present an application of the model discussed above to a broad-leafed savanna located in the Nylsvley region (South Africa), where the available data allow for an adequate testing of the model (Scholes and Walker, 1993).

The broad-leafed sites at Nylsvley are dominated by a vegetation field composed mainly of *Eragostris pallens* (herbaceous) and *Burkea africana* (woody) with a canopy cover of about 30%–40% and a discontinuous grass cover in the rest of the area. This vegetation is generally found on nutrient-poor, acidic soils, while fine-leafed savannas are typically observed on more fertile grounds (Scholes and Walker, 1993). Both in the fertile and in the unfertile sites, soils are sandy and about 1 m deep, though most of the organic matter and nutrients are concentrated in the top 80 cm.

Nitrogen may represent a limiting factor for productivity in the broad-leafed savanna, where the release of mineral nitrogen from the decomposition of litter and plant residues is quite slow (the average turnover time is about five years; Scholes and Walker, 1993). Most of the decomposition is due to soil

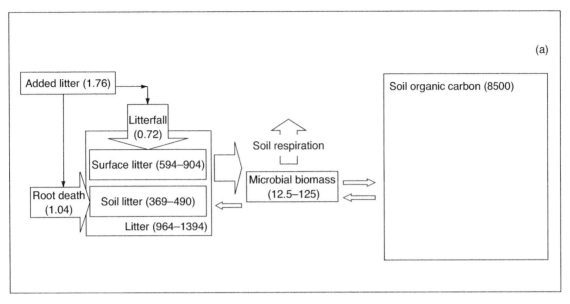

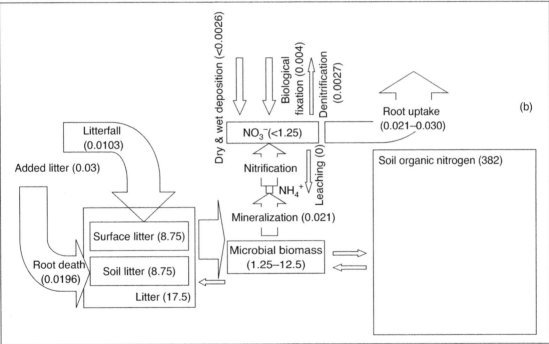

Figure 9.16 Mean carbon (a) and nitrogen (b) cycles in the broad-leafed savanna at Nylsvley (modified from Scholes and Walker, 1993); the concentrations of carbon are expressed as averages in an 80-cm-deep soil layer (in g m^{-3}), while the rates of the processes are expressed as daily means in a 242-day growing season (in g m^{-3} day^{-1}). The box sizes are approximately proportional to the pool sizes. After D'Odorico et al. (2003).

microbes (rather than fire oxidation or termites) and is mostly controlled by the soil water content. Figure 9.16a shows the mean annual carbon cycle in the broad-leafed sites at Nylsvley. The total carbon

Table 9.3 Parameters representative of the nutrient pools at Nylsvley (data from Scholes and Walker, 1993). Modified after D'Odorico et al. (2003).

Parameter	Symbol	Units	Value
Added litter (avg. rate)	ADD	$(\text{gC m}^{-2}\text{day}^{-1})$	1.5
Added litter (C/N ratio)	$(C/N)_{add}$	—	58
Microbial biomass(C/N ratio)	$(C/N)_b$	—	11.5
Humus (C/N ratio)	$(C/N)_h$	—	22

stock of the ecosystem is about $10,000$ g C m^{-3} and most of it is in soil organic matter. Figure 9.16b shows the mean annual cycle of nitrogen; the total nitrogen stock is about 400 g N m^{-3}, and more than 99% is in organic matter, with only a small percentage available to plants as mineral nitrogen. Inorganic nitrogen is present in the soil as nitrate (about 1.0 g N m^{-3}), while ammonium is found in much smaller concentrations, indicating an almost instantaneous nitrification. This suggests that the nitrogen cycle is limited by the slow rate of decomposition (i.e., by the availability of substrate for nitrification), while the process of nitrification does not exert any constraint. From the data in Fig. 9.16, it is possible to estimate the average values of the C/N ratios for the humus and biomass pools as well as the amount and the C/N ratio of the plant residues annually added to the soil as litter fall and dead roots (Table 9.3).

The soil microbes at Nylsvley tend to have short lifetimes and a strong sensitivity to water deficit. For this reason, the microbial biomass experiences water stress earlier than nutrient deficiency. As a consequence, apart from few exceptions reported at the beginning of the growing season, in normal conditions immobilization does not take place. The rate of nitrogen uptake, in the range of 5–7.25 g N m^{-3} y^{-1}, is in approximate equilibrium with the rate of mineralization. Considering also that the overall seasonal leaching losses are negligible, there is neither net seasonal accumulation nor depletion of nitrate in the soil (Scholes and Walker, 1993).

Hypotheses and Calibration of the Model

Since at Nylsvley the nitrogen cycle and plant activity cease almost completely during the dry, dormant, season, the nutrient dynamics was assumed to develop as a sequence of growing seasons in which the final condition at the end of a growing season becomes the initial condition of the following one (D'Odorico et al., 2003). For the sake of simplicity, litter is assumed to be incorporated at a constant rate throughout the growing season. Consistently with Scholes and Walker (1993), dry and wet deposition, nitrogen fixation, ammonia volatilization, and denitrification are neglected, leaving mineralization as the only source of NO_3^- (or NH_4^+), while the outputs of inorganic nitrogen are root uptake and, to a smaller extent, leaching and immobilization (Scholes and Walker, 1993).

The stochastic model of carbon and nitrogen cycles presented in the previous sections was calibrated by D'Odorico et al. (2003) to simulate the soil nutrient dynamics in the broad-leafed savanna at Nylsvley. All model parameters having a clear physical link with the hydrogeochemical processes were directly estimated from the data on soil, climate, vegetation, and nutrients available at Nylsvley (Table 9.3), while the remaining parameters were determined by calibration to reproduce the observed average size of the soil nitrogen and carbon pools as well as the mean rates of decomposition, mineralization, and root uptake (Table 9.4).

Table 9.4 Amounts of carbon and nitrogen in the nutrient pools, and rates of decomposition, mineralization, and root uptake observed at Nylsvley (data from Scholes and Walker, 1993). The carbon and nitrogen concentrations are expressed as the average of an 80-cm-deep soil layer, and the rates as averages throughout a 242-day growing season). After D'Odorico et al. (2003).

Variables	Units	Values
C in humus pool	$(g\,C\,m^{-3})$	8500
C in biomass pool	$(g\,C\,m^{-3})$	12.5–125
C in litter pool	$(g\,C\,m^{-3})$	960–1400
N in ammonium pool	$(g\,N\,m^{-3})$	≈ 0
N in nitrate pool	$(g\,N\,m^{-3})$	≤ 1.25
Average soil moisture	—	0.11
Rate of uptake	$(g\,N\,m^{-3}\,day^{-1})$	0.02–0.03
Rate of mineralization	$(g\,N\,m^{-3}\,day^{-1})$	0.021
Rate of litter decomp.	$(g\,C\,m^{-3}\,day^{-1})$	1.2

Table 9.5 Parameters of the model estimated by calibration for the case of the broad-leafed savanna at Nylsvley. After D'Odorico et al. (2003).

Parameters	Units	Values
a^+	—	0.05
a^-	—	1
DEM^+	$(g\,N\,m^{-3}\,day^{-1})$	0.2
DEM^-	$(g\,N\,m^{-3}\,day^{-1})$	0.5
d	—	3
F	$(m\,day^{-1})$	0.1
k_b	(day^{-1})	8.5×10^{-3}
k_l	$(m^3\,day^{-1}\,g\,C^{-1})$	6.5×10^{-5}
k_i^+	$(m^3\,day^{-1}\,g\,C^{-1})$	1
k_i^-	$(m^3\,day^{-1}\,g\,C^{-1})$	1
k_h	$(m^3\,day^{-1}\,g\,N^{-1}$	2.5×10^{-6}
k_n	$(m^3\,day^{-1}\,g\,C^{-1})$	0.6
r_h	—	$min[0.25;\ (C/N)_h/(C/N)]|_l$
r_r	—	0.6
η	—	1

The values of the constants k_l, k_h, and k_b, for the first-order kinetics of litter and humus decomposition and the death of microbial biomass (Table 9.5) were estimated using the steady-state solution of Eqs. (9.18), (9.21), and (9.24) along with the average litterfall rate (Table 9.3), and the carbon storage in litter (C_l), humus (C_h), and biomass (C_b) pools (Table 9.4). When the temporal derivatives in those equations are set to zero and the values of the other variables are assigned as in Tables 9.3 and 9.4 for the average conditions observed at Nylsvley, the solution of this linear system leads to the values of k_l, k_h, and k_b

reported in Table 9.5. The nondimensional factor $f_s(s)$ expressing the effect of soil water content on the rate of these first-order kinetics was estimated using Eq. (9.8) with an average soil moisture of 0.12.

The estimation of the value of the first-order constant of nitrification, k_n, (Eq. (9.38)) is facilitated by the lack of ammonium and by the high nitrification rates. Being limited by the supply of substrate (i.e., NH_4^+) rather than by microbial activity, nitrification was not very sensitive to k_n as long as an adequately large value is used. For the same reason, owing to the limited amount of ammonium, the values of the parameters k_i^+ and k_i^- (the different susceptibilities of ammonium and nitrate to losses by immobilization, see Eq. (9.31)) are irrelevant to the overall dynamics since most of the immobilization is contributed by NO_3^-.

The parameters F and $DEM^\pm$ for the passive uptake (Eqs. (9.43) and (9.44)) were estimated so as to reproduce the observed rates of root uptake, while the tortuosity factor, d, was assumed to be equal to 3. The isohumic coefficient, r_h, representing the fraction of decomposing litter undergoing humification, was taken as the lower of 0.25 and the ratio $(C/N)_h/(C/N)_l$. Similarly, the constant r_r (i.e., the portion of decomposing carbon that is lost by respiration), was chosen as equal to 0.6. Nitrate is assumed to be completely soluble in the soil solution only when the soil moisture is above field capacity (i.e., $a^- = 1$), while below s_{fc} the solubility coefficient was assumed to decay according to a power law with exponent $g = 0.5$. The same behavior was assumed for the solubility of ammonium, but a smaller value was selected for a^+ to account for the absorption of ammonium by the soil matrix and the consequent low mobility of these ions (Table 9.5). However, owing to the low amounts of NH_4^+ in the soil and the consequently low rates of ammonium uptake and leaching, the particular value used for this parameter is irrelevant to the overall dynamics.

Temporal Dynamics

The input of precipitation was modeled by D'Odorico et al. (2003) as a marked Poisson process with Nylsvley parameters. Depending on the inertia of the various pools and on the degree of dependence on soil moisture, the random fluctuations imposed by precipitation were filtered by the temporal dynamics of the state variables in a very interesting manner. Some variables (e.g., NO_3^-) preserved much of the high-frequency variability imposed by the random forcing of precipitation, while some others (C_h, C_l, and C_b) showed smoother fluctuations (see Fig. 9.17).

Nitrate dynamics is the final product of a number of intertwined processes in which both high- and low-frequency components interact. In particular, the high-frequency component of NO_3^- fluctuations (with a period of days to weeks) can be linked to the direct dependence of mineralization and nitrification on soil moisture, which transfers the random fluctuations of rainfall forcing to the budget of NO_3^-. On the other hand, the low-frequency variability (with a period of seasons to years) resembles that of organic matter (in particular that of the microbial biomass) and depends on the inertia imposed on the dynamics by the dimensions of the soil carbon and nitrogen pools, which are very large compared with the fluxes (see Fig. 9.16).

The different timescales occurring in the carbon and nitrogen cycles were analyzed by means of the power spectra of the relevant variables. Figure 9.18 shows a logarithmic plot of the normalized spectral densities of s, C_b, and NO_3^-. As expected, the energy associated with the high frequencies is higher for soil moisture than for nitrate and microbial biomass, while the converse is true for the low frequencies. The crossing between the C_b (or NO_3^-) and s power spectra is located at frequencies corresponding to periods of seasons to years. The model thus reproduces the change in timescales occurring from the soil moisture dynamics to the nutrient dynamics observed by Schimel et al. (1997). This indicates that, while in semi-arid ecosystems the soil moisture is not able by itself to provide memory to the system at scales

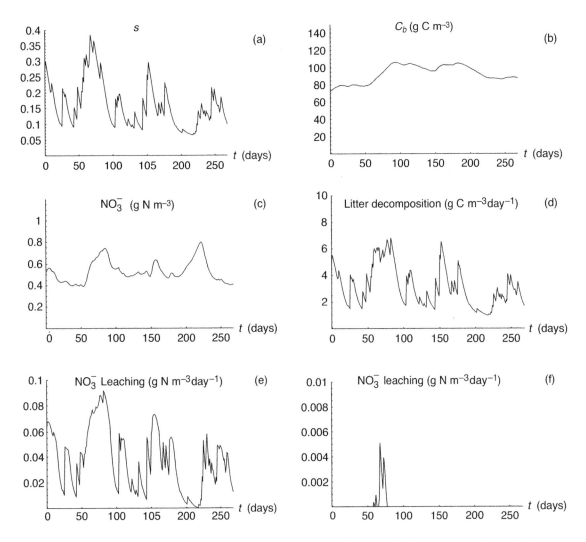

Figure 9.17 Temporal dynamics of soil moisture (a), carbon biomass (b), nitrate (c), litter decomposition (d), nitrate uptake (e), and nitrate leaching (f) simulated during a few growing seasons at Nylsvley. After D'Odorico et al. (2003).

larger than one year (owing to the complete depletion of soil water content by the end of the dry season, e.g., Nicholson, 2000), the nutrient and vegetation dynamics may have a much longer memory, responsible for the interannual persistency observed both in the hydroclimatic and ecosystem processes (Parton et al., 1988; Schimel et al., 1996, 1997).

Analysis of the cross-correlation coefficients of the daily series (Table 9.6) shows that the soil moisture, though it is the driving force of the system, is poorly correlated at lag zero to all the other variables, owing to the different timescales of the fluctuations. For similar reasons, there is a strong correlation between the amounts of nitrate and microbial biomass in the soil, while the aforementioned negative correlation of C_l and C_b propagates to the nitrate content. The correlation coefficients of the carbon humus are very low, because C_h tends to maintain a quasi-constant value with almost no interaction with the other variables. The correlation coefficients for ammonium are not reported because of the extremely low amounts of ammonium characterizing the nitrogen cycle at Nylsvley.

Table 9.6 Cross-correlation coefficients among the state variables relevant to the soil carbon and nitrogen cycles. After D'Odorico et al. (2003).

	Soil moisture	Carbon (litter)	Carbon (humus)	Carbon (biomass)	Nitrate
Soil moisture	1	−0.05	0.03	0.17	0.14
Litter	—	1	0.17	−0.63	−0.64
Humus	—	—	1	0.12	0.16
Biomass	—	—	—	1	0.86

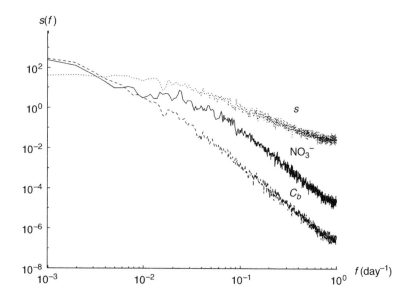

Figure 9.18 Power spectra of soil moisture, nitrate, and microbial biomass. After D'Odorico et al. (2003).

The low cross-correlation coefficient between s and NO_3^- is somewhat surprising, especially when compared with the results in Fig. 9.17. In fact, although the resemblance of the time evolution of the two variables over short timescales (Fig. 9.17) seems to suggest a certain degree of correlation, the s-independent low-frequency fluctuations of nitrate reduce the cross-correlation (Table 9.6). Despite the relatively low values of cross-correlation, such regular behavior is a symptom of the existence of a persistent (though infrequent) link between soil moisture and nitrate dynamics. A closer inspection of the time series of soil moisture and nitrate reveals that such a phenomenon is due to the presence of sudden flushes of nitrate following a prolonged wet period after a drought (see Fig. 9.19). In these conditions, in fact, first the dry soil hinders decomposition and favors SOM accumulation, then the subsequent wet period elicits biomass growth and enhances mineralization (Cui and Caldwell, 1997). Episodic changes in nitrate levels greatly influence plant growth, because the plant response to increased availability of nitrogen tends to be very quick (e.g., Brady and Weil, 1996). These pulses of nitrate are therefore of considerable importance for natural ecosystems. Their modeling would not have been possible using a model based on monthly estimates of soil moisture: the daily temporal resolution is thus fundamental to capturing the impact of soil moisture on nutrient dynamics.

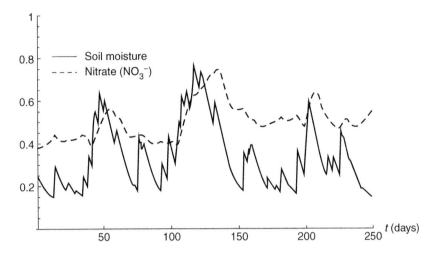

Figure 9.19 Response of the soil nitrate pool to fluctuations in the soil water content. After D'Odorico et al. (2003).

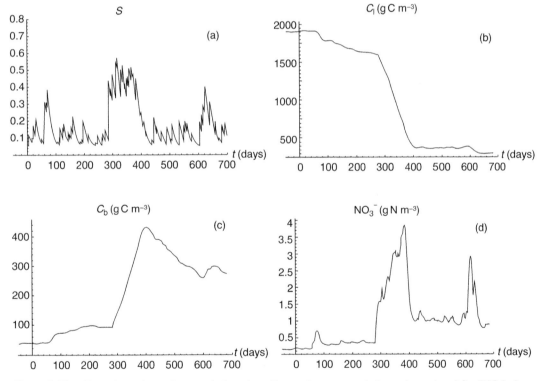

Figure 9.20 Effect of a prolonged wet period on the soil organic matter and nitrate dynamics. After D'Odorico et al. (2003).

A further example of the importance of the stochastic characterization of the model is given in Fig. 9.20, where an exceptionally long wet period is shown to cause a dramatic increase in the microbial biomass activity and nitrate and a consequent decrease in the substrate pools (C_l and, not shown, C_h). However, the high nitrate concentration is soon depleted by plant uptake and leaching and, in the long run, the event

Table 9.7 Cross-correlation coefficients between the state variables and the fluxes of the soil carbon and nitrogen cycles. After D'Odorico et al. (2003).

	Decomposition	Soil respiration	Net mineralization	NO_3^- uptake	Leaching
s	0.81	0.80	0.74	0.78	0.45
C_l	−0.10	−0.16	−−0.43	−0.35	−0.06
C_h	0.22	0.21	0.18	0.15	0
C_b	0.49	0.54	0.66	0.55	0.10
NO_3^-	0.42	0.46	0.62	0.54	0.10

turns to be unfavorable to plants in terms of nitrogen availability because of the marked depletion of the SOM substrate.

The rates of decomposition reported in Fig. 9.17 also present some interesting features. Owing to their direct dependence on soil water content (see Fig. 9.5), they are subject to a high-frequency variability which is closely coupled to that of soil moisture (see Fig. 9.17, and the very high correlation coefficients in the first row of Table 9.7). The decomposition and mineralization fluxes are also correlated with C_b but are negatively correlated to the soil litter content. The average seasonal rates of soil respiration (not shown) are qualitatively similar to the litter decomposition rate, with fluctuations around an average value of 5 g C m^{-3} d^{-1}, which is in good agreement with the Nylsvley observations (Scholes and Walker, 1993).

Since the prevailing arid conditions hamper the diffusive transport of NO_3^- in the soil, its total flux is mainly made up by the passive component, which in turn is mostly controlled by the interaction of the soil moisture and nitrate through the process of transpiration. The nitrogen concentration is always so low that the plant nitrogen demand is never met, in agreement with the fact that the broad-leafed savanna at Nylsvley is a nitrogen-deficient ecosystem (Scholes and Walker, 1993). The dry conditions at Nylsvley are also responsible for the very infrequent occurrence of leaching (Fig. 9.17).

9.4   Outlook

The propagation of intermittent hydrologic forcing through the soil–plant system, with the resulting pulses of microbial activity, is crucial for the C and N cycles and in general for biogeochemical processes. A clear example of the resulting intermittency was seen in Fig. 6.1, with pulses of CO_2 concentration in the soil in response to soil moisture jumps. Throughout this chapter, we focused on the role of soil moisture, and in particular its fluctuations, on the CN dynamics. We saw the sporadic bursting of leaching events (Fig. 9.17), as well as the delayed pulses in the modeled evolution of soil nitrate concentration (Fig. 9.19). Note 9.4 provides further references related to these pulsing dynamics. The impacts of soil water on microbial dynamics and the carbon and nitrogen cycles, in turn, have manifold ramifications not only on vegetation but also on soil evolution, climate dynamics, agricultural production, and water quality.

9.4.1   Water and Soil Microbial Life

In this section, we discuss in more depth the relationship between soil moisture and microbial processes, going from water stress in arid conditions to anoxic conditions in very wet soils. We begin by revisiting

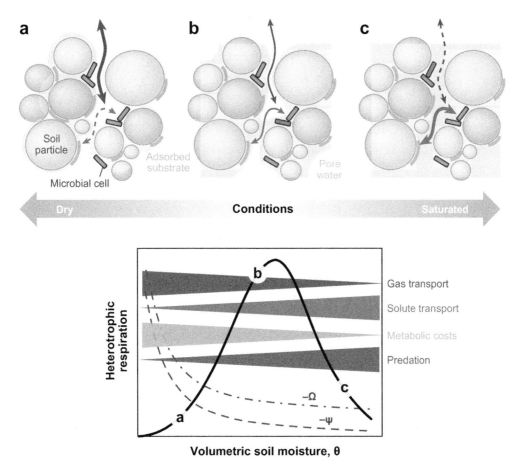

Figure 9.21 Soil moisture effects on microbial activity. In the lower panel, the relationship between soil respiration and moisture (black line) results from interacting effects including diffusion and physiological, biochemical, and ecological processes (Schimel, 2018); ψ indicates the soil water potential and Ω is the cell osmotic potential that would allow a stable turgor pressure to be maintained as ψ declines. After Moyano et al. (2013).

and deconstructing the dependence of microbial activity on soil moisture as shown in Fig. 9.21 (see also Fig. 9.5).

On the dry end of the spectrum, microbial activity is reduced by water stress due to low water potential. In a meta-analysis Manzoni et al. (2012) summarized the soil moisture threshold corresponding to the point when biological activity ceases, determining a water potential of about -14 MPa in mineral soils and -36 MPa in surface litter. These thresholds correspond to soil moisture values for which solute diffusion becomes strongly inhibited, while in litter it is related to the dehydration of microbes rather than diffusion.

Similarly to what we have seen for plants (Secs. 4.9 and 8.1), soil microbes have developed a variety of *strategies* to tolerate or avoid water stress (Manzoni et al., 2014). Drought conditions limit the supply of vital substrates by inhibiting diffusion in dry conditions and these in turn affect carbon and nitrogen cycling in soils. Since drought resistance achieved by active osmoregulation requires large C investment, it tends not to be useful in soils where growth in dry conditions is limited by C supply. In contrast, dormancy

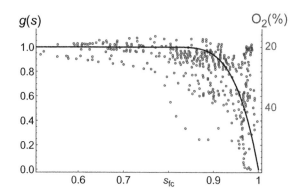

Figure 9.22 The solid line illustrates the control of soil moisture on oxygen, $g(s) = 1 - ((s - s_{fc})/(1 - s_{fc}))^4$ for $s > s_{fc}$; the circles are oxygen/soil-moisture measurements from the Luquillo Experimental Forest. After Calabrese and Porporato (2019).

followed by rapid reactivation upon rewetting seems to be a better strategy in such conditions. Synthesizing more enzymes may also be advantageous because it causes a larger accumulation of depolymerized products during dry periods that can be used upon rewetting (Manzoni et al., 2014).

On the wet end of the spectrum, *oxygen limitation* reduces microbial carbon-use efficiency and microbial growth. While there are some beneficial aspects of anaerobiosis (N_2 fixation is usually more significant, mineralization is retarded, degradation of certain pesticides is accelerated, and certain plant pests are reduced), anaerobiosis typically has detrimental effects, including root stress by anoxia, the loss of N by denitrification, the production of organic acids, H_2S and other plant toxicants, and favored conditions for the development of root pathogens (Tiedje et al., 1984). In general, anaerobic microorganisms have a mechanism to generate energy (and thus grow) other than by coupling electron transport to oxygen reduction, but some anaerobes (called obligate anaerobes) are also inhibited by oxygen because of its toxicity. Anaerobiosis occurs in soil when oxygen consumption rate exceeds the supply rate. The oxygen supply rate is a strongly nonlinear function of moisture content (see Fig. 9.22) and this behavior depends on the physical characteristics of the soil, especially its porosity.

Table 9.8 lists anaerobic processes according to their approximate sequence of occurrence as the redox decreases. It should be clear that, because of soil heterogeneities, some portions of soils are wetter than others and therefore one could expect different moisture conditions within a soil volume and thus the simultaneous presence of some of these processes. We follow Tiedje et al. (1984) in the description of these processes. The first four processes in Table 9.8 take place in soils with temporary anaerobic microsites, e.g. well-drained soils, and are carried out by facultative anaerobes which can shift from aerobic to anaerobic growth. The dissimilatory reduction of NO_3^- to NH_4^+ is carried out by both facultative and obligate anaerobes and is less prevalent. The last four processes are carried out by obligate anaerobes and become significant under flooded or intensively anaerobic conditions. The last three processes in the table occur when organic matter is converted to methane by lithotrophic bacteria that reduce CO_2 with H_2 to form methane, and by bacteria that split acetate into CH_4 and CO_2. Proton reduction is carried out by acetogens that anaerobically oxidize butyrate and propionate to acetate. These can utilize these organic acids in the absence of external electron acceptors. If methane is produced, usually these acetogens are also active; if not, the increased acidity would inhibit methanogenesis.

Table 9.8 Anaerobic microbial processes and their reaction products (OM = organic matter). Only the major reduction products are shown; oxidized products are also produced, usually CO_2 if the electron donor is an organic compound. After Tiedje et al. (1984).

Process	Reaction
Fe^{3+}, Mn^{4+} reduction	$OM + Fe^{3+}, Mn^{4+} \rightarrow Fe^{2+}, Mn^{2+}$
Denitrification	$OM + NO_3^- \rightarrow N_2O, N_2$
Fermentation	$OM \rightarrow$ organic acids (esp. acetate and butyrate)
Nitrate respiration	$OM + NO_3^- \rightarrow NO_2^-$
Dissimilatory NO_3^- reduction to NH_4^+	$OM + NO_3^- \rightarrow NH_4^+$
Sulfate reduction	OM or $H_2 + SO_4^{2-} \rightarrow S^{2-}$
Carbon dioxide reduction	$H_2 + CO_2 \rightarrow CH_4$, acetate
Acetate splitting	Acetate $\rightarrow CO_2 + CH_4$
Proton reduction	Fatty acids and alcohols + $H^+ \rightarrow$ H_2 + acetate + CO_2

To rationalize the observed patterns in the decomposition of SOM, LaRowe and Van Cappellen (2011) included the thermodynamic properties of organic compounds in kinetic models of organic matter degradation. They showed that the persistence of cell-membrane-derived compounds and complex organic compounds in anoxic settings is consistent with their limited catabolic potential under these environmental conditions. Understanding the interactions between these thermodynamics constraints, related to catabolic energy yields and other biotic and abiotic limitations, will help to unravel the mechanisms responsible for the degradation, transformation, and ultimate preservation of organic matter (LaRowe and Van Cappellen, 2011). Because of their implications for climate, the role of these processes on carbon emissions to the atmosphere is currently under intense investigation. We refer to the Notes at the end of the chapter for more references on denitrification, iron cycling, and methane production in wet soils.

9.4.2 From Local Processes to Global Trends: Modeling Across Scales

In spite of the multitude and complexity of the processes that preside over SOM decomposition in different soil moisture conditions, the trends of *global carbon on land* turn out to be surprisingly regular. When plotted as a function of the dryness index (D_I, the ratio of the long-term potential evapotranspiration and rainfall), the global distribution of carbon ($\langle C_{tot} \rangle$, the sum of SOM and biomass) follows a clearly defined monotonic decay (Fig. 9.23). Such a trend is well approximated by a power-law relation:

$$\langle C_{tot} \rangle = aD_I^b, \tag{9.47}$$

where $a = 139.1 \text{ Mg ha}^{-1}$ and $b = -0.4$.

The striking regularity of this interesting trend is reminiscent of the collapse onto the Budyko curve for hydrologic partitioning; see Sec. 7.6. Indeed, some general features can be understood as the combined effect of vegetation and soil microbial response to soil moisture conditions. On the one hand, the

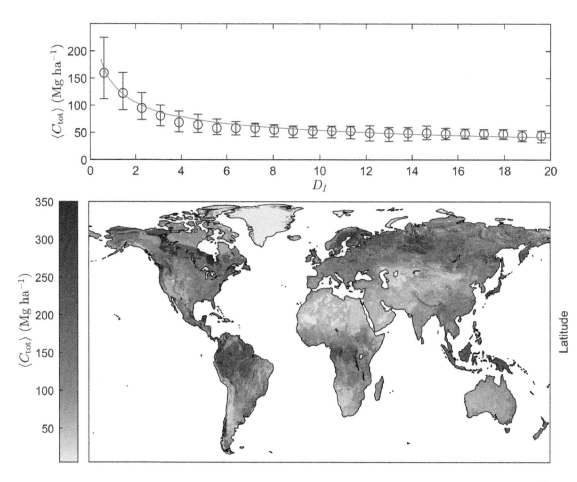

Figure 9.23 Global distribution of carbon (the sum of the carbon in soil organic matter and in biomass, Mg C ha^{-1}) and its dependence on dryness index (D_I). Data were obtained from Hiederer and Köchy (2011) and Spawn et al. (2020). The lines shows the 50th percentile, while the error bars show the 25th and 75th percentiles.

increase in carbon input through photosynthesis increases with soil moisture but tends to plateau in wet conditions (see, e.g., daily timescale assimilation as a function of soil moisture in Sec. 5.4.5 and the long-term average in Sec. 8.1.3). On the other hand, the carbon loss due to decomposition is linked to the soil microbial response, which instead declines as soon as anoxic conditions start appearing, thus favoring the accumulation of carbon.

A more detailed and quantitative justification of this global trend and of its variability would certainly be of great interest; upscaling these dynamics from the pulsing dynamics at the local scales to the global and multi-decadal scales of climatic interest remains an important area for future research. Numerous modeling efforts have been employed towards this goal: in 2009, a review of the literature showed a 6% annual increase in the number of models related to soil carbon and nitrogen cycles (Manzoni and Porporato, 2009). The current spectrum goes from very detailed models at the nanoscale, including genomic information on microbial communities (see Note 9.11 on soil omics), to models for agricultural and environmental engineering applications, and then all the way to models for global simulations. These models have to confront several challenges related to the number of processes and scales involved. However, while it is logical to expect different model architectures and process details, often their mathematical features tend to be linked more to the type of application rather than to the spatial and temporal scales of interest. It is therefore important that these efforts are guided by principles of modeling balance, whereby different processes are described with similar levels of detail and coupled consistently, including the effects of unresolved scales, so that not only is the essential response of interest captured but also distortions (and sometimes exaggerated feedbacks from other processes) are avoided.

9.5   Key Points

- The decomposition of organic matter releases CO_2 to the atmosphere and mineral nitrogen into the soil; it is controlled by both the quality of the organic matter and the environmental conditions, chiefly temperature and soil moisture.
- The rate of microbial activity as a function of soil moisture has a humped shape, initially increasing because of a reduction in water stress, then reaching a maximum at around field capacity, and finally decreasing because of anoxic conditions.
- The soil moisture regime has a profound impact on the dynamics of the soil carbon and nitrogen; in a minimalist predator prey model of substrate microbial dynamics, favorable soil moisture conditions change the stability of the fixed point from a spiral to a node.
- For litter cohorts that remain unmixed, N immobilization and mineralization conditions are described by universal curves; these are parameterized by the litter quality (i.e., its C/N ratio), while the environmental conditions only affect the speed at which these curves are traveled in time.
- When coupled to soil moisture dynamics, simulations of detailed models of carbon and nitrogen cycles reveal a multiplicity of responses at different timescales, some of which have long-term effects.
- In spite of the numerous and complex microbial processes responsible for organic matter decomposition in soils with different levels of saturation, the global trends of organic matter present a regular power-law decay with dryness index.

9.6   Notes, including Problems and Further Reading

9.1 Compute the steady-state solutions of C_s and C_b in the predator–prey model of Eqs. (9.1) and (9.2) and plot them as a function of soil moisture and temperature for the three decomposition functions (i.e., Eqs. (9.4), (9.5), and (9.6)).

9.2 Figure 9.9a shows the eigenvalues of the stability matrix of the predator–prey model with a nonlinear, multiplicative, decomposition function. Plot the corresponding diagrams for the linear and Michaelis–Menten decomposition functions (i.e., Eqs. (9.4) and (9.6)).

9.3 Use the stoichiometric model in Sec. 9.2.3 with the linear decomposition function of Eq. (9.6) to compute the time to reach the end of the immobilization phase (i.e., $C_{tot}(t)/N_{tot}(t) = (C/N)_{CR}$) for $s = 0.2$

and $s = 0.5$ at room temperature. Plot the carbon and nitrogen content as a function of time for these two cases.

9.4 (Water-Induced Soil-Biogeochemistry Pulses) As we have seen in this chapter (see also Fig. 6.1), drying and rapid rewetting events often result in soil respiration pulses with complex biogeochemical responses (Austin et al., 2004). See Daly et al. (2008) and Manzoni et al. (2020) for stochastic analyses of these pulses.

An intermediate pool of dissolved organic matter (DOM) is often required to model the effects of rewetting events on respiration pulses (Lawrence et al., 2009; Manzoni and Katul, 2014; Manzoni et al., 2019). These respiration pulses, known as the Birch effect (Birch, 1958), are attributed to osmolytes being rapidly released to avoid cell lyses, dead microbial cells, and DOM previously inaccessible to microbes due to low diffusivity in dry conditions. Their impact on the overall SOM budget can be large. It has been hypothesized that DOM accumulates in dry conditions because it remains inaccessible to microbes until rewetting; DOM is also involved in other poorly characterized processes (e.g., the plant uptake of organic nitrogen in small-molecular-weight molecules (Schimel and Bennett, 2004) and the leaching of organic carbon and nutrients). See Brangarl et al. (2020) for a detailed model of these processes.

9.5 (Stochastic Substrate-Microbe Water Dynamics)   Couple the minimalist stochastic soil moisture model of Sec. 7.5 to the substrate–microbe model of Sec. 9.2.1. Perform a numerical simulation of the system and analyze the time series and the steady-state PDFs of the state variables of the respiration flux.

9.6 (Substrate-Microbe Water Dynamics) Extend the substrate-microbe model of Sec. 9.2.1 to include two microbial pools decomposing the same substrate: a fast-growing but water-stress-prone bacterial pool and a slow growing but drought-tolerant fungal pool. Assume a linear dependence of the growth-rate constant with soil moisture for the different rates, starting from two different values of microbial wilting point.

9.7 (The Critical Zone: Soil Carbon and Soil Formation) The coupled carbon and water fluxes percolating from the soil–plant system play an important role in the evolution of the deeper soil layers, the saprolite and the fractured bedrock (see Sec. 3.1). By changing the pH and the chemical composition of the soil and bedrock water, the fluxes accelerate mineral dissolution, promoting soil formation, the evolution of different soil orders, and the uptake of atmospheric carbon (Maher and Chamberlain, 2014). Coupling ecohydrological models to reactive transport models is crucial for explaining and predicting several critical-zone processes (Li et al., 2017). Calabrese et al. (2017) presented a dynamical system coupling water- and carbon-balance equations to the law of mass action for weathering reactions.

9.8 (Hydrologic Controls on Soil Nitrogen Oxide Emissions) As mentioned in Sec. 9.1.2, at high soil moisture values denitrification causes nitrogen oxide emissions into the atmosphere, which impact the atmospheric chemical and radiative properties. Nitric oxide (NO) is very reactive in the troposphere and takes part in reactions leading to the formation of acid rain, as well as to the production of ground-level ozone (Logan, 1983). In the troposphere, nitrous oxide (N_2O) is stable and contributes to the greenhouse effect, with a global warming potential 200 times that of carbon dioxide on a per-molecule basis (Williams et al., 1992). Conversely, in the stratosphere N_2O is reactive and participates to the destruction of ozone (e.g., Davidson, 1991).

The process of microbial denitrification consists of anaerobic respiration, leading to the reduction of NO_3^- and NO_2^- to NO, N_2O, and N_2. Soil moisture affects the emission rate and its partitioning between NO, N_2O, and N_2, in at least three ways: (i) through control the soil aeration (i.e., rates of nitrification/denitrification), (ii) through the regulation of the substrate diffusion, and (iii) through control of the physical transport of gases within the soil (Skopp et al., 1990; Davidson, 1991). There is general agreement that the nitrogen oxide emissions are nonmonotonic functions of soil moisture: their rate increases with increasing values of soil water content up to a maximum value and then decreases (Cardenas et al., 1993; Yang and Meixner, 1997; Crill et al., 2000). Conversely, the rate of N_2 emission is an increasing function of s and occurs only with high soil moisture content. The actual values of emissions can in general depend on other ecosystem properties (such as soil temperature and pH, amount and type of microbial biomass, land cover and use, etc.); however, these factors do not significantly affect the shape of the functions representing the dependence on soil water content.

Using the soil moisture model in Chapter 7, coupled to a functional dependence between soil moisture and the normalized rates of emission (Fig. 9.24), Ridolfi et al. (2003) derived probability distributions for the various fluxes from the PDF of soil moisture, providing an indication of the relative importance of hydrological processes on N-oxide emissions at the daily timescale. Their results show a strong dependence of the N-oxide fluxes on the climate, soil, and vegetation characteristics, which are generally consistent with the field measurements in water-controlled ecosystems (Otter et al., 1999). For more detailed estimates, these microbial processes can be coupled to the full dynamic model of Sec. 9.3.

9.9 (Iron-Redox Cycles) Humid tropical forests play a key role in the global carbon cycle, not only for their high levels of primary productivity but also for the high rates of soil respiration (Dubinsky et al., 2010; Scharlemann et al., 2014; Yang and Liptzin, 2015; Barcellos et al., 2018; Calabrese et al., 2020). A peculiarity of humid tropical soils is that the soil C dynamics interacts with iron (Fe) minerals and the Fe-redox cycle (Lovley, 1991; Liptzin and Silver, 2009; Dubinsky et al., 2010).

During periods of high soil wetness, anaerobic processes such as Fe reduction (from Fe^{III} to Fe^{II}) support organic C decomposition. By contrast, low soil wetness favors the formation of Fe oxides (Fe^{III})

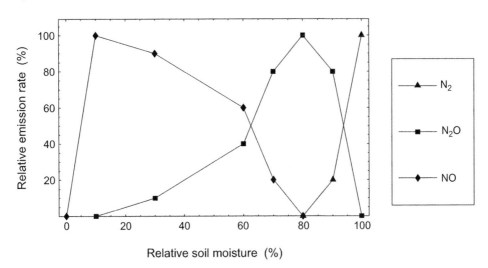

Figure 9.24 Normalized rate of nitrogen emissions as a function of relative soil moisture. Redrawn after Meixner and Eugster (1999).

that protect C and make it inaccessible to soil microorganisms. This link between Fe-redox and C cycles can control C decomposition, owing to the large soil Fe concentration and the frequent occurrence of soil saturated conditions (e.g., Parkin, 1987). As a result, while O_2-limited conditions generally inhibit organic C decomposition, in humid tropical soils O_2 is readily replaced by Fe^{III} as electron acceptor and, once Fe^{III} is reduced, the release of protected C could stimulate further decomposition.

These processes challenge the traditional view of high soil moisture limitations of microbial activity (e.g., Huang and Hall, 2017). A more in-depth understanding of the hydrologic forcing on the Fe-redox cycle is fundamental to predicting how these soils will respond to increasing disturbance, such as warming, land-use change, and droughts (Cusack et al., 2016).

9.10 (Methane Emissions) In very wet soils with anaerobic conditions, the low soil redox potential creates a favorable environment for methanogenesis (Le Mer and Roger, 2001). While part of the CH_4 produced is oxidized, e.g., by oxygen that reaches the soil through the rhizosphere, that remaining is emitted to the atmosphere by diffusion, by ebullition, and especially by plant-mediated transport through the aerenchyma pathway. We refer to Rizzo et al. (2013) and Souza et al. (2021) and references therein for more information on the modeling and the environmental engineering applications of these processes.

9.11 ('Soil Omics') Soil microbial diversity is increasingly analyzed in terms of its genetic differences. Novel techniques, called 'omics', may allow genomics-based metabolic predictions for these microbial communities. Linking environmental gradients and genome-resolved metabolic characterization, these techniques may help unfold specific roles in complex carbon and nitrogen transformations in soils (Diamond et al., 2019). See Pagel et al. (2016) for a modeling effort that starts to incorporate this type of information.

10 Ecohydrology of Agroecosystems

La non-prise en compte des savoirs locaux constitue souvent un frein aux politiques de conservation et d'utilisation durable de la biodiversité dans ces aires.

Failure to take local knowledge into account often hinders policies for the conservation and sustainable use of biodiversity.

Mbayngone and Thiombiano (2011)[†]

> The coupling of ecohydrologic processes and society has led to an acceleration of the biogeochemical cycles often leading to land degradation. In this chapter, a minimalist dynamical system is used to describe the socio-economic pressure on the soil–plant system and its feedbacks causing land degradation and abandonment. We then consider the stochastic water balance for different irrigation methods and the probabilistic dynamics of dissolved minerals, with reference to the problem of soil salinization and phytoremediation. The second part of the chapter deals with examples of stochastic optimization in agroecosystems in the presence of hydrologic forcing, including optimal irrigation and fertilization as well as the economically optimal design of rainwater-harvesting cisterns. We conclude with an overview of optimal control for sustainable soil and water resources, with a focus on aquifer exploitation and sodic soil remediation.

10.1 Coupled Natural–Human Agroecosystems

Human societies are increasingly altering the water and biogeochemical cycles to improve ecosystem productivity and food security while reducing the risks associated with the unpredictable variability of climatic drivers. Agroecosystems are the main stage in this acceleration, comprising 12% of the global land, an area of 16 million square kilometers, equivalent to Brazil and Australia combined. The alterations to ecohydrological processes have the potential to cause dramatic environmental consequences, raising the question how societies can achieve a *sustainable use of natural resources* for the future while ensuring food security for a growing population. In this chapter, we discuss how ecohydrological modeling may help us to better understand and address some of these broad questions; we follow in part Porporato et al. (2015).

[†]E. Mbayngone and A. Thiombiano (2011). "Degradation of protected areas through the use of plant resources: the case of the partial wildlife reserve of Pama, Burkina Faso (West Africa)." *Fruits (Paris)* 66.3, pp. 187–202.

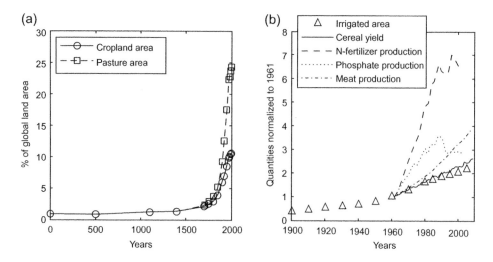

Figure 10.1 (a) Increased human pressure on land in the last two millennia (data from Klein Goldewijk et al., 2011), and (b) acceleration of biogeochemical cycles by agricultural intensification (in terms of cereal yield, irrigated land, fertilization, and meat production) in the last century (irrigated land data from Siebert et al., 2015). After Porporato et al. (2015).

10.1.1 Acceleration of Ecohydrological Processes and Land Degradation

Humans have improved the rate of extraction of food, feed, and fiber from ecosystems by managing vegetation, engineering landscapes, soils, and drainage systems and by intensively controlling the quantity and quality of water, carbon, and nutrient fluxes in and out of ecosystems (Vitousek et al., 1997; Rojstaczer et al., 2001; Haberl et al., 2007), resulting in increasingly stronger human feedbacks on the hydrosphere and ecosystems. This coupling of human and ecohydrological systems is particularly strong in agricultural systems and parallels that between plants and the hydrological cycle in natural ecosystems.

From a physical point of view, ecosystems may be seen as open thermodynamic systems that exchange mass, energy, and entropy in nonequilibrium conditions (Jorgensen and Svirezhev, 2004). Through inputs from solar radiation, water, carbon, and nutrient fluxes, ecosystems build and store organic compounds with high energy availability (i.e., chemical potential), which are used for the growth, maintenance, and reproduction of their constituent members. Human management of water and biogeochemical cycles increases productivity and accelerates the internal cycling of water, energy, and nutrients (see Fig. 10.1), as in the case of irrigation and fertilization, albeit often at the expense of biodiversity, resilience, and other ecosystem services (Altieri, 1999; Lin, 2011). Thus, despite increased productivity, the very efforts to stabilize ecosystem response to small environmental fluctuations may cause over-specialization, over-exploitation, and hence a loss of redundancy, which in turn may increase vulnerability to extreme fluctuations and regime shifts (Scheffer et al., 2001).

Dramatic examples of such *regime shifts* range from large-scale irrigation and deforestation that triggered the loss of fertile land due to erosion and salinization (Hillel, 1998) to desertification induced by land overexploitation (Reynolds et al., 2007). The risk is intensified by severe drought or other temporary lack of resources, such as during the 1930s Dust Bowl in the USA (see Fig. 10.2). As exemplified by the establishment of the US Soil Conservation Service and other emergency relief efforts following the

Figure 10.2 Examples of catastrophic soil loss under intensive land use pervade the agricultural history of the United States. The 1930s Dust Bowl was in part accelerated by severe drought, as described by John Steinbeck, who documented the environmental and economic disaster that eventually displaced hundreds of thousands of Great Plains farmers. "Little by little the sky was darkened by the mixing dust, and the wind felt over the earth, loosened the dust, and carried it away. The wind grew stronger" (Steinbeck, 2006). (a) Dust storm approaching Stratford, Texas, April 18, 1935. (b) Buried machinery in a barn lot in Dallas, South Dakota, May 1936. Much of the southeastern US experienced agricultural degradation due to accelerated water erosion initiated in the nineteenth century with the expansion of cotton. (c) Thousands of farm families, mostly debt-ridden tenants and "share croppers", cultivated cotton with mule and plow and much hand labor. (d) Widespread erosion accompanied cultivation, eventually resulting in substantial losses of surface soils and severe gullying, greatly reducing the land's native productivity. Photograph credits: (a) NOAA George E. Marsh Album, (b) US Department of Agriculture, (c) and (d) USDA Forest Service at Calhoun Critical Zone Observatory (criticalzone.org/calhoun/). After Porporato et al. (2015).

Dust Bowl, substantial resource investments are often required to revert degraded landscapes to their prior productive state.

The Calhoun in South Carolina (see Fig. 10.3) is another example where *land degradation* accelerated from the early nineteenth to the early twentieth century, owing to unsustainable land-use practices for cotton cultivation, and thus drastically altered soils and the geomorphology of the landscape (Ireland et al., 1939; Hoover, 1950; Trimble, 1985; Bastola et al., 2018). The degradation is apparent in the landscape-elevation statistics. Analyzing the first derivatives of the landscape, $\partial_x z$ and $\partial_y z$, describing the change in elevation (z) in the x and y directions, and the local slope, $S = ((\partial_x z)^2 + (\partial_y z)^2)^{0.5}$ from digital

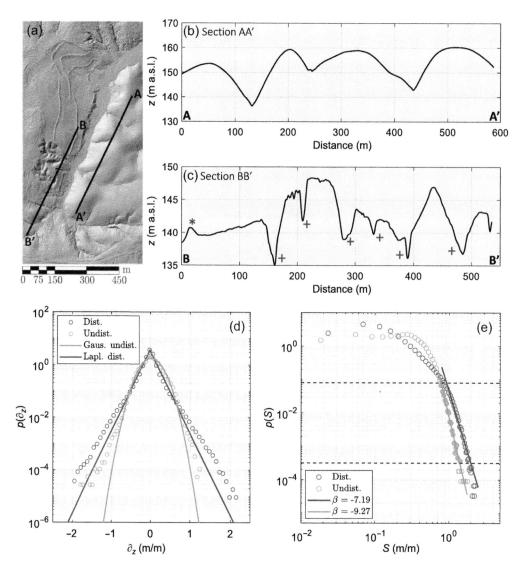

Figure 10.3 (a) Shaded relief image of a subregion of the Calhoun Critical Zone Observatory. On the right, the undisturbed area (hardwood forest) is characterized by a smooth topography with gentle slopes (see cross section AA' in panel (b)); on the left the disturbed area is characterized by the presence of manmade works and deep gullies resulting in steeper slopes (see cross section BB' in panel (c)). This area was partially smoothed in the 1930s with the creation of high (3 m tall) terraces (* in panel (c)). Gullies indicated by + in panel (c) are believed to be rather youthful (many formed in the twentieth century, some subsequent to land smoothing in the 1930s), thus having relatively sharp angles. (d) Gradient and (e) slope distributions of the disturbed and undisturbed areas. In (d) the gradients may be compared with the Gaussian and Laplace distributions for the undisturbed and disturbed cases, respectively. (e) The disturbed areas have distribution tails that decay more slowly, resulting in a higher power-law exponent β. After Bonetti et al. (2019).

elevation models (DEMs), Bonetti and Porporato (2017) showed that the gradient and slope PDFs normally have power-law tails progressively approaching respectively Gaussian and Rayleigh distributions with age. However, with human disturbance, the tails of the PDFs tend to be fatter owing to the observed steeper slopes (Fig. 10.3). The distinctive slope and gradient distribution tails provide direct measures of the effects of accelerated and intensive land use.

10.1.2 A Dynamical Agroecosystem Model

A dynamical system model (see Sec. 2.4) of interacting natural and human-driven agricultural ecosystems is useful to illustrate the qualitative changes induced by including human influence and feedbacks in desertification and land degradation. Following Porporato et al. (2015), we consider a total land area A, partitioned between a managed agricultural land area A_{ag}, yielding food and economic benefits to the population, and a natural land area, $A_{nat} = A - A_{ag}$, which yields associated ecosystem services. In this framework, A_{ag} is a proxy for social dynamics – an increasing A_{ag} indicates a developing social system, whereas a decreasing A_{ag} mirrors a declining society. A dynamic "ecosystem quality" q (with unspecified units Q), which aggregates the effects of ecosystem degradation by agricultural development and ecosystem renewal through natural processes, is sustained by the system's ability to provide ecosystem services η (Q yr^{-1}) (i.e., healthy ecosystems provide more services, which in turn increases their quality). The ecosystem quality q is bounded by a nonlinear function that ensures a steady state for a given level of ecosystem services. The corresponding dynamic equation for q reads

$$\frac{dq}{dt} = \eta - c \left(\frac{q}{q_0} \right)^{\alpha},$$

(10.1)

where the rate of turnover c (Q yr^{-1}) and the exponent α control both the equilibrium value, $q^* = q_0(\eta/c)^{1/(\alpha+1)}$, and how quickly q approaches it. Following Porporato et al. (2015), the reference quality q_0 is assumed to be equal to 1 and α is assumed equal to 2.

The rate of ecosystem services provided by the environment, η, may be assumed to depend on three factors: (i) external inputs from surrounding areas, η_0 (Q yr^{-1}), a measure of non-isolation; (ii) the extent of the natural land area and the quality of the environment; and (iii) negative impacts from adjacent agricultural activities (e.g., nutrient runoff, reduction of biodiversity, etc.):

$$\eta = \eta_0 + k_{nat} q A_{nat} - k_{ag} A_{ag},$$

(10.2)

where k_{nat} (ha^{-1} yr^{-1}) converts quality to ecosystem services and k_{ag} (Q ha^{-1} yr^{-1}) is the degradation rate resulting from nearby agricultural practices. The other state variable in this system, the managed agricultural land area A_{ag}, is assumed to increase with environmental quality, as the associated food and economic benefits of production stimulate human interest in cultivating more land. In addition, we allow for the influence of external markets to drive land cultivation within the system at a rate A_0 (ha yr^{-1}), another measure of non-isolation. We introduce an additional term to ensure that A_{ag} does not exceed the total land area, A. With these assumptions the temporal evolution of the agricultural area is

$$\frac{dA_{ag}}{dt} = (k_h q A_{ag} + A_0) \left(1 - \frac{A_{ag}}{A} \right),$$

(10.3)

where k_h (Q^{-1} yr^{-1}) is the rate at which agricultural land is developed in response to the benefits of production. Equations (10.1)–(10.3) comprise a two-dimensional dynamical system for coupled agricultural area and environmental quality (see Sec. 2.4 for a brief review and references on dynamical systems).

The above dynamical system describes a feedback between agricultural area and environmental quality that involves the negative influence of increased human pressure on ecosystem services and, thus, agricultural productivity. Expanding human pressure to larger areas decreases the quality of natural ecosystems,

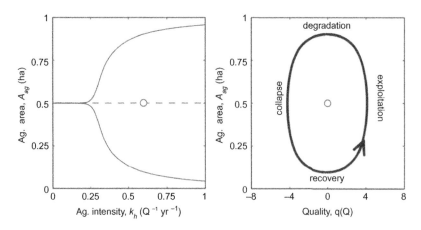

Figure 10.4 Dynamics of the coupled human–agroecosystem described by Eqs. (10.1)–(10.3). Left: Bifurcation diagram for agricultural area as a function of the control parameter k_h represents the intensity with which the social system responds to the benefits of land cultivation. The system undergoes a supercritical Hopf bifurcation from a single, stable, fixed point, approached through decaying spirals, to a stable limit cycle that surrounds an unstable fixed point. Right: System trajectory plotted in quality–agricultural-area phase space for the limit cycle regime. The solid curve traces the stable points (giving a limit cycle), the dashed line depicts an unstable points, and the open circle depicts the unstable fixed point. Parameter values used for the coupled human–environment system are $A = 10^5$ ha, $A_0 = 10^3$ ha yr^{-1}, $k_{\text{nat}} = 4.88 \times 10^{-4}$ m^{-2} yr^{-1}, $\eta_0 = 5$ Q yr^{-1}, $c = 10^3$ Q^{-1} yr^{-1}, $k_{\text{ag}} = 10^{-4}$ Q ha^{-1} yr^{-1}. After Porporato et al. (2015).

which in turn inhibits productivity and further growth and thereby provides a stabilizing mechanism (i.e., a carrying capacity for the system). The presence of an interaction term (qA_{ag}) makes this dynamical system prone to bifurcations as a function of the parameter k_h (Fig. 10.4), which controls the rate at which agricultural development responds to environmental quality and existing agricultural productivity, i.e., the current food and economic output of the land. As the social system responds more rapidly to agricultural production (i.e., higher k_h), the system shifts from one characterized by a single, stable, and productive fixed point approached through decaying spirals to a limit cycle that oscillates between productive and degraded states. This bifurcation is a supercritical Hopf bifurcation, characterized by a loss of stability at the critical point (see Sec. 2.4 and Note 2.15).

As the social system becomes more sensitive to agricultural production and ecosystem quality (i.e., k_h increases), its limit cycle dynamics can be described as a self-sustaining sequence of four phases: exploitation, degradation, collapse, and recovery (Fig. 10.4). In the exploitation phase, the environmental quality is constantly high and the agricultural area expands rapidly. Subsequently, the natural environment is degraded during a period of consistent intensive land use. Once the quality has decreased to a sufficient level, agricultural pressure is released in the collapse phase either through the implementation of conservation practices or due to severe resource impairment. With the release of agricultural pressure, the natural ecosystem then recovers and the system re-enters the exploitation phase. This sequence of phases is conceptually similar to the exploitation–conservation–release–reorganization paradigm introduced by Holling (2001).

The coupled system of Eqs. (10.1)–(10.3) can be seen as an extension of previously proposed models of ecohydrological systems with bistable dynamics (e.g., Scheffer et al., 2001; Runyan et al., 2012), where in moderately stressed conditions the system may present two stable states, a non-degraded and vegetated state along with a degraded and unproductive state (see Note 10.2). In this coupled social–ecological framework, the exploitation–collapse and degradation–recovery phases dynamically link the bistable states, giving rise to an excitable system such as the Fitzhugh–Nagumo model (see Note 2.15). In the following section, a bistability scenario is explored for soil erosion resulting from vegetation impacts on soil production due to excessive agricultural exploitation.

10.1.3 Bistable Soil–Plant Dynamics

Following Pelak et al. (2016), we analyze a dynamical system for soil–plant interactions, where a positive plant–soil feedback leads to a "humped" soil production function, which induces soil-depth bistability, when erosion depends on the vegetation biomass. The soil is modeled as an aggregated soil column of depth h, whose time rate of change is the difference between production (PR) and erosion (ER):

$$\frac{dh}{dt} = \underbrace{(p_0 + p_v b) \exp(-k_s h)}_{\text{PR}(h,b)} - \underbrace{\left[e_0 + e_1 \exp(-k_v b)\right] \Theta(h)}_{\text{ER}(h,b)}, \tag{10.4}$$

where p_0 is the abiotic soil production rate in the absence of vegetation and soil, p_v describes the sensitivity of soil production to vegetation, k_s represents the rate at which increasing soil depth decouples soil production from surface weathering processes, e_0 is the rate of erosion that occurs in the presence of full vegetative cover, e_1 is the range of erosion rates between the fully vegetated and the unvegetated states, k_v controls the sensitivity of erosion to biomass, and $\Theta(\cdot)$ is the Heaviside step function (see Eq. (6.25)). The plant biomass density, b, is balanced by the growth (GR) and harvest (HV) rates,

$$\frac{db}{dt} = \underbrace{r\left[1 - \exp(-k_g h)\right] b - mb^2}_{\text{GR}(h,b)} - \underbrace{f_h(t)b}_{\text{HV}(b,t)}, \tag{10.5}$$

where r is the maximum growth rate, k_g controls the effect of soil depth on growth, m is the plant mortality rate, and $f_h(t)$, which here represents the agricultural pressure, is the fraction of biomass harvested per unit time. Equations (10.4) and (10.5) provide a minimalist model for the impacts of various biotic and abiotic factors on soil production, erosion, plant growth, and harvest (Pelak et al., 2016).

We first analyze the soil formation in the absence of vegetation, in which case the soil formation is driven entirely by abiotic processes. The abiotic case is described by Eq. (10.4) with $b = 0$, which has a single stable fixed point (see Fig. 10.5a):

$$h^* = \begin{cases} -k_s^{-1} \ln(\rho^{-1}) & \rho > 1, \\ 0 & \rho \leq 1, \end{cases} \tag{10.6}$$

where $\rho = p_0/(e_0 + e_1)$ is the ratio of the abiotic production and erosion rates. In the case where the abiotic production rate is greater than the abiotic erosion rate (i.e., $\rho > 1$), there is one stable soil depth at a nonzero value. This is the transport-limited regime, where erosion is dictated by $e_0 + e_1$. In the case where erosion exceeds production (i.e., $\rho < 1$), there is one stable soil depth at $h^* = 0$, and soil production is weathering-limited. Owing to the nature of the step function, the soil production rate in this weathering-limited regime is equal to p_0, but any soil produced is instantaneously eroded.

We now consider the coupled plant–soil dynamics in the absence of harvest. Because the typical time-scale for plants to reach a stable biomass is orders of magnitude longer than that for soil depth, we

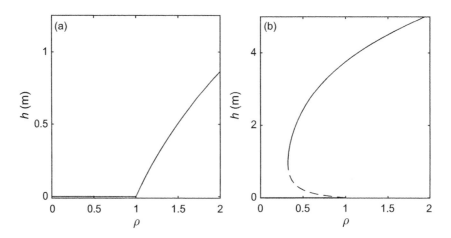

Figure 10.5 Soil depth bifurcation diagram as a function of the control parameter ρ for (a) the abiotic and (b) the biotic cases. The control parameter ρ is varied by varying e_0 and $\sigma = 0.8$. Other default parameters: $p_0 = 0.05$ mm yr^{-1}, $p_v = 0.4$ mm yr^{-1} kg^{-1}, $r = 0.2$ yr^{-1}, $m = 0.05$ yr^{-1} kg^{-1}, $k_g = 0.2$ m^{-1}, $k_s = 0.8$ m^{-1}, $e_0 = 0.02$ mm yr^{-1}, $e_1 = 1.0$ mm yr^{-1}, $k_v = 0.4$ kg^{-1}. After Pelak et al. (2016).

assume the plant biomass to be in instantaneous equilibrium. This equilibrium, b^*, can be found by setting $db/dt = 0$ in Eq. (10.5), so that Eq. (10.4) becomes

$$
\begin{aligned}
\frac{dh}{dt} &= \text{PR}(h, b^*) - \text{ER}(h, b^*) \\
&= \left\{ p_0 + \frac{p_v r}{m} \left[1 - \exp\left(-k_g h\right) \right] \right\} \exp\left(-k_s h\right) \\
&\quad - \left\{ e_0 + e_1 \exp\left(-\frac{k_v r}{m} \left[1 - \exp\left(-k_g h\right) \right] \right) \right\} \Theta(h).
\end{aligned}
\tag{10.7}
$$

Dynamic regimes for Eq. (10.7) with different plant–soil feedbacks are presented in Fig. 10.5b. Similarly to the abiotic case, the coupled system has a single nonzero stable state ($h^* > 0$) when $\rho > 1$. However, when $\rho < 1$ the soil stability depends on the shape of the soil production and erosion functions, which are related to a parameter $\sigma = [p_0 m/(p_v r) + 1]k_s/(k_s + k_g)$. When $\sigma \leq 1$, the soil production function decreases monotonically with h (see the solid curve in Fig. 10.6b), suggesting weak plant–soil feedback and a single stable soil depth of zero (the upper left regime in Fig. 10.6a). When $\sigma < 1$, we have a humped soil production function, which has a maximum at a nonzero soil depth (see the dotted curve in Fig. 10.6b). Thus strong plant–soil feedback allows a transition from a monostable bare state to a bistable regime of both zero and nonzero soil depths (lower left regime in Fig. 10.6a). The transition is located at $\rho_c(\sigma)$ where Eq. (10.7) is tangent to the line $dh/dt = 0$. A comparison of the abiotic and biotic systems demonstrates that positive plant–soil feedback stabilizes a vegetated, soil-mantled state in what would otherwise be a weathering-limited, exposed bedrock state in the absence of vegetation (Pelak et al., 2016).

We finally analyze the soil stability in relation to the removal (*harvest*) of native vegetation in favor of agriculture, an occurrence which greatly increases erosion rates and thus decreases the stability of the system. This stability can be quantified from the potential function (see Sec. 2.4.1) $\mathcal{V}(h) = \int_0^h (\text{PR} - \text{ER}) dh$, as shown in Fig. 10.7a for three harvest levels (see also Sec. 2.4 and Eq. (2.159)). In the absence of harvest, the system is bistable, as indicated by the two local minima (the zero and nonzero stable states) and a local maximum (the potential barrier) in the potential function (thick solid line, Fig. 10.7a). Under the critical

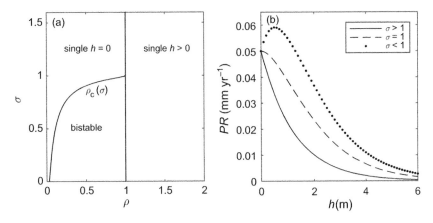

Figure 10.6 (a) Three stability regimes for the biotic system as a function of the control parameters ρ and σ. The parameter ρ is varied by varying the parameter k_g. (b) The soil production function, PR, for various values of σ, which controls the effect of vegetation on soil production (σ is varied by changing the value of p_v). After Pelak et al. (2016).

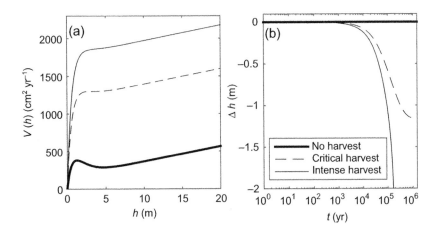

Figure 10.7 (a) Soil potential function $\mathcal{V}(h)$ with no harvest (thick solid line), a critical harvest level at which the stable and unstable states have joined together (dashed line), and intense harvest, which pushes the system toward a bare stable state (thin solid line). The valleys correspond to stable steady states and the peaks correspond to unstable steady states. (b) A time series illustrating the change in soil depth (Δh) for zero, critical, and intense harvest regimes. After Pelak et al. (2016).

harvest level (the dashed line in Fig. 10.7a), the stable and unstable states have merged, resulting in a half-stable state. At this point, any perturbation to a lower soil depth will result in a collapse to the degraded state. Further increases in the harvest rate will shift the system to a monostable regime, driving the system toward a degraded state (thin solid line in Fig. 10.7a). The corresponding soil-depth time series for no harvest, critical harvest, and intense harvest are shown in Fig. 10.7b (Pelak et al., 2016).

This interpretation of the soil production function and its link to biologic activity suggest that increasing the intensity of agricultural pressure or altering other factors could strongly impact geomorphologic processes in agroecosystems. As large portions of the Earth's surface are affected in some way by agricultural activity, such impacts on the interplay between soil production, erosion, and vegetation have the potential to affect geomorphic processes on a global scale.

10.2   Irrigation

Irrigation is the process of applying controlled amounts of water to plants to avoid or reduce water stress, in contrast to *rain-fed agriculture*, which relies only on direct rainfall. Globally, irrigated agriculture is the primary user of freshwater, accounting for nearly 85% of total water consumption (Jury and Vaux Jr., 2007), and providing about 40% of the total food production (Fereres and Soriano, 2007). Water demand for irrigation has grown in recent years (see Fig. 10.1) and is projected to increase as a result of population growth, changing food habits and biofuel production, and projected climate change (Schmidhuber and Tubiello, 2007; De Fraiture et al., 2010). Irrigation has important ecohydrological implications, related to increased evapotranspiration and percolation fluxes and changes in soil moisture fluctuations, which are complicated by rainfall unpredictability. In addition to the implications related to potential soil salinization (Assouline et al., 2006, see also Sec. 10.3.1), water management through irrigation has also profound implications for river flows and related ecosystem services (Eheart and Tornil, 1999; Baron et al., 2002; Falkenmark and Lannerstad, 2005; Miles et al., 2006) as well as for groundwater levels and quality (Schoups et al., 2005; Scanlon et al., 2007a; Caylor et al., 2009; McGuire, 2009).

From a modeling point of view, irrigation appears as an input in the soil water balance equation (1.5), reported here for convenience,

$$nZ_r\frac{ds}{dt} = R + J - C_i - Q - E - L. \tag{10.8}$$

In this section, we follow the analysis of Vico and Porporato (2010, 2011a) to quantify irrigation requirements with different irrigation strategies under stochastic rainfall conditions. Optimal irrigation with consideration of sustainability and profitability will be discussed later in Sec. 10.4.1.

10.2.1   Irrigation Strategies

Irrigation scheduling is an important aspect of water-resource management and optimization (English et al., 2002). There is a need to minimize the amount of irrigation water per cultivated area to harmoniously balance the concurring water demands for industrial and municipal uses with the requirements of natural ecosystems. This is often in contrast with food-security requirements of maximizing crop yields while reducing irrigation costs, and the constraints of water-resources managers who must know in advance the water demand for agriculture to allocate water and plan long-term investments for infrastructure and its maintenance.

We focus on demand-based irrigation, in which a water application is triggered by plant or soil water status reaching a pre-set threshold. On the basis of the multiple constraints and goals, irrigation may either be designed to fully meet crop water requirements for avoiding plant water stress (i.e., *stress-avoidance irrigation*) or to maintain a limited stress level (i.e., *deficit irrigation*; see Chalmers et al., 1981). In the first case, the crop is always maintained under well-watered conditions, while the latter case allows a certain level of water stress to occur. As such, deficit irrigation may result in lower water requirements at the cost of yield reduction (Geerts and Raes, 2009).

Traditional irrigation typically consists of repeated, massive, but infrequent applications of water that fill the root zone up to field capacity (loosely defined here as the point where drainage from the root zone becomes significant) by furrow, flood, or sprinkler irrigation. Conversely, modern irrigation principles (see Fig. 10.8) advocate a more finely tuned irrigation, realized by means of *micro-irrigation* techniques (e.g., drip or trickle irrigation, microsprinklers, microsprayers, or subsurface emitters (Hillel, 2003)), in

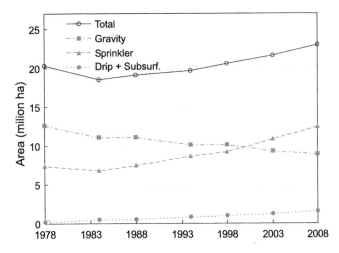

Figure 10.8 Changes in irrigated areas in relation to irrigation methods in the contiguous USA over the period 1978–2008. Data source: US Census of Agriculture, Farm and Ranch Irrigation Survey. After Vico and Porporato (2011a).

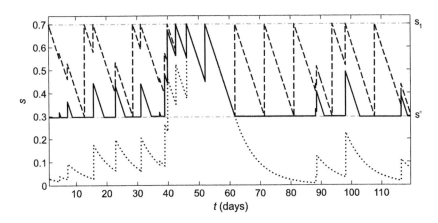

Figure 10.9 Example of soil moisture time series for rainfed agriculture (dotted line), micro-irrigation (solid line), and traditional irrigation (dashed line). The soil is sandy loam, with $n = 0.43$, $Z_r = 25$ cm, $s^* = 0.3$, and $s_1 = 0.7$ (see horizontal dot-dashed lines). The average precipitation depth is $\alpha = 15$ mm and frequency $\lambda = 0.15$ day^{-1}, with interception parameters $\Delta = 1$ mm and $k_{int} = 0.9$. The potential evapotranspiration rate is $E_{max} = 0.45$ cm day^{-1}. After Vico and Porporato (2010).

which water is supplied continuously, to avoid water stress while minimizing runoff and percolation losses. This also results in reduced leaching of soluble chemicals (e.g., pesticides and fertilizers, Böhlke, 2002), although it may also increase the risk of soil salinization (Bras and Seo, 1987) or reduce recharge to groundwater and streams (English et al., 2002).

The two irrigation schemes may be idealized as follows (see Fig. 10.9): (i) a modern micro-irrigation scheme with a continuous supply of water which maintains the root zone soil moisture just above the stress level until the next rainfall event; (ii) a traditional irrigation scheme, consisting in concentrated applications of water, when the soil moisture reaches the same stress level, that bring the soil moisture back to field capacity. These two idealized irrigation schemes are optimal in the sense that they avoid crop

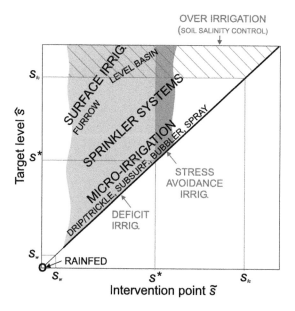

Figure 10.10 Diagram summarizing irrigation schemes and irrigation methods as a function of the irrigation parameters "intervention point" $\tilde{s}$ (x-axis) and "target level" $\hat{s}$ (y-axis). The irrigation parameter space is limited by the condition $\hat{s} \geq \tilde{s}$ (the white areas identify unrealistic combinations of parameters). The parameter space can be further subdivided into deficit-irrigation (light shaded area) and stress-avoidance irrigation (dark shaded area). Target levels above soil field capacity (hatched area) correspond to over-irrigation for soil salinity control. Regarding the irrigation methods, the whole shaded area corresponds to traditional irrigation schemes with impulsive applications. The thick solid line ($\hat{s} = \tilde{s}$) represents micro-irrigation. After Vico and Porporato (2011a).

water stress while minimizing water losses by percolation and runoff. Furthermore, they cover the two extremes cases of continuous and fully concentrated irrigation (Vico and Porporato, 2011a).

More general irrigation schemes, including deficit irrigation, were analyzed in Vico and Porporato (2011a). In fact, depending on the set level of plant or soil water status at which an irrigation application is initiated, either stress-avoidance or deficit irrigation may be performed. Sometimes a minimal stress level (deficit irrigation) may be acceptable when water supply is limited (English, 1990) and/or expensive, or it may even be desirable for quality purposes for specific products such as some fruit trees and vines (e.g., Fereres and Soriano, 2007). Depending on the technology used for water distribution, each irrigation application may either provide a set amount of water, thus restoring an adequate plant or soil water level (furrows or sprinkled systems, i.e., traditional irrigation) or supply enough water to balance current losses through evapotranspiration until the next rainfall event occurs (drip or trickle, bubbler, or micro-spray systems), i.e., modern micro-irrigation. These irrigation strategies can be described by means of two parameters (Vico and Porporato, 2011a): a soil moisture threshold, which triggers the irrigation application (the *intervention point*, $\hat{s}$); and the amount of water applied at each treatment or, equivalently, the soil moisture level to be restored by the irrigation application (the *target level*, $\tilde{s}$). In conditions prone to soil salinization, over-irrigation may be adopted to flush salts from the soil by enhanced percolation; in that case the target level is set above field capacity.

A generalized irrigation scheme is illustrated in Fig. 10.10. For traditional irrigation, each event corresponds to a volume per unit area equal to

$$nZ_r(\tilde{s} - \hat{s}),$$ (10.9)

which for stress-avoidance irrigation becomes

$$nZ_r(s_{fc} - s^*).$$ (10.10)

For micro-irrigation, the rates are simply equal to the soil moisture losses at that level of water stress; in particular, within the framework of Chapter 7, for stress-avoidance irrigation the rate is equal to E_{max}.

From a practical point of view, these irrigation schemes are attained in the field by employing various irrigation methods (e.g., Cuenca et al., 1989), as indicated qualitatively in Fig. 10.10. Traditional irrigation is often carried out by means of surface or sprinkler irrigation. Surface irrigation, where water is redistributed by gravity in basins or furrows, is particularly suited for larger applications (exceeding 50 mm per treatment), even though it generally has relatively low application efficiency and distribution uniformity. Sprinkler or spray systems are able to mimic rainfall events as small as 5 mm in the case of center-pivot systems (Trout and Kincaid, 2007), and hence are effective for light and frequent application. Nevertheless, sprinkler systems tend to have a higher energy demand for operation and higher investment costs when compared with surface irrigation systems (Kruse et al., 1990). Conversely, the micro-irrigation scheme, requiring almost continuous water applications at low rates, needs more sophisticated systems such as drip or trickle irrigation (also termed localized irrigation). These localized irrigation systems are often automated and are capable of high-frequency irrigation, with high application efficiencies, but require high initial investment, frequent maintenance, and proper operation.

Currently, surface irrigation is the most common irrigation system worldwide, while sprinkler irrigation is the most common system, followed by gravity systems, in the USA (Fig. 10.8). Despite its costs, localized irrigation is becoming more and more common. Such systems are being used predominantly for high-value horticultural crops, because of their high efficiency and water control, while they are still at the research stage for row crops (Ayars et al., 1999; Lamm and Trooien, 2003). The use of localized irrigation and its extension to other crops is expected to be reinforced by water scarcity and increase in crop and water prices, as well as the search for higher water productivity.

10.2.2 Stochastic Analysis

To quantify the irrigation water, we consider both random timing and amounts of rainfall and model the soil water in the soil–plant system at the daily level (see Chapter 7). We focus here on stress-avoidance irrigation, hence aiming at crop yield maximization under optimal irrigation. Ideally, unless time scheduling constraints are present, or in the case of reduced water availability, water applications should be made just before the soil moisture reaches the stress level. This level can be reasonably associated with the soil moisture corresponding to incipient stomatal closure, s^* (Hsiao, 1973; Porporato et al., 2001).

Following Vico and Porporato (2010, 2011a), we limit our analysis to stochastic steady-state conditions, i.e., we assume that $\partial p(s, t)/\partial t = 0$ (see Sec. 6.4.2). To obtain the desired PDF of soil moisture under a stochastic steady state, $p(s)$, we analyze the crossing properties of the soil moisture process (see Sec. 6.8). We focus here on the frequency of excursions of the soil moisture process below and above a generic threshold ξ, i.e., on the frequencies of downcrossing, $\nu_\xi^\downarrow$, and upcrossing, $\nu_\xi^\uparrow$. Because irrigation does not alter the crossing during a soil moisture drydown, the frequency of downcrossings of a generic threshold $\xi \geq \tilde{s}$ can be expressed as (see Sec. 6.8)

$$\nu_\xi^\downarrow = \rho(\xi)p(\xi).$$ (10.11)

Conversely, the frequency of upcrossings needs to account for jumps in soil moisture caused by either rainfall events or irrigation applications. Following Sec. 6.8, the frequency of soil moisture jumps caused by the occurrence of rainfall is $\lambda \int_{\tilde{s}}^{\xi} e^{-\gamma(\xi-u)} p(u)du$, while the frequency of soil moisture jumps caused by irrigation is the same as the frequency of downcrossing of the threshold $\xi = \tilde{s}$, i.e., $\rho(\tilde{s})p(\tilde{s})$. Because of the assumed irrigation scheme, the upcrossings caused by irrigation applications can occur only when the threshold ξ is such that $\tilde{s} \leq \xi \leq \hat{s}$. Hence, by combining these two cases, the following frequency of upcrossing for a generic soil moisture level $s \geq \tilde{s}$ is obtained;

$$v_{\xi}^{\uparrow}(\xi) = \Theta(\hat{s} - \xi)\rho(\tilde{s})p(\tilde{s}) + \lambda \int_{\tilde{s}}^{\xi} e^{-\gamma(\xi-u)} p(u)du, \tag{10.12}$$

where $\Theta(\cdot)$ is the Heaviside function. Because under stochastic steady-state conditions the frequency of upcrossing of a generic soil moisture threshold $\xi = s$ equals the frequency of downcrossing of the same threshold, using Eqs. (10.11) and (10.12) the following equation can be written:

$$\rho(s)p(s) = \Theta(\hat{s} - s)\rho(\tilde{s})p(\tilde{s}) + \lambda \int_{\tilde{s}}^{s} e^{-\gamma(s-u)} p(u)du. \tag{10.13}$$

Multiplying Eq. (10.13) by $\exp(\gamma s)$ and differentiating with respect to s, a first-order ordinary linear differential equation is obtained. Its solution is the desired PDF of the soil moisture (Vico and Porporato, 2011a)

$$p(s) = C \frac{e^{-\int_{\tilde{s}}^{s}\left(\gamma - \frac{\lambda}{\rho(u)}\right)du}}{\rho(s)} \left\{1 + \int_{\tilde{s}}^{s}[\gamma\Theta(\hat{s} - u) - \delta(\hat{s} - u)]e^{\int_{\tilde{s}}^{u}\left(\gamma - \frac{\lambda}{\rho(y)}\right)dy} du\right\} \tag{10.14}$$

where $\rho(s)$ is a generic normalized evapotranspiration loss function, $\delta(\cdot)$ is the Dirac delta function, and C is a normalizing constant, which can be obtained by imposing $\int_{\tilde{s}}^{s_{fc}} p(u)du = 1$. For $\tilde{s} \to 0$ and $\hat{s} \to 0$, Eq. (10.14) simplifies to the case of rainfed agriculture (i.e., Eq. (7.28)). For $\hat{s} \to \tilde{s}$, the case of micro-irrigation is retrieved (Vico and Porporato, 2010):

$$p(s) = \frac{C}{\rho(s)} \exp\left(-\gamma s + \lambda \int_{\tilde{s}}^{s} \frac{du}{\rho(u)}\right), \tag{10.15}$$

$$p_0 = \frac{C}{\lambda} e^{-\gamma\tilde{s}}, \tag{10.16}$$

where the finite duration of the irrigation applications produces an atom of probability at $\tilde{s}$ (see Note 7.12).

Examples of steady-state soil moisture PDFs are reported in Fig. 10.11 for different choices of the parameters $\tilde{s}$ and $\hat{s}$. In the most general case, the PDF exhibits different behaviors over the three ranges of soil moisture limited by $\tilde{s}$, s^*, and $\hat{s}$ (clearly, some of these ranges may not exist with particular choices of the irrigation parameters). Above $\hat{s}$, the PDF has the same (though rescaled) shape as that pertaining to rainfed agriculture, because in that range of soil moistures the distribution of probability is not directly impacted by irrigation applications. In this representation each irrigation treatment is instantaneous. However, in the limiting case of micro-irrigation ($\tilde{s} \to \hat{s}$), the treatments become continuous, resulting in irrigation periods of nonzero duration and an atom of probability at $\tilde{s}$ or $\hat{s}$ (see Fig. 10.11b).

Each irrigation treatment needs to supply the volume (per unit area) $nZ_r(\hat{s} - \tilde{s})$ to bring the soil moisture back to the target level $\hat{s}$. The frequency of irrigation is the same as the frequency of downcrossing of the threshold $\tilde{s}$, i.e., $v_{\tilde{s}} = \rho(\tilde{s})p(\tilde{s})$ (see Eq. (10.11)). Hence, the average volume per unit cultivated area required for irrigation purposes over a growing season of duration T_{seas} is

$$V_{ideal}(\tilde{s}, \hat{s}) = nZ_r(\hat{s} - \tilde{s})v_{\tilde{s}}T_{seas} = nZ_r(\hat{s} - \tilde{s})\rho(\tilde{s})p(\tilde{s})T_{seas}. \tag{10.17}$$

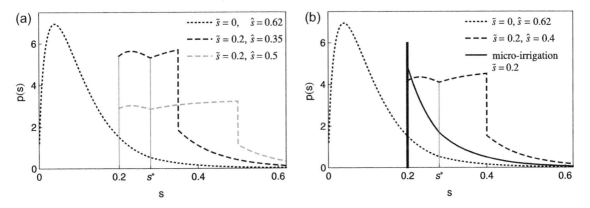

Figure 10.11 Example of steady-state probability density functions for the available plant soil moisture, for different choices of the parameters $\tilde{s}$ and $\hat{s}$. The soil moisture PDF for rainfed agriculture is presented for comparison (dotted curves), along with the case of micro-irrigation (solid curve in (b)). In (b) the atom of probability at $s = 0.2$ (thick vertical line) is not to scale. Other parameters are $n = 0.43$, $s_{fc} = 0.62$, $E_{max} = 0.55$ cm day^{-1}, $s^* = 0.28$, $Z_r = 50$ cm, $\alpha = 15$ mm, and $\lambda = 0.15$ day^{-1}. After Vico and Porporato (2011a)

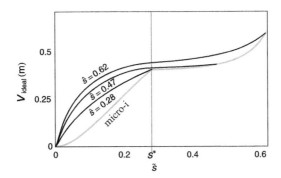

Figure 10.12 Average required irrigation volumes as a function of the intervention point $\tilde{s}$, for different choices of the parameter $\hat{s}$, over a 110-day growing season. The gray line refers to the extreme case of $\hat{s} \rightarrow \tilde{s}$, i.e., micro-irrigation. The thin vertical line corresponds to $\tilde{s} = s^* = 0.28$. All the other parameters are as in Fig. 10.11. After Vico and Porporato (2011a).

This required irrigation volume already accounts for losses to runoff and deep percolation caused by any input of water exceeding s_{fc}, through its dependence on the soil moisture water balance. Nonetheless, the water requirement in Eq. (10.17) represents an ideal situation, with irrigation application efficiency equal to 1. Accounting for typical irrigation application efficiencies is crucial for realistic estimation of water requirements. Figure 10.12 represents the average required irrigation volumes, for different choices of irrigation parameters, when the rainfall statistics are kept constant. In general, when all the other parameters are fixed, the higher is $\hat{s}$ or $\tilde{s}$, the higher the irrigation volumes will also be, due to the larger losses at higher soil moisture levels.

The long-term average of Eq. (10.8) can be found by adding the irrigation component into our analysis of the mean soil water balance, as discussed in Sec. 7.4.2. The inputs to the system are the average daily rainfall, $\langle R \rangle = \alpha \lambda$, and the average daily irrigation, $\langle J \rangle = V_{ideal} T_{seas}^{-1}$. The average daily losses through

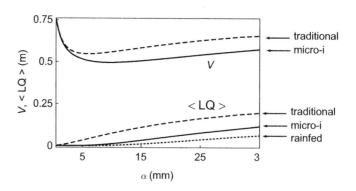

Figure 10.13 Long-term irrigation volumes, V, and deep percolation and runoff losses, $\langle LQ \rangle$, for fixed total growing season rainfall depth ($\langle R \rangle = 400$ mm) and variable α and λ, in the cases of rain-fed agriculture, micro-irrigation ($\tilde{s} = \hat{s} = s^*$), and traditional irrigation ($\tilde{s} = s^*$, $\hat{s} = s_{fc}$). The components of the water balance are expressed as volumes per unit area, i.e., depths. The growing season duration, T_{seas}, is assumed to be 180 days; all the other parameters are as in Fig. 10.9. After Vico and Porporato (2010).

evapotranspiration can be expressed as $\langle E \rangle = nZ_r \int_{\tilde{s}}^{s_{fc}} \rho(u)p(u)du$, while runoff and deep percolation losses can be obtained as $\langle LQ \rangle = \langle R \rangle + \langle I \rangle - \langle E \rangle$.

For the soil and climate parameters in Fig. 10.9 over a growing season of 180 days, the corresponding water balance under various different rainfall timings is presented in Fig. 10.13 (fixed total rainfall amount, with increasing α and decreasing λ). Both irrigation schemes exhibit an increase in irrigation water requirements with increasing mean event depth, with the exception of very small rainfall depths. For small event depths, interception by crop significantly decreases the effective rainfall, thus enhancing the water demand and reducing runoff and deep infiltration. At the other extreme (large α), more water is lost to runoff or deep percolation compared with the case of frequent and small events, thus larger irrigation volumes are required. At intermediate event depths, a minimum in the irrigation volumes is found, occurring for a value which depends on the adopted irrigation scheme. As expected, even micro-irrigation results in higher water losses than rainfed agriculture. In fact, at higher levels of average soil moisture it is more likely that a rainfall event exceeds the soil storage capacity $nZ_r(s_{fc} - s)$, which also explains why in turn the water lost under traditional irrigation is the highest.

This stochastic soil moisture description of irrigation provides the average irrigation requirements for a given soil, crop, and type of climate and the related water balance. This framework can be readily coupled to a minimalist model of yield and an economic balance, allowing the assessment of a variety of irrigation strategies in terms of water conservation, crop productivity, and profitability, under current and future rainfall patterns (see Sec. 10.4.1).

10.3 Dynamics of Soil Water Minerals

Several important problems in the ecohydrology of agroecosystems concern the joint dynamics of soil water and minerals, the latter often being either nutrients (as in Chapter 9) or contaminants. In this section we analyze the problem of soil salinization and the remediation of contaminated soil using plants (so-called phytoremediation). The analysis of these problems can be approached by considering the coupled

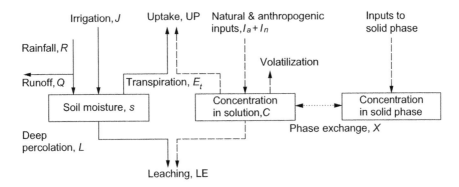

Figure 10.14 Schematic illustration of mass balances of the soil moisture, s, and a generic soil compound in solution with concentration $C = m/(nZ_r s)$ (e.g., nitrate, the soluble fraction of a contaminant, dissolved salts, etc.). The solid lines represent fluxes of water, the dashed lines fluxes of dissolved mineral of mass m, and the dotted lines fluxes exchanged with the adsorbed phase. Equations (10.18) and (10.19) describe mathematically this coupled system forced by rainfall and anthropogenic inputs. After Porporato et al. (2015).

balance equations of soil moisture and the corresponding minerals dissolved in the soil solution. As before, for simplicity we refer to vertically averaged dynamics over a representative soil rooting zone of depth Z_r.

Figure 10.14 offers a schematic illustration of the budgets of the soil moisture s and the mass m of a generic soil mineral in solution (e.g., nitrate, the soluble fraction of a contaminant, dissolved salts, etc.) having concentration $C = m/(nZ_r s)$. The solid lines represent fluxes of water, the dashed lines fluxes of dissolved mineral of mass m, and the dotted lines the fluxes exchanged with the adsorbed phase. The temporal evolution of the solute mass m dissolved in the soil solution per unit area of soil for a generic solute M, can be described by the following mass-balance equation:

$$\frac{dm(t)}{dt} = I_n(t) + I_a(t) - X(t, s(t))$$
$$- \text{UP}(t, s(t), m(t)) - \text{LE}(t, s(t), m(t)) - \text{VOL}(t, s(t), m(t)), \qquad (10.18)$$

where $I_n(t)$ is the rate of natural input by dry and wet deposition, $I_a(t)$ is the anthropogenic input rate, $X(t, s(t))$ is net exchange between the soluble and insoluble fractions (e.g., nutrient mineralization from organic matter, adsorption/desorption, dissolution of salts, etc.), $\text{UP}(t, s(t), m(t))$ is the rate of nutrient or contaminant uptake by plants (linked in part to transpiration), $\text{LE}(t, s(t), m(t))$ is the rate of solute loss due to leaching and runoff (associated with LQ), and $\text{VOL}(t, s(t), m(t))$ is the possible volatilization flux. These generic terms take on specific forms depending on the type of solute analyzed, as will be detailed in the following sections.

While rainfall has the strongest control on soil moisture dynamics, some feedbacks between soil moisture and mineral content may also be present. Including formally such feedbacks, Eq. (10.8) becomes

$$nZ_r \frac{ds(t)}{dt} = R + J - C_i - E(s, m) - L(s, m) - Q(s, m). \qquad (10.19)$$

In this equation, the term E explicitly contains m to indicate the impact of changes in the osmotic potential due to the presence of solutes in the soil solution on evapotranspiration (see Sec. 4.9.2), while the presence of m in the terms related to percolation and runoff, $L(s, m)$ and $Q(s, m)$, is related to the possible changes in soil hydraulic properties and thus infiltration due to high concentrations of solutes, especially in the case of soil sodicity (Sec. 10.5.2; see also Kramer and Mau, 2020).

The external stochastic forcing through rainfall and the nonlinear coupling between soil moisture and solute concentration make the behavior of the stochastic system (10.18) and (10.19) difficult to predict theoretically in the general case. Practical insight can be obtained by a combination of numerical simulations of the full equations and specific analyses on simplified models. The latter approach is possible, for example, when there is a clear separation of timescales between the soil moisture temporal evolution and the duration of leaching events; in such a case, it is possible to treat the two equations separately by first solving the stochastic differential equation for the soil moisture and then approximating the short-duration leaching events as independent, instantaneous events whose frequency is controlled by the probability of reaching percolation thresholds (Suweis et al., 2010; Manzoni et al., 2011b). In the case of negligible feedback on the soil moisture, some progress is also possible by focusing on averaged behaviors of the variables described by so-called "macroscopic equations". This approach is typical for complex systems and has been extensively used in statistical mechanics (e.g., Van Kampen, 1992) and turbulence modeling (e.g., Pope, 2001). Analyzing these systems requires overcoming a so-called "closure problem", wherein the flux terms cannot be expressed just in terms of the means s and m, because of nonlinearities, but involve higher-order joint moments of s and m. We present some examples of these approaches in the following sections.

10.3.1 Soil Salinization

Large areas of cultivated land worldwide are affected by soil salinity. Szabolcs et al. (1989) estimated that 10% of arable land with a total area of over nine million km^2 over 100 countries is salt-affected, especially in arid and semi-arid regions (see Fig. 10.15). Salinity refers to large concentrations of easily soluble salts present in water and soil on a unit volume or weight basis (typically expressed as electrical conductivity (EC) of the soil moisture in dS/m, i.e. decisiemens per meter at 25 °C; for NaCl, 1 mg/l $\sim 15 \times 10^{-4}$ dS/m). High salinity causes both ion-specific and osmotic stress effects (see Sec. 4.9), with important consequences for plant production and quality. Normally, yields of most crops are not significantly affected if EC ranges from 0 to 2 dS/m, while above levels of 8 dS/m most crops show severe yield reductions (e.g., Ayars et al., 1993). Prevention or remediation of soil salinity is usually done by leaching salts by over-irrigation, and has resulted in the concept of leaching requirements (e.g., Hillel, 1998). Alternative amelioration strategies that harvest salt-accumulating plants have also been proposed, but appear to be less effective (Qadir et al., 2001).

Salt accumulation in the root zone may be due to natural factors (primary salinization) or due to irrigation (secondary salinization). In what follows, we consider these dynamics using minimalist models and a simplified macroscopic equation for the mean salt concentration. In line with the previous analyses of the soil water balance, we consider vertically averaged soil moisture and salt balance, a simplified approach used by Allison et al. (1994), Hillel (2000), and Suweis et al. (2010).

Minimalist Models of Soil Primary Salinization

We begin with an analysis of the macroscopic equation, i.e., the equation for the ensemble average, following Porporato et al. (2015). We focus on the long-term dynamics of salt in soils caused by natural inputs, such as wet (rain) and dry (aerosol) deposition (primary salinization). Our aim is to illustrate the main mechanisms underlying the accumulation of salt mass and the concentration increase in the root zone as a function of the main hydro-climatic parameters. For simplicity, we neglect the feedback of m on s and

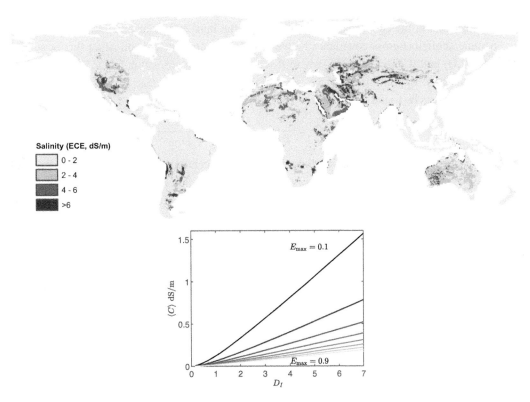

Figure 10.15 (Top) Global distribution of salt-affected soils. Data from Harmonized World Soil Database (Hiederer and Köchy, 2011). (Bottom) The steady-state concentration of salt, Eq. (10.23), plotted against the climatic dryness index D_I. Each line corresponds to changes in the potential evapotranspiration (in mm/day) varying from low (darker lines) to high (lighter lines), while the mean rainfall depth is kept constant at $\alpha = 10$ mm. After Porporato et al. (2015).

plant uptake and instead combine, as in Chapter 9, the percolation and runoff in a single term LQ. As a result, the macroscopic equations for Eqs. (10.19) and (10.18) in conditions without irrigation and at steady state can be written as

$$\langle R \rangle = \langle E(s) \rangle + \langle LQ(s) \rangle \tag{10.20}$$

$$\langle I_n \rangle = \langle UP \rangle + \langle LE \rangle = K_d \kappa \langle C E(s) \rangle + K_d \langle C LQ(s) \rangle, \tag{10.21}$$

where K_d is the partition coefficient between the adsorbed and dissolved salt fractions, κ is the ratio of the concentration in the transpiration stream and the concentration in the soil water (Dietz and Schnoor, 2001),[1] and

$$C = \frac{m}{nZ_r s} \tag{10.22}$$

is the concentration of dissolved salts in the soil. To obtain an explicit solution, we assume the plant uptake to be linearly dependent on the soil moisture, a reasonable assumption over large spatial areas, and $E = E_{\max} \langle s \rangle$, where $E_{\max}$ is the potential evapotranspiration. We also approximate the leakage rate using a modified first term of its Taylor expansion (Laio et al., 2002; Feng et al., 2015) such that $\langle LQ \rangle =$

---

[1] This term represents a simple way to account for the active uptake – see Sec. 9.3.3.

$(\lambda/\gamma) \exp[-\gamma \epsilon(1 - \langle s \rangle)]$, where ϵ (not to be confused with the ratio of the molar masses of water and dry air, used in the early chapters) accounts for the bulk effects of other soil features. Furthermore, we assume negligible cross-covariance between salt concentration and soil water fluxes, such that $\langle C L Q \rangle = \beta \langle C \rangle \langle L Q \rangle$ and $\langle C E \rangle = \beta \langle C \rangle \langle E \rangle$, where β serves as a linear correction factor.

With these approximations, the steady-state ensemble average concentration of salt under these conditions can be found as a function of the dryness index:

$$\langle C \rangle = \frac{\langle I_n \rangle}{K_d E_{\max} \beta \left[\kappa \langle s \rangle + D_I e^{-\gamma \epsilon(1 - \langle s \rangle)} \right]}. \tag{10.23}$$

This synthetic expression, which links the mean concentration to climate parameters, is shown in the graph in Fig. 10.15, where the steady-state salt concentration is plotted against increasing dryness index for different values of the potential evapotranspiration. The increase in D_I comes as a result of decreasing mean rainfall frequency λ, while the mean rainfall depth α is kept constant. As can be clearly seen, primary salinization is especially pronounced in drier climates (high D_I) where the potential evapotranspiration greatly exceeds the rainfall that may leach salt out of the soil. This trend is even more accentuated in locations with lower potential evapotranspiration where, for the same D_I, rainfall rates are comparatively even lower (as seen in some higher latitudes; see the map in Fig. 10.15). These data are in agreement with global trends reported in Hassani et al. (2020).

If one considers that the typical timescales for salt-mass dynamics in the root zone are orders of magnitude larger than those characterizing rainfall (and thus wet deposition) and percolation, it is possible to simplify the problem and obtain a stochastic differential equation for the salt content in the soil, which in turn can be solved to give the steady-state PDF of m for primary salinization (Suweis et al., 2010). Because of this separation of scales, the soil moisture typically reaches steady-state conditions within a growing season (e.g., 5–7 months), while the salt-mass balance only does so on much longer timescales (e.g., decades). Accordingly, at those long timescales, the salt-mass dynamics is characterized by slow growth due to a mean input rate $\Upsilon = (1/T) \int_0^T I_n(t)dt$, and punctuated by sudden and infrequent leaching events induced by percolation, given by the salt concentration times the percolation rate, $C(t)L(s(t))$. As a result, Eq. (10.18) can be described by the simple equation

$$\frac{dm}{dt} = \Upsilon - \frac{m}{nZ_r s} L(s), \tag{10.24}$$

according to which, m steadily grows during periods with no percolation and intermittently jumps down because of leaching events modulated by m itself.

On the basis of this separation of scales, we can consider percolation $L(s)$ to happen instantaneously (i.e., a point process), taking place when the soil moisture reaches the percolation threshold (e.g., s_1; see Sec. 7.5). Suweis et al. (2010) further assumed that percolation events take place independently, namely as a marked Poisson process with frequency λ_p and mean α_p (Botter et al., 2007). Since percolation can be shown to act as a censoring process (similarly to canopy interception, see Sec. 7.3.2), the percolation–depth distribution is identical to the PDF of rainfall depths (i.e., exponential) with mean depth $\alpha_p = \alpha$. Regarding the frequency λ_p, the rate of the percolation process is the frequency at which the soil moisture crosses the threshold s_1, which can be found from Sec. 6.8 for the mean rate of crossing in terms of the soil moisture PDF as $\lambda_p = \rho(s_1)p(s_1)$.

With these assumptions, the salt dynamics in Eq. (10.24) can be simplified to

$$\frac{dm}{dt} = \Upsilon - \frac{mb}{nZ_r s_1} \diamond \mathcal{F}_{\lambda_p, 1/\alpha_p}, \tag{10.25}$$

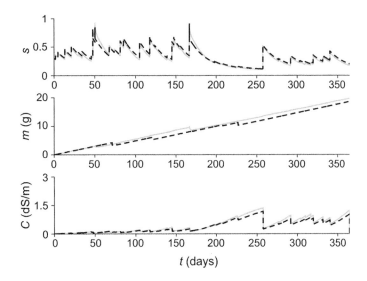

Figure 10.16 For comparison, the soil moisture (top), salt mass (middle), and specific salt concentration (bottom) for a sandy loam from the minimalist model (dashed line, Eq. (10.25)) and from the complete model (solid line, Eq. (10.18)). For both models $\lambda = 0.1$ day^{-1}, $\alpha = 1.8$ cm, $n = 0.45$, $Z_r = 30$ cm, $s_w = 0.1$, $E_{\max} = 0.35$ cm day^{-1}, and $\Upsilon = 84$ mg m^{-2} day^{-1} (typical of coastal regions); for the minimalist model, $s_1 = 0.8$ and $b = 0.6$. After Suweis et al. (2010).

where $\mathcal{F}_{\lambda_p,1/\alpha_p}$ is the formal derivative of a marked Poisson noise (see Sec. 6.6), the leaching-efficiency parameter b is used to account for incomplete salt dissolution, and the typical value of soil moisture during leaching events is approximated by the value s_1. The special multiplication, indicated by the diamond, in $m \diamond \mathcal{F}_{\lambda_p,1/\alpha_p}$ stands as a reminder that the multiplication of the state variable $m(t)$ with a white noise is not a uniquely defined mathematical concept but requires a so-called "prescription". Therefore, when taking the limit to instantaneous jumps, the value of m multiplying the jumps needs to be chosen consistently as a suitable value within the discontinuity before and after the jump. Here it is sufficient to recall that the consistent prescription preserves the rules of calculus; we refer to Suweis et al. (2011) and Bartlett and Porporato (2018) for more information about this problem (see also Sec. 7.7).

We can now use a variable transformation to write the problem in the form seen previously (see Chapter 6). Before doing so, however, it is important to reassure ourselves that the simplifications made provide a realistic representation of primary salinization. To this purpose, Fig. 10.16 gives a comparison of the more complete salinity model given by Eqs. (10.18) and (10.19) versus the simplified model of Eq. (10.25), showing a very good fit. The parameters s_1 and b are fitted to best approximate the complete model (Suweis et al., 2010).

As we said in the previous paragraph, from a mathematical viewpoint, Eq. (10.25) is a stochastic differential equation with multiplicative white (jump) noise. The noise prescription allows us to transform Eq. (10.25) into

$$\frac{dy}{dt} = \Upsilon e^y - \frac{b}{nZ_rs_1}\mathcal{F}_{\lambda_p,1/\alpha_p}, \tag{10.26}$$

where $y(t) = \ln[m(t)]$. In this new variable, the jump term is additive and, therefore, the corresponding Chapman–Kolmogorov equation can be obtained, following Sec. 6.7. In particular, a stationary solution

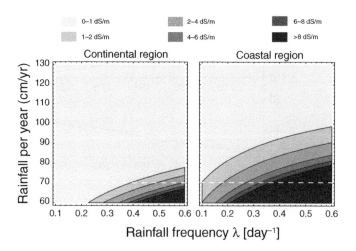

Figure 10.17 Contour plot of the asymptotic mean salt concentration $\langle C \rangle$ as a function of rainfall rate and frequency. The values reported in the legend refer to the corresponding salt concentration values with respect to the average soil moisture. The contour lines represent significant soil salinity values $(1, 2, \ldots, 8 \text{ dS/m})$. $\Upsilon = 84 \text{ mg day}^{-1}\text{m}^{-2}$ for coastal regions and $\Upsilon = 6 \text{ mg day}^{-1}\text{m}^{-2}$ for continental regions. After Suweis et al. (2010).

can be obtained using Eq. (6.89). Then, with the derived distribution approach (Sec. 6.3.8), i.e., $p(m) = p(y)dy/dm$, the PDF of m for the salt mass in the root zone can be obtained as a gamma distribution:

$$p(m) = \mathcal{N}\, m^{1/\mu} e^{-m\lambda_p/\Upsilon}, \tag{10.27}$$

where $\mathcal{N} = \left(\lambda_p/\Upsilon\right)^{(\mu+1)/\mu} / \Gamma[(\mu+1)/\mu]$ with $\mu = b\alpha_p/(nZ_r s_1)$. Equation (10.27) links the soil salinity statistics as a function of climate, soil, and vegetation parameters. The mean and the variance for the gamma distribution are given by Eq. (6.35).

Figure 10.17 represents the mean salt concentration, approximated as $\langle C \rangle = \langle m \rangle/(nZ_r\langle s \rangle)$, as a function of the mean rainfall rate and frequency. The contour lines connect equal values of the mean salt concentration in the soil, for a given input of salt, Υ, for two different geographic regions. Typical salt inputs in coastal areas are 100–200 kg/(ha yr) of salt, while values drop by an order of magnitude in continental regions (Hillel, 2000). Between the black areas and the light gray areas the values of concentration change substantially: above a certain total rainfall per year, the input of salt in relation to rainfall frequency becomes immaterial, as leaching effectively washes out the salt mass from the root zone; for lower total rainfall values, however, the salt in the soil increases with increasing λ. For a given annual precipitation depth, with low rainfall frequencies, rainfall events carry enough water to trigger leaching. Conversely, if λ is high, evapotranspiration dominates and leaching is largely reduced, thereby causing salt accumulation in the root zone. Therefore, $\langle m \rangle$ strongly increases with λ. Relatively small reductions in rainfall at the transition between these two regimes may entail a dramatic increase in long-term soil salinization. Figure 10.17 also shows the threshold ($\langle C \rangle < 2 \text{ dS/m}$) of soil salinity above which regular (e.g., non-halophytic) vegetation is damaged. For coastal areas, soil salinization may occur even in relatively more humid regions, especially when rainfall events are frequent and not very intense. In contrast, in continental regions only arid climates may begin to develop soil salinization (in the absence of irrigation and groundwater input).

One can evaluate the risk of soil salinization in rain-fed agriculture by estimating the typical salt inputs, the total rainfall per year, and the rainfall frequency. For example, a rain-fed crop in a semi-arid climate (e.g., with a rainfall depth of 70 cm/yr) in a continental region risks salinization only when rainfall events are not very intense (e.g., $\alpha < 1$ cm or $\lambda > 0.48$ day^{-1}). If the same crop is located in a coastal area, salinization occurs for a wider range of rainfall parameters (e.g., $\alpha < 1$ cm or $\lambda > 0.18$ day^{-1}).

The present discussion of primary salinization can be extended to include the plant uptake of salt (e.g., Mau et al. (2014) discussed the PDF of m when a linear uptake term is added), as well as the feedback of changes in osmotic potential salinization on evapotranspiration at both local (Bras and Seo, 1987; Perri et al., 2019) and regional scales (Perri et al., 2020).

Stochastic Analysis of Secondary Salinization and Over-Irrigation

Secondary salinization refers to soil (and groundwater) salinization induced by human activities, especially irrigation. The latter in fact often carries non-negligible amounts of salt, which over time may accumulate because of a lack of sufficient drainage and high evaporation. However, with a sufficient input of water at modest concentrations (over-irrigation), it is possible to enhance percolation, thereby reducing salinity. Thus irrigation can play two contrasting roles: introducing additional salt (the term I_a in Eq. (10.18) now designates the concentration of salt in the irrigation water) and increasing leaching and salt flushing.

To understand this dual role of irrigation as providing both water and salt inputs, it is instructive to consider a crude macroscopic approximation, similar to Eq. (10.23), that neglects the primary salt input as well as the effects of cross-correlations between the fluctuations in soil moisture and salt (i.e., only the feedback between averages is retained):

$$\langle C \rangle = \frac{C_J n Z_r \langle J \rangle}{K_d E_{\max} \beta \left[\kappa \langle s \rangle + D_I e^{\gamma \epsilon (1 - \langle s \rangle)} \right]}, \tag{10.28}$$

where the connection between the mean the anthropogenic input of salt $\langle I_a \rangle$ and the average irrigation $\langle J \rangle$ is established by setting $\langle I_a \rangle = C_J n Z_r \langle J \rangle$, with C_J the salt concentration in the irrigation water.

The behavior of the mean salt concentration in the soil given by Eq. (10.28) is shown in Fig. 10.18a as a function of the average irrigation rate $\langle J \rangle$ for various values of the dryness index D_I. As can be seen, the concentration of salt in the soil is determined by both the climate and the irrigation rate. Particularly under drier climates, a maximum for the soil salt concentration exists at an intermediate rate of irrigation, owing to the opposing effects of irrigation and leaching. In drier climates, salt in the irrigation water increases the soil salt concentration. As irrigation is increased, however, enhanced leaching counterbalances the additional salt input, and soon the salt concentration decreases again. In contrast, in wetter climates, leaching events due to naturally occurring rainfall already dominate the system, so that the overall salt mass balance is not as drastically affected by the increase due to irrigation. The maximum observed in drier climates disappears. A comparison of the approximate analytical solutions and the numerical simulation of the complete system, in which the covariances of salt mass and soil water fluxes (e.g., $\langle CL \rangle$ and $\langle CE \rangle$) are explicitly taken into account, reveals that the behavior of the system and the existence of a maximum concentration are consistent across different climate regimes (see Figs. 10.18b and c).

We conclude this section on salinity by noting that some salt-tolerant crops may offer valuable alternatives in regions with soil salinization (e.g., dates and atriplex) and that moderate levels of salinity may actually promote growth in some crops. In most species salinity, even if not toxic, triggers stomatal closure and reduced growth (Munns and Tester, 2008; Perri et al., 2019). Therefore, considering long-term

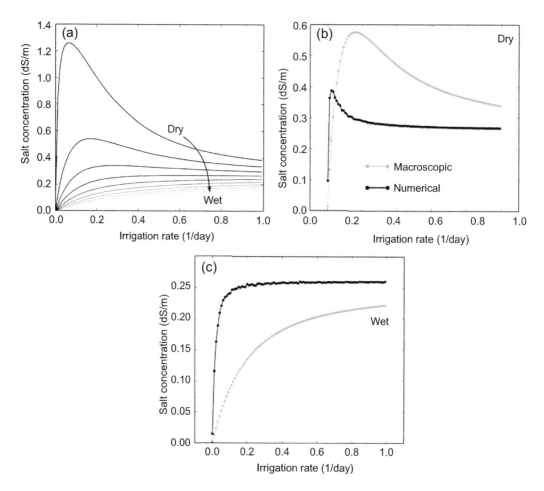

Figure 10.18 (a) Mean salt concentration as a function of irrigation rate in liters per day for various climate conditions (using the macroscopic equation (10.28)), where "Dry" and "Wet" span the range $D_I = 50$ to $D_I = 0.55$. (b) and (c) For comparison, the salt-concentration–irrigation-rate relations obtained from the macroscopic solution of Eq. (10.28) (gray lines) and from the corresponding numerical solution in which the covariance of the salt mass and soil water fluxes are explicitly simulated (black lines); $D_I = 5$ in panel (b) and $D_I = 1$ in panel (c) Other parameters: $C_J = 1000$ mg cm^{-1} m^{-2}, $nZ_r = 15$ cm, $E_{max} = 0.5$ cm day^{-1}, $\alpha = 1.0$ cm, $\kappa = 0.1$, $K_d = 0.6$, $\epsilon = 0.1$, and $\beta = 16$. After Porporato et al. (2015).

salinization trends is crucial, to inform current efforts to avoid salinization and promote a sustainable use of soil and water resources (Hillel, 1998; Jobbagy and Jackson, 2001; Perri et al., 2020).

10.3.2 Phytoremediation

Phytoremediation uses plants to clean up contaminated soils (Salt et al., 1998; Dietz and Schnoor, 2001; Pilon-Smits, 2005; Gerhardt et al., 2009), taking advantage of plants' ability to bioaccumulate chemicals through root uptake (e.g., Sec. 9.3.3). While it is cost-effective, phytoremediation has two major downsides: (i) it has a longer durations than traditional techniques, and (ii) there are potentially high contaminant leaching losses. To quantify the long-term mean extraction duration and leaching losses, here we follow

Manzoni et al. (2011b) to analyze the mass balance for the contaminant under the stochastic framework for the soil water. This allows us to quantify analytically the partitioning of contaminants between leaching and plant uptake as a function of rainfall statistics and soil and vegetation characteristics. This approach is suitable for the optimal design and risk assessment of remediation projects, as it provides a probabilistic representation of both contaminant concentrations through time and remediation duration, under various site management practices (e.g., regarding the choice of plant species and soil type).

We consider the total contaminant concentration (x, mass per unit soil volume, ML^{-3}), which includes the concentration in solution (x_s, mass per unit soil water volume, ML^{-3}) and the adsorbed contaminant concentration (x_a, mass per unit soil mass). We also use a linear adsorption isotherm ($x_a = K_d x_s$) to describe the adsorption kinetics (e.g., Verma et al., 2006; Trapp and Matthies, 1995). This allows us to express x as $ns x_s + \rho_b x_a = (ns + \rho_b K_d) x_s$ and thus x_s becomes a function of s and x,

$$x_s(x, s) = \frac{x}{ns + \rho_b K_d}. \tag{10.29}$$

The dynamics of the soil contaminant follows the minimalist model of Sec. 10.3.1, where percolation and runoff events are assumed to be instantaneous when the soil moisture reaches a threshold s_1. These events are assumed to be a renewal point process with mean frequency $\lambda_p = \rho(s_1) p(s_1)$ and depth $\alpha_p = \alpha$ (see Sec. 10.3.1). Thus, similarly to Eq. (10.25), leaching events are modeled as a multiplicative process forced by marked Poisson events,

$$\text{LE} = x_s(x, s = s_1) \diamond \mathcal{F}_{\lambda_p, 1/\alpha_p}(t) \tag{10.30}$$

where $s = s_1$ indicates that percolation takes place at s_1. The uptake of contaminants may be approximated as a passive process (see Sec. 9.3.3) that is simply proportional to the mean transpiration rate and the contaminant concentration in the soil solution at average soil moisture,

$$\text{UP} = \kappa \langle E \rangle x_s(x, s = \langle s \rangle) \tag{10.31}$$

where κ is a nondimensional coefficient often referred to as the transpiration stream concentration factor, which is nearly constant for low x_s (Vogeler et al., 2001; Mathur, 2004). The long-term transpiration rate, $\langle E \rangle$ (see Sec. 7.6.2), and the mean soil moisture, $\langle s \rangle$, are used to reflect the fact that soil moisture fluctuations are much faster than the contaminant dynamics.

With the leaching and uptake defined in Eqs. (10.30) and (10.31), we can express the mass balance equation for contaminant concentration as (Manzoni et al., 2011b)

$$Z_r \frac{dx}{dt} = -\text{UP} - \text{LE} = -\frac{\kappa \langle E \rangle}{n \langle s \rangle + \rho_b K_d} x - \frac{1}{ns_1 + \rho_b K_d} x \diamond \mathcal{F}_{\lambda_p, 1/\alpha_p}(t). \tag{10.32}$$

An example of a simulated time series of x, shown in Fig. 10.19c (black curve), is close to the one from the fully coupled model (hidden gray curve), supporting the assumption that contaminant dynamics are primarily controlled by the mean soil moisture between leaching events.

Similarly to Eq. (10.26) for soil salinization, the multiplicative noise in Eq. (10.32) can be converted to an additive noise by means of a change of variable, obtaining

$$\frac{dy}{dt} = -\beta - \mathcal{F}_{\lambda_p, 1/\alpha_p}(t), \tag{10.33}$$

where $y = Z_r(ns_1 + \rho_b K_d) \ln x$, and $\beta = \kappa \langle E \rangle (ns_1 + \rho_b K_d)/(n \langle s \rangle + \rho_b K_d)$. This type of stochastic differential equation with constant drift term has already been seen in Sec. 7.5. Here we are interested in the duration

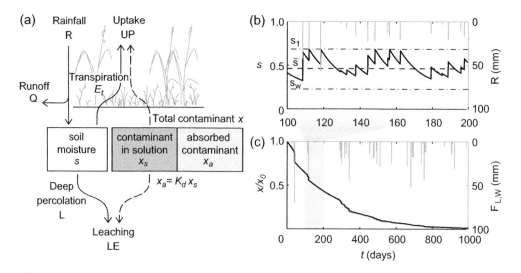

Figure 10.19 (a) Modeling scheme including the soil water and contaminant mass balances. The boxes represent compartments; the arrows indicate the mass fluxes for water (solid arrows) and contaminant (dashed arrows). Time series of (b) daily soil moisture, s (solid black curve), driven by rainfall, R (gray bars) and (c) the long-term contaminant dynamics, without simulating the daily fluctuations in the soil moisture (solid black curve), driven by mean uptake and stochastic leaching events (gray bars), are very close to the fully coupled model (solid gray curve, overlapping the black curve). Model parameters: $\lambda = 0.2$ day^{-1}, $\alpha = 0.015$ m, $K_d = 10^{-6}$ m^3 g^{-1}, $\kappa = 0.75$ (trichloroethylene contamination in a sandy loam soil low in organic matter). After Manzoni et al. (2011b).

of the remediation process $\langle T \rangle$, starting from the initial condition x_0, needed to reach an acceptable level of contaminant, x_c. The mean first passage time $\langle T \rangle$ can be calculated as (see Sec. 6.8)

$$\langle T \rangle = \frac{Z_r \left(ns_1 + \rho_b K_d\right) \ln\left(x_0/x_c\right)}{\langle R \rangle + \langle E \rangle \left[\kappa \left(ns_1 + \rho_b K_d\right) / \left(n\langle s \rangle + \rho_b K_d\right) - 1\right]}, \tag{10.34}$$

which gives the duration and the ratio of initial and target contaminant concentrations as functions of rainfall, contaminant, and plant characteristics.

Plants not only provide a sink for the contaminant but also decrease the soil moisture and hence lower or delay leaching losses (Chen et al., 2004). This important indirect effect of plants on contaminant fate is captured by this model through the coupling of plant transpiration and soil moisture in the soil water balance equation, as higher transpiration reduces the long-term percolation and runoff. To assess the phytoremediation efficiency, Manzoni et al. (2011b) calculated the ratio of total bio-accumulated contaminant and the mean total mass loss. At any given time during the phytoremediation, Eq. (10.32) can be averaged to find the mean plant uptake as $\langle UP \rangle = \kappa \langle x \rangle \langle E \rangle / (n\langle s \rangle + \rho_b K_d)$ and approximate the mean leaching as $\langle LE \rangle = \langle x \rangle \alpha_p \lambda_p / (ns_1 + \rho_b K_d)$. The phytoremediation efficiency then can be expressed as

$$\chi = \frac{\langle UP(t) \rangle}{\langle UP(t) \rangle + \langle LE(t) \rangle} = \left[1 + \frac{\alpha_p \lambda_p \left(n\langle s \rangle + \rho_b K_d\right)}{\kappa \langle E \rangle \left(ns_1 + \rho_b K_d\right)}\right]^{-1}, \tag{10.35}$$

which does not depend on time because the contaminant fluxes scale as the concentrations.

The rainfall regime affects the duration of phytoremediation through its controls on soil moisture and in turn on the frequency and intensity of leaching events. Thus remediation is faster in humid climates,

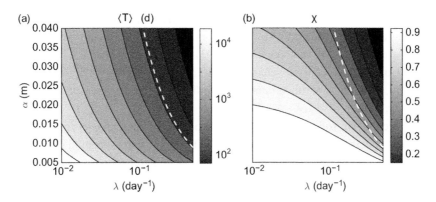

Figure 10.20 Mean remediation (a) duration $\langle T \rangle$ and (b) efficiency χ, as functions of mean rainfall depth α and frequency λ ($x_c/x_0 = 0.1$ and other parameters are listed in Fig. 10.19). The dashed lines indicate climatic conditions for which the mean rainfall equals the potential evapotranspiration, $\langle R \rangle = \alpha \lambda = E_{max}$; thus below the dashed lines the system tends to be more water-limited. After Manzoni et al. (2011b).

because of the increased plant uptake and leaching losses. However, this may lead to the significant contamination of groundwater and rivers downstream because of enhanced leaching. To quantify this tradeoff, in Fig. 10.20, we compare the mean remediation duration $\langle T \rangle$ with the phytoremediation efficiency χ. Drier climates on average need longer durations to achieve the same concentration goal, but under these drier conditions bioaccumulation is the main contaminant export pathway (high χ). For example, durations of the order of few years with relatively high mean efficiencies ($\chi \geq 0.6$) can be achieved for trichloroethylene in climates with a mean growing season rainfall in the range 300–500 mm (the central area in Fig. 10.20).

10.4 Water Resource and Agriculture Management as Stochastic Optimization Problems

The stochastic behaviors of hydroclimatic forcing and socio-economic dynamics have strong impacts on the availability of soil and water resources, resulting in complicated decision-making processes in agroecosystems. To overcome these problems, one may seek solutions by stochastic optimization methods to minimize costs or maximize gains (Spall, 2005). In this section, we offer three examples of optimization under uncertainties due to hydroclimatic forcing. We consider the optimization of irrigation strategies, following the previous developments in Sec. 10.2, optimal fertilization rates for maximum profit, and the optimal design of a rainwater cistern with minimal costs. These examples illustrate the complications brought about by rainfall stochasticity, even for the relatively easier conditions of a steady state.

10.4.1 Optimizing Irrigation for Crop Yield, Profitability, and Sustainability

A sustainable water resource management for agriculture should preserve ecosystems and their services while also guaranteeing the necessary yields and economic profits. Choices of irrigation strategies and water management must therefore simultaneously optimize crop yield, profitability, and sustainability, while considering the conflicting needs of ecosystems, farmers, and society. Sustainability is achieved when local and regional agricultural practices are such that land, water, and other resources are not degraded, thus providing similar opportunities in the future in terms of both available resources and their quality (Oster and Wichelns, 2003). Besides the essential impact of the unpredictability of rainfall on irrigation, discussed in Sec. 10.2, the problem of "optimal" irrigation is further complicated by a number of uncertainties regarding both the economic situation and the actual crop productivity.

In this section, we follow Vico and Porporato (2011b) and consider the total irrigation volume, which includes not only the ideal volume from Eq. (10.17) but also non-beneficial water uses such as evaporation from furrows, runoff from the field, wind drift, and interception by vegetation. The ratio of the ideal irrigation volume and the total irrigation volume is defined as the irrigation application efficiency, η_A. Even though this efficiency may be affected by a number of conditions, it is mainly controlled by the irrigation method and thus may be modeled as a function of the irrigation parameters (i.e., the intervention point $\tilde{s}$ and the target level $\hat{s}$). In most settings, the application efficiency does not exceed 50% for surface systems but can reach 80% for sprinkler irrigation and even be above 90% for drip irrigation (Trout and Kincaid, 2007). Such behavior is modeled as

$$\eta_A(\tilde{s}, \hat{s}) = \left(\eta_{A,m} - \eta_{A,f}\right) \varphi(\tilde{s}, \hat{s}) + \eta_{A,f}, \tag{10.36}$$

where $\eta_{A,f}$ is the minimum application efficiency (here assumed comparable with the value typical of furrow irrigation, e.g., $\eta_{A,f} = 0.5$), and $\eta_{A,m}$ is the maximum application efficiency (corresponding to modern micro-irrigation, e.g., $\eta_{A,m} = 0.9$). The function $\varphi(\tilde{s}, \hat{s})$ ranges from 0, for no irrigation applications, to 1, for micro-irrigation, and gives the relationship between application efficiency and the different irrigation methods. A flexible function to describe this dependence on the irrigation parameters $\tilde{s}$ and $\hat{s}$ is

$$\varphi(\tilde{s}, \hat{s}) = \left(1 + c_2 - e^{-1}\right)^{-1} \left[1 - \exp\left(-\left(\frac{\tilde{s}}{\hat{s}}\right)^{c_1}\right) + c_2 \left(\frac{s}{s}\right)^{c_3}\right]. \tag{10.37}$$

For positive values of the three parameters c_1, c_2, and c_3, this functional form reproduces the observed behavior of the efficiency, with a steep increase from furrow to sprinkler irrigation methods, a plateau covering most of the parameter combinations pertaining to traditional irrigation obtained through sprinklers, and a final increase for modern micro-irrigation. With this application efficiency and the ideal irrigation volumes in Eq. (10.17), we can find the total irrigation volume per unit area over a growing season of duration T_{seas}:

$$V(\tilde{s}, \hat{s}) = \eta_A(\tilde{s}, \hat{s})^{-1} V_{\text{ideal}}(\tilde{s}, \hat{s}) = \eta_A(\tilde{s}, \hat{s})^{-1} n Z_r (\hat{s} - \tilde{s}) \rho(\tilde{s}) p(\tilde{s}) T_{\text{seas}}, \tag{10.38}$$

where $p(\tilde{s})$ can be found from Eq. (10.14).

When adequate soil nutrients are available, and in the absence of pests and diseases, the yields per unit cultivated area are primarily controlled by water availability (see Fig. 1.11). Fitting several empirical relationships of crop yield as a function of plant transpiration (Payero et al., 2006; Igbadun et al., 2007), Vico and Porporato (2011b) gives

$$Y = Y_{\text{max}} \frac{E_{\text{seas}}^a}{E_{\text{seas},50\%}^a + E_{\text{seas}}^a}, \tag{10.39}$$

where Y_{max} represents the maximum yield (i.e., the asymptotic yield for very high E_{seas}), $E_{\text{seas},50\%}$ is the E value corresponding to a yield of $Y_{\text{max}}/2$, and the parameter a defines the steepness of the curve. Next, to minimize the supplied water (rainfall and irrigation) per unit crop yield it is useful to consider a "water productivity", defined as (Vico and Porporato, 2011b)

$$\text{WP} = \frac{Y}{V(\tilde{s}, \hat{s}) + \mathcal{R}_{\text{seas}}}, \tag{10.40}$$

which takes into account the total supplied water (through both rainfall $\mathcal{R}_{\text{seas}}$ and irrigation $V(\tilde{s}, \hat{s})$) rather than the more customary supplied irrigation water alone. Thus, WP allows us to evaluate both the generalized irrigation scheme and rainfed agriculture using the same metric.

To be profitable, an agricultural enterprise needs to balance the investments and costs required to achieve the expected yields and their related incomes. The gross income per unit cultivated area, G, is determined by the yield Y times the crop sale price (i.e., the crop price received by the farmer) c_c, i.e., $G = c_c Y$ (subsidies are not included here). The costs sustained by the farmer can be divided into three main types (Vico and Porporato, 2011b): fixed costs per unit area for crops, C_0, fixed irrigation costs, C_J, and variable irrigation-related costs, C_w. The first type of cost is associated with seeds, pesticides, fertilizer, field machinery, labor, and operation expenses. The second is related to irrigation equipment capital costs, maintenance, and irrigation-related labor. In general, irrigation technologies with higher application efficiencies have higher costs of installation and maintenance. As a first approximation, the fixed irrigation cost can be linked to the irrigation efficiencies in Eq. (10.37) and modeled as $C_J = c_{J,m} \varphi(\tilde{s}, \hat{s})$, in which $c_{J,m}$ is the fixed cost to apply modern micro-irrigation per unit cultivated area. The last type of cost is mainly dependent on the amount of applied water (e.g., as fuel for pumping or price paid for off-farm water; application-specific labor), C_w. Assume a linear dependence of the variable irrigation costs on the applied irrigation water, $C_w = c_w V$, where c_w is the cost per unit applied water. According to the above assumptions, the net income per unit cultivated area, N, may be expressed as

$$N = G - [C_0 + C_J + C_w] = c_c Y - \left[C_0 + c_{J,m} \varphi(\tilde{s}, \hat{s}) + c_w V \right]. \tag{10.41}$$

Vico and Porporato (2011b) parameterized the above irrigation strategies for US corn agriculture (see Table 10.1) and determined water requirements, crop yield, water productivity, and economic gain for given climatic conditions, water costs, and crop sale prices. Figure 10.21 reports these quantities as a function of the irrigation parameters $\tilde{s}$ and $\hat{s}$, for a total seasonal rainfall of 248 mm and an average crop sale price and water cost (see Table 10.1). As expected, there is no single irrigation strategy that simultaneously minimizes water requirements while also maximizing crop yield and economic gain. Hence, the choice of the best strategy depends on the importance associated with each of these criteria. If water supply is extremely limited, rainfed agriculture or extreme deficit irrigation are the best strategies (Fig. 10.21a), while deficit micro-irrigation allows the most efficient use of scarce available water (Fig. 10.21c). The water savings associated with rainfed agriculture and deficit irrigation, however, cause a significant reduction in crop yield (Fig. 10.21b) and net gain, in particular if investment-intensive micro-irrigation is applied (Fig. 10.21d). In contrast, when larger amounts of water are available, stress-avoidance irrigation guarantees higher crop yields and higher gross incomes, with net gain depending on the cost of the irrigation method employed. Hence, if economic gain maximization is the target, a mild deficit traditional irrigation with relatively infrequent but large irrigation applications (i.e., relatively high $\hat{s}$) is the most profitable strategy (Fig. 10.21d), at the same time leading to near-maximum crop yields (Fig. 10.21b). Micro-irrigation is not the most profitable solution in this case, because current water cost and crop sale price are relatively low and thus the high installation costs of micro-irrigation are not offset by the limited savings associated with lower water requirements and by the yields obtained.

Table 10.1 Parameters describing irrigation application to _Zea mays_ grown on a sandy loam soil. After Vico and Porporato (2011b).

Parameter	Unit	Value	Description
a	—	2.57	Eq. (10.39)
C_0	$ ha^{-1}	600	Eq. (10.41)
c_1	—	1/3	Eq. (10.37)
c_2	—	1/2	Eq. (10.37)
c_3	—	8	Eq. (10.37)
c_c	$ kg^{-1}	0.12	Eq. (10.37)
$c_{I,m}$	$ ha^{-1}	658	Eq. (10.41)
c_w	$ m^{-1}ha^{-1}	148	Eq. (10.41)
E_{max}	cm day^{-1}	0.55	max. transpiration
$E_{seas,50\%}$	mm	694	Eq. (10.39)
n	—	0.45	porosity
s^*	—	0.28	onset of crop stress
s_{fc}	—	0.62	field capacity
T_{seas}	days	110	e.g., Eq. (10.38)
Y_{max}	ton ha^{-1}	26	Eq. (10.39)
Z_r	cm	50	root-zone depth
$\eta_{A,f}$	—	0.5	Eq. (10.36)
$\eta_{A,m}$	—	0.9	Eq. (10.36)

The fluctuations in water and crop prices directly influence the optimal irrigation strategies. When water costs are higher, micro-irrigation may be more profitable than traditional irrigation for higher intervention points. For traditional irrigation, the optimal intervention point increases with increasing crop sale price because, when the crop value on the market is high, the higher investments in water required by stress-avoidance irrigation are offset by returns from crop sales, while low crop values may not justify the increased water expenditures associated with stress-avoidance irrigation. Where water costs are elevated, irrigation planning must produce a crop-price forecast for the season, which represents a further source of uncertainty. It should finally be noted that the question of irrigation sustainability may also require the consideration of other environmental factors, such as the risks of soil erosion or soil salinization (see Sec. 10.3.1).

10.4.2 A Dynamical System of Crop Model for Optimal Fertilization

Nitrogen-rich organic amendments and mineral fertilizers have played a pivotal role in the so-called green revolution, making it possible to dramatically increase food production. However, such unprecedented inputs of nitrogen, phosphorous, and potassium to agricultural soils have also altered the natural biogeochemical cycles (Fig. 10.1), resulting in diffuse eutrophication (Vitousek et al., 1997) and contributing to greenhouse gas accumulation. Once in the ecosystems (either natural or managed), the concentrations and fate of nutrients depend on climate and its interactions with vegetation and soil biota. Not only does the soil moisture play a major role in controlling the balance of nutrient uptake by vegetation and leaching (similarly to how it affects the balance of soil contaminants, see Sec. 10.3.2), but it also impacts the mineralization of soil organic matter (see Chapter 9).

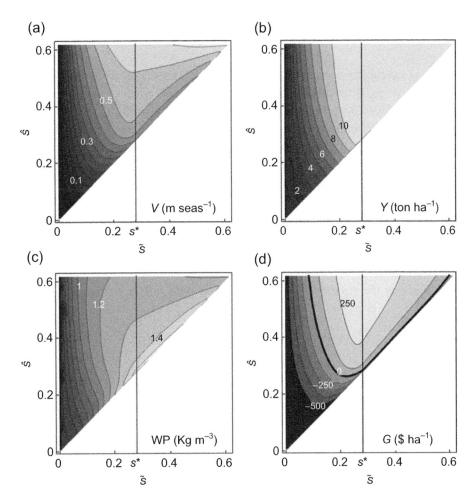

Figure 10.21 Total water requirements (a), average crop yield (b), water productivity (c), and economic gain (d), as a function of the irrigation parameters for *Zea mays*. The parameters are listed in Table 10.1. The vertical lines correspond to $\tilde{s} = s^*$. After Vico and Porporato (2011b).

The amount of fertilizer (here identified with mineral nitrogen for simplicity) that needs to be supplied to maximize profit is strongly linked to hydrologic variability. Several data-driven, plot- and watershed-scale models to predict the soil N fate have been proposed, including crop models (e.g., Birkinshaw and Ewen, 2000; Maggi et al., 2008; Steduto et al., 2009). The latter represent an important tool for studying the impact of different management strategies aimed at maximizing yield, minimizing water and fertilizer use, and reducing the leaching of fertilizers under hydroclimatic variability. Here we follow Pelak et al. (2017), who reinterpreted the AquaCrop model (Steduto et al., 2009) in the form of a dynamical system (Sec. 2.4) to model crop biomass, yield, and profitability under varying climate scenarios, irrigation strategies, and fertilization treatments. The system consists of a set of ordinary differential equations (ODEs) describing the states of crop canopy cover, $c(t)$, soil moisture $s(t)$, and soil nitrogen $N(t)$ at the daily timescale. The explicit inclusion of stochastic rainfall forcing is crucial for long-term design and risk assessment.

The rate of change in canopy cover, modeled as a balance between the increase due to canopy growth and the decrease due to metabolic limitations and senescence, is given by

$$\frac{dc}{dt} = \underbrace{r_G \text{UP}}_{\text{growth}} - \underbrace{\left[r_M + \gamma_m (t - t_{\text{sen}}) \Theta (t - t_{\text{sen}})\right] c^2}_{\text{senescence}} \tag{10.42}$$

where r_G is the canopy-cover increase per amount of nitrogen taken up (UP), r_M is a constant metabolic limitation term, γ_m is the mortality rate after the senescence onset time (t_{sen}), and $\Theta(\cdot)$ is the Heaviside step function (Sec. 6.3.4). The first term gives the canopy growth rate and the second term combines the effects of metabolic limitation and senescence. For unstressed conditions (sufficiently high soil moisture and nitrogen) prior to t_{sen}, Eq. (10.42) is the logistic growth equation (Murray, 2002), which includes the approximately exponential growth of c in the initial growth stage, the slowing of growth as a limit is reached, and the negligible growth rate near the carrying capacity.

The governing equation for soil moisture is adapted from the water balance equation (10.8), where the evapotranspiration is further divided into crop transpiration and soil evaporation, $E = E_t + E_s$. Crop transpiration is modeled as

$$E_t = c K_t(s) E_{t,\max}, \tag{10.43}$$

where $E_{t,\max}$ is the maximum crop transpiration with full crop canopy cover and the water stress coefficient, $K_t(s)$, is expressed as

$$K_t(s) = \begin{cases} 0 & s \le s_w, \\ \dfrac{s - s_w}{s^* - s_w} & s_w < s \le s^*, \\ 1 & s > s^*, \end{cases} \tag{10.44}$$

which captures the plant stomatal response to the soil moisture conditions. The soil evaporation rate, E_s, is given by

$$E_s = (1 - c) K_s(s) E_{s,\max}, \tag{10.45}$$

where $E_{s,\max}$ is the maximum soil evaporation with no canopy cover and $K_r(s)$ is a soil evaporation reduction coefficient,

$$K_s(s) = \begin{cases} 0 & s \le s_h, \\ \dfrac{s - s_h}{1 - s_h} & s \ge s_h. \end{cases} \tag{10.46}$$

A diagram showing K_t and K_s as functions of s is shown in Fig. 10.22a; $E_{t,\max}$ and $E_{s,\max}$ are taken to be of similar magnitude, to capture the dominance of E_s shortly after planting and that of E_t later in the growing season (Kelliher et al., 1995a).

Most crop models include a nitrogen balance owing to its key role in the growth and development of crops (e.g., Stöckle et al., 2003; Brisson et al., 2003). In order to examine crop growth and yield under various management conditions, total mineral nitrogen (no distinction is made here between different mineral forms) is modeled by the balance between deposition and fertilization as inputs and leaching and plant uptake as outputs (also see Sec. 9.3):

$$\frac{dN}{dt} = D + F - \text{LE} - \text{UP}, \tag{10.47}$$

where N is the nitrogen content per unit area of soil, D is the rate of natural nitrogen addition to the soil and is assumed to be constant for a heavily agricultural region, and F is the fertilization rate, assumed here as a constant control parameter for the entire length of the growing season, t_{GS} (i.e., fertilization is

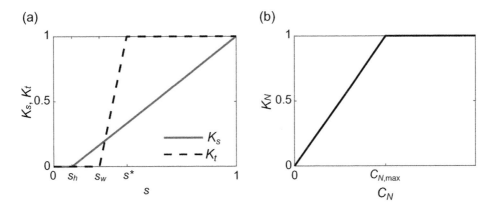

Figure 10.22 (a) Crop water stress coefficient (K_t, dashed line) and soil evaporation reduction coefficient (K_s, solid line) as a function of soil moisture, s. Adapted from Pelak et al. (2017). (b) Nutrient limitation coefficient K_N as a function of nutrient concentration C_N.

assumed to be applied continuously in time either because of the slow release of minerals into the soil solution or because of its application in combination with micro-irrigation).

The leaching of nitrogen, LE, is given in Eq. (9.40), and reported here for convenience, as

$$\text{LE} = \frac{aN}{snZ_r}L = C_N L, \tag{10.48}$$

where $L = K(s) = K_s s^c$ is the deep percolation (see Eq. (3.13) and Sec. 3.3.3), a is the fraction of the nitrogen content N dissolved in the soil water ($a \approx 1$ for nitrate, while $a \leq 1$ for ammonium), and $C_N = aN/(snZ_r)$ is the nitrogen concentration. The plant uptake of nitrogen, UP, is given in Eqs. (9.41)–(9.43) and simplified as

$$\text{UP} = K_N C_{N,\text{max}} E_t, \tag{10.49}$$

where $C_{N,\text{max}}$ is the maximum plant demand and the coefficient K_N is expressed as

$$K_N = \begin{cases} 1 & C_N \leq C_{N,\text{max}}, \\ C_N/C_{N,\text{max}} & C_N > C_{N,\text{max}}, \end{cases} \tag{10.50}$$

which accounts for the nitrogen limitation due to plant demand (see an example in Fig. 10.22b).

Besides c, s, and N, Pelak et al. (2017) considered the crop biomass B and crop yield Y. The accumulation of plant biomass is associated with the crop transpiration and is limited by nutrient availability (e.g., Steduto et al., 2009):

$$\frac{dB}{dt} = w^* E_t K_N. \tag{10.51}$$

where w^* is the water productivity. The yield is assumed to be a fraction of the plant biomass,

$$Y = hB, \tag{10.52}$$

where h is the harvest index.

The complete model is defined by the balance equations for C, s, and N and their component fluxes. The rainfall stochastic forcing adds an important dynamic component to the coupled model. Figure 10.23

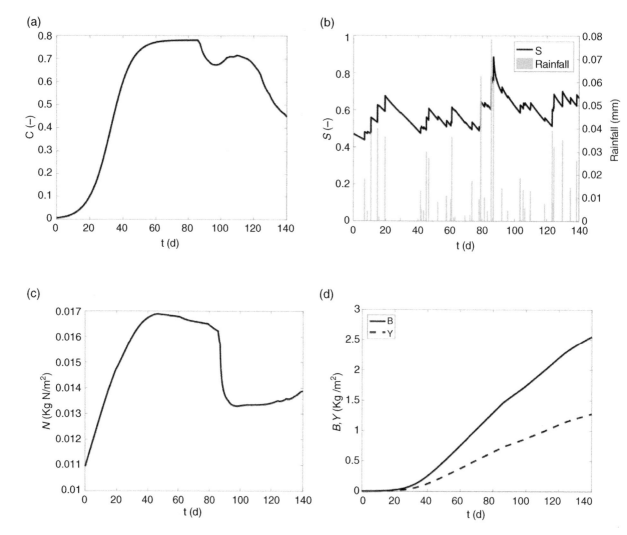

Figure 10.23 Time series of (a) canopy cover C, (b) soil moisture s and rainfall R, (c) soil nitrogen N, (d) crop biomass B and yield Y over a growing season. After Pelak et al. (2017).

shows the development of the main system variables over the course of a growing season with a constant rate of nitrogen fertilization or deposition. The intra-seasonal time dependence of C is associated with the logistic growth term and the time-dependent senescence term in Eq. (10.42); the soil moisture s is driven by the stochastic forcing of rainfall as discussed in Chapter 7; the soil nitrogen content N follows the pattern of C.

To assess the impact of different fertilization strategies under hydroclimatic variability, Pelak et al. (2017) considered the profit from the sale of produce, the costs of fertilizer and irrigation, the "environmental cost" of nitrogen leaching, and the fixed costs, resulting in a net profit

$$P_{net} = p_Y Y (t_{G\,S}) - p_F F_{tot} - p_I I_{tot} - p_L L_{tot} - p_{fix}, \qquad (10.53)$$

where p_Y ($/kg Y) is the unit sale price of the crop yield at the end of the growing season, $Y(t_{GS})$; p_F ($/kg N) and p_I ($/m^3) are the unit prices of fertilizer and irrigation water, respectively. The cumulative

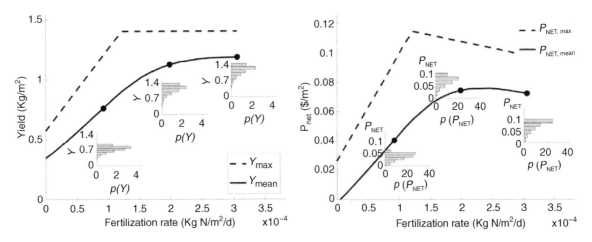

Figure 10.24 Yield and profit as a function of the fertilization rate. The dashed lines represent the theoretical maximum while the solid lines give the mean of many simulations. The inset plots show the numerical probability density functions at the points indicated. After Pelak et al. (2017).

fertilization and the irrigation values are given by $F_{\text{tot}} = \int_0^{t_{\text{GS}}} F(t)dt$ and $I_{\text{tot}} = \int_0^{t_{\text{GS}}} I(t)dt$. The unit environmental cost of leached nitrogen is given by p_L [\$/kg N], conceptualized as the cost necessary to mitigate these losses or to pay associated fines; p_{fix} [\$/m^2] is a fixed amount representing the distribution and energy costs (see Sec. 10.4.1).

Figure 10.24 shows the numerical PDFs of yield and profit for varying fertilization rates. As the fertilization rate increases, both the mean of the yield and its variance increase. The mean increases because the higher fertilization rates lead to a lower likelihood that the crop will experience a shortage of nitrogen; the extra nitrogen raises the maximum possible yield but has a limited impact on the yield in the dry season, leading to the large variance. This uncertainty highlights the importance of considering optimal fertilization under stochastic rainfall conditions. Thus, optimal strategies for a system undergoing stochastic forcing must attempt to maximize profit while also being robust to adverse conditions, such as drought or flood years. As noted by Pelak et al. (2017), this necessarily involves tradeoffs between maximizing yield and profit on the one hand and mitigating risk on the other, which is related to the resilience in the ecological and social systems (see for example Walker et al., 2004).

10.4.3 Design of a Minimal-Cost Rainwater Harvesting Cistern

Rainwater harvesting (RWH) is seen as an increasingly attractive option for a sustainable use of water resources and to reduce the pressure on diminishing water supplies in many regions of the world (Pandey et al., 2003). Its potential for runoff reduction also has benefits even in areas where water is relatively abundant (e.g., Steffen et al., 2013). The volume of the storage is perhaps the most important aspect in RWH design; it is associated with various costs and is complicated by the uncertainties of climate forcing. Here, we follow Pelak and Porporato (2016) in designing an optimal RWH cistern aimed at minimizing costs.

We start from the mass balance of water in an RWH cistern:

$$V\frac{dc}{dt} = \phi AR(t) - H(c) - Q(c), \tag{10.54}$$

where V is the cistern volume, c is the relative cistern storage volume ranging from zero to one, ϕ is a runoff coefficient, A is the area of the roof capable of collecting water, R is the stochastic rainfall rate with mean event depth α and mean frequency λ, Q is the loss rate due to overflow when rainfall completely fills the cistern, and H is the rate of demand of water from the cistern. Demand is assumed constant when the cistern is not empty:

$$H(c) = \begin{cases} 0 & c = 0, \\ h & c > 0. \end{cases} \tag{10.55}$$

When $c = 0$, water is assumed to be provided by other sources, such as a municipal supply.

Noting that the system is mathematically similar to the minimalist model for soil moisture balance developed by Milly (1993) (see Sec. 7.5 and Note 7.12), the solutions (7.39) and (7.40) can be rewritten for cistern storage as

$$p(c) = \frac{\gamma}{D_c} p_0 e^{\gamma(1/D_c - 1)c}, \tag{10.56}$$

$$p_0 = \frac{1 - D_c}{e^{\gamma(1/D_c - 1)} - D_c}. \tag{10.57}$$

where $p(c)$ is the PDF of c, and p_0 is the atom of probability corresponding to an empty cistern. The two dimensionless variables, similar to the storage and dryness indices defined in Eq. (7.46), are $\gamma = V/(\phi A \alpha)$ and $D_c = h/(\phi A \alpha \lambda)$. The long-term partition of rainfall in Eq. (7.52) can be rewritten for the cistern as

$$\langle H \rangle = h (1 - p_0), \tag{10.58}$$

$$\frac{\langle Q \rangle}{\phi A \langle R \rangle} = [1 - D_c(1 - p_0)]. \tag{10.59}$$

where $\langle \cdot \rangle$ refers to a long-term average. Figure 10.25 shows the analytical PDF solutions (right-hand vertical axes) and atom solutions (left-hand vertical axes) along with sample time series. As an analog to the root-zone depth, a larger cistern tends to lower the variability of water storage; cisterns with a larger roof top receive more water at each rainfall event and thus have a lower probability of being empty.

To design the cistern, Pelak and Porporato (2016) used the criterion of minimizing the total costs. The latter typically include the fixed cost of the cistern, G_f, and the distributed cost, G_d, of supplying the water from an external source when the cistern is empty with probability p_0 given in Eq. (10.57):

$$G = G_f + G_d = qV + rhTp_0, \tag{10.60}$$

where q is the unit cost per storage capacity, T is the lifetime of the cistern, and r is the unit water price. Equation (10.60) assumes that the fixed cost of a cistern increases linearly with its volume.

When increasing cistern size, the fixed costs always increase and the distributed costs always decrease, yielding at some point a minimum of the cost. Such a minimum and the corresponding optimal cistern size can be found by solving $dG/dV = 0$. The results are presented in Fig. 10.26 as a function of the rainfall parameters α and λ. For frequent but relatively small events, a larger volume is optimal (upper left), as even though the events are small they are numerous enough to keep a larger cistern full and thereby offset its cost. A larger cistern volume is also optimal for infrequent but large events (lower right), as here the additional volume allows for the system to take advantage of the large size of the events and to store the resulting rainfall during the longer dry periods.

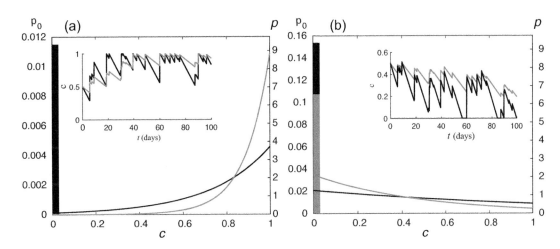

Figure 10.25 Steady-state PDFs of cistern storage, for different cistern volumes (black lines for $V = 7.95$ m^3; gray lines for $V = 19.2$ m^3) and different roof areas (panel (a) for $A = 267$ m^2; panel (b) for $A = 133.5$ m^2). The inset plots contain sample time series for each combination of roof area and cistern volume. Other parameters are: $h = 0.3$ m^3 day^{-1}, $T = 10950$ days, $\alpha = 8.2$ mm, $\phi = 0.8$, $\lambda = 0.31$ day^{-1}. After Pelak and Porporato (2016).

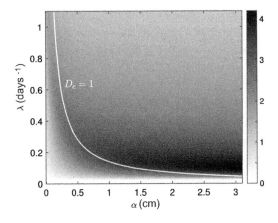

Figure 10.26 The optimal cistern size (m^3) as a function of the climate parameters α and λ. The peak cistern values for particular α and λ values occur near the white curve, which denotes the cistern demand index $D_c = 1$. The cistern parameters are listed in Fig. 10.25, and the cost parameters are $q = 157.4$ \$ m^{-3}, $r = 0.65$ \$ m^{-3}. After Pelak and Porporato (2016).

It should also be noted that in addition to its ability to reduce reliance on municipal supplies, rainwater harvesting may also reduce runoff. Many municipalities assess a storm water fee which may be (partially or wholly) based on the amount of impervious surface on a property (Wang and Zimmerman, 2015). Equation (10.59) can also be interpreted as the ratio of the runoff rates after and before the installation of an RWH system. If this fraction is multiplied by the roof area A, an adjusted roof area is obtained, which could then be used in place of the true roof area when measuring the impervious surface area of a property (Pelak and Porporato, 2016).

10.5 Sustainable Use of Soil and Water Resources as Optimal Control Problems

The sustainable management of soil and water resources requires finding optimal solutions under multiple constraints in systems that are often forced by unpredictable dynamics of a climatic, biological, and socio-economic nature. Control theory offers valuable tools towards this goal; however, the complexity arising from natural–human interactions poses novel challenges for these methods, which have been traditionally confined to more clearly defined engineering and industrial problems.[2]

In general, optimal control aims at finding a strategy that minimizes (respectively maximizes) a given cost (gain) functional, based on generalizations of the calculus of variations by Pontryagin and Bellman among others (Sussmann and Willems, 1997; Pesch et al., 2009; Liberzon, 2011). These methods allow us to rigorously compare a continuum of scenarios and control protocols to find optimal strategies in the entire "strategy space", where control parameters can be continuously changed in time. Applications abound, but only a few examples in the context of environmental problems can be found (see Note 10.19 and the references therein).

In this section, we offer two examples regarding the optimal use of soil and water resources. The first is a classical optimization problem of resource exploitation and refers to the case of groundwater pumping. In its minimalist form, this important problem in water-resource management allows us to present the main concepts of optimal control for transient conditions without excessive mathematical complications. The second problem regards the optimal protocols for sodic soil remediation in minimal time, offering an example of a so-called bang-bang control for a two-dimensional dynamical system in ecohydrology.

10.5.1 Optimal Groundwater Use

As the largest use of water resources globally, irrigation (Sec. 10.2) has led to the over-exploitation of aquifers in several regions of the world (Scanlon et al., 2012). Therefore, determining optimal pumping rates is a problem of considerable interest, linking ecohydrological and socio-economic aspects. In relation to the main tradeoffs and theoretical elements, this problem is also a prototypical example of optimization in resource economics. For more information about this rich field of research, we refer to the original works of Burt (1964) and Brown Jr. and Deacon (1972), which were based on the classical work of Hotelling (1931), models with multiple agents (Negri, 1989), as well as larger-scale analyses including spatial heterogeneities and multiple optimization goals (Cai, 2008).

The problem[3] can be cast by first writing the water balance for a phreatic aquifer (one whose water table is in contact with the unsaturated zone):

$$nA\frac{dh}{dt} = \mathcal{R} - \mathcal{Q} - \mathcal{B} \tag{10.61}$$

where h [L] is the water-table level, n [—] is the aquifer storage coefficient (i.e., the aquifer effective porosity), A [L^2] is the representative aquifer area, $\mathcal{R}$ [L^3/T] is the groundwater recharge, $\mathcal{Q}$ [L^3/T] is the

[2] After trying to ride unsuccessfully the South Sea Bubble in 1719–1720, Isaac Newton, frustrated with his trading experience, concluded (Brunnermeier and Oehmke, 2013): "I can calculate the motions of the heavenly bodies, but not the madness of people."

[3] We thank Sara Cerasoli for help in developing this section.

groundwater withdrawal by pumping, and $\mathcal{B}[L^3/T]$ is the flow out of the aquifer, called the baseflow. A profit or *utility functional* U [$] can be defined as

$$U = \int_{t_0}^{t_0+T} [\mathcal{G}(t) - \mathcal{C}(t)]e^{-\rho t}dt, \tag{10.62}$$

over a period T, where t_0 is the time at which pumping is started, $\mathcal{G}$ [T^{-1}] is the gain or revenue rate related to crop sale and other water uses, $\mathcal{C}$ [T^{-1}] is the loss due to both economic costs (e.g., infrastructure construction and electricity) and, possibly, environmental costs (e.g., soil subsidence, reduction in the baseflow fed by the aquifer), and ρ is the discount rate [T^{-1}] on future utility. The coupling between Eqs. (10.61) and (10.62) stems from the dependence of $\mathcal{G}$ and $\mathcal{C}$ on $\mathcal{Q}$.

Usually, the control problem consists in determining the optimal function $\mathcal{Q}(t)$ that maximizes the utility among its infinite possible values, over a finite or infinite ($T \to \infty$) horizon (i.e., a so-called intertemporal problem). The problem is translated into mathematical terms as the extremization of a utility functional, which depends on a *control variable*, in this case the normalized pumping rate, $q = \mathcal{Q}/(nA)$.

To keep the mathematical complications to a minimum, we assume constant recharge $\mathcal{R}/(nA) = r_0$ and a linear dependence of base flow on aquifer level, $B/(nA) = \beta h$, that gains are proportional to withdrawals, $\mathcal{G} = \kappa q$, and that the cost $\mathcal{C} = \mu (h_{\max} - h)^\delta q^\theta$ is dependent on both the pumping rate and the aquifer depletion $(h_{\max} - h)^\delta$, where $h_{\max}$ is the maximum aquifer level. As a result, the utility maximization over all possible pumping rates $q(t)$ is

$$\max_{q(t)} U = \int_{t_0}^{\infty} [\kappa q(t) - \mu (h_{\max} - h(t))^\delta q(t)^\theta]e^{-\rho t}dt, \tag{10.63}$$

subject to the constraint

$$\frac{dh(t)}{dt} = r_0 - q(t) - \beta h(t). \tag{10.64}$$

Before analyzing the full variational problem in transient conditions, it is instructive to consider the simpler steady-state problem, in which case the optimization is reduced to finding the maximum of a function. Accordingly, for a constant pumping rate $\bar{q}$ applied for a very long time, the problem reaches a steady state ($dh/dt = 0$) at the level

$$\bar{h} = (r_0 - \bar{q})/\beta. \tag{10.65}$$

The utility per unit time, neglecting the initial transient before steady state, is also constant in time,

$$\bar{U} = \int_{t_1 \gg t_0}^{t_1+\infty} [\kappa\bar{q} - \mu\bar{q}^\theta (h_{\max} - \bar{h})^\delta]e^{-\rho t}dt = \frac{\kappa}{\rho}\bar{q} - \frac{\mu}{\rho}\bar{q}^\theta (h_{\max} - \bar{h})^\delta, \tag{10.66}$$

where $\int_0^\infty e^{-\rho t}dt = 1/\rho$ is applied in the last equality. The steady-state utility function is shown in Fig. 10.27. The optimal pumping rate is the maximum given by the condition $d\bar{U}/d\bar{q} = 0$. It shows the optimum as the tangency point between the highest attainable utility $\bar{U}$ and the constraint in Eq. (10.65).

Now we can proceed to study the optimal protocol in transient conditions. It can be shown that over a long period this must lead to the steady-state solution that we have just analyzed (Arrow, 1967). The first step in the variational analysis is to build the Hamiltonian (Liberzon, 2011):

$$\mathcal{H} = e^{-\rho t}[\kappa q(t) - \mu (h_{\max} - h(t))^\delta q(t)^\theta] + \lambda(t)[r_0 - q(t) - \beta h(t)]. \tag{10.67}$$

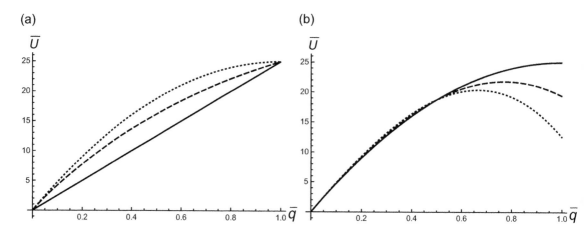

Figure 10.27 Steady-state utility as a function of the aquifer pumping rate for different values of the exponents δ and θ in (10.63). Parameter values: $\kappa = 1$, $\mu = 0.5$, $h_{max} = 2$, $R = 1.5$, $\beta = 1$, $\rho = 0.02$. (a) The role of δ: $\theta = 2$ and $\delta = 0$ (solid line), $\delta = 1/2$ (dashed line), $\delta = 1$ (dotted line); (b) the role of θ: $\delta = 0$ and $\theta = 1$ (solid line), $\theta = 3/2$ (dashed line), $\theta = 2$ (dotted line).

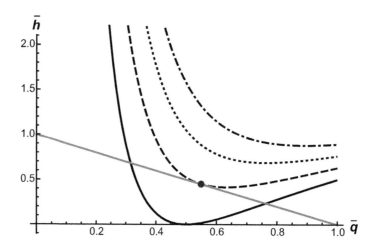

Figure 10.28 Optimal withdrawal at steady state. The optimum is the tangency point between the highest attainable utility $\bar{U}$ (dashed curve) and the constraint (10.65) (diagonal line). Values for parameters: $\kappa = 1$, $\mu = 0.5$, $h_{max} = 2$, $R = 1$, $\beta = 1$, $\rho = 0.02$, $\theta = 2$ $\delta = 1$, and $\bar{U} = 0.25$ (solid line), $\bar{U} = 0.32$ (dashed line), $\bar{U} = 0.38$ (dotted line), $\bar{U} = 0.5$ (dot-dashed line).

Applying the maximum principle (Liberzon, 2011), we obtain the following conditions:

$$\frac{\partial \mathcal{H}}{\partial q} = e^{-\rho t}\left[\kappa - \theta\mu q(t)^{\theta-1}(h_{max} - h(t))^{\delta}\right] - \lambda = 0, \tag{10.68}$$

$$\frac{\partial \mathcal{H}}{\partial h} = \delta\mu e^{-\rho t}q(t)^{\theta}(h_{max} - h(t))^{\delta-1} - \beta\lambda(t) = -\dot{\lambda}, \tag{10.69}$$

$$\lim_{t \to \infty} \lambda h(t) = 0, \tag{10.70}$$

where Eq. (10.70) is the so-called transversality condition, to be satisfied at the end of the period. From the first condition, we can extract

$$\lambda = e^{-\rho t}\left[\kappa - \theta\mu q(t)^{\theta-1}(h_{\max} - h(t))^{\delta}\right],\tag{10.71}$$

which, substituted in the second condition, gives the governing equation for the dynamical system

$$\begin{cases}\dot{q} = \dfrac{\delta q}{(\theta-1)(h_{\max}-h)}\left[\begin{array}{l} -\dfrac{\kappa(\beta+\rho)q^{1-\theta}(h_{\max}-h)^{1-\delta}}{\delta\theta\mu} \\ +\dfrac{(\beta+\rho)(h_{\max}-h)}{\delta} - \beta h + \dfrac{q}{\theta} - q + R\end{array}\right] \\ \dot{h} = r_0 - q - \beta h,\end{cases}\tag{10.72}$$

where both h and q are functions of t.

To study the system's stability, we linearize around the steady state (h^*, q^*) as follows (see Sec. 2.4):

$$\begin{bmatrix}\dot{\hat{h}}(t) \\ \dot{\hat{q}}(t)\end{bmatrix} \approx \mathcal{J}(h^*, q^*) \times \begin{bmatrix}\hat{h}(t) \\ \hat{q}(t)\end{bmatrix},\tag{10.73}$$

where $\mathcal{J}$ is a Jacobian matrix whose determinant can be shown to be negative.[4]

The value of the negative determinant equals the product of the eigenvalues that are opposite in sign, showing that the steady state is a saddle point (see Table 2.4), as shown in the phase diagram (Fig. 10.29). The optimal path is forced to be on the stable arm (or *saddle path*) as all the other paths would violate either feasibility (positive pumping $q > 0$ with fully depleted aquifer $h = 0$) or Eq. (10.70): for any given initial aquifer level h_0, q_0 will jump onto this arm and then travel towards the steady point (Acemoglu, 2008). The expression for the saddle path can be obtained by the time-elimination method (Mulligan and Martin, 1991) as

$$q'(h) = \frac{dq}{dh} = \frac{dq/dt}{dh/dt} = \frac{\dot{q}(t)}{\dot{h}(t)},\tag{10.75}$$

which is a nonlinear ODE in $q(h)$ that usually has no analytical solution. Once it is solved numerically, the optimal pumping rate is defined for each h. The temporal dynamics of the optimal aquifer-withdrawal dynamics is obtained by inserting $q(h)$ into the aquifer balance equation (10.61),

$$\dot{h}(t) = r_0 - q(h(t)) - \beta h(t).\tag{10.76}$$

Let us now consider the case of an aquifer initially with no pumping, i.e. $q = 0$, starting from a generic initial state ■): when left undisturbed with recharge r_0, the water table level tends to the natural steady state of the aquifer, $h_n^* = r_0/\beta$. At a certain time t^*, when the system is at the state □, the aquifer starts to be exploited. If the exploitation is carried out optimally, the pumping rate follows Eq. (10.75) and the water table decreases up to its new steady state (⋆), following Eq. (10.76).

A suboptimal exploitation leads to a loss in the net profit and/or to aquifer depletion. Let us consider an over-exploited aquifer: starting from a state ● that is the result of an excessive pumping rate, it undergoes heavy depletion up to a state ○. At this point, the optimal recovery consists in jumping to the saddle path

[4] This can be shown explicitly for the case $\delta = 1, \theta = 2$, where

$$\det \mathcal{J}(h^*, q^*) = \begin{vmatrix} \partial\dot{h}/\partial h & \partial\dot{h}/\partial q \\ \partial\dot{q}/\partial h & \partial\dot{q}/\partial q \end{vmatrix}$$

$$= \begin{vmatrix} -\beta & -1 \\ \dfrac{q^*(-q^*+2R-2h_{\max}\beta)\mu-2\kappa(\beta+\rho)}{2(h_{\max}-h^*)^2\mu} & \dfrac{-q^*+R+h_{\max}(\beta+\rho)-h^*(2\beta+\rho)}{h_{\max}-h^*} \end{vmatrix} < 0.\tag{10.74}$$

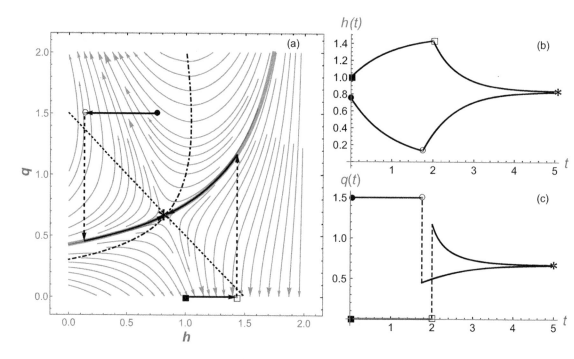

Figure 10.29 (a) Phase diagram for the system (10.72). The intersection between the nullclines $\dot{h} = 0$ (dotted line) and $\dot{q} = 0$ (dot-dashed line) locates the steady point (h^*, q^*) with saddle stability ($\star$). The thick gray solid line represents the saddle path (10.75). The black arrows represent the optimal paths. Parameter values: $\kappa = 1$, $\mu = 0.5$, $h_{\max} = 2$, $R = 1.5$, $\beta = 1$, $\rho = 0.02$, $\theta = 2$, $\delta = 1$. Time evolution of the (b) aquifer level h and (c) pumping rate q, for two cases: an aquifer initially not exploited, with initial conditions $h_0 = 1$, $q_0 = 0$ ($\blacksquare$), then optimally exploited from time $t^* = 2$ ($\square$): and an overexploited aquifer with initial conditions $h_0 = 0.75$, $q_0 = 1.5$ ($\bullet$), with optimal recovery from $t^* = 1.75$ ($\circ$).

(at time t^*) and following it toward the steady state $\star$. Figures 10.29b and c show the time evolution for these cases.

It is important to note that once the optimal path is chosen, control of the pumping rate is continuous until the steady state is reached. This is not always the case, because sometimes the optimal control may be a switching function (i.e., a piecewise continuous function). This happens when the control variable, here the rate of pumping q, is linearly expressed in the utility function and the constraint condition (i.e., $\theta = 1$). In such a case, the highest attainable utility occurs at the minimum or maximum value of the control, depending on the parameters of the problem; these are the so-called corner solutions (Kamien and Schwartz, 2012). The resulting type of control is also know as *bang-bang control*, and it appears in the next section for the case of the optimal remediation of sodic soils.

10.5.2 Sodic Soil Remediation

The problem of the remediation of saline and sodic soils offers an interesting example where optimal control can be applied neatly, with the quality of the irrigation water as the control parameter. Soil salinity refers to high concentrations of total salt ions (e.g., Na^+, K^+, Ca^{2+}, Mg^{2+}), whereas *soil sodicity* refers to conditions where sodium predominates (e.g., a high percentage of Na^+ in the soil). The problem of soil sodicity[5] often arises when brackish or slightly saline water is used for irrigation in arid and semi-arid regions (Assouline et al., 2015). The remediation of soil salinity and sodicity may be achieved by changing the irrigation water properties in time, e.g., by mixing costly freshwater to the available brackish water or by dissolving amendments such as gypsum in the applied irrigation water. Following Porporato et al. (2015) and Mau and Porporato (2015, 2016), here we seek an optimal calcium amendment strategy. which remediates the soil in the least possible time by exchanging sodium ions with calcium ions.

Mau and Porporato (2016) assumed constant soil moisture, maintained by stress-avoidance micro-irrigation with negligible rainfall, and that the salt input occurs exclusively through irrigation and the salt output occurs through leaching. Accordingly, the mass balance equation (10.18), when written for the salt concentration, simplifies to

$$nZ_r\bar{s}\frac{dC}{dt} = JC_J - LC, \tag{10.77}$$

where C is the concentration of total amount of salt cations and JC_J is the rate of salt addition via irrigation (with C_J indicating the salt concentration of the irrigation water, a control function in this problem).

Besides being dissolved in the soil water, the sodium and other ions can be adsorbed on soil particles, whose surfaces have negative charge. The intricate process of adsorption and desorption in the exchange complex creates a buffer zone for sodium cations and influences the dynamic of soil water sodicity, represented by the exchangeable sodium, ES. This results in a nonlinear relation between ES and C, which in turn makes the evolution equation for ES quite involved. Since the exchange reaction of the cations is much faster than the typical timescale of sodification, the cations are considered to be in thermodynamic equilibrium. With appropriate substitutions and mathematical developments (Mau and Porporato, 2015), the dynamical equation for ES is

$$\frac{dES}{dt} = \frac{JC_JES_J - gCL - nZ_r\bar{s}\frac{dC}{dt}\left(g + \frac{\partial g}{\partial C}\right)}{C_{EC}M + nZ_r\bar{s}\frac{\partial g}{\partial ES}}, \tag{10.78}$$

where ES_J is the equivalent fraction of sodium in the irrigation water, dC/dt is given by Eq. (10.77), C_{EC} is the cation-exchange capacity, M is the mass of dry soil per unit area of depth Z_r, and the function g is

$$g = \frac{2}{1 + \sqrt{1 + 8K_g^2C(ES^{-1} - 1)^2}}. \tag{10.79}$$

Equation (10.78) is nonlinear in the variables ES and C, and also depends nonlinearly upon the irrigation control variables C_J and ES_J. Equations (10.77) and (10.78) form a two-dimensional nonlinear system with control variables C_J and ES_J.

---

[5] Soil sodicity is characterized by a relative high concentration of sodium cations in the exchange complex or in the soil water, which causes negative effects on the physical and chemical properties of the soil. Among the physical changes are the breakdown of macroaggregates (slaking), the release of individual clay platelets from aggregates (dispersion), and surface crusting, all of which all have a detrimental impact on the hydraulic conductivity, infiltration rate, seedling emergence, and water-holding capacity (Bernstein, 1975; Abrol et al., 1988); see also Kramer and Mau, 2020. The chemical effects of sodicity include specific ion deficiencies and toxicities (Ghassemi et al., 1995).

The problem of optimal control addressed in Mau and Porporato (2016) was how to dynamically change the quality of the irrigation water, i.e., vary the value of $ES_J(t)$ and $C_J(t)$, in order to rehabilitate a sodic soil to salinity and sodicity levels within a range suitable for plant growth, $C < 40$ meq/l and $ES < 0.15$, in a minimal amount of time.[6] For an addition of calcium cations q (in meq of calcium cations per day per unit area), the salinity of the irrigation water becomes $C_J = C_J^0 + q/J$ and the sodicity becomes $ES_J = ES_J^0/(1 + q/C_J^0 J)$, in which C_J^0 and ES_J^0 denote respectively the sodicity and salinity of irrigation water when no calcium additive is introduced. Substituting C_J and ES_J into Eqs. (10.77) and (10.78), a system of equations that depends on a single control parameter, $q(t)$, is obtained. Thus we are aiming to find an amendment strategy $q(t)$ that takes the system from a sodic soil $x_1 = (ES_1, C_1)$ to a remediated soil $x_2 = (ES_2, C_2)$ in the minimal amount of time, given that the calcium supplement stays in the range $0 < q(t) < q^{max}$, where q^{max} is the maximal rate of calcium cations added to the irrigation water. Minimization of the total time implies minimization of the amount of good quality irrigation water supplied at a fixed rate, as well as an earlier date for making the soil fully productive and ready for cultivation.

To proceed analytically, Mau and Porporato (2016) linearized the system around its stable point and then further linearized the dependence of the system on the control function q around the point $q^0 = 0$. The resulting system is

$$\frac{dC}{dt} = a_1 + a_2 C + a_3 q, \tag{10.80}$$

$$\frac{dES}{dt} = a_4 + a_5 ES + a_6 q ES, \tag{10.81}$$

where the coefficients a_i are constant (see Mau and Porporato (2016) for typical values). The system is subject to the condition of minimizing the cost functional $U(q) = \int_0^{t_f(q)} \mathcal{L}(q) dt$, where the final time is to be minimized and the Lagrangian (the running cost) is simply $\mathcal{L} = 1$. This type of time-optimal, fixed-endpoint control problem is a classical one (e.g., the brachistochrone problem).

The above equations represent a normal linear control system, which means that the control $q(t)$ has a so-called *bang-bang solution*, i.e., its values are restricted to the extremes of the range $0 < q(t) < q^{max}$, with abrupt transitions between them (Stengel, 1994; Liberzon, 2011). It can be shown (Mau and Porporato, 2016) that the switch of the optimized control parameter occurs once, from an "on" state with maximum calcium amendment ($q = q^{max}$) to an "off" state with irrigation of good quality water but no amendment ($q = 0$) at the switching time t_s, until the final optimal time t_f^{opt}. This bang-bang solution is in agreement with existing amelioration techniques, such as high-salt water dilution methods (Qadir et al., 2001).

Figure 10.30 shows the temporal evolution of C and ES for the optimal remediation strategy, in salinity–sodicity phase space, starting from $ES = 0.35$ and $C = 20$ meq/l (denoted by a triangle) with a final target state $ES = 0.08$ and $C = 31$ meq/l (denoted by a square). For the parameter values chosen, the switching time t_s occurs on day 109, and the remediation duration t_f^{opt} is 161 days. The trajectory starts from sodic conditions, transitions to sodic-saline conditions as sodium is substituted by calcium, and reaches saline conditions when the Na^+ concentration decreases below 0.15 meq/l. Finally, as the calcium amendments are stopped at t_s (while irrigation with clean water continues), the soil switches to the desired normal condition suitable for cultivation. The solid line is obtained via the linearized system of Eqs. (10.80) and

---

[6] An meq is a milliequivalent of solute per liter of solution.

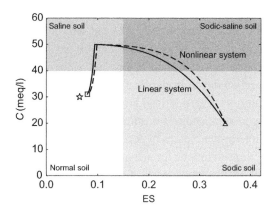

Figure 10.30 Optimal trajectory in salinity–sodicity phase space, defined by the variables C and ES, for remediating a sodic soil in minimal time. The initial condition is marked with a triangle, the goal state with a square, and the steady-state solution ($q = 0$) with a star. The solid trajectory denotes the analytical optimal solution of (10.80) and (10.80), while the dashed trajectory is a simulation of the nonlinear system of equations (10.77) and (10.78), for the same optimal switching strategy. Parameters: $n = 0.43$, $Z_r = 300$ mm, $E_{max} = 5$ mm/day, $M = 450$ kg/m^2, $C_{EC} = 100$ meq/kg, $K_g = 0.01475$, $J = 10$ mm/day, $ES_J^0 = 0.45$ meq/l, $C_J^0 = 15$ meq/l, and $q^{max} = 100$ meq/(m^2/day). Adapted from Porporato et al. (2015).

(10.81), whereas the dashed trajectory is a simulation of the original nonlinear system of Eqs. (10.77) and (10.78) with the same switching strategy as that calculated for the linear system. The comparison in Fig. 10.30 shows that, although simplified, the trajectory of the linearized system captures the dynamics of the nonlinear system reasonably well.

It is interesting to make a connection between the type of optimal protocol found here (bang-bang control, namely a discontinuous switch from maximum calcium amendment to clean water) with the problem of aquifer depletion with no dependence of cost on groundwater level (see Fig. 10.27); the latter problem also leads to a bang-bang solution with a sudden switch between the so-called corner solutions of pumping at a maximum rate or no pumping at all. This similarity stems from the linearity (i.e., the lack of local convexity) of the cost function: a simple linear dependence on the time of control in the sodic soil problem (the Lagrangian equal to 1) and the linear proportionality with pumping rate (the Lagrangian linearly proportional to q).

10.6 Key Points

- The modern acceleration of ecohydrological processes in agroecosystems increases the risk of land degradation; the coupling of bistable soil–plant systems with the socio-economical component may trigger an excitable dynamics characterized by phases of exploitation, degradation, collapse, and recovery.
- Optimizing irrigation, the largest user of water resources, for productivity, profitability, and sustainability requires comparing the stochastic soil moisture dynamics resulting from the different irrigation methods and strategies (traditional versus micro-irrigation and stress-avoidance versus deficit irrigation).

- Primary soil salinization takes place in arid and semi-arid regions where the slow salt input from dry and wet deposition overcomes intermittent outputs due to leaching; the latter takes the form of a state-dependent stochastic jump in the soil salinity equation.
- Secondary salinization due to irrigation inputs may be curbed by over-irrigation aimed at flushing out the excess salts.
- In both phytoremediation and fertilization a successful solution lies in favoring the mineral plant uptake over the intermittent losses by leaching.
- The design of optimal solutions for an agroecosystem is made difficult by the stochastic environmental conditions, as illustrated by the problems of optimal irrigation, fertilization, and rainwater harvesting.
- In transient conditions, the design of optimal solutions is elegantly framed within optimal control theory; this is illustrated by the problems of aquifer exploitation and of sodic soil remediation.

10.7 Notes, including Problems and Further Reading

10.1 (Plants and Soil Formation) With reference to the problem of soil erosion and land degradation, discussed in Sec. 10.1.3, plot the phase diagram of h and b for the system of Eqs. (10.4) and (10.5) under different conditions of harvest pressure.

10.2 (Drivers and Feedbacks of Desertification) D'Odorico et al. (2013) analyzed the drivers and feedbacks related to desertification, namely the changes in soil properties, vegetation, or climate which result in a persistent loss of ecosystem services that are fundamental to sustaining life. In classic desertification theories, involving transitions between bistable ecosystem dynamics, a shift to a "desertified" (or degraded) state is typically sustained by positive feedbacks, which stabilize the system in the new state. Desertification feedbacks may involve land-degradation processes (e.g., nutrient loss or salinization), changes in rainfall regime resulting from land–atmosphere interactions (e.g., precipitation recycling, dust emissions), or changes in plant community composition (e.g., shrub encroachment, decrease in vegetation cover). D'Odorico et al. (2013) also discussed some desertification control strategies, noting that these feedback mechanisms may be enhanced by interactions with socio-economic drivers.

10.3 (Terraces, Diguettes, and Other Soil Conservation Practices) Agricultural terraces are built to provide a larger surface area for cultivation on hillslopes and to aid farming production. They typically increase infiltration and reduce runoff; in Mediterranean areas, the runoff coefficient on abandoned terraces is 20%–40%, but is less in more humid climates (Arnáez et al, 2015). These terraces also reduce soil loss and enhance the retention of nutrients and carbon as well as providing drought mitigation (Kosmowski, 2018), although they require maintenance to avoid problems such as stability and debris flows.

In drylands, and especially in the sub-Saharan regions, the extensive constructions of diguettes (i.e., micro-dams) and similar soil water conservation practices have been advocated and implemented to combat desertification. Assessing their effectiveness and potential large-scale feedbacks on local environmental conditions is a difficult task, which requires not only a combination of theory, experiment, and remote sensing observations, but also a variety of field observations which have important social components (e.g., Thor West et al., 2020).

10.4 (Nonlinear Dynamics and Desertification Control) As mentioned in the previous note, controlling soil–water–vegetation patterns in drylands has the potential to help reduce desertification and reverse land degradation trends. By formulating the problem of infiltration enhancement and vegetation management in semi-arid ecosystems as a spatial resonance of the underlying nonlinear dynamics, the paper Mau et al. (2015) presented an interesting step in the theoretical analysis of these issues.

10.5 (Tele-Coupling of Market Pressure and Land Degradation) The tele-coupling framework (Liu et al., 2013) has been useful for conceptualizing the pathways in which the structure of price and knowledge transmits from world to regional or local markets. Tele-coupling may bring a divergence between private (or household) decisions serving market requirements and social needs to sustain local soil and water resources, resulting in an acceleration of biogeochemical cycling and land degradation. Instances of such long-distance dynamics abound. For example, soybean production in Brazil has been driven by a booming world market in vegetable oils and protein feed for livestock (Bruinsma, 2006; Nepstad et al., 2006). Since 2000, the annual growth of soybean production in Brazil has been 4.5 million tons, which represents an increase of 16% per year. Brazilian production jumped from 39.5 million tons (14 million hectares) in the 2000/2001 harvest to 133 million tons in the 2020/2021 (39 million hectares) harvest and was mainly exported to China and Europe (Nepstad et al., 2006). Examples of policy-driven tele-coupling include the impact of American subsidies of cotton in the past decades on the livelihood of cotton farmers in developing nations, such as Brazil and West Africa (Watkins, 2002; Fadiga et al., 2006). Other examples include quinoa, oil palm, bananas, wine production, and wood-fiber plantations.

Unless coupled with increasingly sustainable management practices, these economic interactions can rapidly drive agricultural land use to adversely alter local environments and social conditions. While the emerging framework of tele-coupling has mainly been employed to understand market and land-use dynamics at very large scales, less attention has been paid to the local social and environmental context, where land users make decisions that can result in land degradation and other environmental effects.

10.6 (Seasonal Stochastic Dynamics of Small Irrigation Dams) Modify the seasonal soil moisture model of Sec. 7.8.1 to analyze both numerically and analytically the probability of failure of the cisterns or small irrigation dams typical of seasonally dry climates (e.g., northeastern Brazil).

10.7 (Model Similarity of Micro-Irrigation and Rainwater Harvesting Problems) Compare the solutions of the micro-irrigation problem (Sec. 10.2.2) and the cistern optimization problem (Sec. 10.4.3) with those of the minimalist soil moisture model of Milly (Sec. 7.5).

10.8 (Cultivos de Vazante) Several interesting adaptations for sustainable agriculture in semi-arid regions have been implemented. In northeastern Brazil, which is characterized by an extreme seasonally dry hydrologic regime, more than 70,000 small dams have been built to retain superficial runoff and underground water flow, creating small reservoirs. Recession agriculture consists of cropping the margins of the reservoirs, on slight slopes, while the water level progressively decreases (Antonino et al., 2005).

10.9 (Impacts of Large-Scale Irrigation on Climate) By changing the soil moisture conditions, large-scale irrigation has the potential to affect local climate, especially in continental regions. Through the various forms of land–atmosphere feedback (see Sec. 5.4.3), it has been argued that irrigation enhances precipitation (Stidd et al., 1975; Barnston and Schickedanz, 1984; DeAngelis et al., 2010) and cools temperatures (Kueppers et al., 2007).

10.10 (Genetically Engineered Drought Tolerant Crops) Use the framework of Sec. 8.6 to quantify the reduction in plant water stress of a genetically modified crop with reduced threshold for water stress s^* in conditions of rainfed agriculture. Discuss ways to extend this analysis to conditions with deficit and stress-avoidance irrigation.

10.11 (Over-irrigation for Salinity Control) Simulate the soil moisture and salt concentration dynamics of Eqs. (10.18) and (10.19), in the case of overirrigation for no-rain conditions. Calculate the time to critical soil salinization as a function of the concentration of salts in the irrigation water and of the other main parameters.

10.12 (PDF of Salt Content) Derive step by step the steady-state PDF of m, Eq. (10.27), and plot it as a function of the main soil and climate parameters.

10.13 (Salinization and Ancient Civilizations) Secondary soil salinization and the related land degradation is not only a pressing modern problem, but has played a role in the decline of ancient civilizations, especially those that thrived in the fertile crescent of Mesopotamia (Hillel, 1992). These lessons from history remind us that ecohydrology has dimensions that reach well beyond the boundaries of science and engineering.

10.14 (Irrigation, Salinization, and Soil Biogeochemical Processes) Irrigation and possibly the related secondary salinization are known to have an impact on soil biogeochemical processes (Li et al., 1994; Rietz and Haynes, 2003). Much remains to be done, on both the experimental and modeling fronts, to understand the impact of irrigation on the carbon and nitrogen cycles, the soil microbial dynamics, and other related processes discussed in Chapter 9.

10.15 (Phytoremediation) Using the Laplace transform in Eq. (6.68), obtain the transient solution of Eq. (10.33) in terms of the contaminant concentration.

10.16 (Vegetation–Groundwater Feedbacks and Soil Salinization) In some regions of Australia, the clearing of deep-rooted salt-tolerant forests has caused a rise in the saline water table (Ruprecht and Schofield, 1991) with dramatic consequences for soil and plant conditions. This shift to degraded conditions may be considered a secondary form of salinization, not directly linked to irrigation. A model for such bistable dynamics was proposed by Runyan and D'Odorico (2010).

10.17 (Aquifer Exploitation and Crop Water-Use Efficiency) For the simple aquifer irrigation model of Sec. 10.5.1, analyze the optimal aquifer level and pumping rate at steady state, as a function of the main coefficients of the utility function.

10.18 (Aquifer Depletion and Ponzi Schemes) For the simple aquifer irrigation model of Sec. 10.5.1, plot the non-optimal solutions resulting from the full system (10.72) without transversality condition. Show that these solutions, the so-called Ponzi schemes, correspond to non-optimal conditions.

10.19 (Optimal Control and Resource Economics) Optimal control has been applied to minimalist models of environmental systems as a way to provide first approximate solutions to the increasingly complex challenges faced by modern environmental sciences and engineering. Besides the cases described in Sec. 10.5, here we mention the references English (1981), Zilberman (1982), McConnell (1983), Mannocchi and Mecarelli (1994), and Sohngen and Mendelsohn (2003), as well as the review by Herman et al. (2020).

10.20 (Wetland Optimal Control for Climate Mitigation) Wetlands provide numerous ecosystem services, including carbon storage, but they are also potential emitters of greenhouse gases. Many constructed wetlands can be managed to optimize climate mitigation, among other goals. Rice fields, which are important contributors to the global methane budget, offer an excellent example of optimal control with regard to their flooding regime. Using an ecohydrological model to analyze the effect of drainage protocols on methane emissions, Souza et al. (2021) showed that the methane emissions can be reduced by more than 60% if drainage is carried out optimally.

References

Aber, J. D., J. M. Melillo, K. J. Nadelhoffer, J. Pastor, and R. D. Boone (1991). "Factors controlling nitrogen cycling and nitrogen saturation in northern temperate forest ecosystems." In: *Ecological Applications* 1.3, pp. 303–315.

Abramowitz, M. and I. A. Stegun (1964). *Handbook of Mathematical Functions With Formulas, Graphs, and Mathematical Tables*. United States Department of Commerce.

Abrol, I., J. S. P. Yadav, and F. Massoud (1988). Salt-affected soils and their management. 39. Food & Agriculture Organization.

Acemoglu, D. (2008). *Introduction to Modern Economic Growth*. Princeton University Press.

Adler, P. B., J. HilleRisLambers, and J. M. Levine (2007). "A niche for neutrality." In: *Ecology Letters* 10.2, pp. 95–104.

Allen, C. D. et al. (2010). "A global overview of drought and heat-induced tree mortality reveals emerging climate change risks for forests." In: *Forest Ecology and Management* 259.4, pp. 660–684.

Allen, M., W. Swenson, J. Querejeta, L. Egerton-Warburton, and K. Treseder (2003). "Ecology of *Mycorrhizae*. A conceptual framework for complex interactions among plants and fungi." In: *Annual Review of Phytopathology* 41.1, pp. 271–303.

Allen, M. F. (2007). "Mycorrhizal fungi: Highways for water and nutrients in arid soils." In: *Vadose Zone Journal* 6.2, pp. 291–297.

Allison, G., G. Gee, and S. Tyler (1994). "Vadose-zone techniques for estimating groundwater recharge in arid and semiarid regions." In: *Soil Science Society of America Journal* 58.1, pp. 6–14.

Altieri, M. A. (1999). "The ecological role of biodiversity in agroecosystems." In: *Invertebrate Biodiversity as Bioindicators of Sustainable Landscapes*. Elsevier, pp. 19–31.

Amenu, G. and P. Kumar (2008). "A model for hydraulic redistribution incorporating coupled soil–root moisture transport." In: *Hydrology and Earth System Sciences* 12.1, pp. 55–74.

Anderegg, W. R. L. et al. (2018). "Hydraulic diversity of forests regulates ecosystem resilience during drought." In: *Nature* 561.7724, pp. 538–541.

Andrews, L. C., (1986). *Elementary Partial Differential Equations with Boundary Value Problems*. Academic Press.

Antonino, A. C. D., C. Hammecker, S. Montenegro, A. M. Netto, R. AnguloJaramillo, and C. Lira (2005). "Subirrigation of land bordering small reservoirs in the semi-arid region in the northeast of Brazil: monitoring and water balance." In: *Agricultural Water Management* 73.2, pp. 131–147.

Archer, S., C. Scifres, C. Bassham, and R. Maggio (1988). "Autogenic succession in a subtropical savanna: conversion of grassland to thorn woodland." In: *Ecological Monographs* 58.2, pp. 111–127.

Argyris, J. H., G. Faust, M. Haase, and R. Friedrich (2015). *An Exploration of Dynamical Systems and Chaos*, completely revised and enlarged second edition. Springer.

Arnáez, J., N. Lana-Renault, T. Lasanta, P. Ruiz-Flaño, and J. Castroviejo (2015). "Effects of farming terraces on hydrological and geomorphological processes. A review." In: *Catena* 128, pp. 122–134.

Arnold, J., K. Potter, K. King, and P. Allen (2005). "Estimation of soil cracking and the effect on surface runoff in a Texas Blackland Prairie watershed." In: *Hydrological Processes* 19.3, pp. 589–603.

Arrow, K. J. (1967). *Application of Control Theory to Economic Growth*. Stanford University.

Assouline, S., M. Möller, S. Cohen, M. Ben-Hur, A. Grava, K. Narkis, and A. Silber (2006). "Soil–plant system response to pulsed drip irrigation and salinity: Bell pepper case study." In: *Soil Science Society of America Journal* 70.5, pp. 1556–1568.

Assouline, S., D. Russo, A. Silber, and D. Or (2015). "Balancing water scarcity and quality for sustainable irrigated agriculture." In: *Water Resources Research* 51.5, pp. 3419–3436.

Augé, R. M. (2004). "Arbuscular mycorrhizae and soil/plant water relations." In: *Canadian Journal of Soil Science* 84.4, pp. 373–381.

Austin, A. T., L. Yahdjian, J. M. Stark, J. Belnap, A. Porporato, U. Norton, D. A. Ravetta, and S. M. Schaeffer (2004). "Water pulses and biogeochemical cycles in arid and semiarid ecosystems." In: *Oecologia* 141.2, pp. 221–235.

Ayars, J., R. Hutmacher, R. Schoneman, S. Vail, and T. Pflaum (1993). "Long term use of saline water for irrigation." In: *Irrigation Science* 14.1, pp. 27–34.

Ayars, J., C. Phene, R. Hutmacher, K. Davis, R. Schoneman, S. Vail, and R. Mead (1999). "Subsurface drip irrigation of row crops: a review of 15 years of research at the Water Management Research Laboratory." In: *Agricultural Water Management* 42.1, pp. 1–27.

Azaele, S., S. Suweis, J. Grilli, I. Volkov, J. R. Banavar, and A. Maritan (2016). "Statistical mechanics of ecological systems: neutral theory and beyond." In: *Reviews of Modern Physics* 88.3, p. 035003.

Ball, J. T., I. E. Woodrow, and J. A. Berry (1987). "A model predicting stomatal conductance and its contribution to the control of photosynthesis under different environmental conditions." In: *Progress in Photosynthesis Research*. Springer, pp. 221–224.

Barcellos, D., C. S. O'Connell, W. Silver, C. Meile, and A. Thompson (2018). "Hot spots and hot moments of soil moisture explain fluctuations in iron and carbon cycling in a humid tropical forest soil." In: *Soil Systems* 2.4, p. 59.

Barenblatt, G. I. (1996). *Scaling, Self-Similarity, and Intermediate Asymptotics: Dimensional Analysis and Intermediate Asymptotics*. Cambridge University Press.

Barnston, A. G. and P. T. Schickedanz (1984). "The effect of irrigation on warm season precipitation in the southern Great Plains." In: *Journal of Applied Meteorology and Climatology* 23.6, pp. 865–888.

Baron, J. S. et al. (2002). "Meeting ecological and societal needs for freshwater." In: *Ecological Applications* 12.5, pp. 1247–1260.

Barraclough, D. (1997). "The direct or MIT route for nitrogen immobilization: a 15N mirror image study with leucine and glycine." In: *Soil Biology and Biochemistry* 29.1, pp. 101–108.

Bartlett, M. S. and A. Porporato (2018). "State-dependent jump processes: ItoStratonovich interpretations, potential, and transient solutions." In: *Physical Review E* 98.5, p. 052132.

Bartlett, M. S., G. Vico, and A. Porporato (2014). "Coupled carbon and water fluxes in CAM photosynthesis: modeling quantification of water use efficiency and productivity." In: *Plant and Soil* 383.1-2, pp. 111–138.

Bartlett, M, E Daly, J. McDonnell, A. J. Parolari, and A. Porporato (2015). "Stochastic rainfall-runoff model with explicit soil moisture dynamics." In: *Proc. Roy. Soc. A* 471.2183, p. 20150389.

Bartlett, M., A. J. Parolari, J. McDonnell, and A. Porporato (2016a). "Beyond the SCS-CN method: a theoretical framework for spatially lumped rainfallrunoff response." In: *Water Resources Research* 52.6, pp. 4608–4627.

— (2016b). "Framework for event-based semidistributed modeling that unifies the SCS-CN method, VIC, PDM, and TOPMODEL." In: *Water Resources Research* 52.9, pp. 7036–7052.

— (2017). "Reply to comment by Fred L. Ogden et al. on Beyond the SCSCN method: a theoretical framework for spatially lumped rainfall-runoff response." In: *Water Resources Research* 53.7, pp. 6351–6354.

Bastola, S., Y. Dialynas, R. Bras, L. Noto, and E. Istanbulluoglu (2018). "The role of vegetation on gully erosion stabilization at a severely degraded landscape: a case study from Calhoun Experimental Critical Zone Observatory." In: *Geomorphology* 308, pp. 25–39.

Bear, J. (1972). *Dynamics of Fluids in Porous Media*. Courier Corporation.

Beck, C. and E. G. Cohen (2003). "Superstatistics." In: *Physica A: Statistical Mechanics and Its Applications* 322, pp. 267–275.

Bejan, A. (2006). *Advanced Engineering Thermodynamics*. Wiley.

Bell, G. (2000). "The distribution of abundance in neutral communities." In: *The American Naturalist* 155.5, pp. 606–617.

Benjamin, J. R. and C. A. Cornell (2014). *Probability, Statistics, and Decision for Civil Engineers*. Courier Corporation.

Bensimon, D., L. P. Kadanoff, S. Liang, B. I. Shraiman, and C. Tang (1986). "Viscous flows in two dimensions." In: *Reviews of Modern Physics* 58.4, p. 977.

Berger-Landefeldt, U. (1936). "Der Wasserhaushalt der Alpenpflanzen." In:

Bernstein, L. (1975). "Effects of salinity and sodicity on plant growth." In: *Annual Review of Phytopathology* 13.1, pp. 295–312.

Betts, A. (1973). "Non-precipitating cumulus convection and its parameterization." In: *Quart. J. Roy. Meteorol. Soc.* 99.419, pp. 178–196.

Beven, K. (2006). "A manifesto for the equifinality thesis." In: *Journal of Hydrology* 320.1-2, pp. 18–36.

Birch, H. (1958). "The effect of soil drying on humus decomposition and nitrogen availability." In: *Plant and Soil* 10.1, pp. 9–31.

Birkinshaw, S. J. and J. Ewen (2000). "Nitrogen transformation component for SHETRAN catchment nitrate transport modelling." In: *Journal of Hydrology* 230.1-2, pp. 1–17.

Bogie, N. A., R. Bayala, I. Diedhiou, M. H. Conklin, M. L. Fogel, R. P. Dick, and T. A. Ghezzehei (2018). "Hydraulic redistribution by native Sahelian shrubs: bioirrigation to resist in-season drought." In: *Frontiers in Environmental Science* 6.

Böhlke, J.-K. (2002). "Groundwater recharge and agricultural contamination." In: *Hydrogeology Journal* 10.1, pp. 153–179.

Bonan, G. (2002). *Ecological Climatology: Concepts and Applications.* Cambridge University Press.

Bonetti, S. and A. Porporato (2017). "On the dynamic smoothing of mountains." In: *Geophysical Research Letters* 44.11, pp. 5531–5539.

Bonetti, S., X. Feng, and A. Porporato (2017). "Ecohydrological controls on plant diversity in tropical South America." In: *Ecohydrology* 10.6, e1853.

Bonetti, S., D. D. Richter, and A. Porporato (2019). "The effect of accelerated soil erosion on hillslope morphology." In: *Earth Surface Processes and Landforms* 44.15, pp. 3007–3019.

Borgogno, F., P. D'Odorico, F. Laio, and L. Ridolfi (2009). "Mathematical models of vegetation pattern formation in ecohydrology." In: *Reviews of Geophysics* 47.1.

Botter, G., A. Porporato, I. Rodríguez-Iturbe, and A. Rinaldo (2007). "Basin-scale soil moisture dynamics and the probabilistic characterization of carrier hydrologic flows: slow, leaching-prone components of the hydrologic response." In: *Water Resources Research* 43.2.

Bradford, K. J. and T. C. Hsiao (1982). "Physiological responses to moderate water stress." In: *Physiological Plant Ecology II.* Encyclopedia of Plant Physiology, pp. 263–324. Springer.

Brady, N. and R. Weil (1996). *The Nature and Properties of Soils.* Pearson Prentice Hall.

Brangarl, A. C., S. Manzoni, and J. Rousk (2020). "A soil microbial model to analyze decoupled microbial growth and respiration during soil drying and rewetting." In: *Soil Biology and Biochemistry* 148, p. 107871.

Brantley, S. L., M. B. Goldhaber, and K. V. Ragnarsdottir (2007). "Crossing disciplines and scales to understand the critical zone." In: *Elements* 3.5, pp. 307–314.

Bras, R. L. and D.-J. Seo (1987). "Irrigation control in the presence of salinity: extended linear quadratic approach." In: *Water Resources Research* 23.7, pp. 1153–1161.

Brill, P. H. et al. (2008). *Level Crossing Methods in Stochastic Models.* Springer.

Brisson, N. et al. (2003). "An overview of the crop model STICS." In: *European Journal of Agronomy* 18.3-4, pp. 309–332.

Brooks, J. R., F. C. Meinzer, J. M. Warren, J.-C. DOMEC, and R. Coulombe (2006). "Hydraulic redistribution in a Douglas-fir forest: lessons from system manipulations." In: *Plant, Cell & Environment* 29.1, pp. 138–150.

Brooks, R. H. and A. T. Corey (1966). "Properties of porous media affecting fluid flow." In: *Journal of the Irrigation and Drainage Division* 92.2, pp. 61–90.

Brown, G., Jr., and R. Deacon (1972). "Economic optimization of a single-cell aquifer." In: *Water Resources Research* 8.3, pp. 557–564.

Bruinsma, J. (2006). *World Agriculture: Towards 2030/50, an FAO Perspective.*

Brunnermeier, M. K. and M. Oehmke (2013). "Chapter 18 – Bubbles, financial crises, and systemic risk." In: eds. G. M. Constantinides, M. Harris, and R. M. Stulz. *Vol. 2. Handbook of the Economics of Finance.* Elsevier, pp. 1221–1288.

Brutsaert, W. (2005). *Hydrology: An Introduction.* Cambridge University Press, 605 pp.

Brutsaert, W. and D. Chen (1995). "Desorption and the two stages of drying of natural tallgrass prairie." In: *Water Resources Research* 31.5, pp. 1305–1313.

— (1996). "Diurnal variation of surface fluxes during thorough drying (or severe drought) of natural prairie." In: *Water Resources Research* 32.7, pp. 2013–2019.

Budyko, M. (1974). *Climate and Life*. International Geophysics Series. Academic Press.

Burdine, N. et al. (1953). "Relative permeability calculations from pore size distribution data." In: *Journal of Petroleum Technology* 5.03, pp. 71–78.

Burgess, T. L. (1995). "Desert grassland, mixed shrub savanna, shrub steppe, or semidesert scrub." In: *The Desert Grassland*, pp. 31–67.

Burke, I. C., W. K. Lauenroth, R. Riggle, P. Brannen, B. Madigan, and S. Beard (1999). "Spatial variability of soil properties in the shortgrass steppe: the relative importance of topography, grazing, microsite, and plant species in controlling spatial patterns." In: *Ecosystems* 2.5, pp. 422–438.

Burt, O. R. (1964). "Optimal resource use over time with an application to ground water." In: *Management Science* 11.1, pp. 80–93.

Cabon, F., G. Girard, and E. Ledoux (1991). "Modelling of the nitrogen cycle in farm land areas." In: *Fertilizer Research* 27.2-3, pp. 161–169.

Cai, X. (2008). "Implementation of holistic water resources – economic optimization models for river basin management – reflective experiences." In: *Environmental Modelling & Software* 23.1, pp. 2–18.

Calabrese, S. and A. Porporato (2019). "Impact of ecohydrological fluctuations on iron-redox cycling." In: *Soil Biology and Biochemistry* 133, pp. 188–195.

Calabrese, S., A. J. Parolari, and A. Porporato (2017). "Hydrologic transport of dissolved inorganic carbon and its control on chemical weathering." In: *Journal of Geophysical Research: Earth Surface* 122.10, pp. 2016–2032.

Calabrese, S., D. Barcellos, A. Thompson, and A. Porporato (2020). "Theoretical constraints on Fe reduction rates in upland soils as a function of hydroclimatic conditions." In: *Journal of Geophysical Research: Biogeosciences*, e2020JG005894.

Caldwell, M. M., T. E. Dawson, and J. H. Richards (1998). "Hydraulic lift: consequences of water efflux from the roots of plants." In: *Oecologia* 113.2, pp. 151–161.

Callen, H. (2006). *Thermodynamics and an Introduction to Thermostatistics*. Student Edition. Wiley India Pvt. Limited.

Campbell, G. S. and J. M. Norman (1998). *An Introduction to Environmental Biophysics*. Springer, New York.

Cárdenas, L., A. Rondón, C. Johansson, and E. Sanhueza (1993). "Effects of soil moisture, temperature, and inorganic nitrogen on nitric oxide emissions from acidic tropical savanna soils." In: *Journal of Geophysical Research: Atmospheres* 98.D8, pp. 14783–14790.

Carson, D. (1973). "The development of a dry inversion-capped convectively unstable boundary layer." In: *Quart. J. Roy. Meteorol. Soc.* 99.421, pp. 450–467.

Caylor, K. K., T. M. Scanlon, and I. Rodríguez-Iturbe (2009). "Ecohydrological optimization of pattern and processes in water-limited ecosystems: a tradeoff-based hypothesis." In: *Water Resources Research* 45.8.

Celia, M. A., E. T. Bouloutas, and R. L. Zarba (1990). "A general massconservative numerical solution for the unsaturated flow equation." In: *Water Resources Research* 26.7, pp. 1483–1496.

Cermák, J., J. Kucera, W. L. Bauerle, N. Phillips, and T. M. Hinckley (2007). "Tree water storage and its diurnal dynamics related to sap flow and changes in stem volume in old-growth Douglas-fir trees." In: *Tree Physiology* 27.2, pp. 181–198.

Chalmers, D. J., P. D. Mitchell, and L. Van Heek (1981). "Control of peach tree growth and productivity by regulated water supply, tree density, and summer pruning." In: *Journal of the American Society for Horticultural Science* 106, pp. 307–312.

Chapman, S. (1976). *Methods In Plant Ecology*.

Chater, C. C., R. S. Caine, A. J. Fleming, and J. E. Gray (2017). "Origins and evolution of stomatal development." In: *Plant Physiology* 174.2, pp. 624–638.

Chave, J. (2004). "Neutral theory and community ecology." In: *Ecology Letters* 7.3, pp. 241–253.

Chave, J., H. C. Muller-Landau, and S. A. Levin (2002). "Comparing classical community models: theoretical consequences for patterns of diversity." In: *The American Naturalist* 159.1, pp. 1–23.

Chen, Y., Z. Shen, and X. Li (2004). "The use of vetiver grass (*Vetiveria zizanioides*) in the phytoremediation of soils contaminated with heavy metals." In: *Applied Geochemistry* 19.10, pp. 1553–1565.

Chesson, P. (2000). "Mechanisms of maintenance of species diversity." In: *Annual Review of Ecology and Systematics* 31.1, pp. 343–366.

Christov, C. (2009). "On frame indifferent formulation of the Maxwell–Cattaneo model of finite-speed heat conduction." In: *Mechanics Research Communications* 36.4, pp. 481–486.

Clapp, R. B. and G. M. Hornberger (1978). "Empirical equations for some soil hydraulic properties." In: *Water Resources Research* 14.4, pp. 601–604.

Clark, J. S. (2020). *Models for Ecological Data: An Introduction.* Princeton University Press.

Coppola, A. (2000). "Unimodal and bimodal descriptions of hydraulic properties for aggregated soils." In: *Soil Science Society of America Journal* 64.4, pp. 1252–1262.

Corless, R. M., G. H. Gonnet, D. E. Hare, D. J. Jeffrey, and D. E. Knuth (1996). "On the Lambert W function." In: *Advances in Computational Mathematics* 5.1, pp. 329–359.

Cosby, B., G. Hornberger, R. Clapp, and T. Ginn (1984). "A statistical exploration of the relationships of soil moisture characteristics to the physical properties of soils." In: *Water Resources Research* 20.6, pp. 682–690.

Cover, T. M. (1999). *Elements of Information Theory.* John Wiley & Sons.

Cowan, I. (1986). "Economics of carbon fixation in higher plants." In: *On the Economy of Plant Form and Function: Proceedings of the Sixth Maria Moors Cabot Symposium on Evolutionary Constraints on Primary Productivity and Adaptive Patterns of Energy Capture in Plants, Harvard Forest, August 1983.* Cambridge University Press.

Cowan, I. and G. Farquhar (1977). "Stomatal function in relation to leaf metabolism and environment." In: *Symposia of the Society for Experimental Biology,* vol. 31, p. 471.

Cox, D. and V. Isham (1986). "The virtual waiting-time and related processes." In: *Advances in Applied Probability* 18.2, pp. 558–573.

Cox, D. and H. Miller (1965). *The Theory of Stochastic Processes.* Methuen.

Crill, P., M. Keller, A. Weitz, B. Grauel, and E. Veldkamp (2000). "Intensive field measurements of nitrous oxide emissions from a tropical agricultural soil." In: *Global Biogeochemical Cycles* 14.1, pp. 85–95.

Cross, M. C. and P. C. Hohenberg (1993). "Pattern formation outside of equilibrium." In: *Reviews of Modern Physics* 65.3, p. 851.

Cruiziat, P., H. Cochard, and T. Amglio (2002). "Hydraulic architecture of trees: main concepts and results." In: *Annals of Forest Science* 59.7, pp. 723–752.

Cuenca, R. H. et al. (1989). *Irrigation System Design. An Engineering Approach.* Prentice Hall.

Cueto-Felgueroso, L. and R. Juanes (2008). "Nonlocal interface dynamics and pattern formation in gravity-driven unsaturated flow through porous media." In: *Physical Review Letters* 101.24, p. 244504.

Cui, M. and M. M. Caldwell (1997). "A large ephemeral release of nitrogen upon wetting of dry soil and corresponding root responses in the field." In: *Plant and Soil* 191.2, pp. 291–299.

Cusack, D. F., J. Karpman, D. Ashdown, Q. Cao, M. Ciochina, S. Halterman, S. Lydon, and A. Neupane (2016). "Global change effects on humid tropical forests: evidence for biogeochemical and biodiversity shifts at an ecosystem scale." In: *Reviews of Geophysics* 54.3, pp. 523–610.

Czernik, T., J. Kula, J. Luczka, and P. Hänggi (1997). "Thermal ratchets driven by Poissonian white shot noise." In: *Physical Review E* 55.4, p. 4057.

Daly, E. and A. Porporato (2006a). "Impact of hydroclimatic fluctuations on the soil water balance." In: *Water Resources Research* 42.6.

— (2006b). "Probabilistic dynamics of some jump-diffusion systems." In: *Physical Review E* 73.2, p. 026108.

— (2007). "Intertime jump statistics of state-dependent Poisson processes." In: *Physical Review E* 75.1.

— (2010). "Effect of different jump distributions on the dynamics of jump processes." In: *Physical Review E* 81.6, p. 061133.

Daly, E., A. Porporato, and I. Rodríguez-Iturbe (2004a). "Coupled dynamics of photosynthesis, transpiration, and soil water balance. Part I: Upscaling from hourly to daily level." In: *Journal of Hydrometeorology* 5.3, pp. 546–558.

— (2004b). "Coupled dynamics of photosynthesis, transpiration, and soil water balance. Part II: Stochastic analysis and ecohydrological significance." In: *Journal of Hydrometeorology* 5.3, pp. 559–566.

Daly, E., A. C. Oishi, A. Porporato, and G. G. Katul (2008). "A stochastic model for daily subsurface CO2 concentration and related soil respiration." In: *Advances in Water Resources* 31.7, pp. 987–994.

Daly, E., S. Palmroth, P. Stoy, M. Siqueira, A. C. Oishi, J.-Y. Juang, R. Oren, A. Porporato, and G. G. Katul (2009). "The effects of elevated atmospheric CO2 and nitrogen amendments on subsurface CO2 production and concentration dynamics in a maturing pine forest." In: *Biogeochemistry* 94.3, pp. 271–287.

Daly, E., S. Calabrese, J. Yin, and A. Porporato (2019a). "Hydrological spaces of long-term catchment water balance." In: *Water Resources Research* 55.12, pp. 10747–10764.

— (2019b). "Linking parametric and water-balance models of the Budyko and Turc spaces." In: *Advances in Water Resources,* p. 103435.

Davidson, E. A. (1991). "Fluxes of nitrous oxide and nitric oxide from terrestrial ecosystems." In: *Microbial Production and Consumption of Greenhouse Gases: Methane, Nitrous Oxide, and Halomethanes,* pp. 219–235.

De Bruin, H. (1983). "A model for the Priestley–Taylor parameter α." In: *J. Clim. Appl. Meteorol.* 22.4, pp. 572–578.

De Fraiture, C., D. Molden, and D. Wichelns (2010). "Investing in water for food, ecosystems, and livelihoods: an overview of the comprehensive assessment of water management in agriculture." In: *Agricultural Water Management* 97.4, pp. 495–501.

De Gennes, P.-G. (1985). "Wetting: statics and dynamics." In: *Reviews of Modern Physics* 57.3, p. 827.

De Rooij, G. (2000). "Modeling fingered flow of water in soils owing to wetting front instability: a review." In: *Journal of Hydrology* 231, pp. 277–294.

DeAngelis, A., F. Dominguez, Y. Fan, A. Robock, M. D. Kustu, and D. Robinson (2010). "Evidence of enhanced precipitation due to irrigation over the Great Plains of the United States." In: *Journal of Geophysical Research: Atmospheres* 115.D15.

Deardorff, J. W. et al. (1970). "A numerical study of three-dimensional turbulent channel flow at large Reynolds numbers." In: *J. Fluid Mech.* 41.2, pp. 453–480.

Del Giorgio, P. A. and J. J. Cole (1998). "Bacterial growth efficiency in natural aquatic systems." In: *Annual Review of Ecology and Systematics* 29.1, pp. 503–541.

Delwiche, M. J. and J. R. Cooke (1977). "An analytical model of the hydraulic aspects of stomatal dynamics." In: *Journal of Theoretical Biology* 69.1, pp. 113–141.

Dewar, R. (1995). "Interpretation of an empirical model for stomatal conductance in terms of guard cell function." In: *Plant, Cell & Environment* 18.4, pp. 365–372.

Diamond, S. et al. (2019). "Mediterranean grassland soil C–N compound turnover is dependent on rainfall and depth, and is mediated by genomically divergent microorganisms." In: *Nature Microbiology* 4.8, pp. 1356–1367.

DiCarlo, D. A. (2013). "Stability of gravity-driven multiphase flow in porous media: 40 years of advancements." In: *Water Resources Research* 49.8, pp. 4531–4544.

Dietrich, W. E. and J. T. Perron (2006). "The search for a topographic signature of life." In: *Nature* 439.7075, pp. 411–418.

Dietz, A. C. and J. L. Schnoor (2001). "Advances in phytoremediation." In: *Environmental Health Perspectives* 109.suppl. 1, pp. 163–168.

Dingman, S. L. (2015). *Physical Hydrology.* Waveland Press.

D'Odorico, P. and A. Porporato (2004). "Preferential states in soil moisture and climate dynamics." In: *Proceedings of the National Academy of Sciences* 101.24, pp. 8848–8851.

D'Odorico, P., L. Ridolfi, A. Porporato, and I. Rodríguez-Iturbe (2000). "Preferential states of seasonal soil moisture: the impact of climate fluctuations." In: *Water Resources Research* 36.8, pp. 2209–2219.

D'Odorico, P., F. Laio, A. Porporato, and I. Rodríguez-Iturbe (2003). "Hydrologic controls on soil carbon and nitrogen cycles. II. A case study." In: *Advances in Water Resources* 26.1, pp. 59–70.

D'Odorico, P., A. Bhattachan, K. F. Davis, S. Ravi, and C. W. Runyan (2013). "Global desertification: drivers and feedbacks." In: *Advances in Water Resources* 51, pp. 326–344.

Domec, J.-C., J. S. King, A. Noormets, E. Treasure, M. J. Gavazzi, G. Sun, and S. G. McNulty (2010). "Hydraulic redistribution of soil water by roots affects whole-stand evapotranspiration and net ecosystem carbon exchange." In: *New Phytologist* 187.1, pp. 171–183.

Doswell, C. A., III, and E. N. Rasmussen (1994). "The effect of neglecting the virtual temperature correction on CAPE calculations." In: *Weather and Forecasting* 9.4, pp. 625–629.

Dubinsky, E. A., W. L. Silver, and M. K. Firestone (2010). "Tropical forest soil microbial communities couple iron and carbon biogeochemistry." In: *Ecology* 91.9, pp. 2604–2612.

Dunne, T. (1978). "Field studies of hillslope flow processes." In: *Hillslope Hydrology,* pp. 227–293.

Dunne, J. et al. (2020). "The GFDL Earth System Model version 4.1 (GFDLESM 4.1): overall coupled model description and simulation characteristics." In: *Journal of Advances in Modeling Earth Systems*, e2019MS002015.

Durner, W. (1994). "Hydraulic conductivity estimation for soils with heterogeneous pore structure." In: *Water Resources Research* 30.2, pp. 211–223.

Dymnikov, V. P. and A. N. Filatov (2012). *Mathematics of Climate Modeling.* Springer Science & Business Media.

Eagleson, P. S. (1978a). "Climate, soil, and vegetation: 1. Introduction to water balance dynamics." In: *Water Resources Research* 14.5, pp. 705–712.

— (1978b). "Climate, soil, and vegetation 3. A simplified model of soil Moisture movement in the liquid phase." In: *Water Resources Research* 14.5, pp. 722–730.

Eagleson, P. S. and R. I. Segarra (1985). "Water-limited equilibrium of savanna vegetation systems." In: *Water Resources Research* 21.10, pp. 1483–1493.

Eheart, J. W. and D. W. Tornil (1999). "Low-flow frequency exacerbation by irrigation withdrawals in the agricultural midwest under various climate change scenarios." In: *Water Resources Research* 35.7, pp. 2237–2246.

Ehleringer, J. R. and R. K. Monson (1993). "Evolutionary and ecological aspects of photosynthetic pathway variation." In: *Annual Review of Ecology and Systematics* 24.1, pp. 411–439.

Elser, J. J. et al. (2000). "Nutritional constraints in terrestrial and freshwater food webs." In: *Nature* 408.6812, p. 578.

Eltahir, E. A. and R. L. Bras (1996). "Precipitation recycling." In: *Reviews of Geophysics* 34.3, pp. 367–378.

Emanuel, K. (1994). *Atmospheric Convection.* Oxford University Press.

Emanuel, K. A. and M. Bister (1996). "Moist convective velocity and buoyancy scales." In: *Journal of the Atmospheric Sciences* 53.22, pp. 3276–3285.

Engelbrecht, B. M., L. S. Comita, R. Condit, T. A. Kursar, M. T. Tyree, B. L. Turner, and S. P. Hubbell (2007). "Drought sensitivity shapes species distribution patterns in tropical forests." In: *Nature* 447.7140, pp. 80–82.

Engels, C. and H. Marschner (1995). "Plant uptake and utilization of nitrogen." In: *Nitrogen Fertilization in the Environment,* pp. 41–81.

English, M. (1990). "Deficit irrigation. I: Analytical framework." In: *Journal of Irrigation and Drainage Engineering* 116.3, pp. 399–412.

English, M. J., K. H. Solomon, and G. J. Hoffman (2002). "A paradigm shift in irrigation management." In: *Journal of Irrigation and Drainage Engineering* 128.5, pp. 267–277.

English, M. (1981). "The uncertainty of crop models in irrigation optimization." In: *Transactions of the ASAE* 24.4, pp. 917–0921.

Ewers, B., R. Oren, and J. Sperry (2000). "Influence of nutrient versus water supply on hydraulic architecture and water balance in *Pinus taeda*." In: *Plant, Cell & Environment* 23.10, pp. 1055–1066.

Fadiga, M., D. E. Ethridge, S. Mohanty, and S. Pan (2006). "The impacts of US cotton programs on the world market: an analysis of Brazilian WTO petition." In: *Journal of Cotton Science* 10, pp. 180–192.

Falkenmark, M. and M. Lannerstad (2005). "Consumptive water use to feed humanity – curing a blind spot." In: *Hydrology and Earth System Sciences* 9-12, pp. 15–28.

Farquhar, G., S. V. von Caemmerer, and J. Berry (1980). "A biochemical model of photosynthetic CO_2 assimilation in leaves of C_3 species." In: *Planta* 149.1, pp. 78–90.

Feder, J. (2013). *Fractals. Physics of Solids and Liquids.* Springer US.

Federer, C. (1979). "A soil–plant–atmosphere model for transpiration and availability of soil water." In: *Water Resources Research* 15.3, pp. 555–562.

Fenchel, T., G. M. King, and T. H. Blackburn, 1998. Bacterial Biogeochemistry: The Ecophysiology of Mineral Cycling, San Diego, CA, Academic Press.

Feng, X., G. Vico, and A. Porporato (2012). "On the effects of seasonality on soil water balance and plant growth." In: *Water Resources Research* 48.5.

Feng, X., A. Porporato, and I. Rodríguez-Iturbe (2013). "Changes in rainfall seasonality in the tropics." In: *Nature Climate Change* 3.9, pp. 811–815.

— (2015). "Stochastic soil water balance under seasonal climates." In: *Proc. Roy. Soc. A* 471.2174, p. 20140623.

Feng, X., D. D. Ackerly, T. E. Dawson, S. Manzoni, B. McLaughlin, R. P. Skelton, G. Vico, A. P. Weitz, and S. E. Thompson (2019). "Beyond isohydricity: the role of environmental variability in determining plant drought responses." In: *Plant, Cell & Environment* 42.4, pp. 1104–1111.

Fereres, E. and M. A. Soriano (2007). "Deficit irrigation for reducing agricultural water use." In: *Journal of Experimental Botany* 58.2, pp. 147–159.

Fernandez-Illescas, C. P. and I. Rodríguez-Iturbe (2003). "Hydrologically driven hierarchical competition-colonization models: the impact of interannual climate fluctuations." In: *Ecological Monographs* 73.2, pp. 207–222.

Findell, K. L. and E. A. Eltahir (1997). "An analysis of the soil moisture–rainfall feedback, based on direct observations from Illinois." In: *Water Resources Research* 33.4, pp. 725–735.

— (2003). "Atmospheric controls on soil moisture-boundary layer interactions. Part I: Framework development." In: *Journal of Hydrometeorology* 4.3, pp. 552–569.

Finnigan, J., K. Ayotte, I. Harman, G. Katul, H. Oldroyd, E. Patton, D. Poggi, A. Ross, and P. Taylor (2020). "Boundary-layer flow over complex topography." In: *Boundary-Layer Meteorology* 177.2, pp. 247–313.

Fisher, E. (1923). "Some factors affecting the evaporation of water from soil." In: *The Journal of Agricultural Science* 13.2, pp. 121–143.

Gandhi, P., S. Iams, S. Bonetti, and M. Silber (2019). "Vegetation pattern formation in drylands." In: *Dryland Ecohydrology*. Springer, pp. 469–509.

Gao, Q., P. Zhao, X. Zeng, X. Cai, and W. Shen (2002). "A model of stomatal conductance to quantify the relationship between leaf transpiration, microclimate and soil water stress." In: *Plant, Cell & Environment* 25.11, pp. 1373–1381.

Gardiner, C. (2004). *Handbook of Stochastic Methods for Physics, Chemistry, and the Natural Sciences.* Springer.

Garnier, P., C. Néel, B. Mary, and F. Lafolie (2001). "Evaluation of a nitrogen transport and transformation model in a bare soil." In: *European Journal of Soil Science* 52.2, pp. 253–268.

Garratt, J. R. (1994). *The Atmospheric Boundary Layer.* Cambridge University Press, 316 pp.

Geerts, S. and D. Raes (2009). "Deficit irrigation as an on-farm strategy to maximize crop water productivity in dry areas." In: *Agricultural Water Management* 96.9, pp. 1275–1284.

Gerhardt, K. E., X.-D. Huang, B. R. Glick, and B. M. Greenberg (2009). "Phytoremediation and rhizoremediation of organic soil contaminants: potential and challenges." In: *Plant Science* 176.1, pp. 20–30.

Ghassemi, F., A. J. Jakeman, H. A. Nix et al. (1995). *Salinisation of Land and Water Resources: Human Causes, Extent, Management and Case Studies.* CAB International.

Givnish, T. J. (1999). "On the causes of gradients in tropical tree diversity." In: *Journal of Ecology* 87.2, pp. 193–210.

Goldenfeld, N. and C. Woese (2011). "Life is physics: evolution as a collective phenomenon far from equilibrium." In: *Ann. Rev. Condens. Matter Phys.* pp. 375–399.

Gollan, T., N. Turner, and E.-D. Schulze (1985). "The responses of stomata and leaf gas exchange to vapour pressure deficits and soil water content." In: *Oecologia* 65.3, pp. 356–362.

Gotsch, S. G., N. Nadkarni, A. Darby, A. Glunk, M. Dix, K. Davidson, and T. E. Dawson (2015). "Life in the treetops: ecophysiological strategies of canopy epiphytes in a tropical montane cloud forest." In: *Ecological Monographs* 85.3, pp. 393–412.

Greaves, J. and E. Carter (1920). "Influence of moisture on the bacterial activities of the soil." In: *Soil Science* 10.5, pp. 361–387.

Green, W. H. and G. A. Ampt (1911). "Studies on soil physics." In: *The Journal of Agricultural Science* 4.1, pp. 1–24.

Greenland, D., D. G. Goodin, and R. C. Smith (2003). *Climate Variability and Ecosystem Response at*

Long-Term Ecological Research Sites. Oxford University Press.

Grime, J. P. et al. (1979). *Plant Strategies and Vegetation Processes.* John Wiley and Sons.

Guswa, A. J. (2008). "The influence of climate on root depth: a carbon cost–benefit analysis." In: *Water Resources Research* 44.2.

— (2010). "Effect of plant uptake strategy on the water – optimal root depth." In: *Water Resources Research* 46.9.

Guswa, A. J., M. Celia, and I. Rodríguez-Iturbe (2002). "Models of soil moisture dynamics in ecohydrology: a comparative study." In: *Water Resources Research* 38.9.

Haberl, H., K. H. Erb, F. Krausmann, V. Gaube, A. Bondeau, C. Plutzar, S. Gingrich, W. Lucht, and M. Fischer-Kowalski (2007). "Quantifying and mapping the human appropriation of net primary production in earth's terrestrial ecosystems." In: *Proceedings of the National Academy of Sciences* 104.31, pp. 12942–12947.

Hadas, A., M. Sofer, J. Molina, P. Barak, and C. Clapp (1992). "Assimilation of nitrogen by soil microbial population: NH4 versus organic N." In: *Soil Biology and Biochemistry* 24.2, pp. 137–143.

Hansen, S., M. Shaffer, and H. Jensen (1995). "Developments in modeling nitrogen transformations in soil." In: ed. P. Bacon, *Nitrogen Fertilization in the Environment.* Taylor & Francis. Chapter 3, pp. 83–107.

Hari, P., A. Makela, E. Korpilahti, and M. Holmberg (1986). "Optimal control of gas exchange." In: *Tree Physiology* 2.1-2-3, pp. 169–175.

Hari, P., A. Mäkelä, F. Berninger, and T. Pohja (1999). "Field evidence for the optimality hypothesis of gas exchange in plants." In: *Functional Plant Biology* 26.3, pp. 239–244.

Harris, I., P. D. Jones, T. J. Osborn, and D. H. Lister (2014). "Updated highresolution grids of monthly climatic observations – the CRU TS3. 10 dataset." In: *International Journal of Climatology* 34.3, pp. 623–642.

Hartzell, S., M. S. Bartlett, L. Virgin, and A. Porporato (2015). "Nonlinear dynamics of the CAM circadian rhythm in response to environmental forcing." In: *Journal of Theoretical Biology* 368, pp. 83–94.

Hartzell, S., M. S. Bartlett, and A. Porporato (2017). "The role of plant water storage and hydraulic strategies in relation to soil moisture availability." In: *Plant and Soil* 419.1-2, pp. 503–521.

Hartzell, S., M. Bartlett, J. Yin, and A. Porporato (2018a). "Similarities in the evolution of plants and cars." In: *PLoS One,* e0198044.

Hartzell, S., M. S. Bartlett, and A. Porporato (2018b). "Unified representation of the C3, C4, and CAM photosynthetic pathways with the Photo3 model." In: *Ecological Modelling* 384, pp. 173–187.

Hassan, M. Mahmood-ul, E. Rafique, and A. Rashid (2013). "Physical and hydraulic properties of aridisols as affected by nutrient and crop-residue management in a cotton-wheat system." In: *Acta Scientiarum Agronomy* 35.1, pp. 127–137.

Hassani, A., A. Azapagic, and N. Shokri (2020). "Predicting long-term dynamics of soil salinity and sodicity on a global scale." In: *Proceedings of the National Academy of Sciences* 117.52, pp. 33017–33027.

Hawkins, B. A. et al. (2003). "Energy, water, and broad-scale geographic patterns of species richness." In: *Ecology* 84.12, pp. 3105–3117.

Hawkins, R. H. (1993). "Asymptotic determination of runoff curve numbers from data." In: *Journal of Irrigation and Drainage Engineering* 119.2, pp. 334–345.

Haynes, R. (1986). *Mineral Nitrogen in the Plant–Soil System.* Academic Press.

Herman, J. D., J. D. Quinn, S. Steinschneider, M. Giuliani, and S. Fletcher (2020). "Climate adaptation as a control problem: review and perspectives on dynamic water resources planning under uncertainty." In: *Water Resources Research* 56.2, e24389.

Hiederer, R. and M. Köchy (2011). "Global soil organic carbon estimates and the harmonized world soil database." In: *EUR* 79, p. 25225.

Hillel, D. (1998). *Environmental Soil Physics: Fundamentals, Applications, and Environmental Considerations.* Elsevier Science.

Hillel, D. (1992). *Out of the Earth: Civilization and the Life of the Soil.* University of California Press.

— (2000). *Salinity Management for Sustainable Irrigation: Integrating Science, Environment, and Economics.* The World Bank.

— (2003). *Introduction to Environmental Soil Physics.* Elsevier.

Hochberg, U., F. E. Rockwell, N. M. Holbrook, and H. Cochard (2018). "Iso/anisohydry: a plant–environment interaction rather than a simple hydraulic trait." In: *Trends in Plant Science* 23.2, pp. 112–120.

Holling, C. S. (2001). "Understanding the complexity of economic, ecological, and social systems." In: *Ecosystems* 4.5, pp. 390–405.

Holtan, H. N. (1945). "Time-condensation in hydrograph-analysis." In: *Eos, Transactions of the American Geophysical Union* 26.3, pp. 407–413.

Holton, J. R. (2004). *An Introduction to Dynamic Meteorology*. Elsevier Academic Press.

Homsy, G. M. (1987). "Viscous fingering in porous media." In: *Annual Review of Fluid Mechanics* 19.1, pp. 271–311.

Hoover, M. D. (1950). "Hydrologic characteristics of South Carolina Piedmont forest soils." In: *Soil Science Society of America Journal* 14.C, pp. 353–358.

Horton, R. E. (1933). "The rôle of infiltration in the hydrologic cycle." In: *Transactions of the American Geophysical Union* 14.1, p. 446.

Hotelling, H. (1931). "The economics of exhaustible resources." In: *Journal of Political Economy* 39.2, pp. 137–175.

Hsiao, T. C. (1973). "Plant responses to water stress." In: *Annual Review of Plant Physiology* 24.1, pp. 519–570.

Huang, W. and S. J. Hall (2017). "Elevated moisture stimulates carbon loss from mineral soils by releasing protected organic matter." In: *Nature Communications* 8.1, pp. 1–10.

Hubbell, S. P. (2001). *The Unified Neutral theory of Biodiversity and Biogeography (MPB-32)*. Princeton University Press.

Igbadun, H. E., A. K. Tarimo, B. A. Salim, and H. F. Mahoo (2007). "Evaluation of selected crop water production functions for an irrigated maize crop." In: *Agricultural Water Management* 94.1-3, pp. 1–10.

Ireland, H. A., C. F. S. Sharpe, and D. H. Eargle (1939). *Principles of Gully Erosion in the Piedmont of South Carolina*. US Department of Agriculture.

Istanbulluoglu, E. and R. L. Bras (2005). "Vegetation-modulated landscape evolution: effects of vegetation on landscape processes, drainage density, and topography." In: *Journal of Geophysical Research: Earth Surface* 110.F2.

Jackson, R., H. Mooney, and E.-D. Schulze (1997). "A global budget for fine root biomass, surface area, and nutrient contents." In: *Proceedings of the National Academy of Sciences* 94.14, pp. 7362–7366.

Jansson, S. L. (1958). "Tracer studies on nitrogen transformations in soil with special attention to mineralization immobilization relationships." In: *Ann. Rev. Agric. Coll. Sweden* 24, pp. 101–306.

Jarvis, P. G. (1976). "The interpretation of the variations in leaf water potential and stomatal conductance found in canopies in the field." In: *Philosophical Transactions of the Royal Society B: Biological Sciences* 273.927, pp. 593–610.

Javaux, M., V. Couvreur, J. Vanderborght, and H. Vereecken (2013). "Root water uptake: from three-dimensional diophysical processes to macroscopic modeling approaches." In: *Vadose Zone Journal* 12.4.

Jenkinson, D. S. (1990). "The turnover of organic carbon and nitrogen in soil." In: *Phil. Trans. Roy. Soc. B* 329.1255, pp. 361–368.

Jensen, K., K. Berg-Sörensen, H. Bruus, N. Holbrook, J. Liesche, A. Schulz, M. Zwieniecki, and T. Bohr (2016). "Sap flow and sugar transport in plants." In: *Reviews of Modern Physics* 88.3, p. 035007.

Jobbagy, E. G. and R. B. Jackson (2001). "The distribution of soil nutrients with depth: global patterns and the imprint of plants." In: *Biogeochemistry* pp. 51–77.

Johnson-Maynard, J., R. Graham, L. Wu, and P. Shouse (2002). "Modification of soil structural and hydraulic properties after 50 years of imposed chaparral and pine vegetation." In: *Geoderma* 110.3, pp. 227–240.

Jones, H. (1992). *Plants and Microclimate: A Quantitative Approach to Environmental Plant Physiology*. Cambridge University Press.

Jorgensen, S. E. and Y. M. Svirezhev (2004). *Towards a Thermodynamic Theory for Ecological Systems*. Elsevier.

Juang, J.-Y., A. Porporato, P. C. Stoy, M. S. Siqueira, A. C. Oishi, M. Detto, H.-S. Kim, and G. G. Katul (2007). "Hydrologic and atmospheric controls on initiation of convective precipitation events." In: *Water Resource Research* 43.3, W03421.

Jury, W. A. and H. J. Vaux Jr (2007). "The emerging global water crisis: managing scarcity and conflict between water users." In: *Advances in Agronomy* 95, pp. 1–76.

Kaimal, J. C. and J. J. Finnigan (1994). *Atmospheric Boundary Layer Flows: Their Structure and Measurement*. Oxford University Press.

Kamien, M. I. and N. L. Schwartz (2012). *Dynamic Optimization: The Calculus of Variations and Optimal Control in Economics and Management*. Courier Corporation.

Katul, G., C. Hsieh, and J. Sigmon (1997). "Energy–inertial scale interaction for temperature and velocity in the unstable surface layer." In: *Boundary Layer Meteorology* 82, pp. 49–80.

Katul, G., R. Leuning, and R. Oren (2003). "Relationship between plant hydraulic and biochemical properties derived from a steady-state coupled water and carbon transport model." In: *Plant, Cell & Environment* 26.3, pp. 339–350.

Katul, G., A. Porporato, and R. Oren (2007). "Stochastic dynamics of plantwater interactions." In: *Annual Review of Ecology, Evolution, and Systematics*, pp. 767–791.

Katul, G. G., S. Palmroth, and R. Oren (2009). "Leaf stomatal responses to vapour pressure deficit under current and CO2-enriched atmosphere explained by the economics of gas exchange." In: *Plant, Cell & Environment* 32.8, pp. 968–979.

Katul, G. G., A. G. Konings, and A. Porporato (2011). "Mean velocity profile in a sheared and thermally stratified atmospheric boundary layer." In: *Physical Review Letters* 107.26, p. 268502.

Kelliher, F. M., R. Leuning, M. R. Raupach, and E. D. Schulze (1995a). "Maximum conductances for evaporation from global vegetation types." In: *Agricultural and Forest Meteorology* 73.1, pp. 1–16.

Kelliher, F., R Leuning, M. Raupach, and E.-D. Schulze (1995b). "Maximum conductances for evaporation from global vegetation types." In: *Agricultural and Forest Meteorology* 73.1, pp. 1–16.

Kemp, P. R. and G. J. Williams III (1980). "A physiological basis for niche separation between *Agropyron smithii* (C3) and *Bouteloua gracilis* (C4)." In: *Ecology* 61.4, pp. 846–858.

Kestin, J. (1979). *A Course in Thermodynamics*. Vol. 1. Taylor & Francis.

Kim, C., J. Stricker, and P. Torfs (1996). "An analytical framework for the water budget of the unsaturated zone." In: *Water Resources Research* 32.12, pp. 3475–3484.

Klausmeier, C. A. (1999). "Regular and irregular patterns in semiarid vegetation." In: *Science* 284.5421, pp. 1826–1828.

Kleidon, A., F. Hall, G. Collatz, B. Meeson, S. Los, E. Brown, D. E. Colstoun, and D. Landis (2011). "ISLSCP II total plant-available soil water storage capacity of the rooting zone." In: *ORNL DAAC*.

Klein Goldewijk, K., A. Beusen, G. Van Drecht, and M. De Vos (2011). "The HYDE 3.1 spatially explicit database of human-induced global land-use change over the past 12,000 years." In: *Global Ecology and Biogeography* pp. 73–86.

Knapp, A. K. et al. (2002). "Rainfall variability, carbon cycling, and plant species diversity in a mesic grassland." In: *Science* 298.5601, pp. 2202–2205.

Kondepudi, D. and I. Prigogine (2005). *Modern Thermodynamics*. Wiley.

Konings, A. G., G. G. Katul, and A. Porporato (2010). "The rainfall–no rainfall transition in a coupled land-convective atmosphere system." In: *Geophysics Research Letters* 37.14, p. L14401.

Konings, A. G., X. Feng, A. Molini, S. Manzoni, G. Vico, and A. Porporato (2012). "Thermodynamics of an idealized hydrologic cycle." In: *Water Resources Research* 48.5.

Kosmowski, F. (2018). "Soil water management practices (terraces) helped to mitigate the 2015 drought in Ethiopia." In: *Agricultural Water Management* 204, pp. 11–16.

Koster, R. D. et al. (2004). "Regions of strong coupling between soil moisture and precipitation." In: *Science* 305.5687, pp. 1138–1140.

Kramer, I. and Y. Mau (2020). "Soil degradation risks assessed by the SOTE model for salinity and sodicity." In: *Water Resources Research* 56.10

Kramer, P. J. and J. S. Boyer (1995). *Water Relations of Plants and Soils*. Academic Press.

Kruse, E., D. Bucks, R. Von Bernuth et al. (1990). "Comparison of irrigation systems." In: *Agronomy* 30, pp. 475–508.

Kueppers, L. M., M. A. Snyder, and L. C. Sloan (2007). "Irrigation cooling effect: regional climate forcing by land-use change." In: *Geophysical Research Letters* 34.3.

Kumagai, T. and A. Porporato (2012). "Strategies of a Bornean tropical rainforest water use as a function of rainfall regime: isohydric or anisohydric?." In: *Plant, Cell & Environment* 35.1, pp. 61–71.

Lai, C.-T. and G. Katul (2000). "The dynamic role of root-water uptake in coupling potential to actual transpiration." In: *Advances in Water Resources* 23.4, pp. 427–439.

Lai, C.-T., G. Katul, R. Oren, D. Ellsworth, and K. Schäfer (2000). "Modeling CO2 and water vapor turbulent flux distributions within a forest canopy." In: *Journal of Geophysical Research: Atmospheres (1984-2012)* 105.D21, pp. 26333–26351.

Laio, F. (2006). "A vertically extended stochastic model of soil moisture in the root zone." In: *Water Resources Research* 42.2.

Laio, F., A. Porporato, L. Ridolfi, and I. Rodríguez-Iturbe (2001a). "Mean first passage times of processes driven by white shot noise." In: *Physical Review E* 63.3, p. 036105.

Laio, F., A. Porporato, C. Fernandez-Illescas, and I. Rodríguez-Iturbe (2001b). "Plants in water-controlled ecosystems: active role in hydrologic processes and response to water stress: IV. Discussion of real cases." In: *Advances in Water Resources* 24.7, pp. 745–762.

Laio, F., A. Porporato, L. Ridolfi, and I. Rodríguez-Iturbe (2001c). "Plants in water-controlled ecosystems: active role in hydrologic processes and response to water stress: II. Probabilistic soil moisture dynamics." In: *Advances in Water Resources* 24.7, pp. 707–723.

Laio, F., A. Porporato, L. Ridolfi, and I. Rodríguez-Iturbe (2002). "On the seasonal dynamics of mean soil moisture." In: *Journal of Geophysical Research* 107.D15, pp. 4272–4272.

Laio, F., S. Tamea, L. Ridolfi, P. D'Odorico, and I. Rodríguez-Iturbe (2009). "Ecohydrology of groundwater-dependent ecosystems: 1. Stochastic water table dynamics." In: *Water Resources Research* 45.5, W05419.

Lambers, H., F. Chapin III, and T. Pons (1998). *Plant Physiological Ecology*. Springer.

Lamm, F. R. and T. P. Trooien (2003). "Subsurface drip irrigation for corn production: a review of 10 years of research in Kansas." In: *Irrigation Science* 22.3-4, pp. 195–200.

Larcher, W. (1995). *Physiological Plant Ecology: Ecophysiology and Stress Physiology of Functional Groups*. Springer Science & Business Media.

— (2003). *Physiological Plant Ecology: Ecophysiology and Stress Physiology of Functional Groups*. Springer Science & Business Media.

LaRowe, D. E. and P. Van Cappellen (2011). "Degradation of natural organic matter: a thermodynamic analysis." In: *Geochimica et Cosmochimica Acta* 75.8, pp. 2030–2042.

Lauenroth, W. and P. L. Sims (1976). "Evapotranspiration from a shortgrass prairie subjected to water and nitrogen treatments." In: *Water Resources Research* 12.3, pp. 437–442.

Lauenroth, W., J. Dodd, and P. Sims (1978). "The effects of water-and nitrogeninduced stresses on plant community structure in a semiarid grassland." In: *Oecologia* 36.2, pp. 211–222.

Lauenroth, W., O. Sala, D. Milchunas, and R. Lathrop (1987). "Root dynamics of *Bouteloua gracilis* during short-term recovery from drought." In: *Functional Ecology*, pp. 117–124.

Lauenroth, W., O. Sala, D. Coffin, and T. Kirchner (1994). "The importance of soil water in the recruitment of *Bouteloua gracilis* in the shortgrass steppe." In: *Ecological Applications* 4.4, pp. 741–749.

Lawlor, D. W. and W. Tezara (2009). "Causes of decreased photosynthetic rate and metabolic capacity in water-deficient leaf cells: a critical evaluation of mechanisms and integration of processes." In: *Annals of Botany*.

Lawrence, C. R., J. C. Neff, and J. P. Schimel (2009). "Does adding microbial mechanisms of decomposition improve soil organic matter models? A comparison of four models using data from a pulsed rewetting experiment." In: *Soil Biology and Biochemistry* 41.9, pp. 1923–1934.

Le Mer, J. and P. Roger (2001). "Production, oxidation, emission and consumption of methane by soils: a review." In: *European Journal of Soil Biology* 37.1, pp. 25–50.

Leetham, J. and D. Milchunas (1985). "The composition and distribution of soil microarthropods in the shortgrass steppe in relation to soil water, root biomass, and grazing by cattle." In: *Pedobiologia (Jena)* 28.5, pp. 311–325.

Lefever, R. and O. Lejeune (1997). "On the origin of tiger bush." In: *Bulletin of Mathematical Biology* 59.2, pp. 263–294.

Leibold, M. A. and M. A. McPeek (2006). "Coexistence of the niche and neutral perspectives in community ecology." In: *Ecology* 87.6, pp. 1399–1410.

Leuning, R. (1990). "Modelling stomatal behaviour and and photosynthesis of *Eucalyptus grandis*." In: *Functional Plant Biology* 17.2, pp. 159–175.

— (1995). "A critical appraisal of a combined stomatal-photosynthesis model for C3 plants." In: *Plant, Cell & Environment* 18.4, pp. 339–355.

Levitt, J. (1980). "Responses of plants to environmental stresses." In: *Water, Radiation, Salt, and other Stresses* 2.

Lhomme, J.-P. (2001). "Stomatal control of transpiration: examination of the Jarvis-type representation of canopy resistance in relation to humidity." In: *Water Resources Research* 37.3, pp. 689–699.

Lhomme, J.-P., E. Elguero, A. Chehbouni, and G. Boulet (1998). "Stomatal control of transpiration: examination of Monteith's formulation of canopy resistance." In: *Water Resources Research* 34.9, pp. 2301–2308.

Li, C., S. Frolking, and R. Harriss (1994). "Modeling carbon biogeochemistry in agricultural soils." In: *Global Biogeochemical Cycles* 8.3, pp. 237–254.

Li, L. et al. (2017). "Expanding the role of reactive transport models in critical zone processes." In: *Earth-Science Reviews* 165, pp. 280–301.

Liang, X., D. P. Lettenmaier, E. F. Wood, and S. J. Burges (1994). "A simple hydrologically based model of land surface water and energy fluxes for general circulation models." In: *Journal of Geophysical Research: Atmospheres* 99.D7, pp. 14415–14428.

Liang, Y., D. Hazlett, and W. Lauenroth (1989). "Biomass dynamics and water use efficiencies of five plant communities in the shortgrass steppe." In: *Oecologia* 80.2, pp. 148–153.

Liberzon, D. (2011). *Calculus of Variations and Optimal Control Theory: A Concise Introduction.* Princeton University Press.

Lighthill, M. J. and G. B. Whitham (1955). "On kinematic waves. I. Flood movement in long rivers." In: *Proceedings of the Royal Society A: Mathematical, Physical and Engineering Sciences* 229.1178, pp. 281–316.

Limbrunner, J. F., R. M. Vogel, and L. C. Brown (2000). "Estimation of harmonic mean of a lognormal variable." In: *Journal of Hydrologic Engineering* pp. 59–66.

Lin, B. B. (2011). "Resilience in agriculture through crop diversification: adaptive management for environmental change." In: *BioScience* 61.3, pp. 183–193.

Lindner, B., J. Garcia-Ojalvo, A. Neiman, and L. Schimansky-Geier (2004). "Effects of noise in excitable systems." In: *Physics Reports* 392.6, pp. 321–424.

Linn, D. M. and J. W. Doran (1984). "Effect of water-filled pore space on carbon dioxide and nitrous oxide production in tilled and nontilled soils 1." In: *Soil Science Society of America Journal* 48.6, pp. 1267–1272.

Liptzin, D. and W. L. Silver (2009). "Effects of carbon additions on iron reduction and phosphorus availability in a humid tropical forest soil." In: *Soil Biology and Biochemistry* 41.8, pp. 1696–1702.

Liu, J. et al. (2013). "Framing sustainability in a telecoupled world." In: *Ecology and Society* 18.2.

Liu, M.-C., J.-Y. Parlange, M. Sivapalan, and W. Brutsaert (1998). "A note on the time compression approximation." In: *Water Resources Research* 34.12, pp. 3683–3686.

Liu, Y., A. J. Parolari, M. Kumar, C.-W. Huang, G. G. Katul, and A. Porporato (2017). "Increasing atmospheric humidity and CO2 concentration alleviate forest mortality risk." In: *Proceedings of the National Academy of Sciences* 114.37, pp. 9918–9923.

Logan, J. D. (2013). *Applied Mathematics.* John Wiley and Sons, 658 pp.

Logan, J. A. (1983). "Nitrogen oxides in the troposphere: global and regional budgets." In: *Journal of Geophysical Research: Oceans* 88.C15, pp. 10785–10807.

Lohammar, T., S. Larsson, S. Linder, and S. Falk (1980). "FAST: simulation models of gaseous exchange in Scots pine." In: *Ecological Bulletins,* pp. 505–523.

Lovley, D. R. (1991). "Dissimilatory Fe (III) and Mn (IV) reduction." In: *Microbiology and Molecular Biology Reviews* 55.2, pp. 259–287.

Lüttge, U. and F. Beck (1992). "Endogenous rhythms and chaos in crassulacean acid metabolism." In: *Planta* 188.1, pp. 28–38.

Maggi, F. and A. Porporato (2007). "Coupled moisture and microbial dynamics in unsaturated soils." In: *Water Resources Research* 43.7.

Maggi, F. et al. (2008). "A mechanistic treatment of the dominant soil nitrogen cycling processes: model development, testing, and application." In: *Journal of Geophysical Research: Biogeosciences* 113.G2.

Maher, K. and C. Chamberlain (2014). "Hydrologic regulation of chemical weathering and the geologic carbon cycle." In: *Science* 343.6178, pp. 1502–1504.

Mahrt, L. (2014). "Stably stratified atmospheric boundary layers." In: *Annual Review of Fluid Mechanics* 46, pp. 23–45.

Mannocchi, F. and P. Mecarelli (1994). "Optimization analysis of deficit irrigation systems." In: *Journal of Irrigation and Drainage Engineering* 120.3, pp. 484–503.

Manzoni, S. and G. Katul (2014). "Invariant soil water potential at zero microbial respiration explained by hydrological discontinuity in dry soils." In: *Geophysical Research Letters* 41.20, pp. 7151–7158.

Manzoni, S. (2017). "Flexible carbon-use efficiency across litter types and during decomposition partly compensates nutrient imbalances? Results from analytical stoichiometric models." In: *Frontiers in Microbiology* 8, p. 661.

Manzoni, S. and A. Porporato (2007). "A theoretical analysis of nonlinearities and feedbacks in soil carbon and nitrogen cycles." In: *Soil Biology and Biochemistry* 39.7, pp. 1542–1556.

— (2009). "Soil carbon and nitrogen mineralization: theory and models across scales." In: *Soil Biology and Biochemistry* 41.7, pp. 1355–1379.

Manzoni, S., R. B. Jackson, J. A. Trofymow, and A. Porporato (2008). "The global stoichiometry of litter nitrogen mineralization." In: *Science* 321.5889, pp. 684–686.

Manzoni, S., J. A. Trofymow, R. B. Jackson, and A. Porporato (2010). "Stoichiometric controls on carbon, nitrogen, and phosphorus dynamics in decomposing litter." In: *Ecological Monographs* 80.1, pp. 89–106.

Manzoni, S., G. Katul, P. A. Fay, H. W. Polley, and A. Porporato (2011a). "Modeling the vegetation-atmosphere carbon dioxide and water vapor interactions along a controlled CO2 gradient." In: *Ecological Modelling* 222.3, pp. 653–665.

Manzoni, S., A. Molini, and A. Porporato (2011b). "Stochastic modelling of phytoremediation." In: *Proc. Roy. Soc. A: Mathematical, Physical and Engineering Sciences* 467.2135, pp. 3188–3205.

Manzoni, S., J. P. Schimel, and A. Porporato (2012). "Responses of soil microbial communities to water stress: results from a meta-analysis." In: *Ecology* 93.4, pp. 930–938.

Manzoni, S., G. Vico, A. Porporato, and G. Katul (2013). "Biological constraints on water transport in the soil–plant–atmosphere system." In: *Advances in Water Resources* 51, pp. 292–304.

Manzoni, S., S. Schaeffer, G. Katul, A Porporato, and J. Schimel (2014). "A theoretical analysis of microbial eco-physiological and diffusion limitations to carbon cycling in drying soils." In: *Soil Biology and Biochemistry* 73, pp. 69–83.

Manzoni, S., M. H. Ahmed, and A. Porporato (2019). "Ecohydrological and stoichiometric controls on soil carbon and nitrogen dynamics in drylands." In: *Dryland Ecohydrology*, pp. 183–199. Springer

Manzoni, S., A. Chakrawal, T. Fischer, J. P. Schimel, A. Porporato, and G. Vico (2020). "Rainfall intensification increases the contribution of rewetting pulses to soil heterotrophic respiration." In: *Biogeosciences* 17.15, pp. 4007–4023.

Martínez-Vilalta, J., R. Poyatos, D. Aguadé, J. Retana, and M. Mencuccini (2014). "A new look at water transport regulation in plants." In: *New Phytologist* 204.1, pp. 105–115.

Mathur, S. (2004). "Modeling phytoremediation of soils." In: *Practice Periodical of Hazardous, Toxic, and Radioactive Waste Management* 8.4, pp. 286–297.

Matthews, T. J. and R. J. Whittaker (2014). "Neutral theory and the species abundance distribution: recent developments and prospects for unifying niche and neutral perspectives." In: *Ecology and Evolution* 4.11, pp. 2263–2277.

Mattson Jr., W. J. (1980). "Herbivory in relation to plant nitrogen content." In: *Annual Review of Ecology and Systematics* 11.1, pp. 119–161.

Mau, Y. and A. Porporato (2015). "A dynamical system approach to soil salinity and sodicity." In: *Advances in Water Resources* 83, pp. 68–76.

— (2016). "Optimal control solutions to sodic soil reclamation." In: *Advances in Water Resources* 91, pp. 37–45.

Mau, Y., X. Feng, and A. Porporato (2014). "Multiplicative jump processes and applications to leaching of salt and contaminants in the soil." In: *Physical Review E* 90.5, p. 052128.

Mau, Y., L. Haim, and E. Meron (2015). "Reversing desertification as a spatial resonance problem." In: *Physical Review E* 91.1, p. 012903.

Maxwell, R. M. et al. (2014). "Surface-subsurface model intercomparison: a first set of benchmark results to diagnose integrated hydrology and feedbacks." In: *Water Resources Research* 50.2, pp. 1531–1549.

McConnell, K. E. (1983). "An economic model of soil conservation." In: *American Journal of Agricultural Economics* 65.1, pp. 83–89.

McDonnell, J. J. (2013). "Are all runoff processes the same?." In: *Hydrological Processes* 27.26, pp. 4103–4111.

McDowell, N. et al. (2008). "Mechanisms of plant survival and mortality during drought: why do some plants survive while others succumb to drought?." In: *New Phytologist* 178.4, pp. 719–739.

McDowell, N. G. et al. (2013). "Evaluating theories of drought-induced vegetation mortality using a multimodel–experiment framework." In: *New Phytologist* 200.2, pp. 304–321.

McGuire, V. L. (2009). "Water-level changes in the High Plains aquifer, predevelopment to 2007, 2005–06, and 2006–07." In:

McNeal, B. and N. Coleman (1966). "Effect of solution composition on soil hydraulic conductivity." In: *Soil Science Society of America Journal* 30.3, pp. 308–312.

Medlyn, B. E. et al. (2011). "Reconciling the optimal and empirical approaches to modelling stomatal conductance." In: *Global Change Biology* 17.6, pp. 2134–2144.

Meinzer, F. C., D. M. Johnson, B. Lachenbruch, K. A. McCulloh, and D. R. Woodruff (2009). "Xylem hydraulic safety margins in woody plants: coordination of stomatal control of xylem tension with hydraulic capacitance." In: *Functional Ecology* 23.5, pp. 922–930.

Meixner, F. X. and W. Eugster (1999). "Effects of landscape pattern and topography on emissions and transport." In: eds. J. D. Techunen and P. Kabat, *Effects of Landscape Pattern and Topography on Emissions and Transport.* pp. 18–23.

Meron, E. (2015). *Nonlinear Physics of Ecosystems.* CRC Press.

Miles, L., A. C. Newton, R. S. DeFries, C. Ravilious, I. May, S. Blyth, V. Kapos, and J. E. Gordon (2006). "A global overview of the conservation status of tropical dry forests." In: *Journal of Biogeography* 33.3, pp. 491–505.

Milly, P. C. D. (1985). "Stability of the Green–Ampt profile in a delta function soil." In: *Water Resources Research* 21.3, pp. 399–402.

Milly, P. (1993). "An analytic solution of the stochastic storage problem applicable to soil water." In: *Water Resources Research* 29.11, pp. 3755–3758.

Mitchell, K. J. (1975). "Dynamics and simulated yield of Douglas fir." In: *Forest Science* 21.suppl. 1, a0001-z0001.

Molina, J., C. Clapp, M. Shaffer, F. Chichester, and W. Larson (1983). "NCSOIL, a model of nitrogen and carbon transformations in soil: description, calibration, and behavior 1." In: *Soil Science Society of America Journal* 47.1, pp. 85–91.

Moore, R. (1985). "The probability-distributed principle and runoff production at point and basin scales." In: *Hydrological Sciences Journal* 30.2, pp. 273–297.

Morowitz, H. J. (1993). *Beginnings of Cellular Life: Metabolism Recapitulates Biogenesis.* Yale University Press.

Moyano, F. E., S. Manzoni, and C. Chenu (2013). "Responses of soil heterotrophic respiration to moisture availability: an exploration of processes and models." In: *Soil Biology and Biochemistry* 59, pp. 72–85.

Mualem, Y. (1976). "A new model for predicting the hydraulic conductivity of unsaturated porous media." In: *Water Resources Research* 12.3, pp. 513–522.

Mulligan, C. B. and X. Sala-i Martin (1991). Tech. rep. National Bureau of Economic Research.

Munns, R. and M. Tester (2008). "Mechanisms of salinity tolerance." In: *Ann. Rev. Plant Biol.* 59, pp. 651–681.

Munson, B., B. Young, and T. Okiishi (2005). *Fundamentals of Fluid Mechanics.* Wiley.

Murray, J. D. (2002). *Mathematical Biology.* Springer.

Negri, D. H. (1989). "The common property aquifer as a differential game." In: *Water Resources Research* 25.1, pp. 9–15.

Nepstad, D. C. et al. (1994). "The role of deep roots in the hydrological and carbon cycles of Amazonian forests and pastures." In: *Nature* 372.6507, pp. 666–669.

Nepstad, D. C., C. M. Stickler, and O. T. Almeida (2006). "Globalization of the Amazon soy and beef industries: opportunities for conservation." In: *Conservation Biology* 20.6, pp. 1595–1603.

Neumann, R. B. and Z. G. Cardon (2012). "The magnitude of hydraulic redistribution by plant roots: a review and synthesis of empirical and modeling studies." In: *New Phytologist* 194.2, pp. 337–352.

Nicholson, S. (2000). "Land surface processes and Sahel climate." In: *Reviews of Geophysics* 38.1, pp. 117–139.

Nilsen, E. T., M. R. Sharifi, P. W. Rundel, W. M. Jarrell, and R. A. Virginia (1983). "Diurnal and seasonal water relations of the desert phreatophyte *Prosopis glandulosa* (honey mesquite) in the Sonoran Desert of California." In: *Ecology* 64.6, pp. 1381–1393.

Nilsen, E. and D. Orcutt (1998). "Physiology of plants under stress: abiotic factors." In:

Nobel, P. S. (1999). *Physicochemical and Environmental Plant Physiology*. Academic Press.

Norman, J. M. (1982). "Simulation of microclimates." In: *Biometeorology in Integrated Pest Management*, pp. 65–99.

Novick, K. A., A. G. Konings, and P. Gentine (2019). "Beyond soil water potential: an expanded view on isohydricity including land–atmosphere interactions and phenology." In: *Plant, Cell & Environment* 42.6, pp. 1802–1815.

Noy-Meir, I. (1973). "Desert ecosystems: environment and producers." In: *Annual Review of Ecology and Systematics,* pp. 25–51.

Ojha, L., M. B. Wilhelm, S. L. Murchie, A. S. McEwen, J. J. Wray, J. Hanley, M. Massé, and M. Chojnacki (2015). "Spectral evidence for hydrated salts in recurring slope lineae on Mars." In: *Nature Geoscience* 8.11, pp. 829–832.

Orcutt, D. M. (2000). *The Physiology of Plants under Stress: Soil and Biotic Factors*. Vol. 2. John Wiley and Sons.

Oren, R., J. Sperry, G. Katul, D. Pataki, B. Ewers, N. Phillips, and K. Schafer (1999). "Survey and synthesis of intra and interspecific variation in stomatal sensitivity to vapour pressure deficit." In: *Plant, Cell & Environment* 22.12, pp. 1515–1526.

Oster, J. and D Wichelns (2003). "Economic and agronomie strategies to achieve sustainable irrigation." In: *Irrigation Science* 22.3-4, pp. 107–120.

Otter, L. B., W. X. Yang, M. C. Scholes, and F. X. Meixner (1999). "Nitric oxide emissions from a southern African savanna." In: *Journal of Geophysical Research: Atmospheres* 104.D15, pp. 18471–18485.

Pacala, S. W., C. D. Canham, J. Saponara, J. A. Silander Jr., R. K. Kobe, and E. Ribbens (1996). "Forest models defined by field measurements: estimation, error analysis and dynamics." In: *Ecological Monographs* 66.1, pp. 1–43.

Pagel, H., C. Poll, J. Ingwersen, E. Kandeler, and T. Streck (2016). "Modeling coupled pesticide degradation and organic matter turnover: From gene abundance to process rates." In: *Soil Biology and Biochemistry* 103, pp. 349–364.

Pandey, D. N., A. K. Gupta, D. M. Anderson et al. (2003). "Rainwater harvesting as an adaptation to climate change." In: *Current Science* 85.1, pp. 46–59.

Panikov, N. S. and M. V. Sizova (1996). "A kinetic method for estimating the biomass of microbial functional groups in soil." In: *Journal of Microbiological Methods* 24.3, pp. 219–230.

Parkin, T. B. (1987). "Soil microsites as a source of denitrification variability." In: *Soil Science Society of America Journal* 51.5, pp. 1194–1199.

Parlange, J.-Y., W. Hogarth, P. Ross, M. B. Parlange, M. Sivapalan, G. C. Sander, and M. C. Liu (2000). "A note on the error analysis of time compression approximations." In: *Water Resources Research* 36.8, pp. 2401–2406.

Parolari, A. J., G. G. Katul, and A. Porporato (2014). "An ecohydrological perspective on drought-induced forest mortality." In: *Journal of Geophysical Research: Biogeosciences* 119.5, pp. 965–981.

Parton, W. J., D. S. Schimel, C. V. Cole, and D. S. Ojima (1987). "Analysis of factors controlling soil organic matter levels in great plains grasslands." In: *Soil Science Society of America Journal* 51.5, pp. 1173–1179.

Parton, W. J., J. W. Stewart, and C. V. Cole (1988). "Dynamics of C, N, P and S in grassland soils: a model." In: *Biogeochemistry* 5.1, pp. 109–131.

Pastor, J., J. D. Aber, C. A. McClaugherty, and J. M. Melillo (1984). "Aboveground production and N and P cycling along a nitrogen mineralization gradient on Blackhawk Island, Wisconsin." In: *Ecology* 65.1, pp. 256–268.

Paul, E. A. (2006). *Soil Microbiology, Ecology and Biochemistry.* Academic Press.

Payero, J. O., S. R. Melvin, S. Irmak, and D. Tarkalson (2006). "Yield response of corn to deficit irrigation in a semiarid climate." In: *Agricultural Water Management* 84.1-2, pp. 101–112.

Pelak, N. and A. Porporato (2016). "Sizing a rainwater harvesting cistern by minimizing costs." In: *Journal of Hydrology* 541, pp. 1340–1347.

— (2019). "Dynamic evolution of the soil pore size distribution and its connection to soil management and biogeochemical processes." In: *Advances in Water Resources* 131, p. 103384.

Pelak, N., A. J. Parolari, and A. Porporato (2016). "Bistable plant–soil dynamics and biogenic controls on the soil production function." In: *Earth Surface Processes and Landforms* 41.8, pp. 1011–1017.

Pelak, N., R. Revelli, and A. Porporato (2017). "A dynamical systems framework for crop models: toward optimal fertilization and irrigation strategies under climatic variability." In: *Ecological Modelling* 365, pp. 80–92.

Penman, H. L. (1948). "Natural evaporation from open water, bare soil and grass." In: *Proc. Roy. Soc. A* 193.1032, pp. 120–145.

Perri, S., G. G. Katul, and A. Molini (2019). "Xylem–phloem hydraulic coupling explains multiple osmoregulatory responses to salt stress." In: *New Phytologist* 224.2, pp. 644–662.

Perri, S., S. Suweis, A. Holmes, P. R. Marpu, D. Entekhabi, and A. Molini (2020). "River basin salinization as a form of aridity." In: *Proceedings of the National Academy of Sciences* 117.30, pp. 17635–17642.

Pesch, H. J., M. Plail, and D. Munich (2009). "The maximum principle of optimal control: a history of ingenious ideas and missed opportunities." In: *Control and Cybernetics* 38.4A, pp. 973–995.

Philip, J. R. (1957a). "The theory of infiltration: 1. The infiltration equation and its solution." In: *Soil Science* 83.5, pp. 345–358.

— (1957b). "The theory of infiltration: 4. Sorptivity and algebraic infiltration equations." In: *Soil Science* 84.3, pp. 257–264.

— (1968). "Steady infiltration from buried point sources and spherical cavities." In: *Water Resources Research* 4.5, pp. 1039–1047.

— (1973). "On solving the unsaturated flow equation: 1. the flux-concentration relation." In: *Soil Science* 116.5, pp. 328–335.

— (1991). "Hillslope infiltration: planar slopes." In: *Water Resources Research* 27.1, pp. 109–117.

— (1997). "Effect of root water extraction on wetted regions from continuous irrigation sources." In: *Irrigation Science* 17.3, pp. 127–135.

Phillips, O. and J. S. Miller (2002). *Global Patterns of Plant Diversity: Alwyn H. Gentry's Forest Transect Data Set.* Missouri Botanical Press.

Phillips, R. and S. R. Quake (2006). "The biological frontier of physics." In: *Physics Today* 59.5, pp. 38–43.

Pilon-Smits, E. (2005). "Phytoremediation." In: *Ann. Rev. Plant Biol.* 56, pp. 15–39.

Pope, S. B. (2001). *Turbulent Flows.* IOP Publishing.

Porporato, A. (2009). "Atmospheric boundary-layer dynamics with constant Bowen ratio." In: *Boundary-Layer Meteorology* 132.2, pp. 227–240.

Porporato, A. and P. D'Odorico (2004). "Phase transitions driven by statedependent Poisson noise." In: *Physical Review Letters* 92.11, p. 110601.

Porporato, A. and I. Rodríguez-Iturbe (2002). "Ecohydrology – a challenging multidisciplinary research perspective/Ecohydrologie: une perspective stimulante de recherche multidisciplinaire." In: *Hydrological Sciences Journal* 47.5, pp. 811–821.

— (2013). "From random variability to ordered structures: a search for general synthesis in ecohydrology." In: *Ecohydrology* 6.3, pp. 333–342.

Porporato, A., F. Laio, L. Ridolfi, and I. Rodríguez-Iturbe (2001). "Plants in water-controlled ecosystems: active role in hydrologic processes and response to water stress: III. Vegetation water stress." In: *Advances in Water Resources* 24.7, pp. 725–744.

Porporato, A., P. D'Odorico, F. Laio, and I. Rodríguez-Iturbe (2003a). "Hydrologic controls on soil carbon and nitrogen cycles. I. Modeling scheme." In: *Advances in Water Resources* 26.1, pp. 45–58.

Porporato, A., F. Laio, L. Ridolfi, K. K. Caylor, and I. Rodríguez-Iturbe (2003b). "Soil moisture and plant stress dynamics along the Kalahari precipitation gradient." In: *Journal of Geophysical Research: Atmospheres* 108.D3, p. 4127.

Porporato, A., E. Daly, and I. Rodríguez-Iturbe (2004). "Soil water balance and ecosystem response to climate change." In: *The American Naturalist* 164.5, pp. 625–632.

Porporato, A., G. Vico, and P. A. Fay (2006). "Superstatistics of hydro-climatic fluctuations and interannual ecosystem productivity." In: *Geophysical Research Letters* 33.15.

Porporato, A., X. Feng, S. Manzoni, Y. Mau, A. J. Parolari, and G. Vico (2015). "Ecohydrological modeling in agroecosystems: examples and challenges." In: *Water Resources Research* 51.7, pp. 5081–5099.

Priestley, C. and R. Taylor (1972). "On the assessment of surface heat flux and evaporation using large-scale parameters." In: *Monthly Weather Review* 100.2, pp. 81–92.

Priestley, M. B. (1981). *Spectral Analysis and Time Series*. Academic Press.

Purves, D. W., J. W. Lichstein, N. Strigul, and S. W. Pacala (2008). "Predicting and understanding forest dynamics using a simple tractable model." In: *Proceedings of the National Academy of Sciences* 105.44, pp. 17018–17022.

Qadir, M., S. Schubert, A. Ghafoor, and G. Murtaza (2001). "Amelioration strategies for sodic soils: a review." In: *Land Degradation & Development* 12.4, pp. 357–386.

Rand, R. H. (1983). "Fluid mechanics of green plants." In: *Annual Review of Fluid Mechanics* 15.1, pp. 29–45.

Ratkowsky, D., J. Olley, T. McMeekin, and A. Ball (1982). "Relationship between temperature and growth rate of bacterial cultures." In: *Journal of Bacteriology* 149.1, pp. 1–5.

Raupach, M. and J. Finnigan (1988). "'Single-layer models of evaporation from plant canopies are incorrect but useful, whereas multilayer models are correct but useless': Discuss." In: *Functional Plant Biology* 15.6, pp. 705–716.

— (1997). "The influence of topography on meteorological variables and surface–atmosphere interactions." In: *Journal of Hydrology* 190.3-4, pp. 182–213.

Reeves, P. C. and M. A. Celia (1996). "A functional relationship between capillary pressure, saturation, and interfacial area as revealed by a pore-scale network model." In: *Water Resources Research* 32.8, pp. 2345–2358.

Refsgaard, J. C. and J. Knudsen (1996). "Operational validation and intercomparison of different types of hydrological models." In: *Water Resources Research* 32.7, pp. 2189–2202.

Reynolds, J. F. et al. (2007). "Global desertification: building a science for dryland development." In: *Science* 316.5826, pp. 847–851.

Richardson, C. W. (1981). "Stochastic simulation of daily precipitation, temperature, and solar radiation." In: *Water Resources Research* 17.1, pp. 182–190.

Richter, D. D. and D. Markewitz (2001). *Understanding Soil Change: Soil Sustainability over Millennia, Centuries, and Decades*. Cambridge University Press.

Ridolfi, L., P. D'Odorico, A. Porporato, and I. Rodríguez-Iturbe (2000). "Duration and frequency of water stress in vegetation: an analytical model." In: *Water Resources Research* 36.8, pp. 2297–2307.

Ridolfi, L., P. D'Odorico, F. Laio, S. Tamea, and I. Rodríguez-Iturbe (2008). "Coupled stochastic dynamics of water table and soil moisture in bare soil conditions." In: *Water Resources Research* 44.1, W01435.

Ridolfi, L., P. D'Odorico, and F. Laio (2011). *Noise-Induced Phenomena in the Environmental Sciences*. Cambridge University Press.

Ridolfi, L., P. D'odorico, A. Porporato, and I. Rodríguez-Iturbe (2003). "The influence of stochastic soil moisture dynamics on gaseous emissions of NO, N2O, and N2." In: *Hydrological Sciences Journal* 48.5, pp. 781–798.

Ridolfi, L., P. D'Odorico, and F. Laio (2006). "Effect of vegetation–water table feedbacks on the stability and resilience of plant ecosystems." In: *Water Resources Research* 42.1.

Rietz, D. and R. Haynes (2003). "Effects of irrigation-induced salinity and sodicity on soil microbial activity." In: *Soil Biology and Biochemistry* 35.6, pp. 845–854.

Rigby, J. R. and A. Porporato (2006). "Simplified stochastic soil-moisture models: a look at infiltration." In: *Hydrology and Earth System Sciences* 10.6, pp. 861–871.

Rigby, J., J. Yin, J. D. Albertson, and A. Porporato (2015). "Approximate analytical solution to diurnal atmospheric boundary-layer growth under wellwatered conditions." In: *Boundary-Layer Meteorology* 156.1, pp. 73–89.

Ritchie, J. and J. Adams (1974). "Field measurement of evaporation from soil shrinkage cracks." In: *Soil Science Society of America Journal* 38.1, pp. 131–134.

Rizzo, A., F. Boano, R. Revelli, and L. Ridolfi (2013). "Role of water flow in modeling methane emissions from flooded paddy soils." In: *Advances in Water Resources* 52, pp. 261–274.

Rockhold, M. L., R. Yarwood, M. R. Niemet, P. J. Bottomley, and J. S. Selker (2002). "Considerations for modeling bacterial-induced changes in hydraulic properties of variably saturated porous media." In: *Advances in Water Resources* 25.5, pp. 477–495.

Rockwell, F. E. and N. M. Holbrook (2017). "Leaf hydraulic architecture and stomatal conductance: a functional perspective." In: *Plant Physiology* 174.4, pp. 1996–2007.

Rodríguez-Iturbe, I., D. Entekhabi, and R. L. Bras (1991). "Nonlinear dynamics of soil moisture at climate scales 1. Stochastic analysis." In: *Water Resourses Research* 27.8, pp. 1899–1906.

Rodríguez-Iturbe, I., P. D'odorico, A. Porporato, and L. Ridolfi (1999a). "On the spatial and temporal links between vegetation, climate, and soil moisture." In: *Water Resources Research* 35.12, pp. 3709–3722.

Rodríguez-Iturbe, I. and A. Porporato (2004). *Ecohydrology of Water-Controlled Ecosystems: Soil Moisture and Plant Dynamics*. Cambridge University Press.

Rodríguez-Iturbe, I., A. Porporato, L. Ridolfi, V. Isham, and D. Cox (1999b). "Probabilistic modelling of water balance at a point: the role of climate, soil and vegetation." In: *Proc. Roy. Soc. A: Mathematical, Physical and Engineering Sciences* 455, pp. 3789–3805.

Rodríguez-Iturbe, I., A. Porporato, F. Laio, and L. Ridolfi (2001). "Intensive or extensive use of soil moisture: plant strategies to cope with stochastic water availability." In: *Geophysical Research Letters* 28.23, pp. 4495–4497.

Rojstaczer, S., S. M. Sterling, and N. J. Moore (2001). "Human appropriation of photosynthesis products." In: *Science* 294.5551, pp. 2549–2552.

Romps, D. M. (2017). "Exact expression for the lifting condensation level." In: *Journal of the Atmospheric Sciences* 74.12, pp. 3891–3900.

Rosindell, J., S. P. Hubbell, F. He, L. J. Harmon, and R. S. Etienne (2012). "The case for ecological neutral theory." In: *Trends in Ecology & Evolution* 27.4, pp. 203–208.

Ross, S. (2014). *Introduction to Probability Models*. Elsevier Science.

Rotach, M. W. and D. Zardi (2007). "On the boundary-layer structure over highly complex terrain: key findings from MAP." In: *Q. J. R. Meteorol. Soc.* 133.625, pp. 937–948.

Runyan, C. W. and P. D'Odorico (2010). "Ecohydrological feedbacks between salt accumulation and vegetation dynamics: role of vegetation–groundwater interactions." In: *Water Resources Research* 46.11.

Runyan, C. W., P. D'Odorico, and D. Lawrence (2012). "Physical and biological feedbacks of deforestation." In: *Reviews of Geophysics* 50.4.

Ruprecht, J. and N. Schofield (1991). "Effects of partial deforestation on hydrology and salinity in high salt storage landscapes. I. Extensive block clearing." In: *Journal of Hydrology* 129.1-4, pp. 19–38.

Russell, J. (1931). *Soil Conditions and Plant Growth*. Cran Green and Co.

Sack, L. and N. M. Holbrook (2006). "Leaf hydraulics." In: *Annual Review of Plant Biology* 57.1, pp. 361–381.

Sala, O., R. Jackson, H. Mooney, R. Howarth, and E. Odum (1976). *Methods in Ecosystem Science*. Springer

Sala, O., W. Lauenroth, and W. Parton (1982). "Plant recovery following prolonged drought in a shortgrass steppe." In: *Agricultural Meteorology* 27.1-2, pp. 49–58.

— (1992). "Long-term soil water dynamics in the shortgrass steppe." In: *Ecology* 73.4, pp. 1175–1181.

Sala, O. E., W. Lauenroth, W. Parton, and M. Trlica (1981). "Water status of soil and vegetation in a shortgrass steppe." In: *Oecologia* 48.3, pp. 327–331.

Salisbury, F. and C. Ross (1969). *Plant Physiology*. Wadsworth.

Salt, D. E., R. Smith, and I. Raskin (1998). "Phytoremediation." In: *Annual Review of Plant Biology* 49.1, pp. 643–668.

Salvucci, G. D. (2001). "Estimating the moisture dependence of root zone water loss using conditionally averaged precipitation." In: *Water Resources Research* 37.5, pp. 1357–1365.

Salvucci, G. D. (1993). "An approximate solution for steady vertical flux of moisture through an unsaturated homogeneous soil." In: *Water Resources Research* 29.11, pp. 3749–3754.

Salvucci, G. D. and D. Entekhabi (1994). "Explicit expressions for Green–Ampt (delta function diffusivity) infiltration rate and cumulative storage." In: *Water Resources Research* 30.9, pp. 2661–2663.

Sarmiento, G. (1984). *The Ecology of Neotropical Savannas*. Harvard University Press.

Scanlon, B. R., I. Jolly, M. Sophocleous, and L. Zhang (2007a). "Global impacts of conversions from natural to agricultural ecosystems on water resources: quantity versus quality." In: *Water Resources Research* 43.3.

Scanlon, B. R., C. C. Faunt, L. Longuevergne, R. C. Reedy, W. M. Alley, V. L. McGuire, and P. B. McMahon (2012). "Groundwater depletion and sustainability of irrigation in the US High Plains and Central Valley." In: *Proceedings of the National Academy of Sciences* 109.24, pp. 9320–9325.

Scanlon, T. M., K. K. Caylor, S. A. Levin, and I. Rodríguez-Iturbe (2007b). "Positive feedbacks promote power-law clustering of Kalahari vegetation." In: *Nature* 449.7159, pp. 209–212.

Scharlemann, J. P., E. V. Tanner, R. Hiederer, and V. Kapos (2014). "Global soil carbon: understanding and managing the largest terrestrial carbon pool." In: *Carbon Management* 5.1, pp. 81–91.

Scheffer, M., S. Carpenter, J. A. Foley, C. Folke, and B. Walker (2001). "Catastrophic shifts in ecosystems." In: *Nature* 413.6856, pp. 591–596.

Schenk, H. J. and R. B. Jackson (2002a). "Rooting depths, lateral root spreads and below-ground/above-ground allometries of plants in water-limited ecosystems." In: *Journal of Ecology*, pp. 480–494.

— (2002b). "The global biogeography of roots." In: *Ecological Monographs* 72.3, pp. 311–328.

Schenzle, A. and H. Brand (1979). "Multiplicative stochastic processes in statistical physics." In: *Physical Review A* 20.4, p. 1628.

Schimel, D. S., B. Braswell, R. McKeown, D. S. Ojima, W. Parton, and W. Pulliam (1996). "Climate and nitrogen controls on the geography and timescales of terrestrial biogeochemical cycling." In: *Global Biogeochemical Cycles* 10.4, pp. 677–692.

Schimel, D. S., B. Braswell, and W. Parton (1997). "Equilibration of the terrestrial water, nitrogen, and carbon cycles." In: *Proceedings of the National Academy of Sciences* 94.16, pp. 8280–8283.

Schimel, J. and S. M. Schaeffer (2012). "Microbial control over carbon cycling in soil." In: *Frontiers in Microbiology* 3, p. 348.

Schimel, J. P. (2018). "Life in dry soils: effects of drought on soil microbial communities and processes." In: *Annual Review of Ecology, Evolution, and Systematics* 49, pp. 409–432.

Schimel, J. P. and J. Bennett (2004). "Nitrogen mineralization: challenges of a changing paradigm." In: *Ecology* 85.3, pp. 591–602.

Schimel, J. P. and M. N. Weintraub (2003). "The implications of exoenzyme activity on microbial carbon and nitrogen limitation in soil: a theoretical model." In: *Soil Biology and Biochemistry* 35.4, pp. 549–563.

Schlesinger, W. (1997). *An Analysis of Global Change: Biogeochemistry*. Academic Press

Schmidhuber, J. and F. N. Tubiello (2007). "Global food security under climate change." In: *Proceedings of the National Academy of Sciences* 104.50, pp. 19703–19708.

Scholes, R. J. and S. R. Archer (1997). "Tree–grass interactions in savannas." In: *Annual Review of Ecology and Systematics* 28.1, pp. 517–544.

Scholes, R. J. and B. H. Walker (1993). *An African Savanna: Synthesis of the Nylsvley Study*. Cambridge University Press.

— (2004). *An African Savanna: Synthesis of the Nylsvley Study*. Cambridge University Press.

Schoups, G., J. W. Hopmans, C. A. Young, J. A. Vrugt, W. W. Wallender, K. K. Tanji, and S. Panday (2005). "Sustainability of irrigated agriculture in the San Joaquin Valley, California." In: *Proceedings of the National Academy of Sciences* 102.43, pp. 15352–15356.

Schulze, E. (1993). "Soil water deficits and atmospheric humidity as environmental signals." In: *Water Deficits: Plant Response from Cell to Community,* pp. 129–145.

Schwantes, A. M., A. J. Parolari, J. J. Swenson, D. M. Johnson, J.-C. Domec, R. B. Jackson, N. Pelak, and A. Porporato (2018). "Accounting for landscape heterogeneity improves spatial predictions of tree vulnerability to drought." In: *New Phytologist* 220.1, pp. 132–146.

Scifres, C. J. and B. H. Koerth (1987). "Climate, soils, and vegetation of the La Copita research area." In: *Miscellaneous Publication MP-Texas Agricultural Experiment Station (USA).*

Selker, J., T. Steenhuis, and J.-Y. Parlange (1992). "Wetting front instability in homogeneous sandy soils under continuous infiltration." In: *Soil Science Society of America Journal* 56.5, pp. 1346–1350.

Semenov, A., A. Van Bruggen, and V. Zelenev (1999). "Moving waves of bacterial populations and total organic carbon along roots of wheat." In: *Microbial Ecology* 37.2, pp. 116–128.

Sherman, L. K. (1943). "Comparison f-curves derived by the methods of sharp and Holtan and of Sherman and Mayer." In: *Eos, Transactions of the American Geophysical Union* 24.2, pp. 465–467. ISSN: 2324-9250.

Shmida, A. and T. L. Burgess (1988). "Plant growth-form strategies and vegetation types in arid environments." In: *Plant Form and Vegetation Structure*, pp. 211–241.

Shuttleworth, W., R. Leuning, T. Black, J. Grace, P. Jarvis, J. Roberts, and H. Jones (1989). "Micrometeorology of temperate and tropical forest." In: *Phil. Trans. Roy. Soc. B: Biological Sciences* 324.1223, pp. 299–334.

Siebert, S., M. Kummu, M. Porkka, P. Döll, N. Ramankutty, and B. R. Scanlon (2015). "A global data set of the extent of irrigated land from 1900 to 2005." In: *Hydrology and Earth System Sciences* 19.3, pp. 1521–1545.

Silvertown, J. (2004). "Plant coexistence and the niche." In: *Trends in Ecology & Evolution* 19.11, pp. 605–611.

Silvertown, J., Y. Araya, and D. Gowing (2015). "Hydrological niches in terrestrial plant communities: a review." In: *Journal of Ecology* 103.1, pp. 93–108.

Siqueira, M., G. Katul, and A. Porporato (2008). "Onset of water stress, hysteresis in plant conductance, and hydraulic lift: scaling soil water dynamics from millimeters to meters." In: *Water Resources Research* 44.1.

— (2009). "Soil moisture feedbacks on convection triggers: the role of soil–plant hydrodynamics." In: *Journal of Hydrometeorology* 10.1.

Skopp, J., M. Jawson, and J. Doran (1990). "Steady-state aerobic microbial activity as a function of soil water content." In: *Soil Science Society of America Journal* 54.6, pp. 1619–1625.

Sohngen, B. and R. Mendelsohn (2003). "An optimal control model of forest carbon sequestration." In: *American Journal of Agricultural Economics* 85.2, pp. 448–457.

Sornette, D. (2006). *Critical Phenomena in Natural Sciences: Chaos, Fractals, Selforganization and Disorder: Concepts and Tools.* Springer

Souza, R., J. Yin, and S. Calabrese (2021). "Optimal drainage timing for mitigating methane emissions from rice paddy fields." In: *Geoderma.*

Spall, J. C. (2005). *Introduction to Stochastic Search and Optimization: Estimation, Simulation, and Control.* John Wiley and Sons.

Spawn, S. A., C. C. Sullivan, T. J. Lark, and H. K. Gibbs (2020). "Harmonized global maps of above and belowground biomass carbon density in the year 2010." In: *Scientific Data* 7.1, pp. 1–22.

Sperry, J., U. Hacke, R. Oren, and J. Comstock (2002). "Water deficits and hydraulic limits to leaf water supply." In: *Plant, Cell & Environment* 25.2, pp. 251–263.

Stanhill, G. (1970). "Some results of helicopter measurements of the albedo of different land surfaces." In: *Solar Energy* 13.1, pp. 59–66.

Stark, J. M. and M. K. Firestone (1995). "Mechanisms for soil moisture effects on activity of nitrifying bacteria." In: *Applied and Environmental Microbiology* 61.1, pp. 218–221.

Steduto, P., T. C. Hsiao, D. Raes, and E. Fereres (2009). "AquaCrop – the FAO crop model to simulate yield response to water: I. Concepts and underlying principles." In: *Agronomy Journal* 101.3, pp. 426–437.

Steffen, J., M. Jensen, C. A. Pomeroy, and S. J. Burian (2013). "Water supply and stormwater management benefits of residential rainwater harvesting in US cities." In: *Journal of the American Water Resources Association* 49.4, pp. 810–824.

Steinbeck, J. (2006). *The Grapes of Wrath*. Penguin.

Stengel, R. F. (1994). *Optimal Control and Estimation*. Courier Corporation.

Stevens, B. (2007). "On the growth of layers of nonprecipitating cumulus convection." In: *Journal of the Atmospheric Sciences* 64.8, pp. 2916–2931.

Stidd, C. K., W. Fowler, and J. Helvey (1975). "Irrigation increases rainfall?." In: *Science* 188.4185, pp. 279–281.

Stöckle, C. O., M. Donatelli, and R. Nelson (2003). "CropSyst, a cropping systems simulation model." In: *European Journal of Agronomy* 18.3-4, pp. 289–307.

Strigul, N., D. Pristinski, D. Purves, J. Dushoff, and S. Pacala (2008). "Scaling from trees to forests: tractable macroscopic equations for forest dynamics." In: *Ecological Monographs* 78.4, pp. 523–545.

Strogatz, S. H. (2001). *Nonlinear Dynamics and Chaos: With Applications to Physics, Biology, Chemistry, and Engineering*. CRC Press.

Stull, R. B. (1988). *An Introduction to Boundary Layer Meteorology*. Kluwer Dordrecht, 666 pp.

Sussmann, H. J. and J. C. Willems (1997). "300 years of optimal control: from the brachystochrone to the maximum principle." In: *IEEE Control Systems Magazine* 17.3, pp. 32–44.

Suweis, S., A. Rinaldo, S. van der Zee, E. Daly, A. Maritan, and A. Porporato (2010). "Stochastic modeling of soil salinity." In: *Geophysical Research Letters* 37.7.

Suweis, S., A. Porporato, A. Rinaldo, and A. Maritan (2011). "Prescriptioninduced jump distributions in multiplicative Poisson processes." In: *Physical Review E* 83.6, p. 061119.

Syvertsen, J., G. Nickell, R. Spellenberg, and G. Cunningham (1976). "Carbon reduction pathways and standing crop in three Chihuahuan Desert plant communities." In: *The Southwestern Naturalist*, pp. 311–320.

Szabolcs, I. et al. (1989). *Salt-Affected Soils*. CRC Press.

Tamea, S., F. Laio, L. Ridolfi, P. D'Odorico, and I. Rodríguez-Iturbe (2009). "Ecohydrology of groundwater-dependent ecosystems: 2. Stochastic soil moisture dynamics." In: *Water Resources Research* 45.5, W05420.

Tang, C., B. Shi, C. Liu, L. Zhao, and B. Wang (2008). "Influencing factors of geometrical structure of surface shrinkage cracks in clayey soils." In: *Engineering Geology* 101.3, pp. 204–217.

Tardieu, F. and W. Davies (1993). "Integration of hydraulic and chemical signalling in the control of stomatal conductance and water status of droughted plants." In: *Plant, Cell & Environment* 16.4, pp. 341–349.

Tarnita, C. E., J. A. Bonachela, E. Sheffer, J. A. Guyton, T. C. Coverdale, R. A. Long, and R. M. Pringle (2017). "A theoretical foundation for multi-scale regular vegetation patterns." In: *Nature* 541.7637, pp. 398–401.

Tartakovsky, D. M. (2013). "Assessment and management of risk in subsurface hydrology: A review and perspective." In: *Advances in Water Resources* 51, pp. 247–260.

Tennekes, H. (1973). "A model for the dynamics of the inversion above a convective boundary layer." In: *Journal of Atmospheric Science* 30.4, pp. 558–567.

Tennekes, H. and J. Lumley (1972). *A First Course in Turbulence*. Pe Men Book Co.

Thompson, S., C. Harman, P. Heine, and G. Katul (2010). "Vegetation–infiltration relationships across climatic and soil type gradients." In: *Journal of Geophysical Research: Biogeosciences (2005–2012)* 115.G2.

Thor West, C., S. Benecky, C. Karlsson, B. Reiss, and A. J. Moody (2020). "Bottom-up perspectives on the re-greening of the Sahel: an evaluation of the spatial relationship between soil and water conservation (SWC) and tree-cover in Burkina Faso." In: *Land* 9.6, p. 208.

Tiedje, J., A. Sexstone, T. Parkin, and N. Revsbech (1984). "Anaerobic processes in soil." In: *Plant and Soil* 76.1, pp. 197–212.

Tilman, D. (1982). *Resource Competition and Community Structure*. Princeton University Press.

— (1994). "Competition and biodiversity in spatially structured habitats." In: *Ecology* 75.1, pp. 2–16.

Trapp, S. and M. Matthies (1995). "Generic one-compartment model for uptake of organic chemicals by foliar vegetation." In: *Environmental Science & Technology* 29.9, pp. 2333–2338.

Trimble, S. W. (1985). "Perspectives on the history of soil erosion control in the eastern United States." In: *Agricultural History* 59.2, pp. 162–180.

Trout, T. J. and D. C. Kincaid (2007). "On-farm system design and operation and land management." In: *Irrigation of Agricultural Crops.* John Wiley & Sons. Chapter 5, pp. 133–179.

Tsonis, A. A. (2002). *An Introduction to Atmospheric Thermodynamics.* Cambridge University Press.

Turner, N. (1986). "Adaptation to water deficits: a changing perspective." In: *Functional Plant Biology* 13.1, p. 175.

Turner, N. C. and M. M. Jones (1980). "Turgor maintenance by osmotic adjustment: a review and evaluation." In: *Turgor Maintenance by Osmotic Adjustment: A Review and Evaluation*, pp. 87–103.

Van den Broeck, C., J. Parrondo, R. Toral, and R. Kawai (1997). "Nonequilibrium phase transitions induced by multiplicative noise." In: *Physical Review E* 55.4, p. 4084.

Van den Honert T. H. (1948). "Water transport in plants as a catenary process." In: *Discussions of the Faraday Society* 3, pp. 146–153.

van Genuchten, M. T. (1980). "A closed-form equation for predicting the hydraulic conductivity of unsaturated soils." In: *Soil Science Society of America Journal* 44.5, p. 892.

Van Kampen, N. (1981). "Itô versus Stratonovich." In: *Journal of Statistical Physics* 24.1, pp. 175–187.

Van Kampen, N. G. (1992). *Stochastic Processes in Physics and Chemistry.* Vol. 1. Elsevier.

Vanmarcke, E. (2010). *Random Fields: Analysis and Synthesis.* World Scientific.

Verma, P., K. George, H. Singh, S. Singh, A. Juwarkar, and R. Singh (2006). "Modeling rhizofiltration: heavy-metal uptake by plant roots." In: *Environmental Modeling & Assessment* 11.4, pp. 387–394.

Verschuren, D., K. R. Laird, and B. F. Cumming (2000). "Rainfall and drought in equatorial east Africa during the past 1,100 years." In: *Nature* 403.6768, pp. 410–414.

Vervoort, R. W. and E. van der Zee (2008). "Simulating the effect of capillary flux on the soil water balance in a stochastic ecohydrological framework." In: *Water Resources Research* 44.8, n/a–n/a.

Vico, G. and A. Porporato (2008). "Modelling C3 and C4 photosynthesis under water-stressed conditions." In: *Plant and Soil* 313.1-2, pp. 187–203.

— (2010). "Traditional and microirrigation with stochastic soil moisture." In: *Water Resources Research* 46.3.

— (2011a). "From rainfed agriculture to stress-avoidance irrigation: I. A generalized irrigation scheme with stochastic soil moisture." In: *Advances in Water Resources* 34.2, pp. 263–271.

— (2011b). "From rainfed agriculture to stress-avoidance irrigation: II. Sustainability, crop yield, and profitability." In: *Advances in Water Resources* 34.2, pp. 272–281.

Vitousek, P. M., H. A. Mooney, J. Lubchenco, and J. M. Melillo (1997). "Human domination of Earth's ecosystems." In: *Science* 277.5325, pp. 494–499.

Vogeler, I., S. R. Green, D. R. Scotter, and B. E. Clothier (2001). "Measuring and modelling the transport and root uptake of chemicals in the unsaturated zone." In: *Plant and Soil* 231.2, pp. 161–174.

Volkov, I., J. R. Banavar, S. P. Hubbell, and A. Maritan (2003). "Neutral theory and relative species abundance in ecology." In: *Nature* 424.6952, pp. 1035–1037.

Von Caemmerer, S. (2013). "Steady-state models of photosynthesis." In: *Plant, Cell & Environment* 36.9, pp. 1617–1630.

Walker, B., C. S. Holling, S. R. Carpenter, and A. Kinzig (2004). "Resilience, adaptability and transformability in social–ecological systems." In: *Ecology and Society* 9.2.

Wan, C. and R. E. Sosebee (1991). "Water relations and transpiration of honey mesquite on 2 sites in west Texas." In: *Rangeland Ecology & Management/Journal of Range Management Archives* 44.2, pp. 156–160.

Wang, R. and J. B. Zimmerman (2015). "Economic and environmental assessment of office building rainwater harvesting systems in various US cities." In: *Environmental Science & Technology* 49.3, pp. 1768–1778.

Waring, R. and S. Running (1978). "Sapwood water storage: its contribution to transpiration and effect upon water conductance through the stems of old-growth Douglas-fir." In: *Plant, Cell & Environment* 1.2, pp. 131–140.

Waring, R. H. and S. W. Running (2010). *Forest Ecosystems: Analysis at Multiple Scales.* Elsevier.

Watkins, K. (2002). "Cultivating poverty: the impact of US cotton subsidies on Africa." In: *Oxfam Policy and Practice: Agriculture, Food and Land* 2.1, pp. 82–117.

Webster, P. J. (1994). "The role of hydrological processes in ocean–atmosphere interactions." In: *Reviews of Geophysics* 32.4, pp. 427–476.

West, A., K. Hultine, J. Sperry, S. Bush, and J. Ehleringer (2008). "Transpiration and hydraulic strategies in a piñon–juniper woodland." In: *Ecological Applications* 18.4, pp. 911–927.

Wilcox, B. P., D. D. Breshears, and H. Turin (2003). "Hydraulic conductivity in a pinon–juniper woodland." In: *Soil Science Society of America Journal* 67.4, pp. 1243–1249.

Williams, E., G. Hutchinson, and F. Fehsenfeld (1992). "NOx and N2O emissions from soil." In: *Global Biogeochemical Cycles* 6.4, pp. 351–388.

Wilson, K. et al. (2002). "Energy balance closure at FLUXNET sites." In: *Agricultural and Forest Meteorology* 113.1, pp. 223–243.

Wong, S., I. Cowan, and G. Farquhar (1979). "Stomatal conductance correlates with photosynthetic capacity." In: *Nature* 282.5737, pp. 424–426.

Wood, E. F., D. P. Lettenmaier, and V. G. Zartarian (1992). "A land-surface hydrology parameterization with subgrid variability for general circulation models." In: *Journal of Geophysical Research: Atmospheres* 97.D3, pp. 2717–2728.

Wooding, R. A. (1968). "Steady infiltration from a shallow circular pond." In: *Water Resources Research* 4.6, pp. 1259–1273.

Yang, W. H. and D. Liptzin (2015). "High potential for iron reduction in upland soils." In: *Ecology* 96.7, pp. 2015–2020.

Yang, W. and F. Meixner (1997). "Laboratory studies on the release of nitric oxide from sub-tropical grassland soils: the effect of soil temperature and moisture." In: eds. S. C. Jarvis and B. F. Pain, *Gaseous Nitrogen Emissions from Grasslands,* pp. 67–71.

Yin, J. and A. Porporato (2017). "Diurnal cloud cycle biases in climate models." In: *Nature Communications* 8.1, pp. 1–8.

— (2019). "Looking up or looking down? Hydrologic and atmospheric perspectives on precipitation and evaporation variability." In: *Geophysical Research Letters* 46.21, pp. 11968–11971.

— (2020). "Radiative effects of daily cycle of cloud frequency in past and future climates." In: *Climate Dynamics* 54.3-4, pp. 1625–1637.

Yin, J., A. Porporato, and J. Albertson (2014). "Interplay of climate seasonality and soil moisture-rainfall feedback." In: *Water Resources Research* 50.7, pp. 6053–6066.

Yin, J., J. D. Albertson, J. R. Rigby, and A. Porporato (2015). "Land and atmospheric controls on initiation and intensity of moist convection: CAPE dynamics and LCL crossings." In: *Water Resources Research* 51.10, pp. 8476–8493.

Yin, J., S. Calabrese, E. Daly, and A. Porporato (2019). "The energy side of Budyko: surface-energy partitioning from hydrological observations." In: *Geophysical Research Letters*.

Zaitsev, V. F. and A. D. Polyanin (2012). *Handbook of Exact Solutions for Ordinary Differential Equations.* CRC Press, 791 pp.

Zea-Cabrera, E., Y. Iwasa, S. Levin, and I. Rodríguez-Iturbe (2006). "Tragedy of the commons in plant water use." In: *Water Resources Research* 42.6.

Zelenev, V., A. Van Bruggen, and A. Semenov (2000). ""BACWAVE," a spatial temporal model for traveling waves of bacterial populations in response to a moving carbon source in soil." In: *Microbial Ecology* 40.3, pp. 260–272.

Zemansky, M. and R. Dittman (1997). *Heat and Thermodynamics: An Intermediate Textbook.* McGraw-Hill.

Zeng, N., J. W. Shuttleworth, and J. H. Gash (2000). "Influence of temporal variability of rainfall on interception loss. Part I. Point analysis." In: *Journal of Hydrology* 228.3-4, pp. 228–241.

Zhang, L., K. Hickel, W. Dawes, F. H. Chiew, A. Western, and P. Briggs (2004). "A rational function approach for estimating mean annual evapotranspiration." In: *Water Resources Research* 40.2.

Zilberman, D. (1982). "The use and potential of optimal control models in agricultural economics." In: *Western Journal of Agricultural Economics* 7.2, pp. 395–405.

Index